T0297805

System Assurances

System Assurances

Modeling and Management

Edited by

Prashant Johri
Professor, School of Computing Science & Engineering, Galgotias University, Greater Noida, India

Adarsh Anand
Assistant Professor, Department of Operational Research, University of Delhi, New Delhi, India

Jüri Vain
Professor, Department of Software Science, Tallinn University of Technology, Tallinn, Estonia

Jagvinder Singh
Assistant Professor, USME, DTU East Delhi Campus, Delhi, India

Mohammad Tabrez Quasim
Assistant Professor, University of Bisha, Bisha, Saudi Arabia

Series Editor

Quan Min Zhu

Academic Press is an imprint of Elsevier
125 London Wall, London EC2Y 5AS, United Kingdom
525 B Street, Suite 1650, San Diego, CA 92101, United States
50 Hampshire Street, 5th Floor, Cambridge, MA 02139, United States
The Boulevard, Langford Lane, Kidlington, Oxford OX5 1GB, United Kingdom

Library of Congress Cataloging-in-Publication Data
A catalog record for this book is available from the Library of Congress

British Library Cataloguing-in-Publication Data
A catalogue record for this book is available from the British Library

ISBN 978-0-323-90240-3

For information on all Academic Press publications
visit our website at https://www.elsevier.com/books-and-journals

Publisher: Mara Conner
Acquisition Editor: Sonnini R. Yura
Editorial Project Manager: Leticia M. Lima
Production Project Manager: Prem Kumar Kaliamoorthi
Cover Designer: Mark Rogers

Typeset by STRAIVE, India

Contents

Contributors xxi
Preface xxv
Acknowledgement xxvii

1. **Statistical analysis approach for the quality assessment of open-source software**
Yoshinobu Tamura and Shigeru Yamada

1.1 Introduction 1
1.2 Correspondence analysis 1
1.3 Estimation procedure based on correspondence analysis 3
1.4 Numerical examples 3
1.5 Concluding remarks 7
Acknowledgment 8
References 8

2. **Analytical modeling and performance evaluation of SIP signaling protocol: Analytical modeling of SIP**
Nikesh Choudhary, Vandana Khaitan (nee Gupta), and Vaneeta Goel

2.1 Introduction 11
2.2 Background work 12
2.2.1 Motivation 13
2.2.2 Main contribution 14
2.3 SIP layered structure and its working 14
2.3.1 Types of responses of SIP 14
2.3.2 Layered structure and working of SIP 15
2.4 Proposed SRN model of SIP INVITE transaction 18
2.4.1 SRN and its attributes 18
2.4.2 Proposed SRN model of SIP INVITE transaction 20
2.5 Performance measures 31
2.5.1 Throughput 32
2.5.2 Latency 34
2.6 Numerical illustration 35
2.6.1 Results and discussion 35
2.7 Model validation 37
2.8 Conclusions 38
Acknowledgments 39
References 39

3. An empirical validation for predicting bugs and the release time of open source software using entropy measures—Software reliability growth models

Anjali Munde

3.1 Introduction 41
3.2 Information measures 44
3.2.1 Predicting bugs and the release time of open source software 44
3.3 Conclusion 48
References 48

4. Risk assessment of starting air system of marine diesel engine using fuzzy failure mode and effects analysis

Rajesh S. Prabhu Gaonkar and Sunay P. Pai

4.1 Introduction 51
4.2 Starting air system 53
4.2.1 Explosion in starting air system 54
4.3 Failure mode and effects analysis (FMEA) 55
4.4 The proposed methodology 56
4.5 An illustrative example: Starting air system 59
4.6 Results 62
4.7 Sensitivity analysis 63
4.8 Conclusions 65
References 65

5. Test scenario generator learning for model-based testing of mobile robots

Gert Kanter and Marti Ingmar Liibert

5.1 Introduction 67
5.2 Related work 68
5.3 Preliminaries 69
5.3.1 Model-based testing 69
5.3.2 TestIt toolkit 70
5.4 Tool architecture 70
5.4.1 State-space exploration 71
5.4.2 Clustering state space 72
5.4.3 KMeans 73
5.4.4 Postclustering analysis 73
5.4.5 Automaton construction 74
5.4.6 Model refinement 75
5.4.7 Model checking 75
5.5 Validation 75
5.5.1 Exploration 76
5.5.2 Clustering and creating the automata 76

5.5.3 Results 79
5.6 Conclusion 83
References 83

6. Testing effort-dependent software reliability growth model using time lag functions under distributed environment
Sudeept Singh Yadav, Avneesh Kumar, Prashant Johri, and J.N. Singh

6.1 Introduction 85
6.2 Software reliability growth modeling 89
6.2.1 Framework for modeling 89
6.2.2 Model assumptions and notations used 90
6.2.3 Modeling testing effort 91
6.2.4 Modeling simple, hard, and complex faults 92
6.3 Parameter estimation 95
6.4 Comparison of SRGM criteria 95
6.4.1 Criteria for goodness of fit 95
6.5 Data description and model validation 96
6.6 Conclusion 101
References 101

7. Design and performance analysis of MIMO PID controllers for a paper machine subsystem
Niharika Varshney, Parvesh Saini, and Ashutosh Dixit

7.1 Introduction 103
7.2 Controller tuning 105
7.3 Result analysis 108
7.4 Conclusion 120
References 120

8. Network and security leveraging IoT and image processing: A quantum leap forward
Ajay Sudhir Bale, S. Saravana Kumar, S. Varun Yogi, Swetha Vura, R. Baby Chithra, N. Vinay, and P. Pravesh

8.1 Introduction 123
8.2 Comparative study 124
8.2.1 Vehicular sensor networks 124
8.2.2 Biometrics 126
8.2.3 Network and security 128
8.2.4 Medical applications (healthcare) 129
8.2.5 Protocols 131
8.2.6 Image processing: 132
8.3 Conclusion 136
References 137

9. **Modeling software patching process inculcating the impact of vulnerabilities discovered and disclosed**
Deepti Aggrawal, Jasmine Kaur, and Adarsh Anand

9.1	**Introduction**	143
9.2	**Literature review**	145
9.3	**Notations**	146
9.4	**Model development**	146
	9.4.1 Vulnerability detection process	146
	9.4.2 Vulnerability disclosure process	147
	9.4.3 Vulnerability patching process	147
9.5	**Model illustration**	148
9.6	**Conclusion**	151
	Acknowledgment	151
	References	151

10. **Extension of software reliability growth models by several testing-time functions**
Yuka Minamino, Shinji Inoue, and Shigeru Yamada

10.1	**Introduction**	155
	10.1.1 Background for development of bivariate SRGMs	155
	10.1.2 Our proposed bivariate SRGMs	157
10.2	**Software reliability assessment using bivariate SRGMs**	157
	10.2.1 Existing nonhomogeneous Poisson process (NHPP) model	157
	10.2.2 Bivariate NHPP models based on testing-time functions	158
	10.2.3 Parameter estimation	160
	10.2.4 Bivariate software reliability assessment measures	161
	10.2.5 Numerical examples	161
10.3	**Bivariate Weibull-type SRGMs and their application**	164
	10.3.1 Bivariate Weibull-type SRGMs based on testing-time functions	164
	10.3.2 Parameter estimation	165
	10.3.3 Bivariate Weibull-type SRGMs under budget constraints	167
	10.3.4 Numerical examples	169
10.4	**Conclusions**	173
	Acknowledgments	173
	References	174

11. **A semi-Markov model of a system working under uncertainty**
R.K. Bhardwaj, Purnima Sonker, and Ravinder Singh

11.1	**Introduction**	175
11.2	**Notations**	176

11.3 **Development of system model** 177
11.3.1 States of system model 177
11.3.2 Transition probability 177
11.3.3 Mean sojourn times 180
11.4 **Performance measures** 180
11.4.1 Reliability and mean time-to-system failure 180
11.4.2 Steady-state availability 180
11.4.3 Busy period analysis for the server 181
11.4.4 Expected number of treatment given to the server 182
11.4.5 Expected number of repairs given to the switch 182
11.4.6 Expected number of repairs given to the unit 183
11.4.7 Expected number of server visits 183
11.4.8 Profit 184
11.5 **Simulation study** 184
11.6 **Concluding remarks** 186
Acknowledgment 186
References 186

12. **Design and evaluation of parallel-series IRM system**
Sridhar Akiri, P. Sasikala, Pavan Kumar Subbara, and VSS Yadavalli

12.1 **Introduction** 189
12.1.1 Lagrangean way to understand equality constraints 190
12.1.2 Lagrangean way to understand unequal factors 191
12.2 **Lagrangean procedure for formulation of the problem function:** $c_j = b_j e^{\left[\frac{a_j}{(1-r_j)}\right]}$ 192
12.2.1 Model analysis 192
12.3 **Case problem** 197
12.3.1 Efficiency design with z_j rounding off 199
12.4 **Dynamic programming approach** 200
12.4.1 Introduction 200
12.4.2 Dynamic way to approach programming 200
12.4.3 Basic features of the dynamic programming approach 202
12.4.4 Classification of dynamic programming problems 203
12.4.5 Computational procedure in dynamic programming 203
12.4.6 Computation in the forward and backward directions 203
12.4.7 Development of an optimal decision policy using dynamic programming 204

12.5 The integrated efficiency model by using the dynamic programming method for the function $c_j = b_j e^{\left[\frac{a_j}{(1-r_j)}\right]}$ 205
12.5.1 Dynamic programming solution 205
12.6 Conclusions 207
Further reading 208

13. Modeling and availability assessment of smart building automation systems with multigoal maintenance

Yuriy Ponochovniy, Vyacheslav Kharchenko, and Olga Morozova

13.1 Introduction 209
13.1.1 Motivation 209
13.1.2 Work-related analysis 210
13.1.3 Goals and contribution 211
13.2 Concept of multigoal maintenance 211
13.2.1 Principles of multigoal maintenance 212
13.2.2 Assessment of availability 213
13.3 Development of models 214
13.3.1 General model 214
13.3.2 Availability model considering reliability and cybersecurity 214
13.3.3 Availability model considering joint maintenance strategy 217
13.3.4 Availability model considering separate maintenance strategy 219
13.4 Research of models 219
13.5 Conclusions 223
References 226

14. A study of bitcoin and Ethereum blockchains in the context of client types, transactions, and underlying network architecture

Rohaila Naaz and Ashendra Kumar Saxena

14.1 Blockchain: An insight 230
14.2 Blockchain architectures 230
14.2.1 Infrastructure layer 232
14.2.2 Platform layer 234
14.3 Blockchain adequacy 236
14.3.1 Data integrity 238
14.3.2 Scalability 238

14.3.3 Data transparency 238
14.3.4 Reliability and availability 239
14.3.5 Blockchain architecture design space guidelines 240
14.4 Bitcoin architecture 240
14.4.1 Bitcoin components 240
14.4.2 Bitcoin network nodes 245
14.4.3 Bitcoin consensus 248
14.4.4 Bitcoin miners 251
14.4.5 Bitcoin transactions 252
14.5 Introduction of Ethereum 254
14.6 Ethereum's evolution 255
14.7 Architecture of Ethereum 256
14.7.1 Ethereum virtual machine 256
14.8 Ethereum components 257
14.8.1 Ethereum accounts 258
14.8.2 Mining in Ethereum 258
14.8.3 Ethereum clients 258
14.8.4 Ethereum transaction 262
14.8.5 Ethereum networks 264
14.9 Conclusion 267
References 267
Further reading 269

15. High assurance software architecture and design
Muhammad Ehsan Rana and Omar S. Saleh

15.1 Introduction 271
15.2 Software architecture patterns 271
15.2.1 Client-server pattern 272
15.2.2 Layered architecture pattern 273
15.2.3 Model-view-controller (MVC) pattern 274
15.3 Software design principles 276
15.3.1 Single responsibility principle 276
15.3.2 Open-closed principle 276
15.3.3 Liskov substitution principle 276
15.3.4 Interface segregation principle 276
15.3.5 Dependency inversion principle 277
15.4 Software design patterns 277
15.4.1 Essential elements of design patterns 277
15.4.2 Classification of design patterns 278
15.4.3 Gang of four design patterns 279
15.4.4 Benefits of using design patterns 279
15.5 Software design antipatterns 281
15.5.1 Some common antipatterns 281
15.5.2 When design patterns turn into antipatterns 282
15.6 Conclusion 283
References 283

16. Online condition monitoring and maintenance of photovoltaic system

Neeraj Khera

16.1	**Introduction**	287
16.2	**Condition monitoring of VRLA battery in PV system**	289
	16.2.1 Implementation of predictive fault diagnosis of VRLA battery in PV system	290
	16.2.2 Results and discussion	294
	16.2.3 Web-based condition monitoring of battery	300
16.3	**Condition monitoring of aluminum electrolytic capacitor and MOSFET of power converter in PV system**	301
	16.3.1 Implementation of in-circuit condition monitoring of aluminum-electrolytic capacitor and MOSFET	302
	16.3.2 Results and discussion	302
16.4	**Conclusions**	303
	References	304

17. Fault diagnosis and fault tolerance

Afaq Ahmad and Sayyid Samir Al Busaidi

17.1	**Introduction**	307
17.2	**Digital systems modeling**	308
	17.2.1 Functional modeling at the logic level	309
	17.2.2 Programs as functional models	310
17.3	**Fault models**	311
	17.3.1 Stuck-at fault model	312
	17.3.2 Multiple stuck-at faults model	313
	17.3.3 Bridge fault model	313
	17.3.4 Open fault model	313
	17.3.5 Path delay fault model	314
	17.3.6 Transition fault model	314
	17.3.7 Cell internal fault model	314
17.4	**Fault diagnosis test procedures**	315
17.5	**Fault diagnosis and fault tolerance**	317
	17.5.1 Fault tolerance	318
17.6	**Conclusions**	319
	Acknowledgment	320
	References	320

18. True power loss diminution by Improved Grasshopper Optimization Algorithm

Lenin Kanagasabai

18.1	**Introduction**	323
18.2	**Problem formulation**	323

18.3 Improved Grasshopper Optimization Algorithm 324
18.4 Simulation study 326
18.5 Conclusions 330
References 330

19. Security analytics

Vani Rajasekar, J Premalatha, and Rajesh Kumar Dhanaraj

19.1 Introduction 333
19.1.1 Phishing 334
19.1.2 Spamming 334
19.1.3 Denial-of-service (DoS) attack 334
19.1.4 Malware attacks 335
19.1.5 Botnets 335
19.1.6 Website threats 335
19.2 Different classes of security analytics 335
19.2.1 Core and structural analytics 336
19.2.2 Probability-based analytics 336
19.2.3 Time-based analytics 336
19.2.4 Lifecycle models of vulnerability management 337
19.3 Framework of cyber security analytics 338
19.3.1 Architecture 338
19.3.2 Representation of models 339
19.4 Big data security analytics 342
19.4.1 Analyzing big data sources and security model 342
19.4.2 Security analytics for threat detection 344
19.4.3 Security analytics solution 346
19.5 Security analytics for IoT 347
19.5.1 Interactive model for IoT 347
19.5.2 Ontology-based security modeling 349
19.5.3 Smart home—Security analytics IoT case study 350
19.6 Security analytics in anomaly detection 351
19.6.1 Techniques in anomaly detection 352
19.6.2 Research challenges in anomaly detection 352
References 353

20. Stochastic modeling of the mean time between software failures: A review

Gabriel Pena, Verónica Moreno, and Néstor Barraza

20.1 Introduction 355
20.2 Mathematical background 356
20.2.1 Birth processes 356
20.2.2 The mean value function 359
20.2.3 Mean time between failures 360
20.3 Application 365

20.3.1 Models considered 365
20.3.2 Experiments 366
20.3.3 Discussion 369
20.4 Conclusions 369
Acknowledgments 369
References 369

21. Inliers prone distributions: Perspectives and future scopes

K. Muralidharan and Pratima Bavagosai

21.1 Introduction 371
21.2 Inliers prone models 372
21.2.1 Instantaneous failure models 372
21.2.2 Early failure model-1 373
21.2.3 Early failure model-2 373
21.2.4 Model with inliers at zero and one 374
21.3 Inferences on 0–1 inliers model 374
21.3.1 Parameter estimation 375
21.3.2 Unbiased estimation 376
21.4 Tests of hypothesis about inliers 379
21.5 Data analysis 381
21.6 Inliers-prone distributions: Issues and problems 384
21.7 Future scopes 385
References 386

22. Integration of TPM, RCM, and CBM: A practical approach applied in Shipbuilding industry

Rupesh Kumtekar, Swapnil Kamble, and Suraj Rane

22.1 Introduction 389
22.2 Maintenance strategies 390
22.2.1 Total productive maintenance 390
22.2.2 Reliability centered maintenance 390
22.2.3 Condition-based maintenance 391
22.2.4 Proposed methodology 391
22.3 Reliability of marine propulsion system: A case study 392
22.3.1 Phase I: Preparation—System selection and identifying boundaries 392
22.3.2 Phase II: Analysis 395
22.3.3 Phase III: Task selection 395
22.3.4 Phase IV: Task comparison, review and control 399
22.3.5 Phase V: Involve and empower all employees 400
22.4 Conclusion 401
References 401

23. Revolutionizing the internet of things with swarm intelligence

Abhishek Kumar, Jyotir Moy Chatterjee, Manju Payal, and Pramod Singh Rathore

23.1 Introduction 403
23.2 Characteristics of IoT 405
23.3 The consumer IoT 406
23.4 The industrial IoT 407
23.5 IoT definitions by various companies 407
23.6 The industrial internet 411
23.7 The internet of everything (IoE) 412
23.8 Cyber physical systems (CPS) and industry 4.0 412
23.9 The internet of services (IoS) 412
23.10 The internet of robotic things (IoRT) 413
23.11 More internet of X terms 413
23.12 Swarm intelligence 413
23.13 Definitions of SI 413
23.14 SI and systems intersections 414
23.15 Forests to ants and smart gadgets 416
23.16 Implants & prosthetics 416
23.17 The swarm behavior of the augmented and quantified human—The next stage of the IoE? 418
23.18 ANT colony-based IoT systems 420
23.19 ANT colony optimization (ACO)-based IoT systems 420
23.20 Particle swarm optimization (PSO)-based IoT systems 421
23.21 Artificial bee colony (ABC)-based IoT systems 423
23.22 Bacterial foraging optimization (BFO)-based IoT systems 424
23.23 BAT optimization (BO)-based IoT systems 424
23.24 More SI-based IoT systems 425
23.25 Towards SI-based IoT systems 426
23.26 SI and its applications 427
23.26.1 ACO (Ant colony optimization) 428
23.26.2 BCO (Bee colony optimization) 428
23.27 SI application for IoT processes 428
23.27.1 Use of SI for connected cars 429
23.27.2 Use of SI for data routing 429
23.27.3 Use of SI in cloud computing for data optimization 430
23.28 Conclusion 430
References 430

24. Security and challenges in IoT-enabled systems

S. Kala and S. Nalesh

24.1 Introduction 437
24.2 Commercialized secure hardware primitive designs 439

24.3 Hardware trojan 441
24.4 Side-channel attack (SCA) 441
24.5 Reverse engineering 443
24.6 Key challenges 443
24.7 Conclusion 444
References 444

25. Provably correct aspect-oriented modeling with UPPAAL timed automata
Jüri Vain, Leonidas Tsiopoulos, and Gert Kanter

25.1 Introduction 447
25.2 Related work 448
25.3 Preliminaries 450
25.3.1 Aspect-oriented modeling 450
25.3.2 UPPAAL timed automata 450
25.4 Provably correct weaving of aspects 451
25.4.1 Join points 452
25.4.2 Advice weaving 453
25.4.3 Weaving correctness 457
25.5 Case study: Home rehabilitation system 465
25.5.1 Model description 465
25.5.2 Verification of the aspects' weaving correctness 467
25.6 Usability of AO modeling and verification 471
25.7 Conclusions and discussion 473
Acknowledgment 474
References 474

26. Relevance of data mining techniques in real life
Palwinder Kaur Mangat and Dr. Kamaljit Singh Saini

26.1 Introduction 477
26.2 Methodology 478
26.3 Need of data mining 485
26.4 Types of data mining 486
26.4.1 Ubiquitous data mining 486
26.4.2 Multimedia data mining 486
26.4.3 Distributed/collective data mining 487
26.4.4 Constraint-based data mining 487
26.4.5 Hypertext data mining 488
26.4.6 Spatial and geographic data mining 489
26.4.7 Phenomenal data mining 489
26.4.8 Social security data mining 490
26.4.9 Educational data mining 490
26.4.10 Time series/sequence data mining 492
26.5 Data mining techniques 492
26.5.1 Clustering 493

26.5.2 Classification 493
26.5.3 Decision trees 494
26.6 Categories of data mining techniques 496
26.6.1 Information systems 496
26.6.2 System optimization 496
26.6.3 Knowledge-based systems 497
26.6.4 Modeling 497
26.7 Applications of data mining methods 498
26.7.1 Text mining and web mining 498
26.7.2 Medical/pharmacy 498
26.7.3 Insurance and health care 499
26.7.4 Finance 499
26.7.5 Telecommunications 499
26.7.6 Education 499
26.7.7 Retail industry 500
26.8 Conclusion 500
References 500

27. D-PPSOK clustering algorithm with data sampling for clustering big data analysis

C. Suresh Gnana Dhas, N. Yuvaraj, N.V. Kousik, and Tadele Degefa Geleto

27.1 Introduction 503
27.2 Related work 504
27.3 Proposed work 505
27.3.1 Implementation of DPPSOK-means algorithm 506
27.3.2 DPPSOK algorithm 506
27.3.3 Evaluation of the solutions 507
27.4 Experimental result and discussion 508
27.4.1 Test data 508
27.4.2 Experimental result 508
27.5 Conclusion 511
References 511

28. A review on optimal placement of phasor measurement unit (PMU)

Ashutosh Dixit, Arindam Chowdhury, and Parvesh Saini

28.1 Introduction 513
28.2 Optimal PMU placement (OPP) problem formulation 514
28.3 Mathematical programming method 516
28.3.1 Integer programming (IP) 516
28.4 Meta-heuristic methods 518
28.4.1 Simulating annealing (SA) 518
28.4.2 Genetic algorithm (GA) 518
28.4.3 Tabu search (TS) 519

28.4.4 Differential evolution (DE) 519
28.4.5 Particle swarm optimization (PSO) 519
28.4.6 Ant colony optimization (ACO) 520
28.5 Heuristic methods 520
28.5.1 Depth first search (DeFS) 520
28.6 Algorithm comparison 520
28.7 Future scope 521
28.8 Conclusion 528
References 528

29. Effective motivational factors and comprehensive study of information security and policy challenges

M. Arvindhan

29.1 Introduction 531
29.2 Key information security policies-related challenges 532
29.2.1 Security policy management and updating 532
29.2.2 Illustration of security 533
29.2.3 Open issues 533
29.3 Organizational approaches to information security 535
29.4 Network architecture and threat model 536
29.5 Policy-based SDN security architecture 536
29.6 Trust 537
29.7 Privacy 538
29.8 Privacy and security in cloud computing 539
29.9 Cloud computing framework 540
29.9.1 Eligible 540
29.9.2 Allowed systems 541
29.9.3 Service availability 541
29.9.4 Privacy security of cloud computing 542
29.10 ABE in cloud computing 542
29.11 Conclusions 543
References 544

30. Integration of wireless communication technologies in internet of vehicles for handover decision and network selection

Shaik Mazhar Hussain, Kamaludin Mohamad Yusof, Afaq Ahmad, and Shaik Ashfaq Hussain

30.1 Introduction 547
30.2 Existing works 549
30.3 Research method 551
30.3.1 Handover decision by dynamic Q-learning 551
30.3.2 Network selection using fuzzy CNN 553
30.3.3 Routing 554

30.4	Simulation setup	557
30.5	Comparative analysis and results	558
30.6	Conclusion	560
	References	561

31. Modeling HIV-TB coinfection with illegal immigrants and its stability analysis

Rajinder Sharma

31.1	Introduction	563
31.2	The mathematical model	564
31.3	The mathematical analysis	565
31.4	Numerical analysis and discussion	566
	Acknowledgments	569
	References	569
	Further reading	569

Index 571

Contributors

Numbers in parentheses indicate the pages on which the authors' contributions begin.

Deepti Aggrawal (143), USME, DTU, East Delhi Campus, Delhi, India

Afaq Ahmad (307, 547), Department of Electrical and Computer Engineering, Sultan Qaboos University, Muscat, Oman

Sridhar Akiri (189), Department of Mathematics, G.I.S, GITAM (Deemed to be University), Visakhapatnam, Andhra Pradesh, India

Sayyid Samir Al Busaidi (307), Department of Electrical and Computer Engineering, Sultan Qaboos University, Muscat, Oman

Adarsh Anand (143), Department of Operational Research, Faculty of Mathematical Sciences, University of Delhi, Delhi, India

M. Arvindhan (531), Galgotias University, Greater Noida, India

R. Baby Chithra (123), Department of ECE, New Horizon College of Engineering, Bengaluru, India

Ajay Sudhir Bale (123), Department of ECE, SoET, CMR University, Bengaluru, India

Néstor Barraza (355), Department of Sciences and Technology, University of Tres de Febrero, Caseros, Argentina; School of Engineering, University of Buenos Aires, Autonomous City of Buenos Aires Argentina

Pratima Bavagosai (371), Department of Statistics, Faculty of Science, The Maharaja Sayajirao University of Baroda, Vadodara, India

R.K. Bhardwaj (175), Department of Statistics, Punjabi University Patiala, Patiala, Punjab, India

Jyotir Moy Chatterjee (403), Department of IT, LBEF, Kathmandu, Nepal

Nikesh Choudhary (11), Department of Operational Research, University of Delhi, India

Arindam Chowdhury (513), Department of Electrical Engineering, Maryland Institute of Technology and Management, Jamshedpur, India

Rajesh Kumar Dhanaraj (333), School of Computing Science and Engineering, Galgotias University, Greater Noida, India

C. Suresh Gnana Dhas (503), Department of Computer Science, Ambo University, Ambo, Ethiopia

Ashutosh Dixit (103, 513), Department of Electrical Engineering, Graphic Era Deemed to be University, Dehradun, Uttarakhand, India

Tadele Degefa Geleto (503), Department of Computer Science, Ambo University, Ambo, Ethiopia

Vaneeta Goel (11), Satyawati College, University of Delhi, India

Shaik Ashfaq Hussain (547), Department of Communications Engineering and Advanced Telecommunications Technology, School of Electrical Engineering, Faculty of Engineering, Universiti Teknologi Malaysia (UTM), Johor Bahru, Malaysia

Shaik Mazhar Hussain (547), Department of Communications Engineering and Advanced Telecommunications Technology, School of Electrical Engineering, Faculty of Engineering, Universiti Teknologi Malaysia (UTM), Johor Bahru, Malaysia

Shinji Inoue (155), Faculty of Informatics, Kansai University, Osaka, Japan

Prashant Johri (85), Galgotias University, Greater Noida, Uttar Pradesh, India

S. Kala (437), Department of Electronics and Communication Engineering, Indian Institute of Information Technology Kottayam, Kottayam, India

Swapnil Kamble (389), Mechanical Engineering Department, Goa College of Engineering, Ponda, Goa, India

Lenin Kanagasabai (323), Department of EEE, Prasad V. Potluri Siddhartha Institute of Technology, Vijayawada, Andhra Pradesh, India

Gert Kanter (67, 447), Department of Software Science, Tallinn University of Technology, Tallinn, Estonia

Jasmine Kaur (143), Department of Operational Research, Faculty of Mathematical Sciences, University of Delhi, Delhi, India

Vandana Khaitan (nee Gupta) (11), Department of Operational Research, University of Delhi, India

Vyacheslav Kharchenko (209), Department of Computer Systems, Networks and Cybersecurity, National Aerospace University “KhAI”, Kharkiv, Ukraine

Neeraj Khera (287), Department of ECE, Amity University, Noida, Uttar Pradesh, India

N.V. Kousik (503), School of Computing Science and Engineering, Galgotias University, Greater Noida, Uttar Pradesh, India

Abhishek Kumar (403), Department of CSE, Chitkara University Institute of Engineering and Technology, Chitkara University, Baddi, Himachal Pradesh, India

Avneesh Kumar (85), Galgotias University, Greater Noida, Uttar Pradesh, India

Rupesh Kumtekar (389), Mechanical Engineering Department, Goa College of Engineering, Ponda, Goa, India

Marti Ingmar Liibert (67), Institute of Computer Science, University of Tartu, Tartu, Estonia

Palwinder Kaur Mangat (477), University Institute of Computing, Chandigarh University, Chandigarh, India

Yuka Minamino (155), Graduate School of Engineering, Tottori University, Tottori, Japan

Verónica Moreno (355), Department of Sciences and Technology, University of Tres de Febrero, Caseros, Argentina

Olga Morozova (209), Department of Computer Systems, Networks and Cybersecurity, National Aerospace University "KhAI", Kharkiv, Ukraine

Anjali Munde (41), Amity University, Noida, Uttar Pradesh, India

K. Muralidharan (371), Department of Statistics, Faculty of Science, The Maharaja Sayajirao University of Baroda, Vadodara, India

Rohaila Naaz (229), Teerthanker Mahaveer University, Moradabad, Uttar Pradesh, India

S. Nalesh (437), Department of Electronics, Cochin University of Science and Technology, Kochi, India

Sunay P. Pai (51), Institute of Maritime Studies, Vasco-Da-Gama, Goa, India

Manju Payal (403), Software Developer Academic Hub, Ajmer, Rajasthan, India

Gabriel Pena (355), Department of Sciences and Technology, University of Tres de Febrero, Caseros, Argentina

Yuriy Ponochovnyi (209), Department of Information Systems and Technologies, Poltava State Agrarian University, Poltava, Ukraine

Rajesh S. Prabhu Gaonkar (51), Indian Institute of Technology Goa (IIT Goa), Ponda-Goa, India

P. Pravesh (123), Department of ECE, SoET, CMR University, Bengaluru, India

J Premalatha (333), Department of IT, Kongu Engineering College, Erode, Tamilnadu, India

Vani Rajasekar (333), Department of CSE, Kongu Engineering College, Erode, Tamilnadu, India

Muhammad Ehsan Rana (271), Asia Pacific University of Technology & Innovation (APU), Technology Park Malaysia, Kuala Lumpur, Malaysia

Suraj Rane (389), Mechanical Engineering Department, Goa College of Engineering, Ponda, Goa, India

Pramod Singh Rathore (403), Department of CSE, ACERC, Ajmer, Rajasthan, India

Dr. Kamaljit Singh Saini (477), University Institute of Computing, Chandigarh University, Chandigarh, India

Parvesh Saini (103, 513), Department of Electrical Engineering, Graphic Era Deemed to be University, Dehradun, Uttarakhand, India

Omar S. Saleh (271), Studies, Planning and Follow-Up Directorate, Ministry of Higher Education and Scientific Research, Baghdad, Iraq

S. Saravana Kumar (123), Department of CSE, SoET, CMR University, Bengaluru, India

P. Sasikala (189), Department of Mathematics, G.S.S, GITAM (Deemed to be University), Bangalore, Karnataka, India

Ashendra Kumar Saxena (229), Teerthanker Mahaveer University, Moradabad, Uttar Pradesh, India

Rajinder Sharma (563), CR Department, University of Technology and Applied Sciences-Sohar, Sohar, Oman

J.N. Singh (85), Galgotias University, Greater Noida, Uttar Pradesh, India

Ravinder Singh (175), Department of Statistics, Central University of Haryana, Narnaul, Haryana, India

Purnima Sonker (175), Department of Statistics, Punjabi University Patiala, Patiala, Punjab, India

Pavan Kumar Subbara (189), Department of Mathematics, G.S.S, GITAM (Deemed to be University), Bangalore, Karnataka, India

Yoshinobu Tamura (1), Yamaguchi University, Yamaguchi, Japan

Leonidas Tsiopoulos (447), Department of Software Science, Tallinn University of Technology, Tallinn, Estonia

Jüri Vain (447), Department of Software Science, Tallinn University of Technology, Tallinn, Estonia

Niharika Varshney (103), Department of Electrical Engineering, Graphic Era Deemed to be University, Dehradun, Uttarakhand, India

S. Varun Yogi (123), Department of ECE, SoET, CMR University, Bengaluru, India

N. Vinay (123), Department of ECE, SoET, CMR University, Bengaluru, India

Swetha Vura (123), Department of ECE, SoET, CMR University, Bengaluru, India

Sudeept Singh Yadav (85), Galgotias University, Greater Noida, Uttar Pradesh, India

VSS Yadavalli (189), Department of Industrial and Systems Engineering, Pretoria University, Pretoria, South Africa

Shigeru Yamada (1, 155), Graduate School of Engineering, Tottori University, Tottori, Japan

Kamaludin Mohamad Yusof (547), Department of Communications Engineering and Advanced Telecommunications Technology, School of Electrical Engineering, Faculty of Engineering, Universiti Teknologi Malaysia (UTM), Johor Bahru, Malaysia

N. Yuvaraj (503), Research and Development, ICT Academy, Chennai, India

Preface

In today's neck to neck competitive era, almost every system is facing tough challenges for survival. These challenges include, maintaining the system performance measures such as; reliability, availability, maintainability, cost & failure factors, and many other factors associated with them. The last decade has witnessed huge advancements and technological changes. As a result, more robust and dynamic systems are required to have an overall better performance. These analyses for a system can be performed by taking care of several areas such as mathematical modeling for reliability maintenance, software reliability engineering and standards, coding and cryptography, artificial intelligence and expert systems, operations research and IT, IT in health care, and several others.

This edited issue of Emerging Methodologies and Applications in Modeling, Identification, and Control on *System Assurances: Modeling and Management* includes invited papers appropriate to the theme and the complex solution approaches to handle research challenges in the related domain. Emphasis has been given to the original and qualitative work relevant to the theme, with particular importance given to system assurance.

Prashant Johri
Galgotias University, India

Adarsh Anand
University of Delhi, India

Jüri Vain
Tallinn University of Technology, Estonia

Jagvinder Singh
Delhi Technological University, India

Mohammad Tabrez Quasim
University of Bisha, Bisha, Saudi Arabia

Acknowledgments

We are grateful to various colleagues both in academia and in industry who have contributed to this book. Our appreciation also goes out to Leticia Lima for her cooperation through the publication process. In addition, we are honored to be associated with Prof. Quan Min Zhu, who showed faith in us for carrying out the book compilation work under his series.

Of course, no author can successfully complete a book project without family support. Our beloved appreciation to our respective families for affording us their patience that allowed the compilation of this edited book.

Prashant Johri
Adarsh Anand
Jüri Vain
Jagvinder Singh
Mohammad Tabrez Quasim

Chapter 1

Statistical analysis approach for the quality assessment of open-source software

Yoshinobu Tamura[a] and Shigeru Yamada[b]
[a]*Yamaguchi University, Yamaguchi, Japan,* [b]*Graduate School of Engineering, Tottori University, Tottori, Japan*

1.1 Introduction

At present, many OSS (open source software) are used in many software development organizations. Also, many commercial software have been developed by embedding several OSS components because of standardization, cost reduction, quick delivery, etc. Various methods for software reliability assessment of OSS system have been proposed by several researchers [1–3].

Moreover, many OSS systems have used the fault tracking systems such as the Bugzilla. Also, many research papers have discussed the methods in terms of the fault tracking system [4–13].

In this chapter, we discuss statistical analysis such as correspondence analysis. In particular, we apply correspondence analysis to the fault large-scale data of OSS. Then the discussed method based on correspondence analysis can be used to understand the whole trend of the fault large-scale data of OSS. Correspondence analysis has been used in many marketing areas. Also, several research papers have been proposed by using correspondence analysis [14–16].

Moreover, several numerical examples based on the fault large-scale data in actual OpenStack cloud projects of OSS are shown by using the statistical method applied in this chapter. Then the numerical examples of visualization based on correspondence analysis are shown in this chapter. Finally, we show that the applied method will be useful to understand the reliability trend of OSS for the quality assessment of OSS developed under the OSS project.

1.2 Correspondence analysis

Correspondence analysis is well known as one of the multivariate analyses for qualitative data. In particular, the main purpose of correspondence analysis is

System Assurances. https://doi.org/10.1016/B978-0-323-90240-3.00001-1

visualization. In correspondence analysis, our methods make a visualization by using the survey items and attribute ones. Then the correlation coefficient is maximized for the data visualization.

This chapter focuses on the visualization of fault large-scale data of OSS by using correspondence analysis. By using correspondence analysis, we can easily understand the whole reliability trend of OSS because of the bird's-eye view of correspondence analysis. Many correspondence analyses are used in the marketing area. However, we consider that the correspondence analysis is helpful to understand the whole reliability trend. We use the fault count encoding for the correspondence analysis of OSS fault large-scale data because the software fault data have been used by many software reliability growth models. Thereby, we can comprehend the whole reliability trend by using the fault collection data based on the fault count encoding one.

$$\gamma_{i*} = \sum_j \gamma_{ij}, \tag{1.1}$$

$$\gamma_{*j} = \sum_i \gamma_{ij}, \tag{1.2}$$

$$\gamma = \sum_i \sum_j \gamma_{ij}, \tag{1.3}$$

where γ_{ij} is the data for the first variable i and second variable j.

Then the mean values of the first variable i and the second variable j are as follows:

$$\overline{p} = \frac{1}{n} \sum_i \gamma_{i*} \cdot p_i, \tag{1.4}$$

$$\overline{q} = \frac{1}{n} \sum_j \gamma_{*j} \cdot q_j, \tag{1.5}$$

where p_i is the weight parameter in the first variable i. Similarly, q_j is the weight parameter in the first variable j, n is the number of variables. Therefore, the correlation coefficient is given by

$$\rho = \frac{\frac{1}{n} \sum_i \sum_j \gamma_{ij} \cdot p_i \cdot q_j}{\sqrt{\frac{1}{n} \sum_i \gamma_{i*} \cdot p_i^2} \sqrt{\frac{1}{n} \sum_i \gamma_{j*} \cdot q_j^2}}, \tag{1.6}$$

The estimate of the correlation coefficient ρ^* is the value making ρ of Eq. (1.6) maximize. The estimate can be given from the solutions of the following

$$\frac{d\rho}{du} = \frac{d\rho}{dv} = 0. \tag{1.7}$$

1.3 Estimation procedure based on correspondence analysis

The estimation steps based on the correspondence analysis are as follows:

1. The raw fault large-scale data is obtained from the fault tracking system.
2. The count encoding data is converted from the raw fault large-scale data.
3. The pivot tables are made for each factor of the fault category.
4. We make a correspondence analysis by using the data sets in the pivot table.
5. We discuss the results of the correspondence analysis.

Moreover, we apply the following fault category data to the correspondence analysis:

- Nickname for fault assignee
- Name for software component
- Name for hardware
- Name for OS
- Nickname for fault reporter
- Fault severity level
- Status for software fault
- Number for software version

Then the data in each factor are encoded from the character data to the numerical values by using the count encoding from the standpoint of the relationship between the OSS reliability and OSS fault.

1.4 Numerical examples

This chapter shows several numerical illustrations for OpenStack Project [17] of OSS development project. In particular, StarlingX included in OpenStack projects are well known as the edge-OSS projects. At present, OpenStack projects that include StarlingX. It is useful to show several numerical illustrations for the cloud software. We show several numerical illustrations by using the StarlingX Project as the edge software.

Figs. 1.1–1.8 show the estimated results based on correspondence analysis for assignee, component, hardware, OS, reporter, severity, status, and version. We discuss the estimated results for each factor as follows:

- Nickname of fault assignee:
  Assignee and component exhibit a similar tendency.
- Name of software component:
  Assignee and component exhibit a similar tendency.
- Name of hardware:
  "ppc641e" and "x84_64" show different trends.
- Name of operating system:
  Hardware and OS exhibit a similar tendency.
- Nickname of fault reporter:

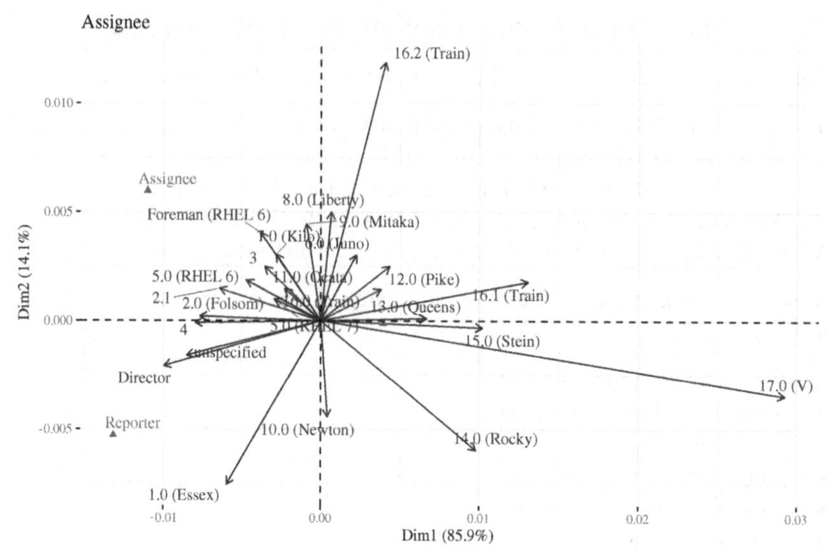

FIG. 1.1 The estimated results based on correspondence analysis for assignee.

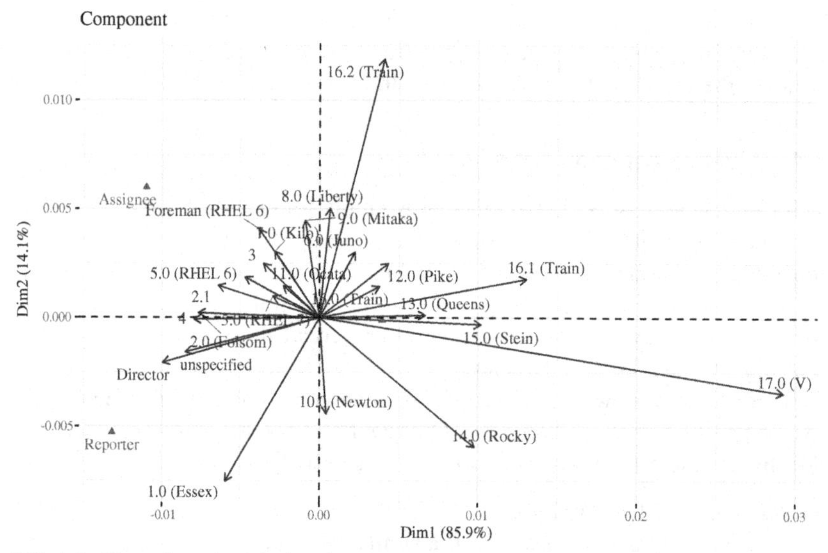

FIG. 1.2 The estimated results based on correspondence analysis for component.

Hardware, OS, and reporter exhibit a similar tendency.

- Fault severity level:
  The low level of fault severity is related to the Reporter.
- Status of software fault:
  The reporter and assignee have a specific role.

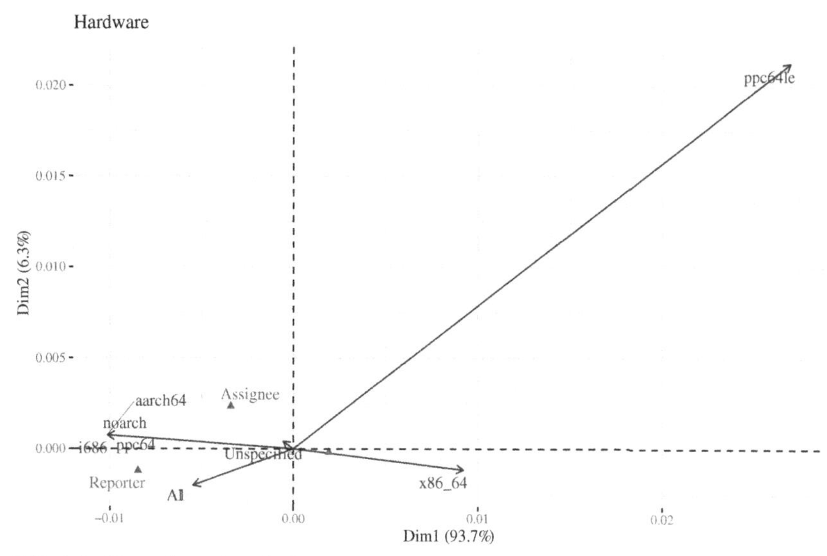

FIG. 1.3 The estimated results based on correspondence analysis for hardware.

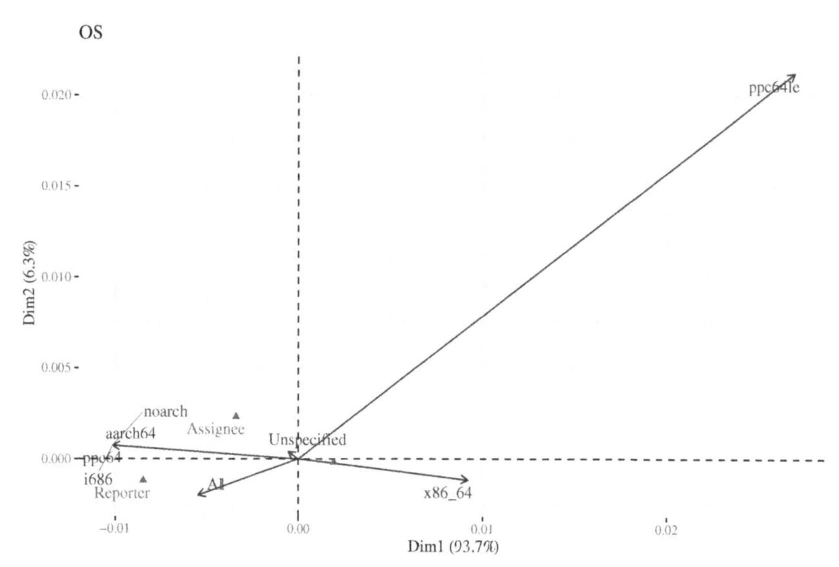

FIG. 1.4 The estimated results based on correspondence analysis for OS.

- Number of software version:
  The old and new versions have a specific role for the assignee and the reporter, respectively.

The correspondence analyses such as Figs. 1.1–1.8 will be highly useful for the edge-OSS developers to understand the reliability of edge-OSS project

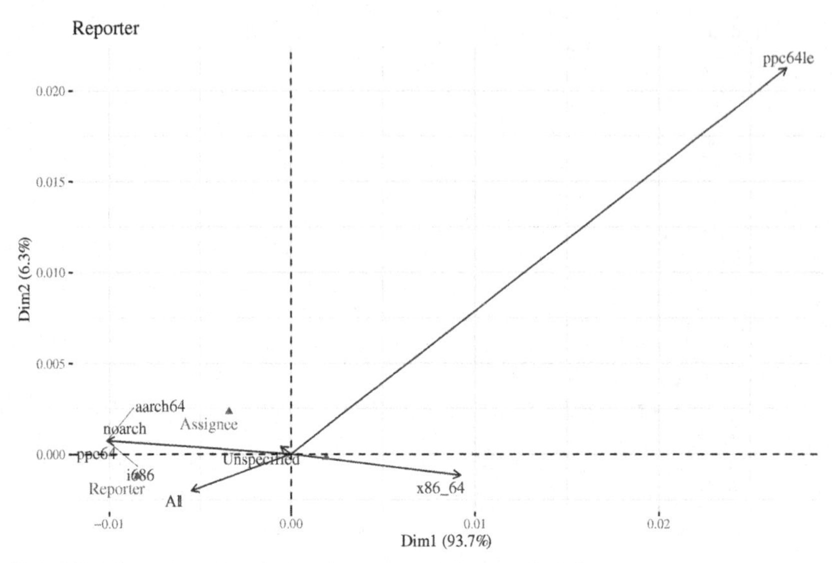

FIG. 1.5 The estimated results based on correspondence analysis for reporter.

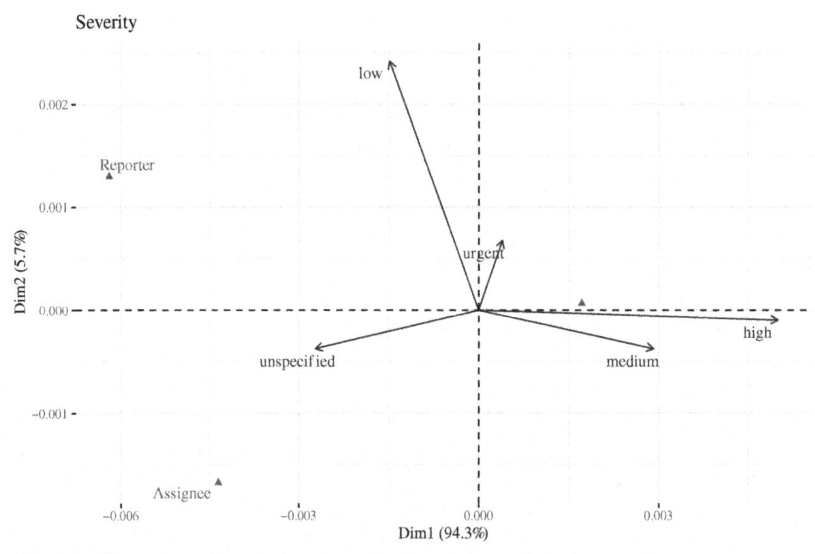

FIG. 1.6 The estimated results based on correspondence analysis for severity.

from various perspectives. Especially, edge-OSS project is structured from assignee, component, hardware, OS, reporter, severity, status, version, etc. By focusing on various factors such as assignee, component, hardware, OS, reporter, severity, status, and version, the edge-OSS developers can understand the reliability of edge OSS from various perspectives.

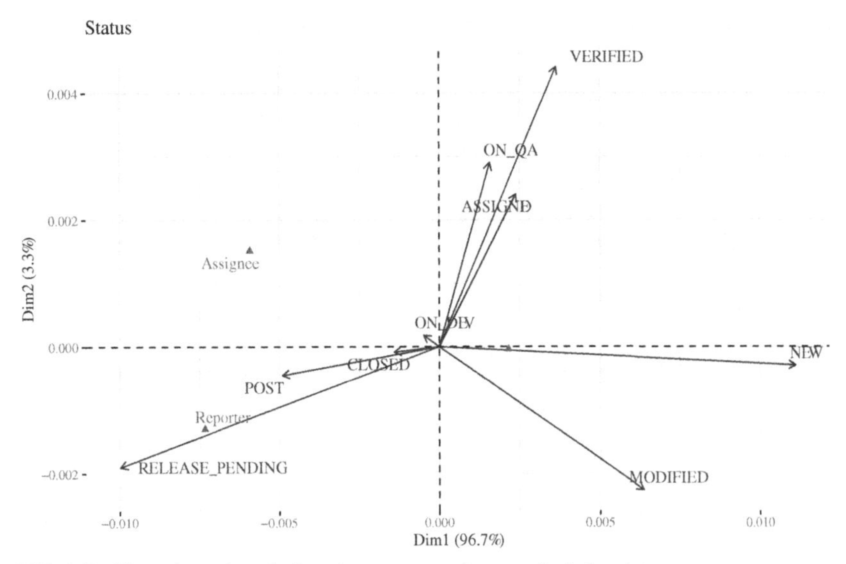

FIG. 1.7 The estimated results based on correspondence analysis for status.

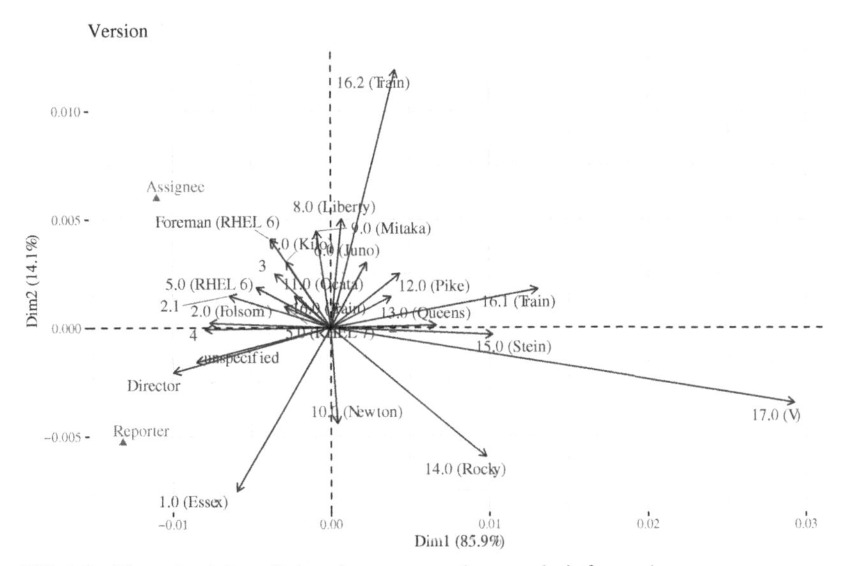

FIG. 1.8 The estimated results based on correspondence analysis for version.

1.5 Concluding remarks

This chapter focuses on statistical analysis by using correspondence analysis for OSS projects. The quality control for OSS is highly important for the edge-OSS developers to manage the edge-OSS quality, software reliability, and cost saving of edge-OSS projects. This chapter has proposed the visualization method based on the correspondence analysis for OSS assessment considering the whole OSS system. It is highly difficult for the edge-OSS project managers to make a visualization of the reliability trend of whole OSS. Then this chapter has discussed several numerical illustrations of visualization based on the correspondence analysis. The discussed method will be available as the reliability visualization method based on the visualization considering the component-based assessment for whole OSS.

Acknowledgment

This work was supported in part by the JSPS KAKENHI Grant no. 20K11799 in Japan.

References

[1] S. Yamada, Y. Tamura, OSS Reliability Measurement and Assessment, Springer International Publishing, Switzerland, 2016.

[2] J. Norris, Mission-critical development with open source software, IEEE Softw. Mag. 21 (1) (2004) 42–49.

[3] Y. Zhou, J. Davis, OSS reliability model: an empirical approach, in: Proceedings of the Fifth Workshop on OSS Engineering, 2005, pp. 67–72.

[4] S. Yamada, Software Reliability Modeling: Fundamentals and Applications, Springer-Verlag, Tokyo/Heidelberg, 2014.

[5] A. Lamkanfi, J. Pérez, S. Demeyer, The Eclipse and Mozilla defect tracking dataset: A genuine dataset for mining bug information, in: Proceedings of 2013 10th Working Conference on Mining Software Repositories (MSR), San Francisco, CA, 2013, pp. 203–206, https://doi.org/10.1109/MSR.2013.6624028.

[6] A. Kaur, S.G. Jindal, Bug report collection system (BRCS), in: Proceedings of 2017 7th International Conference on Cloud Computing, Data Science & Engineering—Confluence, Noida, 2017, pp. 697–701, https://doi.org/10.1109/CONFLUENCE.2017.7943241.

[7] T. Zhang, B. Lee, A bug rule based technique with feedback for classifying bug reports, in: Proceedings of 2011 IEEE 11th International Conference on Computer and Information Technology, Paphos, Cyprus, 2011, pp. 336–343, https://doi.org/10.1109/CIT.2011.90.

[8] S. Just, R. Premraj, T. Zimmermann, Towards the next generation of fault tracking systems, in: Proceedings of 2008 IEEE Symposium on Visual Languages and Human-Centric Computing, Herrsching am Ammersee, 2008, pp. 82–85, https://doi.org/10.1109/VLHCC.2008.4639063.

[9] T. Zimmermann, R. Premraj, J. Sillito, S. Breu, Improving fault tracking systems, in: Proceedings of 2009 31st International Conference on Software Engineering-Companion Volume, Vancouver, BC, 2009, pp. 247–250, https://doi.org/10.1109/ICSE-COMPANION.2009.5070993.

[10] S. Gujral, G. Sharma, S. Sharma, Diksha, Classifying bug severity using dictionary based approach, in: Proceedings of 2015 International Conference on Futuristic Trends on Computational Analysis and Knowledge Management (ABLAZE), Noida, 2015, pp. 599–602, https://doi.org/10.1109/ABLAZE.2015.7154933.

[11] A.W. Khan, S. Kumar, An expert system framework for bug tracking and management, in: Proceedings of 2020 International Conference on Information Science and Communication Technology (ICISCT), KARACHI, Pakistan, 2020, pp. 1–9, https://doi.org/10.1109/ICISCT49550.2020.9080048.

[12] A.F. Otoom, D. Al-Shdaifat, M. Hammad, E.E. Abdallah, Severity prediction of software bugs, in: Proceedings of 2016 7th International Conference on Information and Communication Systems (ICICS), Irbid, 2016, pp. 92–95, https://doi.org/10.1109/IACS.2016.7476092.
[13] Y. Yuk, W. Jung, Comparison of extraction methods for fault tracking system analysis, in: Proceedings of 2013 International Conference on Information Science and Applications (ICISA), Pattaya, Thailand, 2013, pp. 1–2, https://doi.org/10.1109/ICISA.2013.6579462.
[14] M. Ida, Sensitivity analysis for correspondence analysis and visualization, in: Proceedings of 2009 ICCAS-SICE, Fukuoka, 2009, pp. 735–740.
[15] H. Ahn, C. Park, K.P. Kim, A correspondence analysis framework for workflow-supported performer-activity affiliation networks, in: Proceedings of 16th International Conference on Advanced Communication Technology, Pyeongchang, 2014, pp. 350–354, https://doi.org/10.1109/ICACT.2014.6778980.
[16] X. Qin, Q. Wei, Y. Hu, A modified method on economic capital distribution of commercial banks based on correspondence analysis, in: Proceedings of 2011 2nd International Conference on Artificial Intelligence, Management Science and Electronic Commerce (AIMSEC), Dengleng, 2011, pp. 6503–6508, https://doi.org/10.1109/AIMSEC.2011.6010440.
[17] The OpenStack project, Build the future of Open Infrastructure. https://www.openstack.org/.

Yoshinobu Tamura received the BSE, MS, and PhD degrees from Tottori University in 1998, 2000, and 2003, respectively. From 2003 to 2006, he was a research assistant at Tottori University of Environmental Studies. From 2006 to 2009, he was a Lecturer and Associate Professor at Hiroshima Institute of Technology, Hiroshima, Japan. From 2009 to 2017, he was an Associate Professor at the Graduate School of Sciences and Technology for Innovation, Yamaguchi University, Ube, Japan. From 2017 to 2019, he has been working as a professor at the Faculty of Knowledge Engineering, Tokyo City University, Tokyo, Japan. Since 2020, he has been working as a professor at the Faculty of Information Technology, Tokyo City University, Tokyo, Japan. Since 2021, he has been working as a professor at the Graduate School of Sciences and Technology for Innovation, Yamaguchi University, Ube, Japan. The research interests of Yoshinobu Tamura include reliability assessment for edge computing, cloud computing, big data, optimization, and reliability. He is a regular member of the Institute of Electronics, the Information and Communication Engineers of Japan, the Operations Research Society of Japan, the Society of Project Management of Japan, the Reliability Engineering Association of Japan, and the IEEE. He has authored the book entitled OSS Reliability Measurement and Assessment (Springer International Publishing, 2016). Dr. Tamura received the Presentation Award of the Seventh International Conference on Industrial Management in 2004, the IEEE Reliability Society Japan Chapter Awards 2007, the Research Leadership Award in Area of Reliability from the ICRITO 2010, the Best Paper Award of the IEEE International Conference on Industrial Engineering and Engineering Management in 2012, the title of honorary professor from Amity University of India in 2017, the Best Paper Award of the 24th ISSAT International Conference on Reliability and Quality in Design 2018.

Shigeru Yamada received the BSE, M.S, and PhD degrees from Hiroshima University, Japan, in 1975, 1977, and 1985, respectively. Since 1993, he has been working as a professor at the Department of Social Management Engineering, Graduate School of Engineering, Tottori University, Japan. He has published over 500 reviewed technical papers in the area of software reliability engineering, project management, reliability engineering, and quality control. He has authored several books entitled Introduction to Software Management Model (Kyoritsu Shuppan, 1993), Software Reliability Models: Fundamentals and Applications (JUSE, Tokyo, 1994), Statistical Quality Control for TQM (Corona Publishing, Tokyo, 1998), Software Reliability: Model, Tool, Management (The Society of Project Management, 2004), Quality-Oriented Software Management (Morikita Shuppan, 2007), Elements of Software Reliability Modeling Approach (Kyoritsu Shuppan, 2011), Project Management (Kyoritsu Shuppan, 2012), Software Engineering: Fundamentals and Applications (Suurikougaku Publishing, 2013), Software Reliability Modeling: Fundamentals and Applications (Springer-Verlag 2014), and OSS Reliability Measurement and Assessment (Springer International Publishing 2016). Dr. Yamada received the Best Author Award from the Information Processing Society of Japan in 1992, the TELECOM System Technology Award from the Telecommunications Advancement Foundation in 1993, the Best Paper Award from the Reliability Engineering Association of Japan 1999, the International Leadership Award in Reliability Engineering Research from the ICQRIT/SREQOM 2003, the Best Paper Award at the 2004 International Computer Symposium, the Best Paper Award from the Society of Project Management in 2006, the Leadership Award from the ISSAT 2007, the Outstanding Paper Award at the IEEE International Conference on Industrial Engineering and Engineering Management (IEEM2008), the International Leadership and Pioneering Research Award in Software Reliability Engineering from the SREQOM/ICQRIT 2009, the Exceptional International Leadership and Contribution Award in Software Reliability at the ICRITO' 2010, 2011 Best Paper Award from the IEEE Reliability Society Japan Chapter 2012, the Leadership Award from the ISSAT 2014, the Project Management Service Award from the SPM 2014, the Contributions Award for Promoting OR from the ORSJ 2017, and the Research Award from the ISSAT 2017. He is a regular member of the IEICE, the Information Processing Society of Japan, the Operations Research Society of Japan, the Reliability Engineering Association of Japan, the Japan Industrial Management Association, the Japanese Society for Quality Control, the Society of Project Management, and the IEEE.

Chapter 2

Analytical modeling and performance evaluation of SIP signaling protocol: Analytical modeling of SIP

Nikesh Choudhary[a], Vandana Khaitan (nee Gupta)[a], and Vaneeta Goel[b]
[a]Department of Operational Research, University of Delhi, India, [b]Satyawati College, University of Delhi, India

2.1 Introduction

Voice over Internet Protocol (VoIP) (also known as IP telephony) is an approach and collection of technologies for transmitting voice and multimedia sessions over IP networks. It has been actively used everywhere due to its economical cost and noted flexibility related to Public Switched Telephone Networks (PSTN) [1]. One of the most powerful VoIP signaling protocols is the Session Initiation Protocol (SIP) [2]. It is an application-layer signaling protocol introduced by the Internet Engineering Task Force (IETF) considering Request for Comments (RFC) 3261 in 2002. It was formerly designed by the Network Working Group (NWG) for creating, modifying, and tear-down multimedia sessions. This protocol can work on both Transmission Control Protocol (TCP) and User Datagram Protocol (UDP). It also works with other protocols such as session description protocol (SDP), real-time transmission (RTP), resource reservation protocol (RSVP), session announcement protocol (SAP), lightweight directory access protocol (LDAP), real-time streaming protocol (RTSP), and remote authentication dial-in user service protocol (RADIUS).

SIP is also known as a *transaction-oriented protocol* by which messages are carried out through different transactions. The transaction mechanism in SIP controls the exchange of messages between the client and the server and reliably delivers these messages. The two major SIP transactions are the INVITE transaction and the non-INVITE transaction. For establishing a session, the INVITE transaction is used, and for modifying and terminating a

System Assurances. https://doi.org/10.1016/B978-0-323-90240-3.00002-3

session, the non-INVITE transaction is used [3]. In the INVITE transaction, the client sends the INVITE request or acknowledgments to the server, and the server sends different types of responses corresponding to these requests and acknowledgments. Predicting its importance and magnificent potential to grow as a major signaling protocol, IETF chose to operate SIP over H.323 protocol. Despite its potential to succeed H.323, many years of implementation and experimentation have been conducted in the SIP specification, RFC 3261 [4]. The SIP architecture considered in the chapter is adopted from Ref. [4]. Modeling and interpreting SIP specifications using approved methods can assist in ensuring that the phrasing of RFC 3261 is accurate, certain, and user friendly. IETF has published a revised version of the SIP specification, RFC 6026 [5], in September 2010. It also invites people for unicast and multicast sessions.

2.2 Background work

Since put into operation in 2002, SIP has been an interesting research topic. Most of the research articles in SIP are based on its working and services. Suryawanshi et al. [2] described the architecture of SIP, its message types, its applications, and many more. The authors in Ref. [3] discussed the call flow setup in SIP, different servers as a proxy server, network server, registrar server, and SIP security issues. The authors in Ref. [6] determined the potential association between two objects in a social network represented by a graph including edges and nodes. Howie et al. [7] described the advanced form of SIP and its use in collaborating peer-to-peer applications on cellular phones. The authors in Ref. [8] presented the communication protocol design to secure SIP regulated communication of VoIP.

However, we observed that when it comes to the analysis and modeling of SIP, only a few research articles are available in the literature [9, 10]. Lin [9] suggested colored petri nets (CPNs) to design and analyze the INVITE transaction of SIP over the unreliable medium in which messages are reordered and lost. Barakovic et al. [10] considered the simulation model of SIP INVITE transaction designed using CPNs. According to this model, they concluded that the SIP INVITE transactions are free from live locks and dead codes and also observed that these dead codes are both desirable and undesirable at the same time. They also suggested that the INVITE transaction of SIP ought to be suppressed for further improvement to reduce unwanted deadlock. Filal and Bouhdadi [11] considered the reliable and unreliable medium of the SIP INVITE transaction and verified it by using the simulation technique of "Event-B" using the Rodin platform. Kizmaz and Kirci [12] accomplished that the timed colored petri nets (TCPNs) can be used to analyze the functional properties of SIP INVITE transactions.

In the literature discussed earlier, we observe that most of the work on SIP is based on simulation. This inspired us to propose an analytical model for the performance analysis of the protocol.

2.2.1 Motivation

There is plenty of simulation-based research on SIP, but very less literature is available on the analytical modeling of the SIP INVITE transaction. This encouraged us to develop an analytical model for the SIP invite transaction because analytical models provide an absolutely precise connection between the inputs and the outputs. On the contrary, results obtained from simulation models are required to be statistically explained. Moreover, the results of the simulation model are time consuming, error prone, and expensive. The analytical model explains and concentrates on the research procedure "mathematically and statistically." In contrast, the simulation model compares the research work to another related work following the equivalent simulator's situation having similar parameters to test its performance. Hence, in this chapter, we construct an analytical model that provides an abstract representation of the SIP INVITE transaction and helps in analyzing its performance.

Keeping in mind the dynamic behavior of SIP, a stochastic process, specifically a continuous-time Markov chain (CTMC), seems appropriate to develop the analytical model. However, the CTMC model is quite far away from the conventional feel of the system being modeled. Moreover, because of the complex functionality of the SIP, the corresponding CTMC is expected to have a very large state space. To overcome this issue of large state space, we consider proposing a stochastic reward net (SRN) model which is more crisp in its specifications and is almost same as the designer's perception of the system. Also, many software packages are available that can spontaneously transform the SRN model into the underlying CTMC and solve it automatically. In addition to this, the SRN modeling technique has been effectively used in the literature to analyze the performance measures of different protocols, such as TCP [13], TCP NewReno [14], IEEE802.11 MAC [15], RLC protocol of LTE [16], etc. The SRN has proven to be the most suitable tool to describe the dynamic behavior of these protocols by providing a graphical understanding of the packet or message transmission. Hence, we propose an SRN model for the SIP INVITE transaction. Major features of SIP such as various timers, INVITE retransmission, response retransmission, together with the arrival of extra acknowledgments during retransmission are considered in the proposed model. SHARPE [17] software is then used for the numerical analysis of the suggested model. A brief introduction of SRN and its components is given in Section 2.3. For a detailed study on the SRN modeling technique, the reader can refer to Ref. [17].

2.2.2 Main contribution

The important contributions of this chapter are as follows:

1. The chapter proposes an analytical model that provides an abstract representation of SIP INVITE transaction.
2. The analytical model is developed for the SIP INVITE transaction over the unreliable medium using SRN modeling technique.
3. All important features of SIP INVITE transaction such as timers, re-INVITEs, and retransmissions are incorporated in the SRN model.
4. Performance measures such as throughput and latency of the SIP INVITE transaction are determined.
5. The analytical results are validated with the help of discrete-event simulation performed using MATLAB.

The remainder of this chapter is structured in the following manner. In Section 2.3.1, various types of SIP responses are discussed. Section 2.3.2 briefly presents the SIP layered structure and its working. The introduction of SRN and its basic components are discussed in Section 2.4.1. Section 2.4.2 presents the proposed SRN of the SIP INVITE transaction. In Section 2.5, the performance measures of the protocol are discussed. Section 2.6 represents the numerical observations corresponding to the measures of performances discussed in Section 2.5. In Section 2.7, the model validation details are presented. Finally, the chapter is concluded in Section 2.8.

2.3 SIP layered structure and its working

In this section, we discuss the layered structure of SIP INVITE request and its working in correspondence with the client and the server transactions. The SIP architecture we are considering in this chapter is adopted from Ref. [4].

2.3.1 Types of responses of SIP

Before beginning with SIP working, we first need to understand the various types of *responses* used in the protocol. When the client transaction transfers the INVITE request to the server transaction, the server transfers some responses to this INVITE request (see Table 2.1). These responses are described as follows:

1xx (100–199) responses: These responses are called provisional responses. No or numerous provisional responses may appear before single or multiple successful or unsuccessful responses are accepted. Provisional responses meant for an INVITE request can produce "early messages." These early messages in the ongoing session are represented by the *100 trying* response to quench the INVITE request retransmission. These early messages will be required if the server requires to transmit a response to its client inside the message before the primary INVITE transaction ends. Provisional responses

TABLE 2.1 Types of SIP response messages.

Responses	Functions
1*xx*	Provisional response (request received despite that not approved)
2*xx*	Success response (request efficiently received and approved)
3*xx*	Redirection response (some actions are intended to fulfill the request)
4*xx*	Request failure response (some actions are needed to complete the request)
5*xx*	Server error response (server is unsuccessful to respond to the request)
6*xx*	Global failure response (not a single server is available to respond to the request)

are also called "informational responses." 1xx contains different responses such as *180-ringing*, *181-call is being forwarded*, *182-queued*, etc.

2xx (200–299) responses: These responses are called *successful responses*. There is a response in *2xx*, which is represented as *200 OK*. If the client receives *200 OK* response from the server, it represents that the server successfully accepts the request. After receiving this successful response, the client sends an acknowledgment to the server.

3xx (300–399) responses: *3xx* responses provide details regarding the server's position or optional settings that can probably complete the call. *3xx* contains different responses such as *300-multiple responses*, *301-moved permanently*, *302-moved temporarily*, etc.

4xx (400–499) responses: *4xx* responses are the server's complete failure responses. The client cannot retransmit a similar request unless required corrections are done (e.g., attaching relevant authorization). *4xx* includes responses such as *400 bad request*, *401 unauthorized*, etc.

5xx (500–599) responses: These are the failure responses generated when the server encounters some errors. *5xx* contains responses namely *500-server internal error*, *501-not implemented*, *502-bad gateway*, etc.

6xx (600–699) responses: *6xx* responses show that the server contains comprehensive details regarding an appropriate client, and not merely the particular situation shown in the request. Different responses in *6xx* are *600 busy everywhere*, *603 decline*, *604 does not exist anywhere*, etc.

All the responses from *3xx* to *6xx (3xx–6xx)* are called *unsuccessful responses*.

2.3.2 Layered structure and working of SIP

We discuss the layered structure and functioning of the SIP protocol in this section. SIP is structured into four layers, each performing a special task

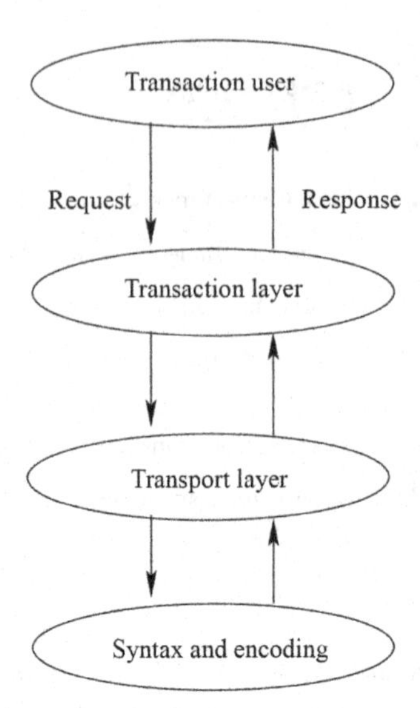

FIG. 2.1 SIP layered architecture.

(see Fig. 2.1). *Syntax and encoding* is the lowest layer of the SIP layered structure. Augmented Backus-Naur Form (ABNF) grammar is used to specify the encoding of the lowest layer. The *transport layer* is the second layer from bottom to top. As mentioned earlier, this layer represents how a client transmits requests and accepts responses, as well as how a server accepts requests and transmits responses across the network. The third layer is the *transaction layer* which lies above the transport layer. This layer manages retransmission by the application layer and matches the responses to the requests and time-outs by the application layer in case of initiation and termination of a session. The topmost layer of the SIP layered structure is called the *transaction user* (*TU*). The TU generates and abandons transactions of SIP and makes use of services provided by the transaction layer.

SIP is a *transaction-oriented protocol* that performs tasks from one transaction to another transaction. Therefore, among all the layers of SIP, the *transaction layer* is the most significant layer. There is only one INVITE request message in the SIP transaction, and corresponding to this INVITE request message, there are many response messages. The transaction is carried out through two sides: the client side and server side. The transactions that occur on the client side are known as *client transactions*, and the transactions that occur on the server side are known as *server transactions*. Requests are sent through client transactions, and responses are sent through server transactions. The objective of a client transaction is to create the INVITE request and transmit the request message to the server transaction. On the other hand, the server transaction accepts a request from the *transaction layer* and transfers it to its TU.

Also, it takes responses from the TU and transfers these responses to the *transaction layer*. Depending on the pattern of the requests and responses received by the client TU and the server TU, there exist two kinds of client transactions, that is, INVITE client transaction and non-INVITE client transaction. INVITE client transaction is used to handle the INVITE requests, and the non-INVITE client transaction is used to handle all types of requests other than the INVITE requests and acknowledgments. Similarly, a server transaction is also of two kinds, INVITE server transaction and non-INVITE server transaction. The INVITE server transaction handles the responses regarding INVITE requests, and the non-INVITE server transaction is used to handle all types of responses other than the INVITE requests and acknowledgments. Fig. 2.2 presents a standard SIP message transaction to establish a session. The illustration comprises two user agents, client and server, and a proxy server. The client transmits an "INVITE" request showing that the client is ready to set up a session with the server. The server then sends different responses, a *100-trying* response before sending a *200-OK*, or *3xx–6xx* response, which is used to quench the requested transmission. The *100-trying* response symbolizes that the server has taken the INVITE request; the server sends the *200-OK* to represent that he/she is ready to participate in this multimedia session, whereas *300–600* responses are

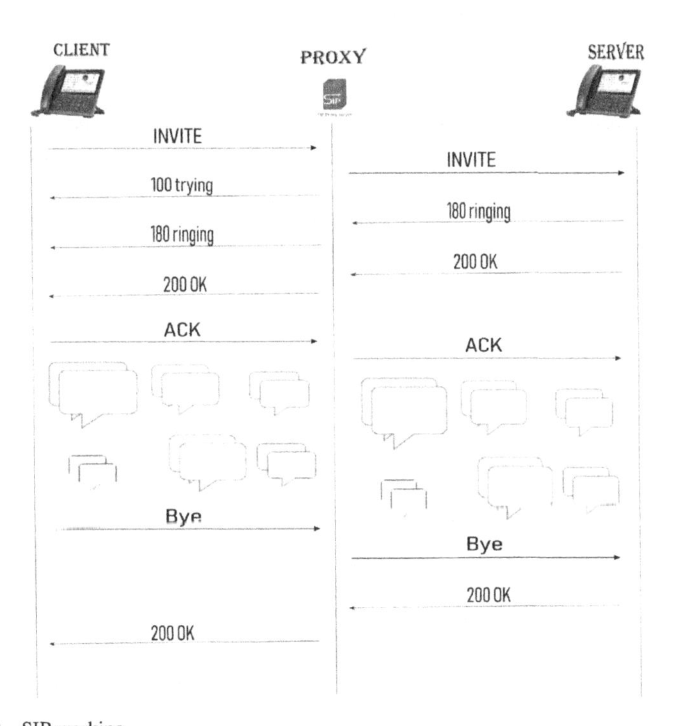

FIG. 2.2 SIP working.

comprised of many responses which represent that there is some error in the request. Upon acceptance of the *200-OK* or *300–600* responses, the client forwards an "ACK" to acknowledge the acceptance of these responses. SIP manages bidirectional handshakes to set up a session: "INVITE request-*100-Trying* response" and "*200-OK (or 300–600)* responses-ACK" handshake. The former handshake is among adjacent SIP nodes such as the client and the proxy, and the latter handshake is among the proxy and its ultimate destination, that is, server. The session is set up by receiving *200 OK* responses from the server. Then the client begins the audio/video conversation. The audio/video data packets are transmitted by following transport protocols such as TCP, UDP, etc. When the client wants to terminate the call, a BYE message is transmitted to the server. The server approves the BYE message by sending a *200 OK* response that tears down a session.

2.4 Proposed SRN model of SIP INVITE transaction

Before presenting the SRN model for the SIP-INVITE-transaction, we first discuss the basics of the SRN modeling technique.

2.4.1 SRN and its attributes

SRN is an enhanced version of stochastic petri nets (SPNs), and SPN is an extension of petri net (PN). PN is a bipartite directed graph in which the nodes are segregated into two disjoint sets, namely, *places* and *transitions*. PN is extended to SPN with time involvement, and a further extension of SPN is known as SRN. For complex systems, it is a top-level descriptive language. SRN has been established as a strong modeling tool for the performance analysis and reliability and availability computation of communication networks. Fig. 2.3 shows the general attributes of an SRN model, which are described as given below:

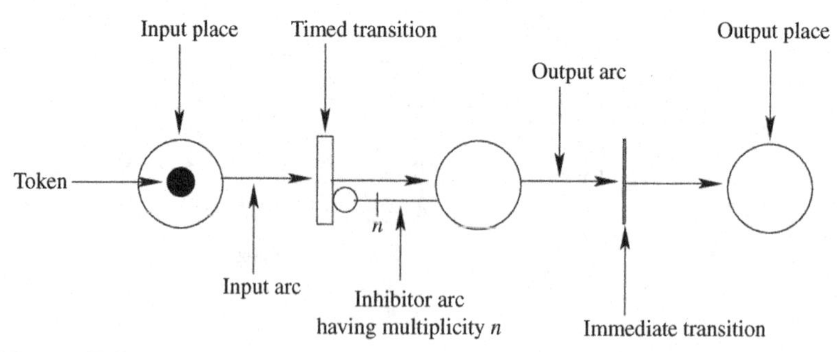

FIG. 2.3 SRN components.

1. *Places*: Places are represented by circles that indicate the state of the system. It can hold tokens and has infinite capacity by default. Tokens are illustrated as dots or numbers in the places.
2. *Transitions*: A transition is represented by a rectangular bar and indicates an action to be performed. It has no capacity and cannot store tokens at all.
3. *Types of places and arcs*:
   - Places and transitions are linked through directed arcs. An arc links a place to a transition and a transition to a place, but no arc can link a transition to a transition or a place to a place. The arcs directed toward the transitions are known as *input arcs* and their corresponding places are known as *input places*. Each arc has some multiplicity, which is represented as a number.
   - The arcs directed out of the transitions are called as *output arcs* and their equivalent places are called as *output places*.
4. *Enabling of transition*: If every input place of a transition holds as many tokens, which are at least equal to the input arcs multiplicities (a positive integer associated with each input arc), then the transition is said to be enabled.
5. *Inhibitor arc*: An arc connecting a place and a transition along with the tiny hollow ring rather than the arrows at the terminating end, having multiplicity $k \geq 1$ is called an *inhibitor arc*. If no multiplicity is mentioned on the arc, then it is assumed to be 1. If an inhibitor arc has multiplicity n, and its input place contains tokens which are greater than or equal to n, then the transition, in spite of being enabled, is inhibited. Firing of a transition can take place only when it is enabled and not inhibited. On firing of a transition, amount of tokens same as the multiplicity of the input arcs is eliminated from all the input places, and amount of tokens same as the multiplicity of the output arcs is transferred to all its output places.
6. *Types of transition*:
   - The transition having some random firing time (distributed exponentially) other than zero is called a *timed transition*.
   - The transition with zero firing time is called an *immediate transition*. Immediate transition possesses higher priority than the timed transition.
7. *Guard function*: The *Guard function* "g" is a Boolean expression associated with a transition, say T. A transition T is enabled if its guard function, g is satisfied (i.e., $g = 1$) and disabled if the guard function $g = 0$.
8. *Marking*: The state of a PN having N places is represented as a marking $N(t)$ which contains N tuples, $(\eta_1(t),\eta_2(t),\ldots,\eta_N(t))$ of positive integers (including zero), where $\eta_i(t)$; $(1 \leq i \leq M)$, indicates the amount of tokens in the ith place at any given time, t. The starting state of the PN is represented by the *initial marking* denoted by "$M(t_0)$" which is at time $t = 0$.

9. *Reward rate*: The *reward* is a positive (including zero) weight allocated to every single marking, and the *reward rate* is defined as the overall marking of the weighted mean of the amount of tokens in the specified place.
10. *Firing probability*: "If multiple transitions struggle for firing, then some weights are needed to be specified to resolve these conflicts" [18]. These weights are known as *firing probabilities*. The probabilities associated with the firing rates of the transition sum up to 1.

$$\sum_{i=1}^{n} p_i = 1.$$

2.4.2 Proposed SRN model of SIP INVITE transaction

We now develop an SRN model for the SIP INVITE client transaction and SIP INVITE server transaction based on its work discussed earlier. For a comprehensive study on modeling using SRN, one can refer to Ref. [18]. Though we have worked with SRN as a modeling tool, we do not use rewards in this model. However, we use other notions such as guard functions that are assisted only by an SRN model but not by a generalized stochastic petri net (GSPN) model. The SRN model depicting the SIP INVITE transaction is shown in Fig. 2.4. This model consists of mainly two parts, namely the client side and the server side. The meanings of each timed transition, place, guard functions, and immediate transitions used in the SRN model are illustrated in Tables 2.2–2.5, respectively. Further, it is to be noted here that even though some functions in the SIP INVITE transaction can be deterministic such as different types of timers such as *timer M*, *timer H*, *timer L*, etc., it is assumed that the firing times of each and every timed transition are distributed exponentially to facilitate modeling using SRN.

The INVITE client transaction and server transaction are initiated by the TU at client and server sides, respectively. When the client's TU wishes to start a session, it creates an INVITE request depicted by the firing of the timed transition T_{client} (see Fig. 2.4). This transition has exponentially distributed firing time with mean λ_1. The INVITE client transaction goes through five different stages one by one after initiating a session. These five stages are:

1. Calling 2. Proceeding 3. Accepted 4. Completed 5. Terminated

These five stages are represented by the places $P_{calling}$, P_{ProcC}, $P_{acceptC}$, P_{compC}, and $P_{terminated}$.

When a transition T_{client} fires, a token is transferred to the place $P_{calling}$. This means that the client creates an INVITE request, and the initial *calling* state is reached. A token in the place $P_{calling}$ enables the transition T_{timerA} which represents that the INVITE request is delivered to server side through the transport

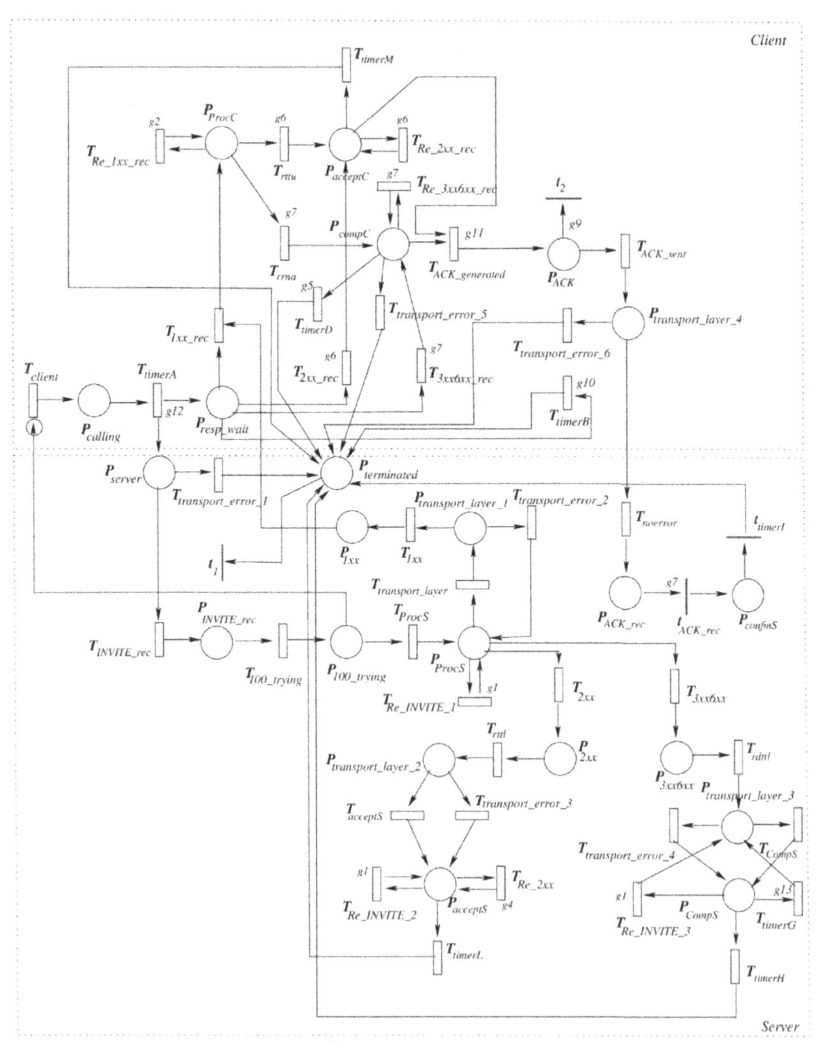

FIG. 2.4 SRN model of SIP INVITE transaction.

layer. As soon as the transition T_{timerA} gets enabled, the guard function with the probability g_{12} checks whether the place P_{INVITE_rec} is empty or not. If the place is empty, then the transition T_{timerA} will fire and the token moves from $P_{calling}$ to either P_{server} or P_{resp_wait} with probabilities p and $(1 - p)$, respectively. A token in the place P_{server} represents that the INVITE request is successfully received by the server TU, and a token in the place P_{resp_wait} describes that the client is waiting for the response while the INVITE request is handled by the server transaction.

TABLE 2.2 Transitions and their meaning.

Transition	Meaning	Average time ($Rate^{-1}$) (ms)
T_{100_trying}	Provisional response needed to quench INVITE request retransmission	0.000048
T_{1xx}	1xx response is generated by the server	0.000055
T_{1xx_rec}	1xx response is received by the client transaction	0.000060
T_{2xx}	2xx response is generated by the server transaction	0.000058
T_{2xx_rec}	2xx response is received by the client transaction	0.000058
T_{3xx6xx}	3xx6xx responses are generated by the server transaction	0.000060
T_{3xx6xx_rec}	3xx6xx responses are received by the client transaction	0.000060
$T_{acceptS}$	Server start entering in the accepted state	0.00004
$T_{ACK\_generated}$	Acknowledgment is generated by the client transaction	0.00003
T_{ACK_sent}	Acknowledgment is sent to server transaction	0.00003
T_{client}	Client transaction creates an INVITE request	0.00001
T_{CompS}	Server transaction start entering in the completed state	0.00004
$T_{transport\_error\_2}$	Transport error occur in the proceeding state	0.000065
$T_{transport\_error\_5}$	Transport error occur in the completed state	0.000065
T_{INVITE_rec}	INVITE request is received by the server transaction	0.000045
$T_{noerror1}$	Successful transmission of acknowledgment to server transaction	0.000060
$T_{noerror2}$	Successful transmission of acknowledgment to server transaction	0.000060
T_{ProcS}	Server start entering in the proceeding state	0.00005
T_{Re_INVITE}	INVITE request retransmission	0.0002
$T_{Re\_1xx\_rec}$	1xx response is again received by the client transaction	0.000062
T_{rttl}	2xx response sent to transport layer	0.00004
T_{re_INVITE}	INVITE request retransmission	0.0002

TABLE 2.2 Transitions and their meaning—cont'd

Transition	Meaning	Average time ($Rate^{-1}$) (ms)
T_{rttu}	2XX response received by the client TU after entering in the proceeding state	0.00004
T_{Re_2xx}	Retransmission of 2xx response	0.000062
T_{rdttl}	3xx6xx responses are delivered to transport layer	0.00004
$T_{Re\_2xx\_rec}$	Rereceiving of 2xx response by the client transaction	0.000060
$T_{Re\_3xx6xx\_rec}$	3xx6xx responses again received by client transaction	0.000062
T_{rrna}	3xx6xx responses received by client TU after entering in the proceeding state	0.00004
$T_{ReINVITE}$	INVITE request retransmission	0.0002
T_{server}	INVITE request is delivered to server transaction	0.00004
$T_{transport\_error\_6}$	Transport error occurred while in the transport layer	0.000065
$T_{transport\_layer}$	Request is delivered to transport layer after proceeding state	0.000052
T_{timerB}	Transaction time-out	0.00003125
$T_{transport\_error\_3}$	Transport error occur before entering the accepted state	0.000065
T_{timerL}	Waiting time for retransmission of 2xx response	0.00003125
T_{timerM}	Amount of time that the TU will wait for retransmission of 2xx response	0.00003125
$T_{transport\_error\_1}$	Transport error occur when acknowledgment is sent to server transaction	0.000065
$T_{transport\_error\_4}$	Transport error occur before entering in the completed state	0.000065
T_{timerG}	Response is passed to transport layer for retransmission	0.000032
T_{timerD}	Amount of time that the server TU will remain in the completed state	0.000032
T_{timerH}	Server transaction abandons retransmitting the 3xx6xx responses	0.00003125

TABLE 2.3 Places and their meaning.

Place	Meaning
P_{100_trying}	100 Trying response is generated to quench request retransmission
P_{1xx}	1xx response is generated by the server transaction
P_{2xx}	2xx response is generated by the server transaction
P_{3xx6xx}	3xx6xx response is generated by the server transaction
P_{ACK}	Acknowledgment is generated from the client transaction
$P_{acceptS}$	Accepted state is entered when the server transaction generates 2xx response successfully
$P_{acceptC}$	Accepted state is entered when the client transaction receives 2xx response from the server transaction
P_{ACK_rec}	Acknowledgment received by the server transaction from the transport layer
$P_{calling}$	Initial calling state is entered when the INVITE client transaction is created
P_{compC}	Completed state is entered when the client transaction receives 3xx6xx response from the server transaction
P_{compS}	Completed state is entered when the server transaction generates 3xx6xx response successfully
P_{confmS}	Confirmed state is entered to absorb additional acknowledgment messages that arrive from retransmission of final response
P_{INVITE_rec}	INVITE request is received by the server transaction
P_{ProcC}	Proceeding state is entered when the client transaction receives 1xx response from the server transaction
P_{ProcS}	Initial proceeding state is entered when INVITE server transaction is created
P_{resp_wait}	Response awaited after sending INVITE request
P_{server}	INVITE request is sent to server transaction
$P_{transport\_error\_1}$	Generated acknowledgment is sent to server transaction through transport layer
P_{trans_layer}	Response sent to transport layer (1xx)
$P_{transport\_error\_2}$	Response sent to transport layer (2xx)
$P_{transport\_error\_3}$	Response sent to transport layer (3xx6xx)
$P_{terminated}$	Finally the INVITE request gets terminate

TABLE 2.4 Guard functions.

Guard function	Condition
g_1	$(P_{INVITE\_rec}) \geq 1$
g_2	$\#(P_{1xx}) \geq 1$
g_3	$\#(P_{acceptS}) == 0$ or $\#(P_{1xx}) == 0$
g_4	$\#(P_{2xx}) \geq 1$
g_5	$\#(P_{CompS}) == 0$
g_6	$\#(P_{acceptS}) \geq 1$
g_7	$\#(P_{CompS}) \geq 1$
g_8	$\#(P_{CompS}) == 0$ or $\#(P_{1xx}) == 0$
g_9	$(P_{ACK\_rec}) \geq 1$
g_{10}	$\#(P_{acceptS}) == 0$ or $\#(P_{1xx}) == 0$ or $\#(P_{CompS}) == 0$
g_{11}	$\#(P_{acceptS}) \geq 1$ or $\#(P_{CompS}) \geq 1$
g_{12}	$(P_{INVITE\_rec}) == 0$
g_{13}	$\#(P_{3xx6xx}) \geq 1$ or $\#(P_{ACK}) == 0$

TABLE 2.5 Immediate transition and their meaning.

Transition	Meaning	Guard function
t_1	Flushing of terminated place	–
t_2	Flushing of extra generated acknowledgments	g_9
t_{ACK_rec}	Server transaction receives acknowledgment and send it to the confirmed state	g_7
t_{timerI}	Determines that the server changes its state from confirmed to terminated	–

A token in the place P_{resp_wait} enables four other transitions, namely T_{1xx_rec}, T_{2xx_rec}, T_{3xx6xx_rec}, and T_{timerB} with probabilities p_1, p_2, p_3, and p_4, respectively. The firing of the transition T_{1xx_rec} implies that the client transaction receives *1xx* response from the server side through transport layer. Similarly, the firing of the transition T_{2xx_rec} and T_{3xx6xx_rec} represent that the client transaction receives *2xx* and *3xx6xx* responses, respectively, from the server side through its transport layer. If, however, the transitions T_{timerB} fires first, this indicates that the client has not received any response from the server transactions, and a time-out has occurred.

If the place P_{server} has a token, then the transitions $T_{transport\_error\_1}$ and T_{INVITE_rec} get enabled with probabilities p_5 and $(1 - p_5)$, respectively. The firing of the transition $T_{transport\_error\_1}$ represents that when an INVITE request is sent to the server transaction, it encounters an error, and due to this error, the request enters the *terminated state*, that is, if the transition $T_{trans\_error\_1}$ fires first with the probability p_5, a token is transferred to the place $P_{terminated}$. Instead, the firing of the transition T_{INVITE_rec} describes that the server receives the INVITE request successfully from the client transaction and a token is transferred to the place P_{INVITE_rec} with the probability $(1 - p_5)$.

When the place P_{INVITE_rec} has a token, it enables the transition T_{100_trying}. The transition T_{100_trying} fires after an exponential time and a token is transferred to the place P_{100_trying} which indicates that the server transaction sends the 100 trying response to the client side to quench the INVITE request retransmission. The INVITE server transaction enters these five different states one by one after receiving an INVITE request from the client's TU:

1. Proceeding 2. Accepted 3. Completed 4. Confirmed 5. Terminated

These five stages are represented by the places P_{ProcS}, $P_{acceptS}$, P_{CompS}, P_{confmS}, and $P_{terminated}$. If the place P_{100_trying} contains a token, it enables the transition T_{ProcS}, and on firing of this transition a token is transferred to the place P_{ProcS}. A token in the place P_{ProcS} indicates that the server enters into the state *proceeding* by receiving the INVITE request, and by sending the *100 trying* response within 200 ms. When a token is transferred to the place P_{ProcS}, it enables four other transitions, namely $T_{Re\_INVITE\_1}$, $T_{Transport\_layer}$, T_{2xx}, and T_{3xx6xx} with probabilities p_6, p_7, p_8, and p_9, respectively. If the transition $T_{Transport\_layer}$ fires first, it implies that the *1xx* response is transmitted to the transport layer and a token is transferred to the place $P_{Transport\_layer\_1}$. If the place $P_{Transport\_layer\_1}$ contains a token, it enables two other transitions $T_{Transport\_error\_2}$ and T_{1xx} with probabilities p_{10} and $(1 - p_{10})$, respectively. If the transition $T_{Transport\_error\_2}$ fires, a token is transferred to the place P_{ProcS} again (depicting that the transport error occurs when a *1xx* response is sent to the transport layer). If the transition T_{1xx} fires, a token is transferred to the place P_{1xx}. If the place P_{1xx} contains a token, it represents that the *1xx* response is successfully generated by the server transaction. If the place P_{1xx} already contains a token and a token enters in the place P_{resp_wait}, then the transition T_{1xx_rec} gets enabled. When the transition T_{1xx_rec} fires, it represents that the client has received *1xx* response from the server transaction. Now, when the transition T_{1xx_rec} fires, a token gets transferred to the place P_{ProcC}, representing that after receiving *1xx* response from the server transaction the client enters the *proceeding state*. If the place P_{ProcS} already has a token and another token enters in the place P_{INVITE_rec}

simultaneously, then the transition $T_{Re\_INVITE\_1}$ fires first, representing that the server accepts the INVITE request repeatedly while it is in the *proceeding state*, and a token is again transferred to the place P_{ProcS}. This condition is managed by the guard function g_1.

The firing of the transition T_{2xx} transfers a token to the place P_{2xx}, representing that the server generates a *2xx* (successful) response. As a token is transferred to the place P_{2xx}, the transition T_{rttl} is enabled representing that the response is transferred to the transport layer. When the transition T_{rttl} fires, a token is transferred to the place $P_{Transmit\_layer}$. If the place $P_{transport\_layer\_2}$ has a token, it enables two other transitions namely, $T_{transport\_error\_3}$ and $T_{acceptS}$ with probabilities p_{11} and $(1 - p_{11})$, respectively. If the transition $T_{Transport\_error\_3}$ fires first, a token is transferred to the place $P_{acceptS}$ which implies that a transport error has occurred while the response is delivered to the transport layer. On the other hand, if the transition $T_{acceptS}$ fires first, a token is transferred to the same place $P_{acceptS}$ indicating that the *2xx* response is successfully delivered by the transport layer. Now, the place $P_{acceptS}$ contains a token which enables three other transitions, namely $T_{Re\_INVITE\_2}$, T_{Re_2xx}, and T_{timerL} with probabilities p_{12}, p_{13}, and p_{14}, respectively. If there is a token in the place $P_{acceptS}$ and at the same time if there are one or more tokens in the place P_{resp_wait} in the client side, the transition T_{2xx_rec} gets enabled in the client side. Simultaneously, if there is a token in the place $P_{acceptS}$ and if there are some tokens in the place P_{procC} of the client side, this enables the transition T_{rttu} (representing that the client has received *2xx* response after entering in the *proceeding state*) in the client side. Also, if the place $P_{acceptS}$ already contains a token and another token enters the place P_{INVITE_rec} at the same time, then the transition $T_{Re\_INVITE\_2}$ will fire, representing that the server transaction again receives the INVITE request while in the *accepted state*, and a token is again transferred to the place $P_{acceptS}$. This condition is managed through the guard function g_1 on $T_{Re\_INVITE\_2}$. Similarly, if in the *accepted state* the server generates *2xx* responses repeatedly, it remains in the *accepted state*. That is, if the place $P_{acceptS}$ already has a token and another token enters in the place P_{2xx} at the same time, then the transition T_{Re_2xx} will fire, representing the repeated generation of *2xx* responses on the server side while in the *accepted state*, and a token is again transferred to the place $P_{acceptS}$. This condition is managed through the guard function g_4 on T_{Re_2xx}. Also, if the transition T_{timerL} fires, a token enters the place $P_{terminated}$, where the firing time of the transition T_{timerL} represents the waiting time for the retransmission of *2xx* (success) response.

When the place P_{ProcS} has a token, the transition T_{3xx6xx} gets enabled and fires, and then a token is transferred to the place P_{3xx6xx} indicating that the server transaction generates the *3xx6xx* (unsuccessful) responses. These unsuccessful responses are generated because the server is moved temporarily or permanently or maybe the server is busy anywhere else as there could be other reasons. If the place P_{3xx6xx} contains a token, it enables the transition T_{rdttl} and when this transition fires, a token is transferred to the place $P_{transport\_layer\_3}$

(describing that response is sent to the transport layer). If the place $P_{transport\_layer\_3}$ contains a token, it enables two other transitions, namely $T_{Transport\_error\_4}$ and T_{CompS} with probabilities p_{15} and $(1 - p_{15})$, respectively. If the transition $T_{Transport\_error\_4}$ fires first, a token is transferred to the place P_{CompS}, representing that some transport error has occurred while moving through the transport layer. If the transition T_{CompS} fires, a token is transferred to the place P_{CompS}. When the transition T_{CompS} fires, it represents that the transport layer successfully sends the *3xx6xx* response to the client side and the server enters into the *completed state*. If the place P_{CompS} contains a token three other transitions, namely $T_{Re\_INVITE\_3}$, T_{timerH}, and T_{timerG} get enabled with probabilities p_{16}, p_{17}, and p_{18}, respectively. Also, if there is a token in the place P_{CompS} and if the place P_{resp_wait} contains a token in the client side, the transition T_{3xx6xx_rec} gets enabled. Simultaneously, if there is a token in the place P_{CompS} and there are one or more token in the place P_{procC}, then the transition T_{rrna} gets enabled. Now, if the place P_{CompS} already contains a token and an extra token enters in the place P_{INVITE_rec}, then the transition $T_{Re\_INVITE\_3}$ at the server side will fire and a token is transferred to the place $P_{transport\_layer\_3}$ for transmission. This condition is managed through the guard function g_1 on $T_{Re\_INVITE\_3}$. The firing of the transition $T_{Re\_INVITE\_3}$ represents that the server receives the INVITE request repeatedly from the client transaction. If the transition T_{timerG} (representing that the *3xx6xx* response is passed to the transport layer again for retransmission) fires, then the token gets transferred to the place $P_{transport\_layer\_3}$. This condition is managed through the guard function *g13* on T_{timerG}. If the transition T_{timerH} fires indicating that time-out has occurred, a token gets transferred to the place $P_{terminated}$.

In the *proceeding state* of client transaction, there are three possibilities: First, the client can accept any number of *1xx* (provisional) responses and remain in *proceeding state*. Second, it can receive *2xx* (successful) responses and enter the *accepted state*, create and send acknowledgment for successful response. And lastly, the client can receive *3xx6xx* (unsuccessful) responses and move to the *completed state*, and create and send acknowledgment for the unsuccessful response. We now consider these three cases one by one. Case 1: A token remains in the place P_{ProcC} and it enables three other transitions, namely, $T_{Re\_1xx\_rec}$, T_{rttu}, and T_{rrna} with probabilities p_{19}, p_{20}, and p_{21}, respectively. If there is already one token in the place P_{ProcC} and simultaneously another token arrives in the place P_{1xx}, then the transition $T_{Re\_1xx\_rec}$ will fire, representing that the client receives the *1xx* responses repeatedly from the server side, and a token is again transferred to the place P_{ProcC}. This condition is managed through the guard function g_2 on $T_{Re\_1xx\_rec}$. Case 2: If the place P_{ProcC} already contains a token and simultaneously another token enters the place $P_{acceptS}$ on the server side, then the transition T_{rttu} will fire, and a token is then transferred to the place $P_{acceptC}$ (depicting that the client receives *2xx* response after receiving *1xx* response from the server transaction). This condition is managed through the guard function g_6 on T_{rttu}. The firing of the transition T_{rttu}

represents that the client is waiting for the response from the server transaction and the server sends *1xx* response along with the *2xx* response. Case 3: When the place P_{ProcC} contains a token and another token enters the place P_{CompS} at the server side, then the transition T_{rrna} will fire, representing that the client receives 3xx6xx responses along with the *1xx* response from the server side, and a token is transferred to the place P_{compC}. This condition is managed through the guard function g_7 on T_{rrna}. Now, if the place P_{resp_wait} contains a token and at the same time another token enters the place $P_{acceptS}$, then the transition T_{2xx_rec} will fire and a token is transferred to the place $P_{acceptC}$. This condition is managed through the guard function g_6 on T_{2xx_rec}. The firing of the transition T_{2xx_rec} represents that the client is waiting for response and receives *2xx* response from the server transaction.

If the place $P_{acceptC}$ contains a token, it enables three other transitions, namely $T_{Re\_2xx\_rec}$, $T_{ACK\_generated}$, and T_{timerM} with probabilities p_{22}, p_{23}, and p_{24}, respectively. If the place $P_{acceptC}$ contains a token and an extra token enters in the place $P_{acceptS}$ simultaneously, then the transition $T_{Re\_2XX\_rec}$ will fire, representing that the client receives *2xx* response repeatedly from the server transaction, and a token is again transferred to the place $P_{acceptC}$. This condition is managed through the guard function g_6 on $T_{Re\_2XX\_rec}$. Also, if the place $P_{acceptC}$ contains a token and simultaneously extra token enters the places $P_{acceptS}$, then the transition $T_{ACK\_generated}$ will fire and a token is transferred to the place P_{ACK}. This condition is managed through the guard function g_{11} on $T_{ACK\_generated}$. The firing of the transition $T_{ACK\_generated}$ represents that the client transfers an acknowledgment for *2xx* response to the server transaction. On the other hand, when the transition T_{timerM} fires first, a token is transferred to the terminated place $P_{terminated}$. The firing time of T_{timerM} is same as the period of time that the TU will hold on for retransmission of *2xx* response.

When the place P_{resp_wait} contains a token and simultaneously another token arrives in the place P_{compS}, then the transition T_{3xx6xx_rec} will fire, and a token is transferred to the place P_{compC} (depicting that the client transaction receives *3xx6xx* response from the server transaction and enters the *completed state*). This condition is managed through the guard function g_7 on T_{3xx6xx_rec}. The firing of the transition T_{3xx6xx_rec} represents that the client is waiting for some response and the server transition transfers *3xx6xx* response to the client transaction. The client's TU will enter the *completed state* directly by getting the response *3xx6xx* from the server transaction or indirectly by getting *3xx6xx* response after the provisional (*1xx*) response. If the place P_{compC} contains a token, it enables four other transitions, namely $T_{Re\_3xx6xx\_rec}$, $T_{ACK\_generated}$, $T_{transport\_error\_5}$, and T_{timerD} with probabilities p_{25}, p_{26}, p_{27}, and p_{28}, respectively. If the place P_{CompC} contains a token and an extra token enters the place P_{CompS} at the same time, then the transition T_{Re_3xx6xx} will fire, representing that the client receives *3xx* response repeatedly from the server transaction. This condition is managed through the guard function g_7 on T_{Re_3xx6xx}. Also, if there is one token in the place P_{compC} and simultaneously another token arrives in the places

P_{CompS}, then the transition $T_{ACK\_generated}$ will fire and one token is transferred to the place P_{ACK}. This situation is managed by the guard function g_{11} on $T_{ACK\_generated}$. When the transition $T_{ACK\_generated}$ fires, it represents that the client sends an acknowledgment for *3xx6xx* response to the server transaction. If transition $T_{transport\_error\_5}$ fires, representing that the acknowledgment from the client transaction encounters a transport error before reaching the server transaction, one token is transferred to the place $P_{terminated}$. When the place P_{compC} contains one token and the place P_{CompS} is emptied, the transition T_{timerD} gets enabled. This condition is managed through the guard function g_5 imposed on the transition T_{timerD}. The firing time of this transition represents the waiting time for *3xx6xx* response retransmission.

Now, a token is there in the place P_{ACK} and this enables two other transitions: the transition T_{ACK_sent} and the immediate transition t_2 with probabilities p_{29} and $(1 - p_{29})$, respectively. The acknowledgments that are created by the client transaction is then sent to the transport layer to transfer them to the server transaction, that is, if the transition T_{ACK_sent} fires, one token is transferred to the place $P_{transport\_layer\_4}$ and this enables two other transitions, namely $T_{transport\_error\_6}$ and $T_{noerror}$ with probabilities p_{30} and $(1 - p_{30})$, respectively. But if the place P_{ACK_rec} already contains a token and an extra token enters the place P_{ACK} simultaneously, then the firing of immediate transition t_2 will start, and the token gets flushed from the place P_{ACK} because only one acknowledgment is required for one INVITE request and more than one acknowledgments get flushed out. This condition is managed through the guard function g_9 on t_2. If a transport error has occurred at the transport layer, the acknowledgments generated by the client transaction enters the *terminated state* and gets destroyed, that is, if the transition $T_{transport\_error\_6}$ fires, a token is transferred to the place $P_{terminated}$.

When the transition $T_{noerror\_6}$ fires, a token is transferred to the place P_{ACK_rec} (implying that the acknowledgment is delivered to the transport layer to transfer it to the server transaction, and no error has occurred at the time of transmission). As a token enters the place P_{ACK_rec} and simultaneously another token arrives in the place P_{CompS}, then the firing of immediate transition t_{ACK_rec} will start and a token is transferred to the place P_{ConfmS}. When the immediate transition t_2 fires, it represents that as the acknowledgment is accepted by the server transaction, the server TU changes its state from *completed* to *confirmed*. This condition is managed through the guard function g_7 on t_{ACK_rec}. In the *confirmed state*, there is an absorption of any additional acknowledgments that are triggered by the retransmission of the final response. When the token enters the place P_{ConfmS}, it enables the immediate transition t_{timerI}. The firing of the immediate transition t_{timerI} will transfer a token to the place $P_{terminated}$. When the immediate transition t_{timerI} fires, it represents that the server changes its state from *confirmed* to *terminated* and the multimedia session is over. On the contrary, if the client does not receive any response from the server transaction, then the transition T_{timerB} will fire representing that the client transaction informs its TU that a time-out has occurred, and then the request enters the *terminated state*. That is, the timed transition T_{timerB} gets enabled when the place P_{resp_wait} contains a token and the places P_{1xx}, $P_{acceptS}$, and P_{CompS} are emptied simultaneously. The transition T_{timerB} will fire and a token is transferred to the place $P_{terminated}$ (implying that the INVITE request gets canceled due to no response from the server TU). This condition is managed through the guard function g_{10} on T_{timerB}.

When the INVITE server transaction sends a response, it should not abandon the transaction only on the basis of encountering an unrecoverable transport error. Rather, the INVITE server transaction state machine continues to exist in the same state. Thus, the INVITE server transaction enables the absorption of the extra INVITE requests rather than treating them as a new INVITE request.

Ultimately, as the client transaction enters into the *Terminated* state ($P_{terminated}$), all the tokens get flushed out and the states are reset again because the round trip time of the data packet has been completed.

In the next part, we present the performance measures of the SIP INVITE request obtained from the above-discussed SRN model.

2.5 Performance measures

In this present section, we address different performance measures of SIP, such as throughput and latency, for the transmission of data packets from the client transaction to the server transaction. Further in the section, we describe how these measures can be evaluated from the proposed SRN model. Concerning the largeness of the SRN model proposed in Section 2.4.2, its analysis can become very complex. However, the largeness problem of such a complex model can be avoided by following some approaches such as state aggregation [19], model decomposition [20, 21], state truncation [22], model composition [23, 24], and state exploration [25, 26]. In this chapter, we implement state aggregation technique to avoid the largeness of the proposed SRN model. "The core idea of state aggregation approach is to (i) partition a very large monolithic SRN into subnets, (ii) analyze the subnets with required analysis metrics of interest, (iii) represent the subnets with simple and equivalent SRN model (normally, a timed transition), and (iv) aggregate the above results into the original partitioned SRN and analyze the aggregated model with required analysis metrics of interest" [27]. This approach helps in reducing the size of the SRN model, and thus prevents the state space from explosion.

For the purpose of numerical analysis of the proposed SRN model, the model is divided into two SRN submodels, namely submodel 1 for the successful transmission (see Fig. 2.5) and submodel 2 for the unsuccessful transmission (see Fig. 2.6) following the state aggregation approach.

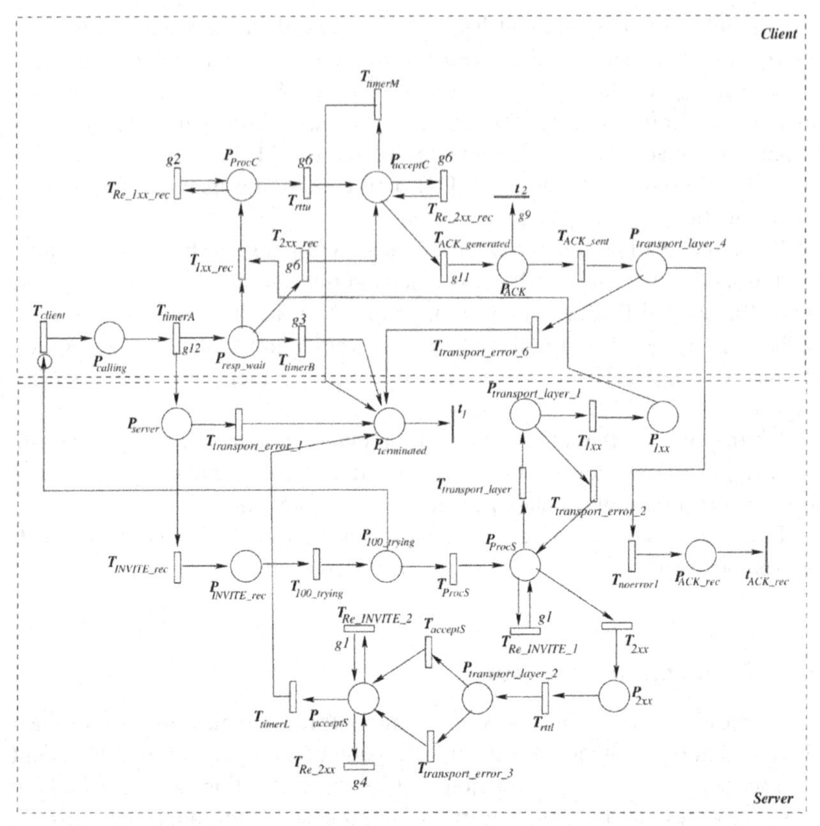

FIG. 2.5 SRN submodel 1 for successful transmission.

2.5.1 Throughput

Throughput is "the average rate at which the data is successfully acknowledged" [28]. The acknowledged data are further divided into two parts: acknowledgment for the successful response *2xx* (known as goodput) and acknowledgment for the unsuccessful response *3xx–6xx* (known as badput).

2.5.1.1 Goodput

Goodput is defined as the number of valuable information bits transmitted by the network to a particular endpoint per unit of time [14]. Goodput refers to "the bandwidth the user actually receives" [29]. For the proposed SRN model, goodput is defined as "the average rate at which successful acknowledgment of the data is received."

In this chapter, the goodput of the SIP INVITE transaction is evaluated from the submodel 1 given in Fig. 2.5. In the model, as the transition ($T_{noerror1}$) fires, the client sends acknowledgment for the successful receiving of 2xx responses,

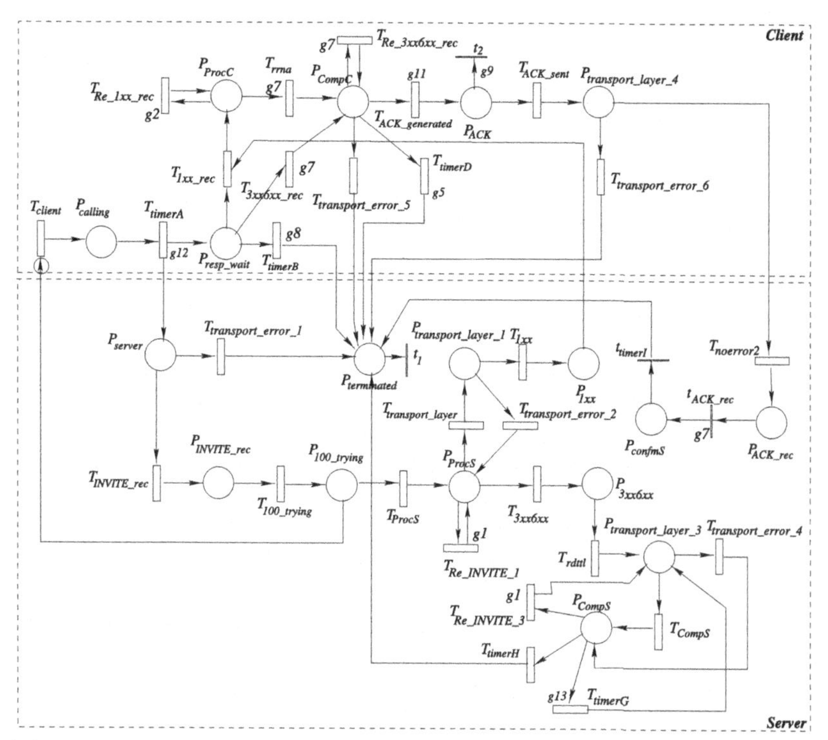

FIG. 2.6 SRN submodel 2 for unsuccessful transmission.

which are a successful response. This means that every time this transition fires, a packet is successfully transmitted. Hence, the throughput of the transition ($T_{noerror1}$) provides the measure of the goodput of the protocol. The goodput of the protocol is hence given by

$$Goodput = \eta_{T_{noerror1}}$$

where η is the throughput of the transition $T_{noerror1}$.

2.5.1.2 Badput

Badput is defined as "the throughput of corrupted packets which provides a measure of wasted network resources" [27, 30]. It includes both necessary and unnecessary retransmissions [31]. For the proposed SRN model, badput is defined as "the average rate at which acknowledgments of unsuccessful transmissions of data packets are received."

In this chapter, the badput of the SIP INVITE transaction is evaluated from the submodel 2 given in Fig. 2.6. In the model, as the transition ($T_{noerror2}$) fires, the client sends acknowledgment for the successful receiving of 3xx6xx responses, which are unsuccessful responses. This means that every time this transition fires, a packet is not successfully transmitted. Hence, the throughput of the transition ($T_{noerror2}$) provides the measure of the badput of the protocol. The badput of the protocol is hence given by

$$Badput = \eta_{T_{noerror2}}$$

where η is the throughput of the transition $T_{noerror2}$.

2.5.2 Latency

Latency is the estimated amount of time needed for an appropriate task to be completed [13]. The completion of the task is represented by the extent of "conceptualization to achievement (delivery)." It is the total amount of time for which a data packet stays in the waiting line and the time needed for its delivery [28]. A data packet can either be forwarded successfully, or it will require retransmission. Hence, the latency starting from the arrival time of the requests till the time of receiving of either the successful responses or the unsuccessful responses can be evaluated as follows.

To find the latency, first we have to find the queueing delay and then the average service time for the successful responses (see Fig. 2.5) and unsuccessful responses (see Fig. 2.6). The queueing delay for both the successful and unsuccessful transmission is same.

The queueing delay represented as $Q_{transmission}$ is obtained as

$$Q_{transmission} = \frac{(P_{calling})}{\eta_{T_{client}}} \tag{2.1}$$

where $\#(P_{calling})$ are the count of tokens in the place $P_{calling}$ and η is the throughput of the transition T_{client}.

The average time taken for the successful transmission, represented as $1/\mu_{success}$ is obtained as

$$\begin{aligned}1/\mu_{success} = \text{Average firing time of } \Big(& T_{timerA} + T_{INVITE\_rec} + T_{100trying} + T_{ProcS} \\ & + T_{transport\_layer} + T_{1xx} + T_{1_{xx\_rec}} + T_{rrna} + T_{ACK\_generated} \\ & + T_{ACK\_sent} + T_{noerror1} \Big)\end{aligned} \tag{2.2}$$

Therefore, the average latency for successful transmission represented by $Latency_{success}$ is obtained as

$$Latency_{success} = Q_{transmission} + \frac{1}{\mu_{success}} \tag{2.3}$$

Similarly, the average time taken for the unsuccessful transmission, represented as $1/\mu_{unsuccess}$ is obtained as

$$1/\mu_{unsuccess} = \text{Average firing time of } \Big(T_{timerA} + T_{INVITE\_rec} + T_{100\_trying} + T_{ProcS} + T_{transport\_layer} + T_{1xx} + T_{1_{xx\_rec}} + T_{rrna} + T_{ACK\_generated} + T_{ACK\_sent} + T_{noerror2}\Big) \quad (2.4)$$

Therefore, the average latency for unsuccessful transmission, represented by $Latency_{unsuccess}$ is obtained as

$$Latency_{unsuccess} = Q_{transmission} + \frac{1}{\mu_{unsuccess}} \quad (2.5)$$

Now to find the latency of the SIP INVITE request, we compare the $Latency_{success}$ and $Latency_{unsuccess}$ and select the maximum of the two as it gives the maximum time for the SIP INVITE transaction to complete its task.

2.6 Numerical illustration

In this section, we perform numerical experiments to obtain the results from the proposed SRN model for the above-discussed measures of performance of the SIP INVITE transaction. SHARPE software is used to obtain the analytical results from the SRN model. For the purpose of demonstration, we assume n (represents number of calls) $= 3$, $\lambda_{success} = \lambda_{unsuccess}$ varying from 0.00001 to 0.0001 m/s. All the timed transition have certain firing rates which are indexed in Table 2.2. The SIP architecture which we are considering in the chapter is adopted from Ref. [4] only. Accordingly, most of the parameter values shown in Table 2.2 are adopted from the technical paper RFC 3261 [4]. The parameter values that were not available are chosen proportionately accordant with the round trip time (RTT) of the SIP INVITE transaction, which is 500 ms [11]. Note that these parameter values are picked only for numerical illustration. We do not evaluate the values of these parameters as evaluating them from some empirical data set is beyond the scope of this chapter.

2.6.1 Results and discussion

The performance measures obtained in Section 2.5 are plotted graphically and the results can be interpreted as follows.

Fig. 2.7 displays the variation in goodput of this SIP INVITE transaction versus the arrival rate of the data packets. We observe that as the arrival rate is increased, there is an increase in the goodput. That is, more data packets are delivered successfully when the rate of arrival of packets is increased.

Fig. 2.8 represents the variation in goodput and badput of the SIP INVITE transaction versus the arrival rate of the data packets. We observe that as the arrival rate of the data packets increases, both the goodput and badput of the SIP INVITE transaction also increase. But the increase in the goodput is more as compared to the increase in the badput, which implies that the overall

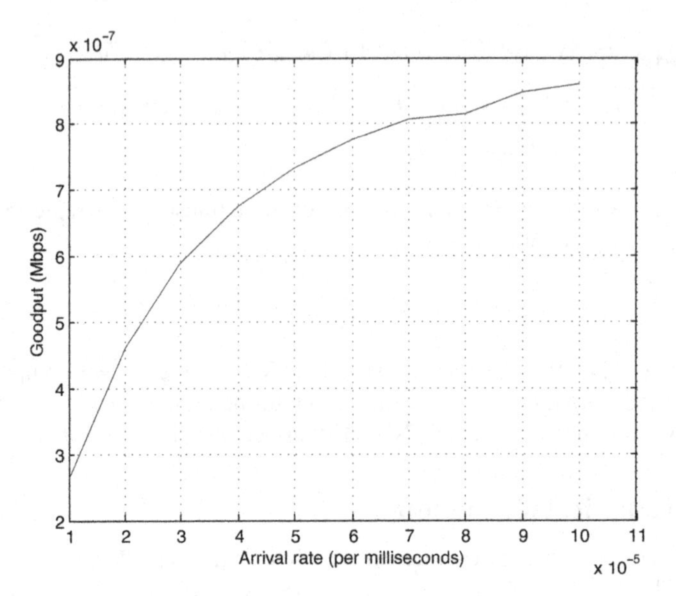

FIG. 2.7 Goodput versus arrival rate of data packets.

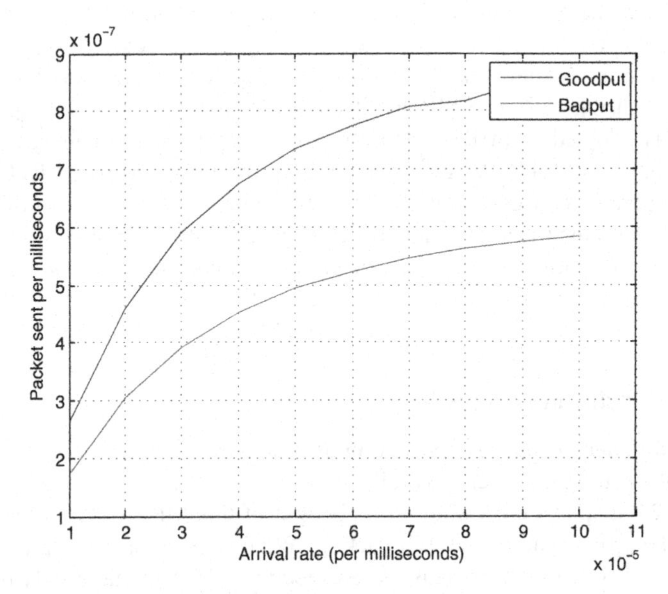

FIG. 2.8 Goodput and badput versus arrival rate of data packets.

throughput of the protocol (rate of successful transmission) increases as the arrival rate of data packets increases. Therefore, the chance of successful delivery of the data packets increases with an increase in the arrival rate of data packets.

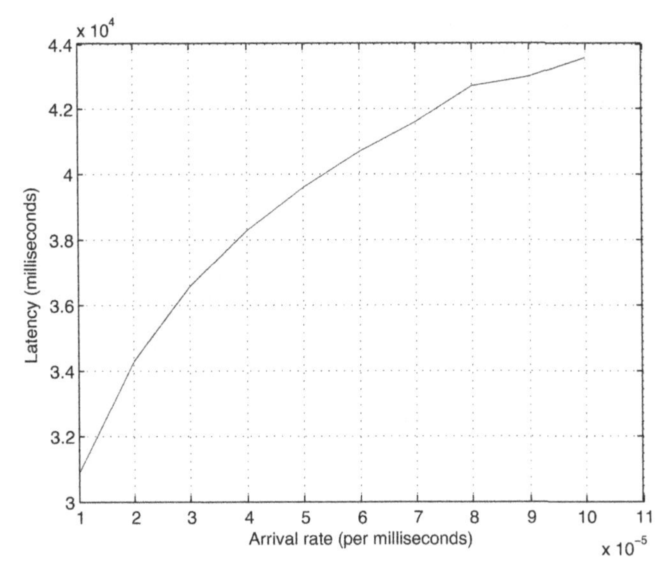

FIG. 2.9 Latency versus arrival rate of data packets.

Fig. 2.9 illustrates the variation in latency of SIP INVITE transaction versus the arrival rate of the data packets. Here also we observe that as the arrival rate of the data packets increases, the latency also increases.

From the graphical results presented earlier, it can be observed that the behavior of the various measures of performance are in accordance with the expected behavior of the SIP INVITE transaction, which proves the correctness of the proposed SRN model.

2.7 Model validation

In this section, simulation is used to further support the analytical results obtained by the proposed SRN model. By using a MATLAB program, we perform discrete-event simulation to simulate the dynamics of the SIP INVITE request. To study the precision of the simulation results, the t-distribution test is used. We use the following mechanism:

We perform the simulation approximately 30 times for the exact same group of parameters as used for obtaining the analytical results and attain 30 values for the latency of the SIP INVITE transaction. We then evaluate the mean and standard deviation of these 30 values. Using the mean and standard deviation, we obtain the confidence interval for 99% accuracy of the results with the help of the formula given below:

$$\text{Confidence Interval (C.I.)} = \overline{\text{x}} \text{-+} \frac{s^{*}t_{0.01,k-1}}{\sqrt{k}}$$

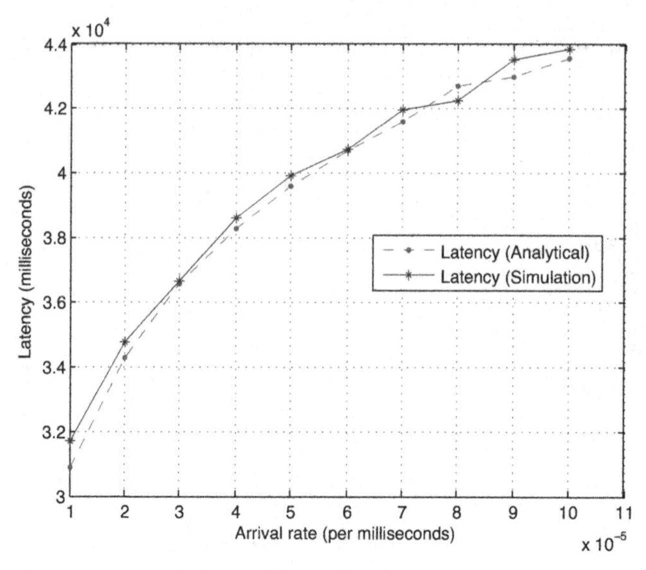

FIG. 2.10 Latency versus arrival rate of data packets.

where

- [] $\overline{x}$ = sample mean,
- [] s = standard deviation,
- [] $t_{0.01,\ k-1}$ = 99% confidence interval tabulated value,
- [] $k - 1$ = degrees of freedom (d.f.), and
- [] k = sample size.

The comparability between the analytical results and simulation results for the latency of the SIP INVITE transaction is presented in Fig. 2.10. The simulation result is represented using a solid line and 99% confidence interval in the figure. As predicted, with an increase in the arrival rate, the latency of the system also increases. Moreover, it is undeniably visible from the plot that the analytical results attained from the SRN model are almost same as the simulation results, thereby validating the proposed analytical model.

2.8 Conclusions

This chapter evaluates the performance of SIP INVITE transaction through analytical modeling. By considering exponential times for all the actions (such as time-outs and transmission), the INVITE transaction flow in SIP is modeled using a CTMC. Howbeit, rather than constructing the CTMC, an SRN model is designed as it is more concise in specifications and as a matter of course produces the underlying CTMC. Some particular measures of performance such as throughput and latency of the SIP INVITE transaction are obtained from the proposed SRN model using the SHARPE software package. Numerical experiments are carried out and graphical results for the goodput, badput, and latency of the SIP are produced. It is observed that all the results are in accord with the anticipated behavior of the SIP displaying the precision of the proposed model. Analytical results are also justified through discrete-event simulation using MATLAB.

Acknowledgments

This research work was supported by a minor research grant (loE/FRP/PCMS/2020/27) under the Faculty Research Programme (FRP) of the IoE scheme of University of Delhi, Delhi, India. One of the authors (N.C.) would like to thank UGC, India for providing her financial support through non-NET fellowship.

References

[1] S. Jalendry, S. Verma, A detail review on voice over internet protocol (VoIP), Int. J. Eng. Trends Technol. 23 (4) (2015) 161–166.

[2] A. Suryawanshi, M. Bhati, R. Dubey, Session initiation protocol (SIP), Int. J. Comput. Sci. Eng. 10 (10) (2013) 18–21.

[3] A. Ali, N. Ahmad, M.S. Akhtar, A. Srivastava, Session initiation protocol, Int. J. Sci. Eng. Res. 4 (1) (2013) 1–6.

[4] E. Schooler, J. Rosenberg, R. Sparks, M.J. Handley, SIP: session initiation protocol, RFC Tech. Rep. 3261 (2002) 1–269.

[5] R. Sparks, T. Zourzouvillys, Correct transaction handling for 2xx responses to session initiation protocol (SIP) INVITE requests, RFC Tech. Rep. 6026 (2010) 1–20.

[6] T.M. Tuan, P.M. Chuan, M. Ali, T.T. Ngan, M. Mittal, Fuzzy and neutrosophic modeling for link prediction in social networks, Evol. Syst. 10 (4) (2019) 629–634.

[7] D. Howie, M. Ylianttila, E. Harjula, E.J. Sanvola, State-of-the-art SIP for mobile application supernetworking, Project: All-IP Application Supernetworking 1 (2008) 1–5.

[8] I. Alsmairat, R. Shankaran, M. Orgun, E. Dutkiewicz, Securing session initiation protocol in voice over IP domain, in: Eighth IEEE International Conference on Dependable, Autonomic and Secure Computing, DASC 2009, Chengdu, China, 2009, pp. 78–83.

[9] L. Lin, Verification of the SIP transaction using coloured petri nets, in: Thirty-Second Australasian Computer Science Conference, Wellington, New Zealand, vol. 91, 2009, pp. 63–72.

[10] S. Barakovic, D. Jevtic, J.B. Husic, Modeling of session initiation protocol invite transaction using colored petri nets, Int. J. Comput. Sci. Eng. 6 (1) (2012) 43–50.

[11] R. Filal, M. Bouhdadi, A mechanically proved and an incremental development of the session initiation protocol INVITE transaction, Int. J. Comput. Netw. Commun. 2014 (2014) 1–11.

[12] S. Kizmaz, M. Kirci, Verification of session initiation protocol using timed colored petri net, Int. J. Commun. Netw. Syst. Sci. 4 (3) (2011) 170–179.

[13] V. Gupta, S. Dharmaraja, M. Gong, Analytical modeling of TCP flow in wireless LANs, Int. J. Math. Comput. Model. 53 (5–6) (2011) 684–693.

[14] R. Vinayak, D. Krishnaswamy, S. Dharmaraja, Analytical modeling of transmission control protocol new reno using generalized stochastic petri nets, Int. J. Commun. Syst. 27 (12) (2014) 4185–4198.

[15] V. Gupta, M. Gong, S. Dharmaraja, C. Williamson, Analytical modeling of bidirectional multichannel IEEE802.11 MAC protocols, Int. J. Commun. Syst. 24 (5) (2011) 647–665.

[16] S. Gupta, V. Gupta, Analytical modeling of RLC protocol of LTE using stochastic reward nets, Int. J. Commun. Syst. 32 (6) (2018) 1–18.

[17] R.A. Sahner, K.S. Trivedi, A. Puliato, Performance and reliability analysis of computer systems: an example-based approach using the SHARPE software package, IEEE Trans. Reliab. 46 (3) (1997) 441–845.

[18] K.S. Trivedi, Probability and Statistics With Reliability, Queuing, and Computer Science Applications, Wiley & Sons, New York, 2001.

[19] P. Buchholz, An adaptive aggregation/disaggregation algorithm for hierarchical Markovian models, Eur. J. Oper. Res. 116 (3) (1999) 545–564.

[20] S.M. Koriem, Fast and simple decomposition techniques for the reliability analysis of interconnection networks, Int. J. Syst. Softw. 45 (2) (1999) 155–171.

[21] M.A. Marsan, R. Gaeta, M. Meo, Accurate approximate analysis of cell-based switch architectures, Perform. Eval. 45 (1) (2001) 33–56.

[22] G.P. Katerina, K.S. Trivedi, Stochastic modeling formalisms for dependability, performance and performability, Perform. Eval. Orig. Dir. 1 (2000) 403–422.

[23] P. Ballarini, S. Donatelli, G. Franceschinis, Parametric stochastic well-formed nets and compositional modelling, in: Proceedings of the 21st International Conference on Application and Theory of Petri Nets, 2000, pp. 43–62.

[24] W.J. Knottenbelt, P.G. Harrison, M.A. Mestern, P.S. Kritzinger, A probabilistic dynamic technique for the distributed generation of very large state spaces, Perform. Eval. 39 (1–4) (2000) 127–148.

[25] I. Davies, J. Knottenbelt, Symbolic methods for the state space exploration of GSPN models, in: Proceedings of 12th International Conference on Modelling Techniques and Tools for Computer Performance Evaluation, 2002, pp. 188–199.

[26] A. Miner, Efficient state space generation of GSPNs using decision diagrams, in: Proceedings International Conference on Dependable Systems and Networks, 2002, pp. 637–646.

[27] D.S. Kim, J.B. Hong, T.A. Nguyen, F. Machid, J.S. Park, K.S. Trivedi, Availability modeling and analysis of a virtualized system using stochastic reward nets, in: IEEE International Conference on Computer and Information Technology, 2016, pp. 210–218.

[28] D. Bertsekas, R. Gallager, Data Networks, second ed., Prentice Hall, Englewood Cliffs, NJ, 1987.

[29] H. Wu, S. Cheng, Y. Peng, K. Long, J. Ma, IEEE distributed coordination function (DCF): analysis and enhancement, in: Proceedings ICC, New York, USA, vol. 1, 2002, pp. 605–609.

[30] Y. Kim, S. Li, Performance analysis of data packet discarding in ATM networks, IEEE/ACM Trans. Netw. 7 (2) (1999) 216–227.

[31] J.J. Sydir, N.T. Plotkin, N. Akar, Using ATM services for (in)efficient support of TCP modeling, in: 7th International Symposium on Analysis and Simulation of Computer and Telecommunication Systems, 1998, pp. 1–22.

Chapter 3

An empirical validation for predicting bugs and the release time of open source software using entropy measures—Software reliability growth models

Anjali Munde
Amity University, Noida, Uttar Pradesh, India

3.1 Introduction

Software is vital in the modern world. The progression of software requires definite advanced models. In earlier times, due to the rising development in software utilities, load grew, putting pressure on software companies to develop reliable software in a shorter time. Recently, an important modification has occurred in software development due to advancements in communication devices, which has escalated the advancement of open source software.

Fault extrapolation has had extensive significance for a substantial period of time. The central issue is reserves distribution: time and manpower being limited reserves, it creates awareness to allocate workforces to parts of a software system with a greater likely amount of faults.

The occurrence of faults in any printed software is predictable, although the cost of mending these faults varies notably based on when the fault is identified. If software inventors are able to distinguish faults at an initial phase, the cost acquired in correcting the fault would be considerably less. Current developments rotate around the fact that faults can now be anticipated before they are discovered. Considerable corpuses of preceding fault data are essential to be able to forecast faults with sufficient precision. Software analytics has released infinite potential for tapping data analytics and rationalizing to

System Assurances. https://doi.org/10.1016/B978-0-323-90240-3.00003-5

enhance the condition of software. Analytics presents the outcomes of the software investigation as real time data, to create informative extrapolations. By establishing the existence or non - existence of a fault in a software edition, developers can envisage the achievement of the software edition even prior to its announcement, established on limited attributes of the announcement edition. If this forecast is implemented at an earlier stage in the software growth phase, it will slash the cost of correcting the fault. Furthermore, by integrating numerous software analytical methods, a fault estimation prototype that is responsive and proficient to be consumed by the software growth business can be established.

The presence of software faults considerably influences software consistency, features, and conservation cost. Attaining fault-free software is hard work—even software applied prudently for long periods can contain unknown faults. Therefore, developing a software fault estimation prototype that could envisage the defectives components in the initial period is an important task in software engineering. Software fault estimation is a fundamental action in software expansion. Software fault estimation is a fundamental action in software expansion since forecasting the fault segments previous to software utilization realizes the worker compensation and hence recovers the complete software functioning.

Existing investigation related to the next announcement comprises the requirements in the software. The amount of constraints are needed to be evaluated for profit expansion and are also consumed as an amount in determining the next announcement of the software. In a closed source situation, the announcement period of the software is most rigid during the prerequisite investigation period of the project. Subsequent to the primary announcement, following announcements happen established on a number of conditions and faults repaired. In the open source scenario, this is verified by an adequate quantity of alterations as well as number of faults repaired in the software. An open source announces its software edition on a regular basis through innovation, necessity, vital area, and time.

Kapur et al. [1] estimated an accurate method for determining when to withdraw software investigation, considering failure attention and cost as two fundamentals simultaneously. The recommended prototype is based on multiattribute efficiency investigation, which aids businesses by generating a cautious outcome on the perfect method of the software. Pachauri et al. [2] generated an attempt to validate the imprecision involved in estimating the limit of SRGM (software reliability growth model) in an insufficient fixing state utilizing fuzzy numbers. They also calculated the uniformity and the ideal announcement period of the software created on cost consistency principles employing a fuzzy environment.

Peng et al. [3] suggested a framework to obtain interesting ways based on bug identification and bug modification under numerous settings.

Hassan [4] suggested the concept of information theory to determine the amount of imprecision or entropy of the distribution in order to calculate the complexity of code changes. He attempted to predict the faults present in the software determined on previous limitations employing entropy measures. Ambros and Robbes [5,6] suggested a standard for approximation of faults and exhibited an extensive calculation of recognized fault approximation techniques.

Kumari et al. [7] recommended defect need arithmetical models by reviewing the immediate description of defects and explanations planned over operators in situations involving entropy-based quantities.

An essential characteristic of open source software is determining when to declare the latest edition. The chief purpose is declaring the software at a suitable time and within limited plan restrictions. The approach following announcing software is a challenge. Announcement planning performs a fundamental function in controlling widespread declarations to the users. It is principally a collection of modern features that are calculated for the current invention environment. The newest declaration follows the development process and subsequent broadcasts are based upon the previous statements and the number of quantities and defects corrected. Each statement encompasses a set of features satisfying specific limitations of the business as required by the customers. Determining what features are to be included in which statement edition is a difficult job. An open source software scheme declares its software edition frequently based on the target, time, patch etc.

Geer and Ruhe [8] recommend their attempt to establish a method to enhance software announcements by means of a genetic algorithm and Baker et al. [9] conducted the investigation with element ordering, applying a greedy algorithm and suggested strengthening for the resulting announcement obstruction. The most pertinent difficulty of NRP (next release planning) is the procedure of release planning (RP), which focuses particular optimum announcements to fulfil software condition limits [10] or announcement times [11,12]. If fault alterations are formulated too close to the announcement date, the complete condition of the fault correcting procedure can be damaging, because of the possible imprecision of interpolations and limited documentation of variations [13]. In fact, if the alterations are confirmed and required in advance of the arranged target, they can be involved in the announcement; otherwise, they would move to the subsequent announcement [14]. A technique to control the following announcement difficulty utilizing support founded on a multilevel algorithm is planned, but was stopped to study the faults and the related variations that happened in the code [15]. Announcement period obstruction created on corrected faults has been recommended in literature [16].

In this chapter, the defects present in main releases of the Bugzilla project are taken into consideration with evident data preprocessing. Then, definite amounts of entropy, specifically, Shannon entropy and Kapur entropy, are calculated for the alterations in various software defects in the period 2011–2019. A method is recommended with the provision of linear regression to accomplish the release period of the Bugzilla open source software through the entropy measures and defects discovered. Regression analysis has been used in two stages. In the first stage, simple linear regression has been applied between the entropy calculated and investigational defects for each release across every period to get the regression coefficients. The coefficients are utilized validating the predicted defects in every release in each time period. In the second phase, multiple linear regression is applied to predict the release time of the software product for each time period by means of the entropy measure and the defects predicted in stage one.

3.2 Information measures

Information theory is an innovative domain of mathematics that was first developed in the 1940s. Several vital concepts were implicit by 1948, when information theory was set on a definite foundation. Shannon [17] broadcast a research paper on "A mathematical theory of communication" and created huge interest in the outcome of research allied to the conduction and conservancy of information seen in the environment and technology. Ambiguity plays an important part in the contrary observations in the surrounding world. A subject that aids in estimating, evaluating, maximizing or minimizing, and eventually regulating it, should definitely be regarded as a critical input to the technical consideration of complex occurrences. The model of entropy was initiated to determine the numerical quantity of ambiguity.

Shannon [17] initiated the notion of entropy in communication theory. The Shannon entropy, H_n, is stated as:

$$H(P) = -\sum_{s=1}^{n} p_s \log p_s \tag{3.1}$$

where $p_s \geq 0$ and $\sum_{s=1}^{n} p_s = 1$.

Kapur formulated a methodical line to achieve a generalization of Shannon entropy. He denoted entropy of order α and degree β, explained as follows:

$$H_{\alpha,\beta}(P) = \frac{1}{1-\alpha} \log \left(\sum_{i=1}^{n} p_i^{\alpha+\beta-1} / \sum_{i=1}^{n} p_i^{\beta} \right), \quad \alpha \neq 1, \alpha > 0, \beta > 0, \alpha+\beta-1 > 0 \tag{3.2}$$

3.2.1 Predicting bugs and the release time of open source software

An assignment is chosen and then subsystems are selected. The CVS (concurrent versions system) logs are looked at and the fault details of all subsystems are collected. In this part, the following approaches are applied. Mostly the data is accumulated and predirected and then entropy quantities are calculated.

An approach is determined to establish the release interval of the software founded on defects remaining and changes in the data. A technique is recommended to determine the estimated period of all releases of the OSS-Bugzilla employing regression analysis. The information is accumulated from the Bugzilla website, www.Bugzilla.org, for every software announcement.

This includes choosing the defects existing in important releases of the Bugzilla plan with particular data preprocessing. Different amounts of entropy, specifically, Shannon entropy [17] and Kapur entropy [18], for the variations in various software announcements have been examined (Tables 3.1 and 3.2).

The SLR (simple linear regression) model continues to be the essential model, including two variables around which one variable is predicted with the other variable. The regression equation yielded by Eq. (3.3) involves being significantly exploited to restrain the dependent inconstant, Y, employing the independent inconstant, X, across the following equality:

$$Y = a + bX \tag{3.3}$$

where a is the y intercept and b is the slope of the line.

To uphold the faults for the next year, employing numerous quantities of entropy is predicted. This fault approximation remains beneficial to the software manager in controlling the supply, through fixing the faults, to maintain the good performance of the software.

Subsequently, Shannon and Kapur entropy have been calculated for many releases of Bugzilla software in every interval. Multiple linear regression is a

TABLE 3.1 Shannon entropy aimed at every time period.

Year	Shannon entropy
2011	0.2441
2012	0.1955
2013	0.3684
2014	0.4320
2015	0.4579
2016	0.3727
2017	0.6041
2018	0.5533
2019	0.5717

TABLE 3.2 Kapur entropy aimed at every time period for $\alpha = 0.1$ and $\beta = 1.0$.

Year	Kapur entropy
2011	0.2947
2012	0.2873
2013	0.4602
2014	0.4675
2015	0.4638
2016	0.4637
2017	0.6875
2018	0.5961
2019	0.5986

regularly employed arithmetical data examination technique, in which two or more independent variables are exploited to predict the value of the dependent variable. The regression examination has been employed in two periods. Thus, the succeeding linear regression prototype is recommended:

$$Y = a + bX_1 \quad (3.4)$$

$$Y_1 = c + dX_1 + eY \quad (3.5)$$

where a, b, c, d, and e are regression coefficients.

In the first stage, the linear regression has been utilized for the exogenous variable X_1 and endogenous variable Y with Eq. (3.4) to obtain the regression coefficients a and b. These coefficients are utilized to reach the estimated amount of defects, i.e., Y for each release in each interval. In the second stage, we study X_1 and Y as independent variables and Y_1 as a dependent variable. Consequently, multiple linear regression in this stage has been employed for the independent variables X_1 and Y and dependent variable Y_1 employing Eq. (3.4) to achieve the regression coefficients c, d, and e. When the calculation of these regression coefficients is estimated, we can estimate the period of all releases. Here, X_1, i.e., entropy, is exclusive in all conditions, as two entropy quantities, specifically, Shannon entropy and Kapur entropy, are developed to estimate the announcement of each release for every interval.

The subsequent record represents the envisaged bugs and the predicted release time of software using the Shannon entropy measure (Table 3.3).

The subsequent record represents the envisaged bugs and the predicted release time of software using the Kapur entropy measure for values of α and β (Table 3.4).

TABLE 3.3 Shannon entropy aimed at every time period.

Year	X_1	Y_o	Y	Observed release time	Y_1
2011	0.2441	4	7.0866	7	10.8923
2012	0.1955	12	5.5639	13	10.8405
2013	0.3684	9	10.9812	11	11.0247
2014	0.432	9	12.9740	11	11.0925
2015	0.4579	5	13.7855	12	11.1201
2016	0.3727	13	11.1160	14	11.0293
2017	0.6041	16	18.3663	5	11.2758
2018	0.5533	22	16.7746	21	11.2217
2019	0.5717	24	17.3511	15	11.2413

TABLE 3.4 Kapur entropy aimed at every time period.

Year	X_1	Y_o	Y	Observed release time	Y_1
2011	0.2947	4	6.0789	7	11.9290
2012	0.2873	12	5.8157	13	11.9909
2013	0.4602	9	11.9648	11	10.5434
2014	0.4675	9	12.2244	11	10.4823
2015	0.4638	5	12.0929	12	10.5133
2016	0.4637	13	12.0893	14	10.5141
2017	0.6875	16	20.0487	5	8.6405
2018	0.5961	22	16.7981	21	9.4057
2019	0.5986	24	16.8870	15	9.3847

Therefore, a method for determining the number of faults documented in numerous announcements of Bugzilla software is established and distinct amounts of entropy for the alterations in various forms of software informed by these intervals are computed. Initially, simple linear regression is employed to predict faults that are still to come in the future. By means of anticipated faults and entropy measures in multiple linear regression, the announcement period of the software has been forecast.

3.3 Conclusion

In this chapter, a methodology is established to determine the announcement time of the software and the forecasted time of every announcement of the OSS-Bugzilla utilizing regression analysis.

The study includes the selection of defects present in foremost announcements of the Bugzilla project with certain data preprocessing. The measures of entropy, Shannon entropy, and Kapur entropy for the corrections in several software updates are examined. In the first stage, linear regression is associated with entropy, calculating and identifying defects for all releases over every period to obtain the regression coefficients. Then these regression coefficients are worked to determine the anticipated faults for each announcement in each time period. In multiple linear regression, the amount of faults envisaged and entropy computed are measured as exogenous variables and duration in months to be anticipated as a endogenous variable for Bugzilla software (www.bugzilla.org). Several measures of entropy, namely, Shannon entropy and Kapur entropy, are computed independently for numerous announcements of Bugzilla software and one entropy at a time is chosen for anticipation of the announcement time in multiple linear regressions.

References

[1] P.K. Kapur, J.N.P. Singh, N. Sachdeva, V. Kumar, Application of multi attribute utility theory in multiple releases of software, in: International Conference on Reliability, Infocom Technologies and Optimization, 2013, pp. 123–132.

[2] B. Pachauri, A. Kumar, J. Dhar, Modeling Optimal Release Policy Under Fuzzy Paradigm in Imperfect Debugging Environment, Elsevier, 2013, https://doi.org/10.1016/j.infsof.2013.06.001.

[3] R. Peng, Y.F. Li, W.J. Zhang, Q.P. Hu, Testing effort dependent software reliability model for imperfect debugging process considering both detection and correction, Reliab. Eng. Syst. Saf. 126 (2014) 37–43.

[4] A.E. Hassan, Predicting faults using the complexity of code changes, in: Proceedings of the 31st International Conference on Software Engineering, Vancouver, BC, Canada, 16–24 May 2009; IEEE Computer Society: Washington, DC, USA, 2009, pp. 78–88.

[5] M.D. Ambros, M. Lanza, R. Robbes, An extensive comparison of bug prediction approaches, in: MSR'10: Proceedings of the 7th International Working Conference on Mining Software Repositories, 2010, pp. 31–41.

[6] M. Ambros, M. Lanza, R. Robbes, Evaluating defect prediction approaches: a benchmark and an extensive comparison, Empir. Softw. Eng. 17 (4–5) (2012) 571–737.

[7] M. Kumari, A. Misra, S. Misra, L. Sanz, R. Damasevicius, V. Singh, Quantity quality evaluation of software products by considering summary and comments entropy of a reported bug, Entropy 21 (1) (2019) 91.

[8] D. Greer, G. Ruhe, Software release planning: an evolutionary and iterative approach, Inf. Softw. Technol. 46 (4) (2003) 243–253.

[9] P. Baker, M. Harman, K. Steinhofel, A. Skaliotis, Search based approaches to component selection and prioritization for the next release problem, in: Proceedings of 22nd International

Conference on Software Maintenance, 2006, pp. 176–185, https://doi.org/10.1109/ICSM.2006.56.

[10] G. Ruhe, M.O. Saliu, The art and science of software release planning, IEEE Softw. 22 (6) (2005) 47–53, https://doi.org/10.1109/MS.2005.164.

[11] B. Yang, H. Hu, L. Jia, A study of uncertainty in software cost and its impact on optimal software release time, IEEE Trans. Softw. Eng. 34 (6) (2008) 813–825, https://doi.org/10.1109/TSE.2008.47.

[12] J. McElroy, G. Ruhe, When-to-release decisions for features with time-dependent value functions, Requir. Eng. 15 (3) (2010) 337–358, https://doi.org/10.1007/s00766-010-0097-5.

[13] E. Capra, C. Francalanci, F. Merlo, The economics of open source software: an empirical analysis of maintenance costs, in: Proceedings of International Conference on Software Maintenance, 2007, pp. 395–404.

[14] C. Francalanci, F. Merlo, Open source development, communities and quality, in: IFIP International Federation for Information Processing, vol. 275, 2008, pp. 187–196.

[15] J. Xuan, H. Jiang, Z. Ren, Z. Luo, Solving the large scale next release problem with a backbone based multilevel algorithm, IEEE Trans. Softw. Eng. 38 (5) (2012) 1195–1212.

[16] P.K. Kapur, V.B. Singh, O.P. Singh, J.N.P. Singh, Software release time based on multi attribute utility functions, Int. J. Reliab. Qual. Saf. Eng. 20 (4) (2013).

[17] C.E. Shannon, A mathematical theory of communication, Bell Syst. Tech. J. 27 (379–423) (1948) 623–656.

[18] J.N. Kapur, Generalized entropy of order α and β, in: The Maths Semi, 1967, pp. 79–84.

Chapter 4

Risk assessment of starting air system of marine diesel engine using fuzzy failure mode and effects analysis

Rajesh S. Prabhu Gaonkar[a] and Sunay P. Pai[b]
[a]*Indian Institute of Technology Goa (IIT Goa), Ponda-Goa, India,* [b]*Institute of Maritime Studies, Vasco-Da-Gama, Goa, India*

4.1 Introduction

The ship is a complex entity with many systems. Operation and safe transportation of the ship depends very much on the reliable operations of all these systems. These can be achieved by carrying out a risk assessment, identifying critical failure modes, and attending to those critical failure modes on top priority. As shown in Table 4.1, the major reason for marine accidents is the marine main engine failure [1]. Lazakis et al. [2] suggested prioritizing the maintenance activities to cope with maintenance issues of any type increasing the reliability and operability of the ship. Maintenance of ship systems is to be undertaken in a systematic and planned manner based on the practical experience and knowledge of the senior staff of the ship. The primary focus of maintenance is to avoid unexpected failures and minimize the operational time and increase the reliability and safety of transportation [3]. This requires a planned maintenance schedule and best maintenance practices [4]. Some of the factors which can affect the performance of maintenance activities include rough weather conditions, the fatigue of crew members, temperature, humidity, ship motion, and vibration. Rabiul et al. [5] developed a model to assess the human error probability during maintenance activities. Kimera et al. [6] used a machine learning approach for predictive maintenance of ballast pumps on the ship. Lazakis et al. [7] proposed an enhanced ship inspection model for improving the maintenance and decision-making in ship operations. Emovon et al. [8] carried out a risk assessment of marine diesel using the risk priority number (RPN) and risk matrix

System Assurances. https://doi.org/10.1016/B978-0-323-90240-3.00004-7

TABLE 4.1 The failure rate distribution for marine accidents.

Sr. no.	Reason	Failure frequency
1	The main engine	232 (41.58%)
2	Manipulate gear	66 (11.83%)
3	Auxiliary diesel engine	120 (21.50%)
4	The boiler	65 (11.65%)
5	The propulsion shaft	63 (11.29%)
6	Others	12 (2.15%)
	Total	558 (100%)

(RM)-based methods. Jeon et al. [9] demonstrated how the failure mode and effects analysis (FMEA) method can be effective in evaluating the reliability of fuel cell-based hybrid power systems for ships.

Among the many systems onboard the ship, the starting air system is prone to explosion if maintenance is not undertaken with proper care. The faults in the starting air system and the resulting corresponding loss cannot be neglected. Several accidents and damages are caused by failures in the starting air system. A reliable operation of the starting air system is very much needed to reduce the chances of accidents and from the standpoint of the safety of personnel. The starting air system may be said to be reliable if it performs its design functions without any explosion for a specified period. The reliability is measured using the probabilities of failures of the key components of the system. It is complementary to the failure probability and can be increased by performing the risk assessment of the system and targeting the critical failure modes to reduce the chances of failures. This requires proper risk assessment strategies and ways to implement the suggested options.

Even if there is no explosion, air start valves and other accessories may sometimes burn away due to carbon deposits from high-temperature air from compression and flame from the cylinder. Excess lubrication may also lead to the accumulation of grease which causes the air start valve to get stuck. Severe explosions in the starting air manifold may even lead to rupture of the manifold causing serious injuries to the crew members and damage to the accessories. The failure probability of the starting air system is influenced by the failure probabilities of its key components such as the automatic starting air valve, air distributor, turning gear, bursting disc, relief valve, and flame arrestor. FMEA is appropriate to carry out the risk assessment of the starting air system. FMEA is at the core of the most frequently used tools for assessing risk [10–14].

The faults in the starting air system can be events that are linked with component failure, human errors, or any other event leading to undesired events. FMEA focuses on identifying which are the probable failure modes of the system, how each failure mode can adversely affect the functioning of the entire system, and the feasible techniques for the early detection of the failure modes.

4.2 Starting air system

High-pressure compressed air is used to start large marine diesel engines. The high-pressure compressed air is required to provide a large torque and overcome the inertia of large reciprocating masses. The air should have enough pressure to provide the necessary speed to the piston to compress the charged air and attain the temperature required for combustion. To facilitate the starting of the engine in any crank position, two starting air valves are opened simultaneously during the starting air sequence. The air is admitted into the cylinder just at the beginning of the power stroke when the piston is moving from the top dead center to the bottom dead center to provide a positive torque. The starting air valves are closed when the exhaust valves (ports) are about to open. During the admission of air, care is taken not to inject fuel to minimize the chances of an air start explosion in the manifold. The design of the air start system is quite complex. The main diesel engines are started by employing manual/automatic, electrical, hydraulic, and pneumatic systems.

The block diagram of a typical starting air system is shown in Fig. 4.1. It consists of an air bottle, a manual valve, a pilot valve, a turning gear interlock, an automatic air start valve, an air distributor, and a cylinder start air valve. Compressed air required for the main engine is provided by opening the air bottle valve. Air passes through the pilot valve. It acts on top of the automatic air start valve. The automatic air start valve is positively closed by this air. The other line is blocked by the turning gear engagement. To start the main engine, the turning gear is disengaged, which allows the supply of air to the automatic air start valve. This air does not go beyond the automatic air start valve due to spring pressure and the pressure exerted by the air supplied through the pilot valve. Once the air start command is given, the pilot valve stops the air flow to the automatic air start valve. The line is vented and there is no positive closing of the automatic air start valve. It allows the air from the turning gear to go to the cylinder air start manifold and air distributor. The set of pilot valves arranged radially around a cam and placed in the air start distributor controls the opening of the main air start valves and maintains the correct sequence of operation. To safeguard the system, safety devices and interlocks are provided as follows:

Bursting disc: It is provided in the starting air line. It bursts and prevents the air line from damage in case of excessive pressure in the line.

Relief valve: It is bolted at the end of the air start manifold. It is used to relieve excess pressure in the start air manifold.

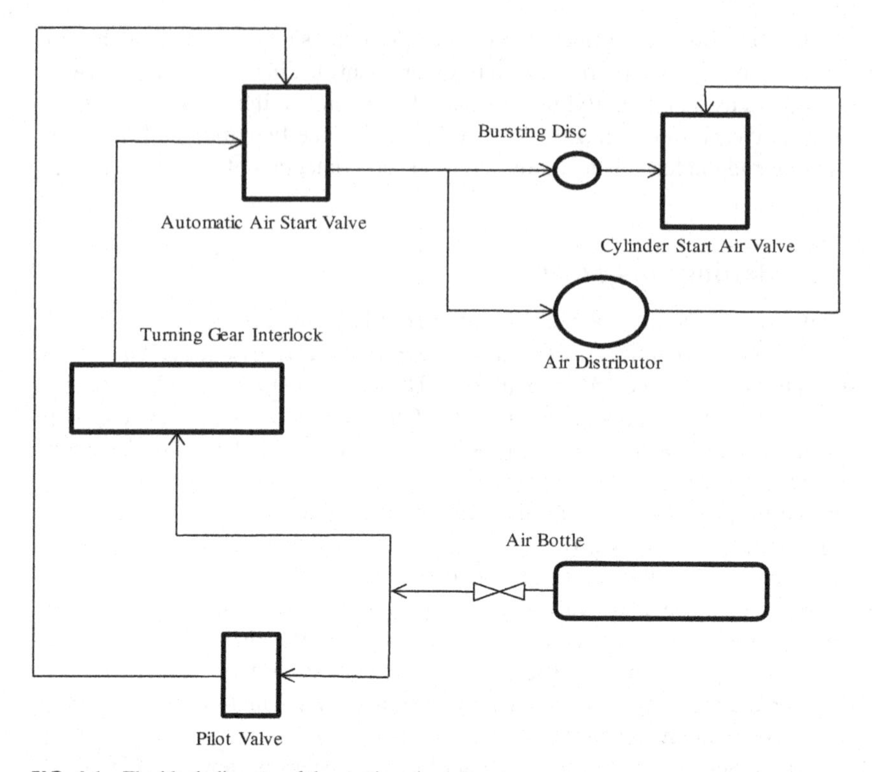

FIG. 4.1 The block diagram of the starting air system.

Flame arrestor: It is made of brass or aluminum and fitted in the branch line before the cylinder air start valve which allows only air to pass through it and arrests the flame traveling to the main air start manifold.

4.2.1 Explosion in starting air system

Explosions in starting air systems are linked to the source of ignition that may be from the following.

1. Leaking of unit cylinder air start valve: The valve may start leaking if any foreign particles get trapped between the valve and the valve seat preventing it from closing fully. Excess lubrication may also lead to an accumulation of grease at the valve seating area causing the air start valve to get stuck. If incorrect clearance is provided between operating parts, the valve will operate sluggishly leading to leakage. The hot gases from the cylinder leak past these valves into the air start manifold overheating the line.
2. Leaking of fuel into the cylinder when the engine is started: During the start sequence, the fuel leaked into the cylinder vaporizes and gets ignited due to

its compression when the piston reaches the TDC. At the opening of the air start valve, these high-pressure combustion gases enter the air start manifold.

3. Accumulating lubricating oil from compressed air: lubricating oil from compressed air gets deposited on the inner surface of the manifold. This oil can autoignite on the admission of high-pressure starting air.

4.3 Failure mode and effects analysis (FMEA)

FMEA is a risk assessment approach wherein the knowledge about risk is increased by identifying the failure modes, their likelihood of occurrence, and chances of early detection. It can have a bottom-up (hardware) or top-down (functional) approach. FMEA is used to identify the prospective failure modes of the system, analyze the effects of these failures on the functioning of the system, and suggest the ways and means to mitigate their effects on the system. It was one of the first systematically driven techniques developed by the US military. FMEA is a qualitative approach and invites innovative inputs from a multidisciplined team of experts. The method can identify the critical failure modes and suggest economical ways of avoiding failures [15]. FMEA improves system reliability by eliminating the faults that are detected in the system. FMEA uses two stages. First, it consists of enumerating all the failure modes with associated consequences. Second, it analyzes the risk by determining the failure rate using the risk priority number (RPN) [16]. RPN is used to prioritize the failure modes to take corrective action. FMEA finds significant contribution toward safety analysis, risk assessment, and decision-making applications [17–19].

$$\text{RPN} = S \times O \times D \tag{4.1}$$

$S \rightarrow$ the consequence severity if a failure occurs.

$O \rightarrow$ the occurrence probability of failure.

$D \rightarrow$ probability of early detection of failure mode.

The major weakness of this method is that different combinations of S, O, and D can give the same RPN. Moreover, the evaluation being subjective, there is a lot of uncertainty in the collected data. The inclusion of fuzzy logic [20,21] in FMEA can reduce this data uncertainty. It is seen that the results obtained by fuzzy logic are better than the ones obtained using the traditional FMEA method [22]. This chapter will analyze the failure modes of the starting air system by applying the fuzzy failure mode and effects analysis (FFMEA) method. The results might vary depending on the knowledge and experience of the expert. To even out and minimize the unpredictable epistemic uncertainty from subjective judgments, one needs to improve the reliability of measures used to collect the data. To provide ease to the expert, linguistic expressions are generally provided.

4.4 The proposed methodology

The prime requirement is to identify various failure modes that trigger the starting air explosion. After identifying the failure modes, the data on their severity, occurrence, and detection chances are collected. Most of the accidents in any industry, such as the marine industry, are either underreported or not reported at all due to a lot of regulations, and hence a lot of uncertainty is involved in estimating the likelihood of an incident. The data for likelihood is gathered by experts either through their experience or from the database. Less knowledge of information about a particular incident results in higher uncertainty and finally does not provide the correct picture of the risk involved. The effect of this is felt in the measures taken to reduce the risk level. The risk assessment objective is to obtain the risk levels and take appropriate actions to reduce the risk level and also explore measures to reduce the uncertainty involved in collecting prior information. Generally, the data are collected from experts using linguistic expressions. This is subjective and carries a lot of uncertainty. To reduce such type of uncertainty, a team of experts is formed with varying experience, knowledge, and expertise. The values of fuzzy FMEA inputs, namely severity, occurrence, and detection, are grouped into five categories of very low (VL), low (L), moderate (M), high (H), and very high (VH). These are assessed on a 1–10 scale. The inputs collected from experts' knowledge being subjective and imprecise, to use in fuzzy logic, they are described by membership functions. The membership function assigns a degree of belongingness $\mu(x) \in [0, 1]$ to every number x elicited from the expert. Different types of membership functions such as triangular, trapezoidal, piecewise linear, Gaussian, and singleton can be designed. But triangular and trapezoidal membership functions are formed using straight lines and are simple. The triangular function is a special case of the trapezoidal membership function where the core of a function is a degenerate interval. These membership functions are simple to use and work well in most practical applications. In this chapter, three input variables are fitted using straight-line membership functions (i.e., triangular and trapezoidal) to decide the degree of membership of each input. Consequently, five sets of overlapping triangular and trapezoidal curves for severity, occurrence, and detection, respectively, are produced using linguistic variables. The description for the same is given in Tables 4.2–4.4, while the fuzzy membership functions are shown in Figs.4.2–4.4.

The proposed steps of the models are described below.

Step 1: Team of m experts $E = \{E_1, E_2, \ldots, E_m\}$ evaluates the system under consideration and identifies all the n failure modes of the system, $FM = \{FM_1, FM_2, \ldots, FM_n\}$.

Step 2: Team of experts with their expertise, knowledge, and experience provides the data on the severity, occurrence, and detection chances in linguistic terms using Tables 4.2–4.4.

TABLE 4.2 Parameter and description of membership function of severity.

Sr. no.	Categories	Parameter	Description
1	Very low (VL)	(0, 0, 1, 2.5)	Failure results in insignificant system damage or loss and may not be detected
2	Low (L)	(1, 2.5, 4)	Failure does not result in any kind of injury or allow exposure/release of any kind of harmful chemicals
3	Moderate (M)	(2.5, 4.5, 5.5, 7.5)	Failure results in lower level of exposure to personnel or slight deterioration of system performance
4	High (H)	(6, 7.5, 9)	Failure results in slight injury to personnel and also exposure to harmful chemicals/radiations/fire
5	Very high (VH)	(7.5, 9, 10, 10)	Failure results in major injury or death of a personnel

TABLE 4.3 Parameter and description of membership function of occurrence.

Sr. no.	Categories	Parameter	Description
1	Very low (VL)	(0, 0, 1, 2.5)	Once per 1000 years or very scarcely
2	Low (L)	(1, 2.5, 4)	Once per 100 years
3	Moderate (M)	(2.5, 4.5, 5.5, 7.5)	Once per 10 years
4	High (H)	(6, 7.5, 9)	Once per year
5	Very high (VH)	(7.5, 9, 10, 10)	Once per month or very frequently

Step 3: Aggregate the experts' judgment using the geometric average approach.

$$\tilde{r}_i = \left(\tilde{a}_{i1} \times \tilde{a}_{i2} \times \cdots \times \tilde{a}_{ik}\right)^{1/k} \tag{4.2}$$

where $\tilde{a}_{ik}$ represents the value of the input variable for the ith failure mode given by the kth expert.

TABLE 4.4 Parameter and description of membership function of detection.

Sr. no.	Categories	Parameter	Description
1	Very low (VL)	(0, 0, 1, 2.5)	Very high chances of detecting the failure or existence of any type of deficiency
2	Low (L)	(1, 2.5, 4)	High chances of detecting the failure or existence of any type of deficiency
3	Moderate (M)	(2.5, 4.5, 5.5, 7.5)	Moderate chances of detecting the failure or existence of any type of deficiency
4	High (H)	(6, 7.5, 9)	Low chances of detecting the failure or existence of any type of deficiency
5	Very high (VH)	(7.5, 9, 10, 10)	Very low chances of detecting the failure or existence of any type of deficiency

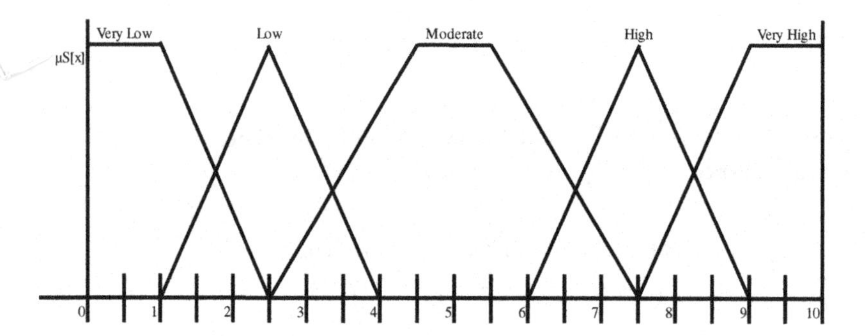

FIG. 4.2 Fuzzy membership function for input severity level.

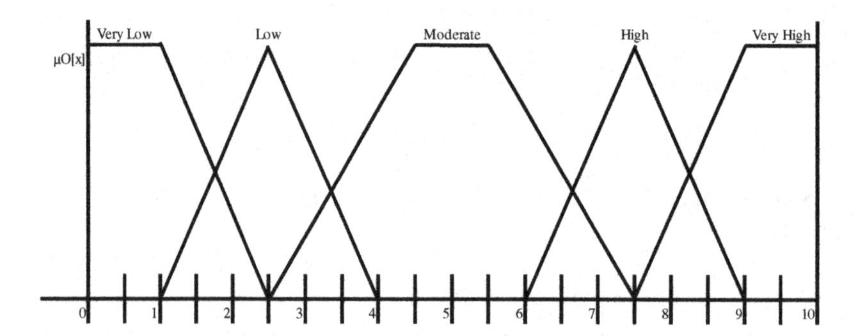

FIG. 4.3 Fuzzy membership function for input occurrence level.

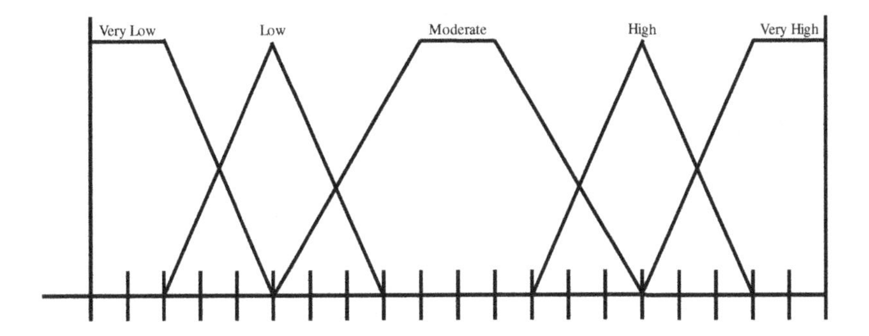

Figure 4: Fuzzy membership function for input detection level

FIG. 4.4 Fuzzy membership function for input detection level.

Step 4: Defuzzify each fuzzy number $\tilde{r}_i$ into a crisp number r_i using the Yager [23] ranking method.

$$r_i = \int_0^1 \frac{1}{2}\left(\left(\tilde{r}_i\right)_\alpha^l + \left(\tilde{r}_i\right)_\alpha^u\right) d\alpha \tag{4.3}$$

Step 5: Calculate the RPN value for each failure mode

$$\text{RPN}_i = r_{iS} \times r_{iO} \times r_{iD} \tag{4.4}$$

r_{iS}, r_{iO}, and r_{iD} are the crisp values of severity, occurrence, and detection of the ith failure mode.

Step 6: Rank the failure modes as per the RPN value. Higher value of RPN indicates that the failure mode is critical and needs urgent attention.

The flowchart for the model is given in Fig. 4.5.

4.5 An illustrative example: Starting air system

As an illustration, risk assessment of the starting air system of the ship is presented in this section. A team of six experts ($m = 6$) is formed consisting of people who are quite knowledgeable and having sizable experience in the maintenance and operation of ship machinery onboard the ship. The team identified seven failure modes ($n = 7$) and the corresponding input parameters. Table 4.5 summarizes the results of the expert team's causes and effects of the explosion in the starting air system. The data on input parameters as given by experts in linguistic expressions are presented in Tables 4.6–4.8.

Aggregated experts' judgment obtained using the geometric average approach (Eq. 4.2) on input parameters is given in Table 4.9. Crisp input parameter values obtained using Eq. (4.3) and RPN values of each failure mode obtained using Eq. (4.4) are given in Table 4.10.

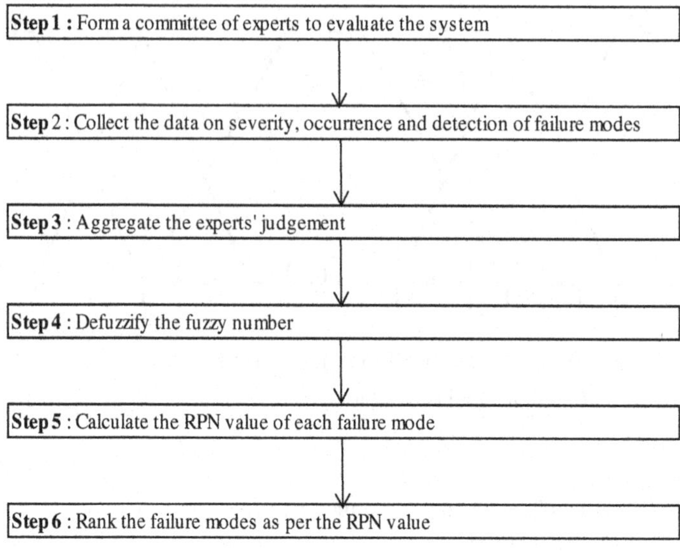

FIG. 4.5 Flowchart for the model.

TABLE 4.5 Cause-effect of starting air system explosion.

Sr. no.	Failure modes	Cause	Effect
1	Bursting disc failing to operate (FM1)	Fitted bursting disc of higher specification /Not original equipment manufacturer spare	May initiate explosion and rupture the air manifold
2	Relief valve failing to operate (FM2)	Spring is jammed or spring with high stiffness/ not tested	May initiate explosion and rupture the air manifold
3	Leaking air start valve (FM3)	Foreign particles trapped or excessive lubrication/ seat clearance not there/ overdue PMS	Overheating of line, hot gases leak past the valve and overheat air start manifold
4	Fuel leaking into the cylinder (FM4)	Leaky fuel injector	Fuel leaked vapourizes and gets ignited due to compression and enter air start manifold
5	Flame arrestor failing to arrest the flame (FM5)	Perforations are damaged, improper maintenance	Flame travels backward in manifold leading to explosion

TABLE 4.5 Cause-effect of starting air system explosion—cont'd

Sr. no.	Failure modes	Cause	Effect
6	Lubricating oil getting deposited in inner surface of the manifold (FM6)	Oil separator at compressor discharge not working, pipelines not properly drained	Ignition of oil leading to explosion/Ramming effect
7	Turning gear interlock not working (FM7)	Improper maintenance	Turning gear and motor fly off

TABLE 4.6 Severity of failure modes.

Failure modes	E1	E2	E3	E4	E5	E6
FM1	H	VH	VH	H	VH	H
FM2	VH	H	VH	VH	H	H
FM3	H	VH	VH	H	H	H
FM4	VH	H	VH	VH	H	VH
FM5	H	H	H	VH	VH	VH
FM6	VH	H	H	VH	VH	VH
FM7	H	M	M	H	H	M

TABLE 4.7 Occurrence probabilities of failure modes.

Failure modes	E1	E2	E3	E4	E5	E6
FM1	VL	L	VL	VL	VL	VL
FM2	L	L	L	VL	L	L
FM3	H	M	M	H	L	M
FM4	M	L	M	M	L	M
FM5	L	VL	VL	L	L	VL
FM6	M	H	H	M	M	L
FM7	VL	L	VL	VL	L	L

TABLE 4.8 Detection chances of failure modes.

Failure modes	E1	E2	E3	E4	E5	E6
FM1	M	M	L	M	M	M
FM2	L	M	M	VL	L	M
FM3	M	M	M	L	M	M
FM4	H	M	M	H	H	M
FM5	VL	L	L	VL	VL	VL
FM6	L	H	M	H	M	H
FM7	VL	VL	VL	VL	L	L

TABLE 4.9 Aggregated experts' judgment on input parameters.

Failure modes	Severity	Occurrence	Detection
FM1	(6.71, 8.22, 8.66, 9.49)	(0, 0, 1.16, 2.7)	(2.15, 4.08, 4.82, 6.75)
FM2	(6.71, 8.22, 8.66, 9.49)	(0, 0, 2.15, 3.7)	(0, 0, 3.18, 5.06)
FM3	(6.46, 7.97, 8.25, 9.32)	(2.87, 4.84, 5.35, 7.18)	(2.15, 4.08, 4.82, 6.75)
FM4	(6.96, 8.47, 9.09, 9.65)	(1.84, 3.7, 4.23, 6.08)	(3.87, 5.81, 6.42, 8.22)
FM5	(6.71, 8.22, 8.66, 9.49)	(0, 0, 1.58, 3.16)	(0, 0, 1.36, 2.92)
FM6	(6.96, 8.47, 9.09, 9.65)	(2.87, 4.84, 5.35, 7.18)	(3.32, 5.27, 5.63, 7.4)
FM7	(3.87, 5.81, 6.42, 8.22)	(0, 0, 1.58, 3.16)	(0, 0, 1.36, 2.92)

4.6 Results

The RPN of FM6 is the highest with a value of 233.78. It indicates that the lubricating oil getting deposited on the inner surface of the manifold is the main cause of the starting air explosion and requires attention on top priority among all the identified failure modes. The RPN of FM6 is the highest because of the

TABLE 4.10 Crisp values of input parameters and RPN of failure modes.

Failure modes	Severity	Occurrence	Detection	RPN
FM1	8.27	0.97	4.45	35.697
FM2	8.27	1.46	2.06	24.873
FM3	8	5.06	4.45	180.14
FM4	8.54	3.96	6.08	205.62
FM5	8.27	1.19	1.07	10.53
FM6	8.54	5.06	5.41	233.78
FM7	6.08	1.19	1.07	7.7417

high consequence severity associated with it. Moreover, there are very low chances of detecting such deposition of lubricating oil. The chances of fuel leaking into the cylinder (FM4) is at the second criticality position (RPN = 205.62) because it is very difficult to detect leakage of fuel into the cylinder. Other failure modes in decreasing order of their criticality are

$$FM6 > FM4 > FM3 > FM1 > FM2 > FM5 > FM7$$

4.7 Sensitivity analysis

From the results of an illustrated example, it is seen that the reliability of a system can be increased by minimizing the chances of failures. This can be achieved by controlling any one or more of the parameters (severity, occurrence, or detection) of RPN. The likelihood of occurrence of any failure can be reduced by its early detection. If an occurrence cannot be avoided, then we should try to keep the severity of an incident to a low level.

The sensitivity analysis is performed in this section to analyze the effects of these parameters on the critical failure modes. Table 4.11 shows the new ranking order after the parameters (occurrence and detection) of the highest ranked failure modes are revised while the severity values for critical failure modes are kept unchanged.

The sensitivity analysis performed on the illustrated example showed that attending the most critical failure mode on priority can lower the chances of its occurrence and thereby increase the reliability of the system.

TABLE 4.11 Revised ranking order after change in occurrence and detection parameters of critical failure mode.

Sr. no.	Failure modes	Occurrence	Detection	RPN	New raking order
1	FM6	(M, M, M, M, M, L)	(L, M, M, M, M, M)	169.11	FM4 > FM3 > FM6 > FM1 > FM2 > FM5 > FM7
2	FM4	(L, L, M, M, L, M)	(M, M, M, M, M, M)	150.73	FM3 > FM6 > FM4 > FM1 > FM2 > FM5 > FM7
3	FM3	(M, M, M, H, L, M)	Unchanged	168.74	FM6 > FM3 > FM4 > FM1 > FM2 > FM5 > FM7

4.8 Conclusions

The chapter demonstrated the use of the FMEA technique in the risk assessment of the starting air system of a marine diesel engine. FMEA is an effective and structured method to identify critical failure modes that may lead to failure of the system. The FMEA tool provides us with insights about knowledge of risk and helps improve the reliability of the system. To reduce the effect of uncertainty in experts' judgment, fuzzy logic is included in FMEA.The FFMEA methodology prioritized the failure modes that need immediate attention. The reliability and operability of the system can be increased by targeting the most critical failure modes during routine maintenance. The results show the suitability of the FFMEA technique in risk assessment in such types of complex systems.

References

[1] T.V. Ta, N.H. Vu, M.A. Triet, D.M. Thien, V.T. Cang, Assessment of marine propulsion system reliability based on fault tree analysis, Int. J. Transp. Eng. Technol. 2 (4) (2016) 55–61, https://doi.org/10.11648/j.ijtet.20160204.14.

[2] I. Lazakis, O. Turan, S. Aksu, Increasing ship operational reliability through the implementation of a holistic maintenance management strategy, Ships Offshore Struct. 5 (4) (2010) 337–357, https://doi.org/10.1080/17445302.2010.480899.

[3] K. Dikis, I. Lazakis, O. Turan, Probabilistic Risk Assessment of Condition Monitoring of Marine Diesel Engines, International Conference on Mechatronics Technology, Glasgow, UK, July 7–July 9, 2014.

[4] J.P. Mileski, G. Wang, L.L. Beacham IV, Understanding the causes of recent cruise ship mishaps and disasters, Res. Transp. Bus. Manag. 13 (2014) 65–70, https://doi.org/10.1016/j.rtbm.2014.12.001.

[5] R. Islam, F. Khan, R. Abbassi, V. Garaniya, Human error probability assessment during maintenance activities of marine systems, Saf. Health Work 9 (2018) 42–52, https://doi.org/10.1016/j.shaw.2017.06.008.

[6] D. Kimera, F.N. Nangolo, Predictive maintenance for ballast pumps on ship repair yards via machine learning, Transp. Eng. 2 (2020) 100020, https://doi.org/10.1016/j.treng.2020.100020.

[7] I. Lazakis, K. Dikis, A.L. Michala, G. Theotokatos, Advanced ship systems condition monitoring for enhanced inspection, maintenance and decision making in ship operations, Transp. Res. Proc. 14 (2016) 1679–1688, https://doi.org/10.1016/j.trpro.2016.05.133.

[8] I. Emovon, M.O. OKWU, Risk assessment tools for categorisation of failure modes of Marine Diesel Engine: a comparative study, J. Adv. Eng. Comput. 2 (1) (2018) 30–43, https://doi.org/10.25073/jaec.201821.80.

[9] H. Jeon, K. Park, J. Kim, Comparison and verification of reliability assessment techniques for fuel cell- based hybrid power system for ships, J. Mar. Sci. Eng. 74 (8) (2020), https://doi.org/10.3390/jmse8020074.

[10] A. Pillay, J. Wang, Modified failure mode and effects analysis using approximate reasoning, Reliab. Eng. Syst. Saf. 79 (1) (2003) 69–85, https://doi.org/10.1016/S0951-8320(02)00179-5.

[11] K. Cicek, M. Celik, Application of failure modes and effects analysis to main engine crankcase explosion failure on-board ship, Saf. Sci. 51 (1) (2013) 6–10, https://doi.org/10.1016/j.ssci.2012.06.003.

[12] X. Su, Y. Deng, S. Mahadevan, Q. Bao, An improved method for risk evaluation in failure modes and effects analysis of aircraft engine rotor blades, Eng. Fail. Anal. 26 (2012) 164–174, https://doi.org/10.1016/j.engfailanal.2012.07.009.

[13] M. Braglia, MAFMA: multi-attribute failure mode analysis, Int. J. Quality Reliab. Manag. 17 (9) (2000) 1017–1034, https://doi.org/10.1108/02656710010353885.

[14] K. Maheswaran, T. Loganathan, A novel approach for prioritisation of failure modes in FMEA using MCDM, Int. J. Eng. Res. Appl. 3 (4) (2013) 733–739.

[15] J. Zhu, B. Shuai, R. Wang, K.S. Chin, Risk assessment for failure mode and effects analysis using the Bonferroni mean and TODIM method, Mathematics 7 (2019) 536, https://doi.org/10.3390/math7060536.

[16] D.H. Stamatis, Six Sigma for Financial Professionals, John Wiley & Sons Inc., 2003. ISBN 0-471-45951-8.

[17] M. Cunbao, G. Zi, Y. Lin, Safety analysis of airborne weather radar based on failure mode, effects and criticality analysis, Proc. Eng. 17 (2011) 407–414, https://doi.org/10.1016/j.proeng.2011.10.048.

[18] H. Liu, X. Deng, W. Jiang, Risk evaluation in failure mode and effects analysis using fuzzy measure and fuzzy integral, Symmetry 9 (2017) 162, https://doi.org/10.3390/sym9080162.

[19] S. Peyghami, M.F. Firuzabad, P. Davari, F. Blaabjerg, Failure Mode, Effects and Criticality Analysis (FMECA) in Power Electronic Based Power Systems, 21st European Conference on Power Electronics and Applications, Aalborg University, Denmark, November, 2019, https://doi.org/10.23919/EPE.2019.8915061.

[20] L.A. Zadeh, Fuzzy Sets and their Applications to Cognitive and Decision Processes, Academic Press, Macon, GA, 1975.

[21] L.A. Zadeh, Fuzzy sets as a basis for a theory of possibility, Fuzzy Set. Syst. 1 (1) (1978) 3–28.

[22] G.A. Keskin, C. Ozkan, An alternative evaluation of FMEA: fuzzy art algorithm, J. Int. Qual. Reliab. Eng. 25 (6) (2009) 647–661, https://doi.org/10.1002/qre.984.

[23] R. Yager, A procedure for ordering fuzzy subsets of the unit interval, Inform. Sci. 24 (1981) 143–161.

Chapter 5

Test scenario generator learning for model-based testing of mobile robots

Gert Kanter[a] **and Marti Ingmar Liibert**[b]
[a]*Department of Software Science, Tallinn University of Technology, Tallinn, Estonia,* [b]*Institute of Computer Science, University of Tartu, Tartu, Estonia*

5.1 Introduction

Robotic systems are continuously growing more complex along with the expectations to their performance and reliability. The software components in autonomous, semiautonomous, and automated systems are constantly becoming larger and more sophisticated, which drives the need for high-quality assurance of these systems. Testing robotic systems is a difficult task which usually entails integration testing the whole robotic system in various conditions. Integration testing robotic systems is usually achieved by tasking the system to perform its intended functions in the conditions in which the system is designed to operate. Testing systems, which are composed of both hardware and software components in the real-world conditions, requires extensive resources, which motivates the need for developing additional methods that reduce the cost and time of testing. In order to achieve this, computer simulation is often used to test software components in addition to testing on real hardware since software defects can account for a large proportion of the total defects in robotic systems. In Refs. [1, 2], simulation was effectively used in uncovering software defects.

Testing robotic systems can be divided into white-box and black-box testing [3]. In white-box testing, the internal structure of the software is analyzed and the tests are created specifically based on that analysis. The downside of using white-box testing is its cost due to complexity and the lack of automation. The alternative to white-box testing is black-box testing which does not rely on source code analysis and treats it as an opaque system that is only interacted with using predefined interfaces. White-box testing and black-box testing can also be combined [4]. Black-box testing approach is well suited in the case of robot operating system (ROS)-based [5] robotic systems as the software

System Assurances. https://doi.org/10.1016/B978-0-323-90240-3.00005-9

components in ROS communicate using topics and services which are clearly defined. This allows testing to be done at ROS topic and service level which is highly suitable for integration testing. Testing at this level can be performed by sending random input signals to the system and then evaluating the resulting behavior. This approach is quite easy to implement but relies on test oracles to discern correct behavior from incorrect. Regardless of having test oracles, it is always possible to automatically measure code coverage of test execution. The approach proposed in this chapter capitalizes on this allowing to automatically generate test scenarios and by using TestIt toolkit to optimize tests based on code coverage. The main benefit of optimizing for code coverage is test efficiency as the optimization strives to minimize test duration while maximizing code coverage.

Since integration testing using black-box approach on ROS-based robotic systems is done using topics and services, these input signals (e.g., navigation goals for mobile robots) can be chained together to form test cases. If these test cases are not fixed before execution, they can be referred to as test scenarios since these tests are then guided during execution to cover specific test goals. In the literature, scenario-based testing using simulation has been described as a viable way to test autonomous systems in Ref. [6]. Designing tests manually is often complex and time consuming since robotic systems state spaces can be vast leading to an enormous number of possible cases. In Ref. [6], the author also emphasizes the need for systematic variation of the test environment to validate correct behavior in all operating conditions. For example, changing simulation parameters (e.g., weather) has been explored in Ref. [7]. As robotic systems are often designed to operate in dynamic environments and for long periods of time, it is exceedingly difficult to design tests manually to account for all variables and over a long period, which motivates the need for automated tool support.

In this chapter, we propose a tool that generates test scenarios automatically using timed automata learned from log traces. We introduce a novel way of reducing the size of the scenario generating automaton by clustering the state space before test generation. The proposed tool is validated on a mobile-robot security system case study. The demonstration case study features a simulation of two patrol robots and an intruder. The goal of using the toolkit is to test the patrol robots' ability of detecting an unknown object which has been left behind by the intruder. The tool is developed as an extension to TestIt [8], a toolkit for model-based testing of ROS-based robotic systems.

5.2 Related work

Model inference is generally divided into two categories: active model inference and passive model inference. Active model inference includes inference techniques that actively interact with black-box systems to extract knowledge about the system, for example, L*-based techniques [9]. L* is the basis of an

MDP-based model learning algorithm in Ref. [10]. L*-based techniques require a minimally adequate teacher (MAT) for the algorithm to query for membership and equivalence. The case studies presented in Ref. [10] require an exceedingly large number of queries due to accounting for erroneous behavior probabilities and their scope is strictly limited to model inference.

According to Ref. [11], learning-based testing (LBT) is an emerging paradigm in automated black-box testing. Similarly to our approach to use LBT in combination with machine learning techniques, constrained active learning techniques have been applied in Ref. [12] to use case test simulated autonomous driving scenarios. The authors present an architecture for a constrained active machine learning (CAML) approach to automate use case testing.

Passive model inference works by inferring models from a fixed set of samples, such as log traces. Algorithms such as kTail, kBehaviour, gkTail, and kLFA are used for passive model inference [13]. A special case of passive model inference, called incremental learning, is a technique where instead of querying the system, we expect receiving observations, one after another [14].

Our proposed approach builds upon these techniques by using passive model inference using already existing log traces to infer models, but also has the option to continue active exploration to increase the number of log traces. We create a simple correct automaton where each of the log traces is a separate state and then try to generalize the model into a scenario generator by using clustering to merge the states.

A somewhat similar approach is presented in Ref. [15], which is demonstrated on a case study featuring a production system. This system was tested using an offline passive model-based testing technique. The tool described in Ref. [15] has a similar general architecture to ours, which incorporates a logger, an explorer, a clusterer, and the model generator. Their tool also uses clustering to segment a trace set into several subsets, but it creates only one subset per entry point of the system. Our solution uses clustering more generally to combine similar states so that they preserve their testing properties. Another key difference is that their tool creates the model as a symbolic transition system whereas the tool presented in this chapter uses UPPAAL timed automata (UTA) [16] as the specification for the model which allows to specify time constraints. Time constraints are especially important in testing robotic systems as some actions might be time constrained and formalisms without the notion of time are not well suited for such modeling.

5.3 Preliminaries

5.3.1 Model-based testing

Model-based testing is a testing technique where formal models are used to guide the testing process [17]. The relations between model, the tester, and system under test (SUT) can be seen in Fig. 5.1.

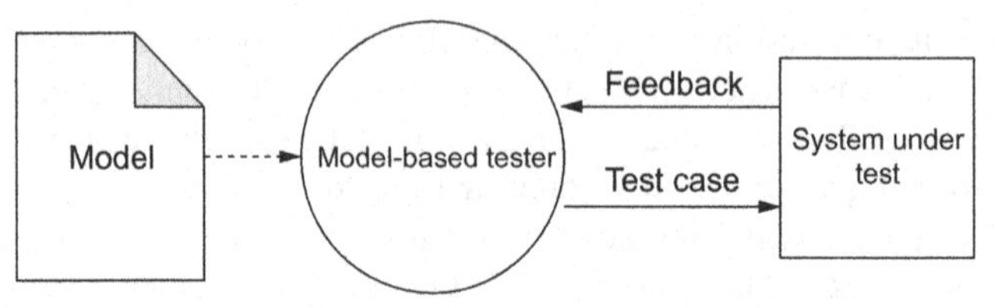

FIG. 5.1 Model-based testing.

The tool proposed in this chapter enables us to learn timed automaton which can be used as a model for generating scenarios for testing using TestIt [8] and UPPAAL tools [16].

Learning the automaton from traces allows us to generate scenarios that let us regression test the system or find new scenarios that may break the system. Timed automaton is useful here because it allows us to test if the robot system is working within its time limits and allows conformance testing also for timed systems. Exploring the state space can be continued and the model can be complemented with new information, which lets us to use it for long-term autonomous testing.

5.3.2 TestIt toolkit

TestIt [8] is an open-source toolkit for model-based testing of robotic systems. Its core features are test optimization, scalability, and long-term autonomy testing using model-based testing technique.

The high-level overview of TestIt can be seen in Fig. 5.2. The general workflow of TestIt starts with the developer making changes to the software code repository which triggers the continuous integration (CI) server. The CI server executes TestIt which initializes the testing infrastructure, performs the tests, and shuts down the testing infrastructure to decrease expenses. The results and feedback of the tests are available to the developer via CI or the TestIt interface.

TestIt test optimization relies on information gathered during previous test execution in exploration mode. The optimized test forms a test scenario with a goal to maximize some criterion (e.g., code coverage, robot-specific measurement value such as localization uncertainty). In order to improve the test scenarios, the test traces should be as concise as possible without losing valuable data due to compression. The addition to this workflow presented in this chapter allows the test cases to be compressed (i.e., clustered) so that the test optimization can focus on salient points in the state space.

5.4 Tool architecture

The task of learning a test scenario generator consists of exploring the state space generating log traces, learning a model from log traces and using that model to generate test scenarios. The tool's components, as seen in Fig. 5.3

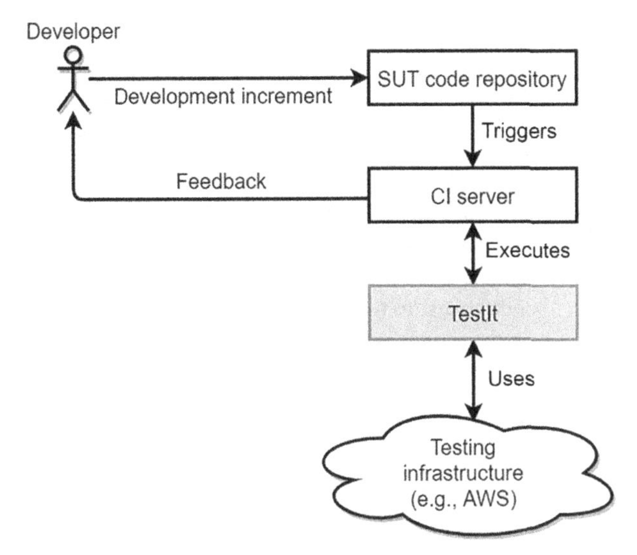

FIG. 5.2 TestIt high-level overview.

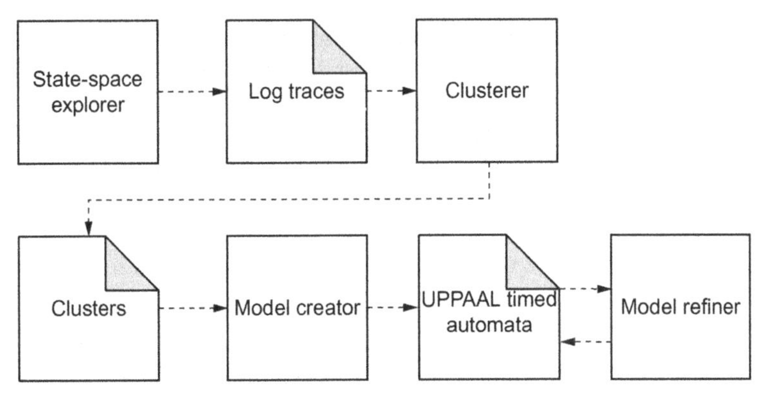

FIG. 5.3 Components and artifacts of the tool.

are the explorer, the clusterer, and the model generator. The explorer produces log traces, which form the input for the clusterer. The clusterer produces clusters, which are used in combination with log traces to create the UTA. The model refiner takes the model and removes deadlocks by trying to create transitions between nearby states.

5.4.1 State-space exploration

Exploring is implemented as a depth-first search in the discrete space defined by the user, as seen in Algorithm 5.1. The explorer increases the values at each search step by the amount defined by the user. This allows the user to explore any continuous or discrete space.

Algorithm 5.1 Explore

Input: Configuration of the explorer; initial robot system state v
Output: Log traces
Label v as visited;
for *all neighbours w from v in distance d sorted by distance in descending order* **do**
  if *vertex w is not visited* **then**
    Send request to the robot system to transition into state w;
    Log the request and its result;
    Explore w;
  end
end

More specifically, the explorer allows the user to define certain ROS topics and the value of the topic fields to be increased or decreased during exploring. It also allows defining constants, initial values, and whether to continue from the last state in the log. Explorable topics can be synced or set to be explored separately. These options also allow the exploration of multirobot systems' state space.

5.4.2 Clustering state space

The log traces created from exploring the environment allows the state space to be defined. A timed automaton is constructed from the state space combined with time stamps. However, generating scenarios from the timed automaton created directly from log traces would lead to visiting many states in said scenarios. This concern gives rise to a need to combine these states somehow to generate better and more optimal scenarios for testing. In order to achieve this, we propose to cluster similar states together to achieve more optimal tests w.r.t. number of stimuli sent to the system.

Clustering is done using the log traces generated by exploring the state space, as seen in Algorithm 5.2. The user can configure what information is logged—for example, the commands sent to the robot systems and responses from the robot systems, but also code coverage at each step. Clustering gives the user the options of choosing which fields to cluster by and how to transform them before clustering. Each synchronized topics' set are clustered together and separate from all other synchronized topics' sets. Only successfully fulfilled commands are being considered as states while clustering.

Algorithm 5.2 Clustering

```
Input: Log traces, User defined coefficients a, b specify the state space
       reduction coefficient
Output: Clusters
state_space := Extract states from log traces;
best_silhouette_coef := -inf;
best_clusters;
for n in range of size(state_space)/a to size(state_space)/b do
    clusters := k_means(state_space, n);
    new_silhouette_coef := silhouette(clusters);
    if new_silhouette_coef > best_silhouette_coef then
        best_silhouette_coef := new_silhouette_coef;
        best_clusters := clusters;
    end
end
for cluster in clusters do
    cluster.center := find_real_center(cluster, log_traces);
    Separate unsuccessfully reached states from successfully reached
      states into a separate cluster;
end
```

5.4.3 KMeans

We used KMeans clustering [18] as a default for combining states but any type of clustering can be defined to be used by the user. KMeans allows the user to use the number of clusters as a hyperparameter. Silhouette analysis [19] was used to optimize the number of clusters. This combination of KMeans clustering and silhouette analysis gave the best result of symmetrical and appropriate sized clusters for the state space. The user can specify the interval of state-space reduction cocfficicnt from which the range of number of clusters will be calculated to be considered for optimization.

5.4.4 Postclustering analysis

In the postclustering step, each system input from each cluster is analyzed separately and separated from nonsuccessful commands to form separate clusters. The clustering algorithm is shown in Algorithm 5.2.

5.4.5 Automaton construction

UTA is used as the formalism for the timed automaton constructed from the clustered state space. Formally, a timed automaton is a tuple

$$\mathcal{A} = \langle \Sigma, L, L_0, C, F, E \rangle \tag{5.1}$$

that consists of the following components:

- Σ is a finite set called the alphabet or actions of $\mathcal{A}$.
- L is a finite set. The elements of L are called the locations or states of $\mathcal{A}$.
- $L_0 \subseteq L$ is the set of start locations.
- C is a finite set called the clocks of $\mathcal{A}$.
- $F \subseteq L$ is the set of accepting locations.
- $E \subseteq L \times \Sigma \times \mathcal{B}(C) \times \mathcal{P}(C) \times L$ is a set of edges, called transitions of $\mathcal{A}$, where
  - $\mathcal{B}(C)$ is the set of clock constraints involving clocks from C, and
  - $\mathcal{P}(C)$ is the powerset of C.

An edge $(\ell, \sigma, g, r, \ell')$ from E is a transition from locations ℓ to ℓ' with action σ, guard g and clock resets r [20].

The automaton is constructed by creating transitions between log traces' clusters, as seen in Algorithm 5.3. Each cluster becomes a state in the timed automata with the center as the value. The alphabet of the automaton is a set of encoded commands that take the robot system from one state to the other. Commands are communicated to the robot system through an adapter that translates them to an understandable format for the robot system.

Algorithm 5.3 Automaton creation

Input: Clusters, Log traces
Output: Timed automaton
prev_cluster := The cluster of the first log trace;
add_start_location(prev_cluster);
for *point in rest of the log traces* **do**
    cluster := Find cluster of the point;
    if *cluster != prev_cluster* **then**
        add_timed_transition(prev_cluster, cluster);
    end
    prev_cluster := cluster;
end

Each transition also has a time constraint based on the time it took for the transition to happen in the log traces. As the transition takes place probably between the borders of the clusters, the time constraints have to be buffered with the average time it takes to reach from the center of the cluster to the edge.

5.4.6 Model refinement

The goal of model refinement is to find and remove possible deadlocks to enable long-term testing based on the learned model. The implemented model refinement strategy tries to find close pairs between which transitions could be possible and tries to use the automaton to move the system through the states and when it encounters a member of a close pair, it tries to transition the system to the other state in the pair and also back to the original state. After it has reached the end of the automaton, it starts to pass through it in reverse. This technique tries to ensure that maximum amount of transitions are bidirectional, which prevents deadlocks.

5.4.7 Model checking

The properties of the obtained model can be verified using UTA model checking [21] to verify that the model fulfills the requirements for the test purpose. After model refinement, we can run a query *A*[] *not deadlock* on the model to see if there are no deadlocks in the model, meaning that we can use it for long-term testing running the test generator indefinitely.

5.5 Validation

The tool was validated on a robotic system that included two patrol robots and one intruder robot. The intruder robot had to plant an unknown object in the building and the patrol robots had to find it. The goal was to test the patrol robots through leading the intruder. The model learning can be considered successful if the automatically created model enables testing the patrol robots as well as a handmade model. The testing ability can be measured by the lines of code covered during testing in the relevant files. The main functionality that is being tested in this case study is the patrol robot unknown object detection. We want to see if leading the intruder by both models enables the testing of the patrol robots' object detection.

The patrol robots have a predetermined path they follow. The intruder has two commands: one that tells the intruder where to go and one that asks the intruder to plant the unknown object. The intruder can plant the unknown object only at a certain location.

5.5.1 Exploration

The explorer sends commands to the intruder for exploring the space of the building depth first and increasing coordinates by 1. It also sends the command of trying to plant the unknown object at each location. Both topics are synced, which means that the planting commands are sent at each location. The explored state space of the robot system can be seen in Fig. 5.5 which corresponds to the floor plan in Fig. 5.4.

5.5.2 Clustering and creating the automata

The log traces acquired via exploring the state space were clustered and used to create the timed automaton. The clustering was performed by x and y coordinates of the first command, the success of the second command and the timestamp.

As can be seen from Fig. 5.6, the event where the plant unknown object command was successful is clustered separately from other command events.

The created UPPAAL automaton without labels can be seen in Fig. 5.7. The refined version of the UPPAAL automaton without labels can be seen in Fig. 5.8. As can be seen, there are more transitions between the states in the refined model and deadlocks have been eliminated. The most notable changes in the automaton structure are marked with circles.

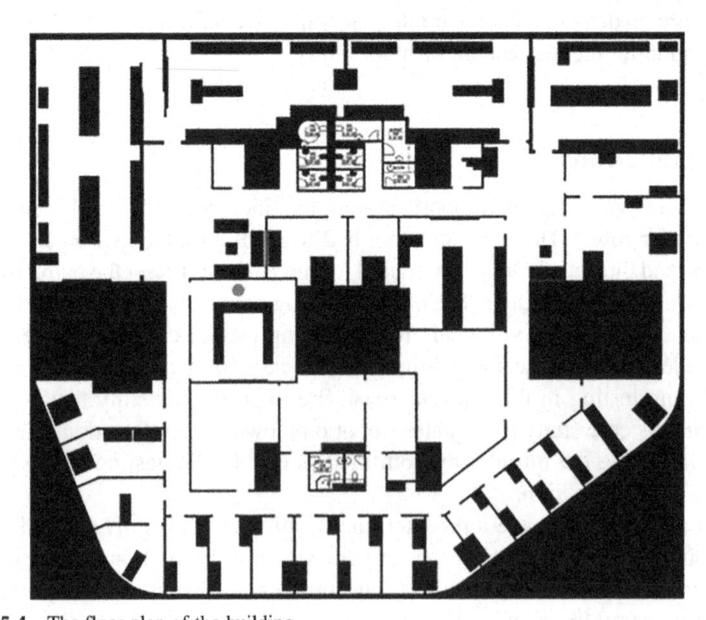

FIG. 5.4 The floor plan of the building.

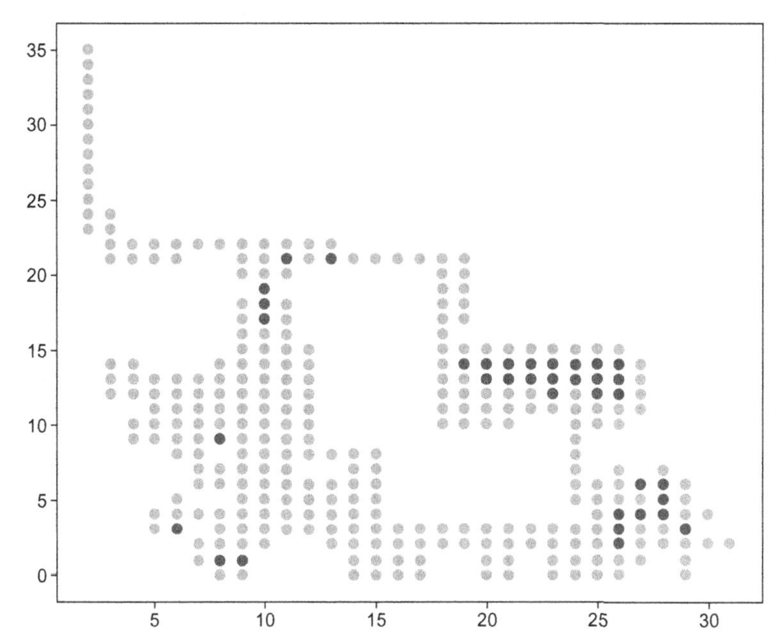

FIG. 5.5 The explored state space of the building.

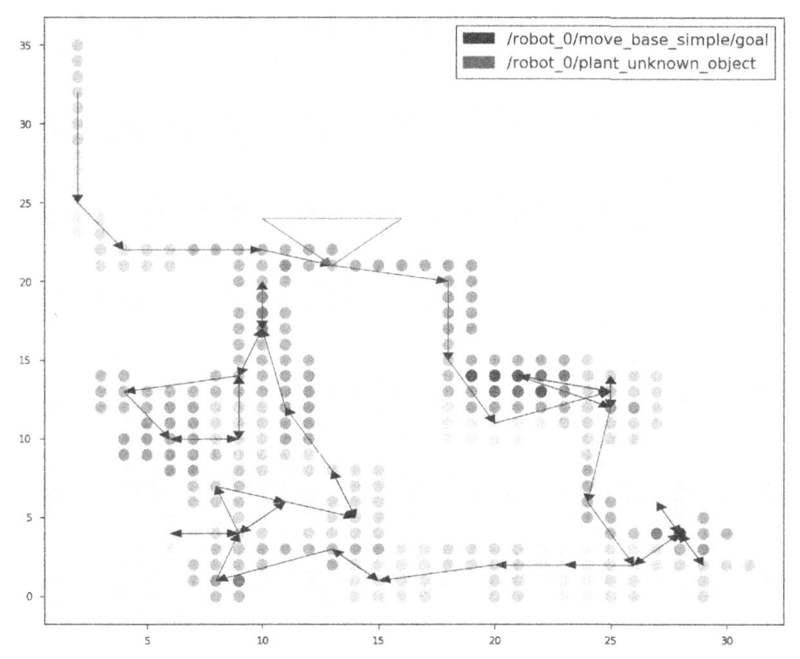

FIG. 5.6 Clusters and transitions of the automata.

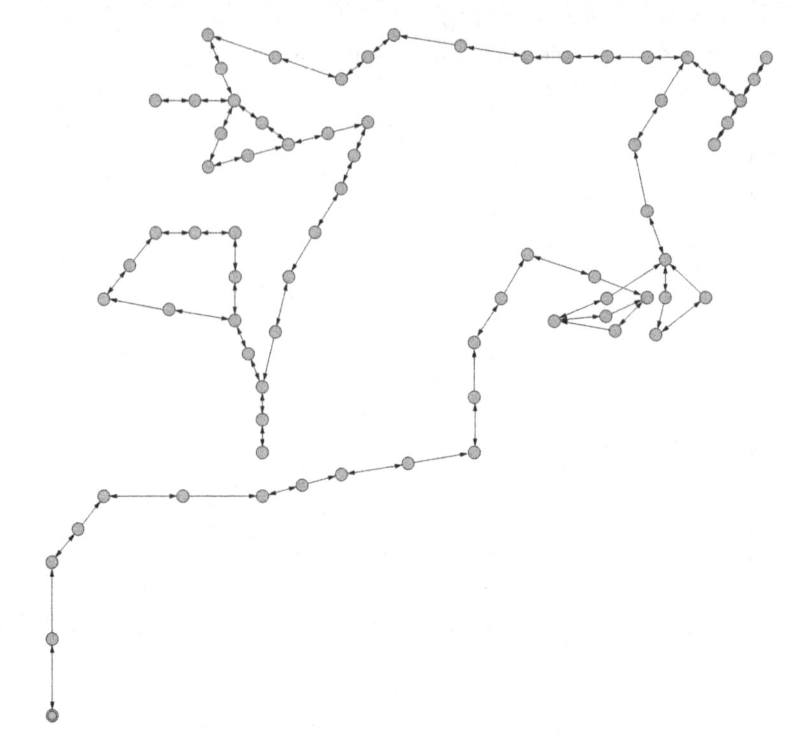

FIG. 5.7 UPPAAL automaton without labels.

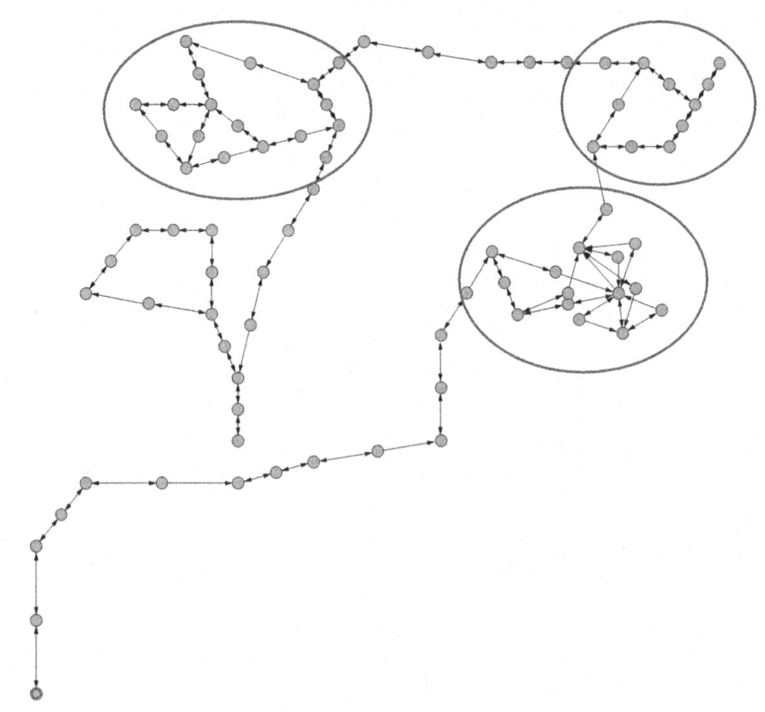

FIG. 5.8 Refined UPPAAL automaton without labels.

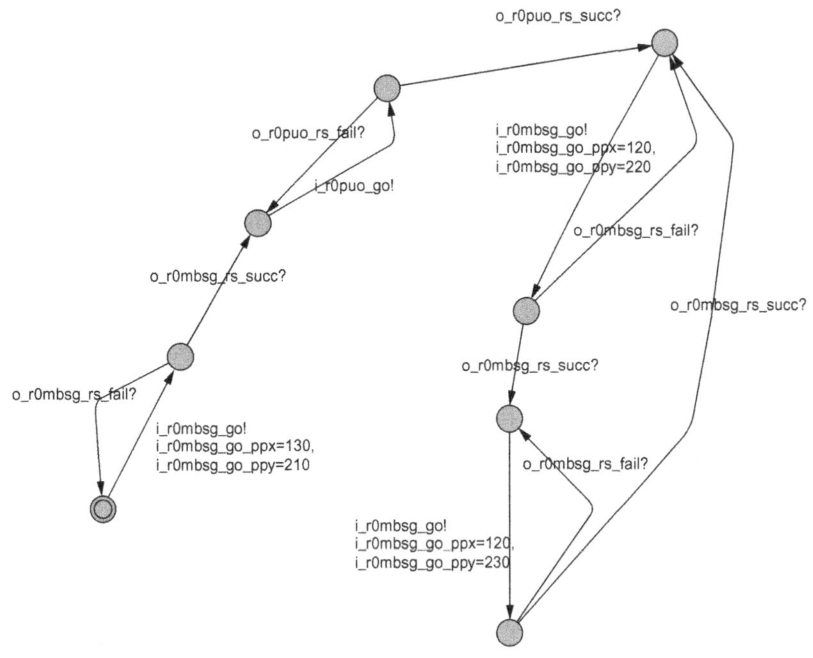

FIG. 5.9 Manually created automaton.

The handmade UPPAAL automaton to test the system can be seen in Fig. 5.9. The handmade model is a simple one which directs the robot directly into the room in which the unknown object can be planted. The handmade model is the fastest possible one to test the scenario of detecting the planted unknown object, since the object will be planted before the first time the patrols reach the room.

5.5.3 Results

The code coverage of the patrol robots was measured during testing execution to assess the test efficiency of the created automaton. We also compared the automatically created automaton's testing efficacy to a manually created automaton's testing efficacy. The main objective was to test the software component responsible for detecting the unknown object, but since the number of lines of code was relatively small, we used a weight of 5 for the object detection file to increase its importance in test optimization.

The relative code coverage percentage of the first patrol robot can be seen in Fig. 5.10. The term relative code coverage refers to the ratio of number of lines

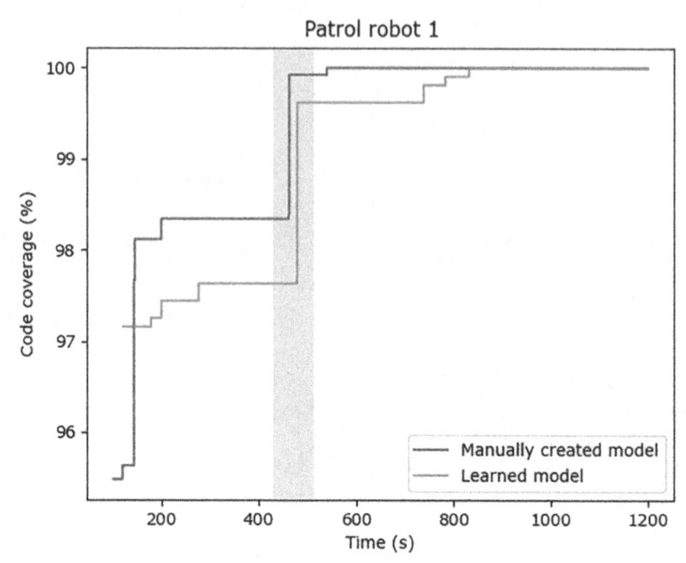

FIG. 5.10 Patrol robot 1 code coverage.

executed over the total number of lines witnessed being executed by the system. At roughly $t = 500$ s in the highlighted interval, the relative code coverage percentage jumps up, which coincides with the time the patrol robot detected the unknown object. In Table 5.1, the unweighted number of new lines discovered by timestamps achieved by running the manual and automatic scenarios can be seen. For each time interval, we have manual scenario lines of code covered and automatic scenario lines of code covered with percentages for each general package ran. The percentages are out of all code lines ran. The package we were interested in is *testit-patrol-obj* **Table 5.1** which is responsible for detecting the unknown object. It can be seen that the *testit-patrol-obj* package's lines of code count goes up to 100% in the interval [451, 600] s, where the detecting of the unknown object occurred.

For both manually and automatically created models the detection of the unknown object occurs roughly at the same time. Testing the system using the learned model was sufficiently good, meaning that the model's state space was small enough for the intruder to place the object before the patrol robots reached the location in their patrol path. This demonstrates that automatic model learning and clustering is a viable method for reducing the state space for scenario generation in model-based testing.

TABLE 5.1 Patrol robot 1 manual scenario lines of code (Man. LoC) compared to automatic scenario lines of code (Aut. Loc) covered out of total by timestamps.

Time interval (s) →	0 .. 150		151 .. 300		301 .. 450		451 .. 600	
Package ↓	*Man. LoC*	*Aut. LoC*	*Man. LoC*	*Aut. LoC*	*Man. LoC*	*Aut. LoC*	*Man. LoC*	*Aut. LoC*
actionlib	135 (95%)	140 (99%)	135 (95%)	141 (100%)	135 (95%)	141 (100%)	136 (96%)	141 (100%)
actionlib_msgs	71 (100%)	71 (100%)	71 (100%)	71 (100%)	71 (100%)	71 (100%)	71 (100%)	71 (100%)
genpy	84 (100%)	84 (100%)	84 (100%)	84 (100%)	84 (100%)	84 (100%)	84 (100%)	84 (100%)
geometry_msgs	61 (100%)	61 (100%)	61 (100%)	61 (100%)	61 (100%)	61 (100%)	61 (100%)	61 (100%)
move_base_msgs	153 (100%)	150 (98%)	153 (100%)	150 (98%)	153 (100%)	150 (98%)	153 (100%)	150 (98%)
rosgraph	126 (100%)	56 (44%)	126 (100%)	56 (44%)	126 (100%)	56 (44%)	126 (100%)	56 (44%)
rosgraph_msgs	60 (100%)	60 (100%)	60 (100%)	60 (100%)	60 (100%)	60 (100%)	60 (100%)	60 (100%)
rospy	431 (100%)	223 (51%)	431 (100%)	225 (52%)	431 (100%)	225 (52%)	431 (100%)	225 (52%)
std_msgs	11 (91%)	11 (91%)	11 (91%)	11 (91%)	11 (91%)	11 (91%)	12 (100%)	12 (100%)
testit-patrol-sut	33 (91%)	31 (86%)	36 (100%)	33 (91%)	36 (100%)	33 (91%)	36 (100%)	33 (91%)

Continued

TABLE 5.1 Patrol robot 1 manual scenario lines of code (Man. LoC) compared to automatic scenario lines of code (Aut. Loc) covered out of total by timestamps.—cont'd

Time interval (s) →	0 .. 150		151 .. 300		301 .. 450		451 .. 600	
Package ↓	*Man. LoC*	*Aut. LoC*	*Man. LoC*	*Aut. LoC*	*Man. LoC*	*Aut. LoC*	*Man. LoC*	*Aut. LoC*
testit-patrol-obj	13 (76%)	13 (76%)	13 (76%)	13 (76%)	13 (76%)	13 (76%)	**17 (100%)**	**17 (100%)**
tf	5 (100%)	5 (100%)	5 (100%)	5 (100%)	5 (100%)	5 (100%)	5 (100%)	5 (100%)
tf2_msgs	54 (100%)	54 (100%)	54 (100%)	54 (100%)	54 (100%)	54 (100%)	54 (100%)	54 (100%)
tf2_ros	14 (100%)	14 (100%)	14 (100%)	14 (100%)	14 (100%)	14 (100%)	14 (100%)	14 (100%)

5.6 Conclusion

In this chapter, we introduced the technique of clustering the state space and creating a timed automaton of the state space to generate scenarios for testing the robot system.

The technique is implemented in the TestIt toolkit as a customizable and flexible tool that enables the exploration, clustering, and creation of the timed automaton used in model-based testing of robotic systems. For exploration, we used depth-first search with user defined parameters, for clustering we used KMeans and for the timed automaton we used UTA formalism.

The tool was validated on a case study of patrol robots, where the objective was to test the patrol robots' software component responsible for detecting an unknown object. The automatically created model was used to guide the environment-controlled actor to place the unknown object. The testing efficacy of the learned model was compared with a manually constructed model to assess the learned model's ability to test the system compared to the manually constructed model. The analysis of the results show that the proposed technique enables creating timed automata suitable for generating test scenarios that are effective at testing the robotic system using model-based testing.

References

[1] C.S. Timperley, A. Afzal, D.S. Katz, J.M. Hernandez, C. Le Goues, Crashing simulated planes is cheap: can simulation detect robotics bugs early? in: 2018 IEEE 11th International Conference on Software Testing, Verification and Validation (ICST), 2018, pp. 331–342, https://doi.org/10.1109/ICST.2018.00040.

[2] T. Sotiropoulos, H. Waeselynck, J. Guiochet, F. Ingrand, Can robot navigation bugs be found in simulation? An exploratory study, in: IEEE International Conference on Software Quality, Reliability and Security (QRS), 2017, pp. 150–159, https://doi.org/10.1109/QRS.2017.25.

[3] G.J. Myers, T. Badgett, C. Sandler, Test-case design, in: The Art of Software Testing, John Wiley & Sons, Ltd, 2012, pp. 41–84, https://doi.org/10.1002/9781119202486.ch4 (Chapter 4).

[4] F. Howar, B. Jonsson, F. Vaandrager, Combining black-box and white-box techniques for learning register automata, in: B. Steffen, G. Woeginger (Eds.), Computing and Software Science: State of the Art and Perspectives, Springer International Publishing, Cham, 2019, pp. 563–588, https://doi.org/10.1007/978-3-319-91908-9_26.

[5] M. Quigley, ROS: an open-source robot operating system, in: ICRA 2009, 2009.

[6] T. Linz, Testing autonomous systems, in: S. Goericke (Ed.), The Future of Software Quality Assurance, Springer International Publishing, Cham, 2020, pp. 61–75, https://doi.org/10.1007/978-3-030-29509-7_5.

[7] H. Abbas, M. O'Kelly, A. Rodionova, R. Mangharam, Safe at any speed: a simulation-based test harness for autonomous vehicles, in: 7th Workshop on Design, Modeling and Evaluation of Cyber Physical Systems (CyPhy17), October, 2017.

[8] G. Kanter, J. Vain, Model-based testing of autonomous robots using TestIt, J. Reliab. Intell. Environ. 6 (2020), https://doi.org/10.1007/s40860-019-00095-w.

[9] D. Angluin, Learning regular sets from queries and counterexamples, Inf. Comput. 75 (2) (1987) 87–106, https://doi.org/10.1016/0890-5401(87)90052-6.

[10] M. Tappler, B.K. Aichernig, G. Bacci, M. Eichlseder, K.G. Larsen, L*-based learning of Markov decision processes (extended version), Form. Asp. Comput. (2021), https://doi.org/10.1007/s00165-021-00536-5.
[11] K. Meinke, Learning-based testing: recent progress and future prospects, in: A. Bennaceur, R. Hähnle, K. Meinke (Eds.), Machine Learning for Dynamic Software Analysis: Potentials and Limits: International Dagstuhl Seminar 16172, Dagstuhl Castle, Germany, April 24–27, 2016, Revised Papers, Springer International Publishing, Cham, 2018, pp. 53–73, https://doi.org/10.1007/978-3-319-96562-8_2.
[12] K. Meinke, H. Khosrowjerdi, Use case testing: a constrained active machine learning approach, in: 15th International Conference on Tests and Proofs, TAP21, 2021.
[13] D. Lo, L. Mariani, M. Santoro, Learning extended FSA from software: an empirical assessment, J. Syst. Softw. 85 (2012), https://doi.org/10.1016/j.jss.2012.04.001.
[14] P. Dupont, Incremental regular inference, in: L. Miclet, C. de la Higuera (Eds.), Grammatical Interference: Learning Syntax From Sentences, Springer, Berlin, Heidelberg, 1996, pp. 222–237.
[15] W. Durand, Automated Test Generation for Production Systems With a Model-Based Testing Approach (Theses), Université Blaise Pascal—Clermont-Ferrand II, 2016, https://tel.archives-ouvertes.fr/tel-01343385.
[16] G. Behrmann, A. David, K.G. Larsen, A tutorial on UPPAAL, in: M. Bernardo, F. Corradini (Eds.), Formal Methods for the Design of Real-Time Systems: International School on Formal Methods for the Design of Computer, Communication, and Software Systems, Bertinora, Italy, September 13–18, 2004, Revised Lectures, Springer, Berlin, Heidelberg, 2004, pp. 200–236, https://doi.org/10.1007/978-3-540-30080-9_7.
[17] M. Utting, B. Legeard, Practical Model-Based Testing: A Tools Approach, Morgan Kaufmann Publishers Inc., San Francisco, CA, 2006.
[18] J.A. Hartigan, M.A. Wong, Algorithm AS 136: a K-means clustering algorithm, J. R. Stat. Soc. C (Appl. Stat.) 28 (1) (1979) 100–108.
[19] P.J. Rousseeuw, Silhouettes: a graphical aid to the interpretation and validation of cluster analysis, J. Comput. Appl. Math. 20 (1987) 53–65, https://doi.org/10.1016/0377-0427(87)90125-7.
[20] D. D'Souza, P. Shankar, Modern Applications of Automata Theory, Co-Published with Indian Institute of Science (IISc), Bangalore, India, 2012, https://doi.org/10.1142/7237.
[21] J. Bengtsson, K. Larsen, F. Larsson, P. Pettersson, W. Yi, UPPAAL—a tool suite for automatic verification of real-time systems, in: R. Alur, T.A. Henzinger, E.D. Sontag (Eds.), Hybrid Systems III, Springer, Berlin, Heidelberg, 1996, pp. 232–243.

Chapter 6

Testing effort-dependent software reliability growth model using time lag functions under distributed environment

Sudeept Singh Yadav, Avneesh Kumar, Prashant Johri, and J.N. Singh
Galgotias University, Greater Noida, Uttar Pradesh, India

6.1 Introduction

Our society's day-to-day operations are becoming largely focused on software-based systems, and the tolerance for system failures is declining. Software engineering is expected to assist not only in the timely and cost-effective delivery of a software product with necessary features, but also in the fulfillment of particular quality requirements. The most noticeable is dependability. Software Reliability Engineering (SRE) is described as the "applied science of predicting, evaluating, and controlling the reliability of software-based systems to improve customer satisfaction" [1,2].

In the preparation and management of software development programs, a very critical role is played by software reliability engineering techniques. During the design and development stages, as well as during the formal testing process, it is important to track the duration and scope of bug incidents as well as the time it takes to repair them. With this information, it is possible to create acceptable models for reliability that can be used for predicting when the desired level of reliability of a software product will be achieved as well as strategies to shorten that time.

According to the IEEE, "the quality of a device in the form of component to do its necessary functions for a specified duration of time under specified conditions." Often project and software development managers equate correctness with dependability, putting a premium on testing and the number of "bugs" discovered fixed with time. While it is critical to find and fix bugs discovered during testing in order to ensure reliability, creating a reliable, high-quality product over the software life cycle is a smarter strategy [3].

System Assurances. https://doi.org/10.1016/B978-0-323-90240-3.00006-0

Although the words mistake, fault, and failure are often interchanged, they have distinct definitions. Programming errors are usually caused by a programmer's action, reduction, and omission, resulting in a fault. A malfunction is caused by a software fault or defect, and a failure occurs when a computer process deviates from the program's requirements in an unacceptable way. When we assess dependability, we look at the following factors; we generally just look at defects that have been detected and defects that have been removed [2].

Software, on the other hand, has a different rate of fault or error detection. When it comes to applications, the error rate is highest during integration and testing. Errors are detected and corrected as the system is checked. During its operational usage, this elimination happens at a slower rate. If no new errors are introduced, the number of errors will gradually decrease. Software, unlike hardware, does not have moving parts or wear out and becomes obsolete [1].

To improve reliability, the emphasis should be on a detailed testing plan and robust requirements that ensure all standards are being tested. The software's maintainability must also be considered, as there will be a "useful-life" period during which sustaining engineering will be needed. As a consequence, in order to prevent software errors, we must take the following steps:

1. Start with the specifications, making sure that the product produced matches the specifications and that all criteria accurately and precisely describe the final product's functionality.
2. Check to see if the code can assist with engineering maintenance without adding new errors.
3. There is a robust evaluation program that verifies all of the features specified in the specifications.

The fault is an error in the code that can result in one or many failures. The faults that are associated with a software product as it was originally written, or as it has been changed, are known as inherent faults. Modified faults are faults that have been implemented as a result of fault correction or design improvements. Software flaws, technical inadequacies in software, and user errors (for example, a lack of user knowledge) may all cause failures. Failure and how to use mistakes, as well as their frequency, have been shown to have a clear correlation with consumer loyalty and perceptions of product quality. Faults, on the other hand, are more developer-oriented since they are usually converted into the amount of work necessary to restore and maintain the device.

The effect of a malfunction or error on the operation of a software-based device is measured in severity. The threat posed by a problem in terms of functionality (service), economics (cost), or, in the case of critical failures, human existence is typically closely linked to its severity. Essentially, major and minor failures are examples of service effect classifications. Failure severity (or faults) is often used to partition operational failure data and thus notify decisions about failures of a specific severity, or to weight data used in reliability and availability calculations [4].

The aim of software testing is to assess and measure the quality of work performed at each point of the software development process. Running a program with the aim of finding an error is known as software testing process. The testing process concentrates on the software's logical internals, ensuring that all claims have been tested, as well as the functional externals, ensuring that specified feedback will produce real results that meet the requirements. While it might seem that way at times, the aim of testing is not to exhaust all budget or schedule resources at the end of a development project. The aim of testing is to ascertain that the software performs as planned and to improve the consistency, reliability, and maintainability of the software. In the production and maintenance of software, testing is an important factor. Testing takes up more time in certain organizations than any other aspect of software development. The adoption of well-defined research procedures and processes cuts down on testing time while enhancing return on investment [5,6].

Integration testing incorporates individual units that were tested and coded during the development process, precedes "acceptance testing" of the completed software system. Acceptance testing is done to help a consumer or client decide whether or not to use a software product. Both complaints and modification requests should be addressed in accordance with the documentation for the software issue procedure. The following forms of testing are used in software testing methodology: integration and regression testing, acceptance testing, functional testing, and system testing. All interfaces, if new or updated, are extensively tested during integration and functional testing. Each feature, command for user input and command choice, is tested. Regression testing is the method of reviewing modifications to computer programs to ensure the older code also performs with the new changes. It is a type of system testing in which the developer's software testing searches for communication flaws between modules, such as information not being passed or being passed incorrectly. The use of case scenarios is the most common method of acceptance testing [6].

Many program software reliability growth models (SRGMs) [7] have been proved to accurately predict the reliability for software and also the number of residual defects. It has been identified that the link between the testing period and the number of flaws removed is exponential, S-shaped, or a mixture of both [5,8–11]. It has been discovered that the program contains multiple types of bugs, each of which usually requires different methods and levels of testing effort to be eliminated. The hyper-exponential SRGM was suggested by Ohba [11]. Assume that software is comprised of various components. Each module has its own set of features, and each module's flaws have their own set of characteristics. As a result, each module's fault removal rate (FRR) differs. He advocated that the defect elimination procedure for each module be simulated independently, considering the concept for phenomenon of fault removal being the addition of fault removal processes of all the modules. In the case of applications with faults of two types, Yamada et al. [12] suggested an updated exponential SRGM. The concept is founded on the premise that for the software development process during the primary stages, the team for testing eliminates a significant count of minor errors for the removal process is very easy. In the later stages of the testing process, the hard faults are eliminated. As a result, they concluded that the fault removal mechanism was a combination of two non-homogenous poison process (NHPP). For the removal of easy defects, the first NHPP models are used, while the second NHPP simulates the removal of the hard faults. For easy types, the FRR per remaining fault is expected to be larger than that for hard types. In the early stages of the testing process, the overall FRR is equal to the fault removal rate of simple faults, and in the upcoming stages of the testing phase, it is equal to the fault removal rate of hard faults. The period between the malfunction observation and the fault removal in the subsequent process is considered to be an indicator of the fault's severity. The longer the time gap, the more severe the fault [13]. If the time delay between failure observation and fault isolation and fault removal is negligible, the fault is classified as easy. A hard fault is one in which there is a time delay. A complex fault is one in which extracting a fault after it has been isolated takes an even longer period In Ohba's [11] and Yamada et al.'s models [12] for each fault type, the assumption removal rate per remaining fault for constant fault still holds true. However, the cumulative fault removal rate for each remaining fault is a function of time. Kapur et al. [5,10] suggested a three-type fault SRGM. For each sort, the fault removal rate per remaining faults is believed to be independent from time. The first kind is modeled using Goel and Okumoto's [9] exponential model. The second sort is represented by Yamada et al.'s delayed S-shaped model [14]. Khoshogoftaar [15] proposed a three-stage Erlang model to model the third form. The superposition for the three SRGMs [16] is used to model the complete removal phenomenon once again. They later expanded their model to include more forms of flaws [5]. During the removal of a newly developed portion, the proposed model in this chapter integrates various debugging time lag functions.

The NHPP-based SRGM for distributed software system with m newly generated and n reused components of software proposed has been by Yamada et al. [17]. It is assumed that many used software components contain software system and follows an exponential curve to the cumulative number of identified faults. If the software system contains many newly generated software components, then the total number of detected faults is represented by the S-shaped growth curve. The SRGM based on NHPP for a distributed development environment with testing effort can be formulated as follows under this assumption:

$$H(t)=a\left[\underbrace{\sum_{i=1}^{n}p_i\left(1-e^{-b_iW(t)}\right)}_{\text{Exponentid}}+\underbrace{\sum_{j=1}^{m}p_{n+j}\left\{1-\left\{1+b_{n+j}W(t)\right\}e^{-b_{n+j}W(t)}\right\}}_{\text{S-Shaped}}\right] \tag{6.1}$$

where

a is the expected number of inherent defects at the start.
b_i is the ith software module, the rate of software failure due to an intrinsic flaw.
p_i is the software components, proportion of the load of total testing is expressed by the weight parameter ($\sum p_i = 1 \quad p_i > 0$).
$H(t)$ is the predicted counted numbers of error captured interval (0, t] time of span.
Throughout the process of testing.
$W(t)$ cumulative testing effort at the end of interval (0, t], i.e., $\int_0^t w(x)dx$.

In this chapter, for distributed systems, we present a software reliability growth model (SRGM) that includes time lag functions for multiple debugging because, based on the non-homogenous Poisson process, for this estimate and forecast the software product's reliability (NHPP) model can be used. The suggested goodness-of-fit model is compared to the Erlang software reliability growth model [1995]. The proposed models in terms of goodness-of-fit results are marginally better. The proposed models are discussed in Section 6.2. The parameter estimation procedure as well as the validation and assessment criteria for the proposed model are described in Sections 6.3–6.5. Section 6.6 concludes this chapter.

6.2 Software reliability growth modeling

6.2.1 Framework for modeling

NHPP is the foundation for the SRGM discussed in this chapter. Based on the assumption of the NHPP models software systems malfunction at random times as a result of the system's remaining faults. As a consequence, the word "NHPP" has been used to define the phenomenon of failure that occurs in the testing phase. The counting $\{N(t), t \geq 0\}$ of an NHPP process is given as follows:

$$\Pr\{N(t) = k\} = \frac{(m(t))^k}{k!} e^{-m(t)}, \quad k = 0, 1, 2 \tag{6.2}$$

and

$$m(t) = \int_0^t \lambda(x)dx$$

The mean value function $m(t)$ (or the intensity function) is the cornerstone of all NHPP models in the literature on software reliability engineering. These models consider a variety of testing scenarios such as the difference between removal and failure procedures, testing manpower learning, the likelihood of inaccurate debugging and error generation, and so on. Models suggested by

Yamada et al. [18] and Trachtenberg [19] investigated the influence of assessing effort strength on the failure phenomenon. If the program has any problems, they are revealed while it is being run. A proposal for a broad framework for model growth has been made.

6.2.2 Model assumptions and notations used

6.2.2.1 *Proposed model assumptions*

Following are the explicit assumptions of the proposed model:

1. The NHPP model simulates the failure observation/fault removal phenomenon.
2. Software will fail during execution as a result of bugs that haven't been fixed.
3. A software system is made up of a number of finite subsystems that have been reused and those that have been constructed from scratch.
4. Software flaws can be classified into three categories: simple, hard, and complex. "First-stage," "second-stage," and "third-stage" processes are used to model them, and they differ in the amount of research effort required to eliminate them.
5. If failure is noted, a delayed (an immediate) attempt is made to determine the reason for the failure and eliminate it. The time between observing a defect and removing it is thought to reflect the magnitude of the fault. The longer the gap, the more severe the fault. The process for fault removal, i.e., the process of perfect debugging, is used.
6. In reused components, fault removal follows an exponential curve, but fault removal in fresh components follows an S-shaped curve and incorporates multiple debugging time lag functions.
7. In $(t, t + \Delta t)$, the anticipated counts of faults removed are proportional to the remaining counts of faults to be removed.

6.2.2.2 *Notations for model*

m, $m(t)$: During the testing process, the anticipated number of faults found in the time span in the interval $(0,t]$.

a: Cumulative or total number of faults.

a_i: Form i fault content was reused in a variable with simple faults.

a_j: Form j newly formed hard fault components have a high initial fault material.

a_k: The initial fault content of a newly formed part of type k with complex faults.

b_i: The proportionality constant for fault type i, reflects the isolation of fault, rate of failure, and the rate of fault removal per fault.

b_j: The proportionality constant for fault type j reflects the fault isolation, rate of failure, and the rate of fault removal per unit fault.

b_k: The proportionality constant for fault type k reflects the rate of failure, isolation of fault, and rate of fault removal per fault.
$b_j(t)$: The fault removal rate per unit fault for hard faults.
$m_{ir}(t)$: By time t, the average number of type i faults from reused components removed has decreased.
$m_{jf}(t)$: By time t, the average number of type j failures caused by newly formed components.
$m_{jr}(t)$: By time t, the average number of type j faults on newly designed components that have been eliminated.
$m_{kf}(t)$: By time t, the average number of type k failures caused by newly formed components.
$m_{kr}(t)$: By time t, the average number of type k faults in newly formed components that have been removed.
p_i: $i = 1, 2$ as constant
p: Components that have been reused but have minor faults.
q: Components that have been established yet have a hard type of faults.
s: Components that have been established yet have faults of complex type.
$\lambda(x)$: Function for intensity.
$w(t)$: Intensity for testing effort.
$w(t)$ or W: Cumulative testing effort at the end of interval $(0, t]$, i.e., $\int_0^t w(x)dx$.

6.2.3 Modeling testing effort

As the amount of time passes, the amount of testing work increases. When the amount of time spent testing grows, so does the amount of effort required. The amount of work put into testing is related to the amount of resources available for testing:

$$\frac{dW(t)}{dt} = v(t)[\alpha - W(t)] \tag{6.3}$$

where testing resource consumption rate is denoted by $v(t)$. $v(t)$ depends on time, applied to the remaining resources as available. Solving Eq. (6.3) with the basic condition $W(0) = 0$.

Situation One:

If $v(t) = v$, Eq. (6.3) gives an exponential type curve, then

$$W(t) = \alpha(1 - e^{-vt}) \tag{6.4}$$

The density function for the above function is $\alpha v e^{-vt}$.

Situation Two:

If $v(t) = v \cdot t$, Eq. (6.3) gives a Rayleigh-type curve, then

$$W(t) = \alpha\left(1 - e^{-vt^2/2}\right) \tag{6.5}$$

The density function for the above function is αvte^{-vt^2}.

Situation Three:

If we have

$$\frac{dW(t)}{dt} = v\frac{W(t)}{\alpha}[\alpha - W(t)] \tag{6.6}$$

then after solving, in the interval (0,t), the cumulative testing effort consumed is given by

$$W(t) = \frac{\alpha}{1 + le^{-vt}}, \quad \text{where} \quad W(0) = \frac{\alpha}{1 + l} \tag{6.7}$$

It is a function for logistic testing effort, where α, v, and l are constants.

The density function for the above function is $\frac{\alpha vle^{-lt}}{(1+ve^{-lt})^2}$.

Situation Four:

Yamada et al. [14] proposed an SRGM describing the time-dependent behavior of "testing effort" expenditure with a Weibull curve. If $\nu(t) = \nu \cdot l \cdot t^{l-1}$ the Eq. (6.3) gives a Weibull-type curve:

$$W(t) = \alpha\left(1 - e^{-vt^l}\right) \tag{6.8}$$

The density function for the above function is $\alpha vlt^{l-1}e^{-vt^l}$.

Weibull curves become special instances of exponential and Rayleigh curves for l = 1, 2, respectively.

6.2.4 Modeling simple, hard, and complex faults

6.2.4.1 *Modeling reused components for fault removal*

Simple fault modeling

First-stage process of reused components for fault removal is modeled as

$$\frac{\frac{d}{dt}m_{ir}(t)}{w(t)} = b_i W(a - m_{ir}(t)) \tag{6.9}$$

The defect removal method is described by first-stage method, as shown in Eq. (6.9). Using the boundary condition to solve the differential equation $m_{ir}(t = 0) = 0$, we obtain

$$m_{ir}(t) = a_i\left(1 - e^{-b_i W*(t)}\right) \tag{6.10}$$

6.2.4.2 *Newly developed component modeling for fault removal*

Software defects in a newly built software component will vary in severity. The magnitude of a fault determines the length of time it would take to eliminate it.

The defects can be modeled as a second-stage or third-stage procedure for fault removal depending on time lag.

Hard fault modeling

When compared to faults of reused parts, hard faults need more testing effort. This suggests that the research team have to devote additional time for determining the root cause of the malfunction, which will necessitate more effort to resolve. Hence

$$\frac{\frac{d}{dt}m_{jk}(t)}{w(t)} = b_j W\left(a - m_{jk}(t)\right) \tag{6.11}$$

and

$$b_j W(t) = \frac{b_j}{1 + \beta e^{-b_j W(t)}}$$

The above failure observation rate shows that as the testing progresses the knowledge gained by the testing team increases, i.e., testing team learning grows. After solving Eq. (6.11) with the initial conditions at $t = 0$, $m_{jf}(t) = 0$, we have

$$m_{jf}(t) = \frac{a_j\left(1 - e^{-b_j W^*(t)}\right)}{1 + \beta e^{-b_j W^*(t)}}$$

Let

$$m_{jr}(t) = m_{jf}(W(t) - \Delta W(t)) \tag{6.12}$$

where

$$\Delta W(t) = \frac{1}{b_j}\log\left(1 + b_j W^*(t)\right)$$

Substituting the value of $\Delta W(t)$ in Eq. (6.12), we obtain

$$m_{jr}(t) = a_j \frac{\left\lfloor 1 - \left(1 + b_j W^*(t)\right) e^{-b_{jr} W^*(t)} \right\rfloor}{1 + \left(1 + b_j W^*(t)\right) \beta e^{-b_{jr} W^*(t)}} \tag{6.13}$$

Modeling complex faults

There may still be components with hard or complex defects. After isolation, these problems may require greater work to remove faults. As a result, they must be characterized with a longer time lag between removal and failure detection. Hence

$$\frac{\frac{d}{dt}m_{kf}(t)}{w(t)} = b_k W(a - m_{kf}(t)) \tag{6.14}$$

$$b_k W(t) = \frac{b_k}{1 + \beta e^{-b_k W^*(t)}}$$

Solving Eq. (6.14) under the initial conditions at $t = 0$, $m_{kf}(t) = 0$, we have

$$m_{kr}(t) = \frac{a_k\left(1 - e^{-b_k W^*(t)}\right)}{1 + \beta e^{-b_k W^*(t)}} \tag{6.15}$$

Let

$$m_{kr}(t) = m_{kr}(t - \Delta W(t)) \tag{6.16}$$

where

$$\Delta W(t) = \frac{1}{b_k}\log\left(1 + b_k W^*(t) + \frac{b_k{}^2 W^*(t)^2}{2}\right)$$

Putting the value of $\Delta W(t)$ in Eq. (6.16), we obtain

$$m_{kr}(t) = a_k \frac{\left[1 - \left(1 + b_k W^*(t) + \frac{b_k^2 W^*(t)^2}{2}\right) e^{-b_k W^*(t)}\right]}{1 + \left(1 + b_k W^*(t) + \frac{b_k^2 W^*(t)^2}{2}\right) \beta e^{-b_k W^*(t)}} \tag{6.17}$$

6.2.4.3 *Total fault removal phenomenon modeling*

The suggested model based on the NHPP is a superposition of repurposed "p" components with freshly generated "q" and "s" components. The mean value functions of the relevant NHPPs [5] are Eqs. (6.10), (6.13), (6.17). As a result, the superposed NHPP mean value function is

$$m(t) = \sum_{i=1}^{p} m_{ir}(t) + \sum_{j=p+1}^{p+q} m_{jr}(t) + \sum_{k=p+q+1}^{p+q+s} m_{kr}(t) \tag{6.18}$$

$$m(t) = \sum_{i=1}^{p} a_i\left(1 - e^{-b_i W*(t)}\right) + \sum_{j=p+1}^{p+q} m_{jr}(t)$$

$$= a_j \frac{\lfloor 1 - (1 + b_j W*(t)) e^{-b_j W*(t)} \rfloor}{1 + (1 + b_j W*(t)) \beta e^{-b_j W*(t)}}$$

$$+ \sum_{k=p+q+1}^{p+q+s} \left(a_k \frac{\left[1 - \left(1 + b_k W*(t) + \frac{b_k^2 W*(t)^2}{2}\right) e^{-b_k W*(t)}\right]}{1 + \left(1 + b_k W*(t) + \frac{b_k^2 W*(t)^2}{2}\right) \beta e^{-b_k W*(t)}} \right)$$

where

$$a_i + a_j + a_k = a \tag{6.19}$$

6.3 Parameter estimation

For reliability evaluation to succeed for the mathematical modeling approach, "the failure data" obtained must be accurate. These results are used to calculate the SRGMs' approximate parameter values. As a result, efforts should be made to improve the transparency and to collect data scientifically. In most cases, data are gathered in two ways: to begin with case, the time between successive failures is stated. Although this is the preferred method of data collection, it is not always practical. For each fault measurement, the testing effort may be difficult, and keeping track of the time spent on each failure report might not be feasible. Grouped data are a more straightforward and commonly collected data form. These data show the number of failures experienced during each cycle as well as the testing intervals. For each of these data types, the maximum likelihood methods and the least squares for estimating SRGM parameters have been presented and are frequently utilized.

6.4 Comparison of SRGM criteria

The capacity of SRGMs to fit historical software failure data is used to judge their performance or show goodness of fit.

6.4.1 Criteria for goodness of fit

a. MSE (mean-square fitting error)
The fault data are simulated using the model under comparison, and the difference between the expected values is calculated. $\hat{m}(t_i)$ and the observed data y_i are measured by MSE [5] as follows:

$$\text{MSE} = \sum_{i=1}^{k} \frac{(\hat{m}(t_i) - y_i)^2}{k} \tag{6.20}$$

where the number of observations is denoted by k. The lower the MSE, the less the fitting errors, and consequently the better is the fit.

b. PE (prediction error)
PE_i is the difference between the observed and predicted number of failures at any given point in time. The lower the prediction error value, the greater the goodness of fit [20].

c. Bias
Bias is defined as the average of PEs. The lower the Bias number, the greater the goodness of fit [20].

d. Variation
Variation is defined as the standard deviation of PE.

$$\text{Variation} = \sqrt{\left({}^{1}/_{N-1}\right)\sum (\text{PE}_i - \text{Bias})^2} \tag{6.21}$$

The lower the variation value, the better the fit [20].

e. RMSPE (root mean square prediction error)
It is a metric for how well a model predicts a given observation.

$$\text{RMSPE} = \sqrt{\left(\text{Bias}^2 + \text{Variation}^2\right)} \tag{6.22}$$

Lower the value of root mean square prediction error better is the goodness of fit [20].

6.5 Data description and model validation

The proposed model is tested on the two, DS-I and DS-II, data sets to determine its validity and software reliability development. On the data sets DS-I and DS-II, SRGMs obtained by exponential, logistic, Rayleigh, and Weibull research effort functions have been validated.

DS-I

Brooks and Motley Brooks [8] provided the information. The fault data set is for a 124-kLOC (kilo lines of code) radar system that was evaluated for 35 months and found to have 1301 faults.

DS-II

The second data set (DS-II) on a real-time system was obtained over the course of 38 weeks of testing, during which 231 faults were discovered. This information can be obtained from Ref. [1] (Tables 6.1 and 6.2).

TABLE 6.1 For DS-I.

a. Parameter results

Models	a	b*1*	b*2*	b*3*	p*1*	p2	p*3*	β*1*	β2
Erlang1-2-3 [1995]	1344	0.0016	0.0035	0.0037	0.723	0.012	0.264	–	–
Erlang-Logistic1-2-3 [1995]	1347	0.0015	0.0062	0.0061	0.839	0.071	0.089	132.05	111.02
Proposed	1308	0.0077	0.0062	0.0053	0.223	0.607	0.168	6.08	19.71

b. Comparison results

Models	R^2	*MSE*	*Bias*	*Variation*	*RMSPE*
Erlang1-2-3[1995]	0.99659	726.60	−4.85	26.90	27.33
Erlang-Logistic1-2-3[1995]	0.99728	578.98	1.03	24.39	24.41
Proposed	0.99933	142.29	−0.48	12.00	12.10

TABLE 6.2 For DS-II.

a. Parameter results

Models	a	b*1*	b2	b*3*	p*1*	p2	p*3*	β*1*	β2
Erlang1-2-3	212	0.0012	0.0016	0.0022	0.596	0.010	0.393	–	–
Erlang-Logistic1-2-3	233	0.001	0.0025	0.0025	0.720	0.011	0.268	28.24	30.48
Proposed	229	0.001	0.0035	0.1617	0.629	0.341	0.029	33.51	7.63

b. Comparison results

Models	R^2	*MSE*	*Bias*	*Variation*	*RMSPE*
Erlang1-2-3	0.96982	112.07	−1.95	10.54	10.72
Erlang-Logistic1-2-3	0.98902	40.78	0.73	6.42	6.47
Proposed	0.99069	34.58	0.15	5.84	5.95

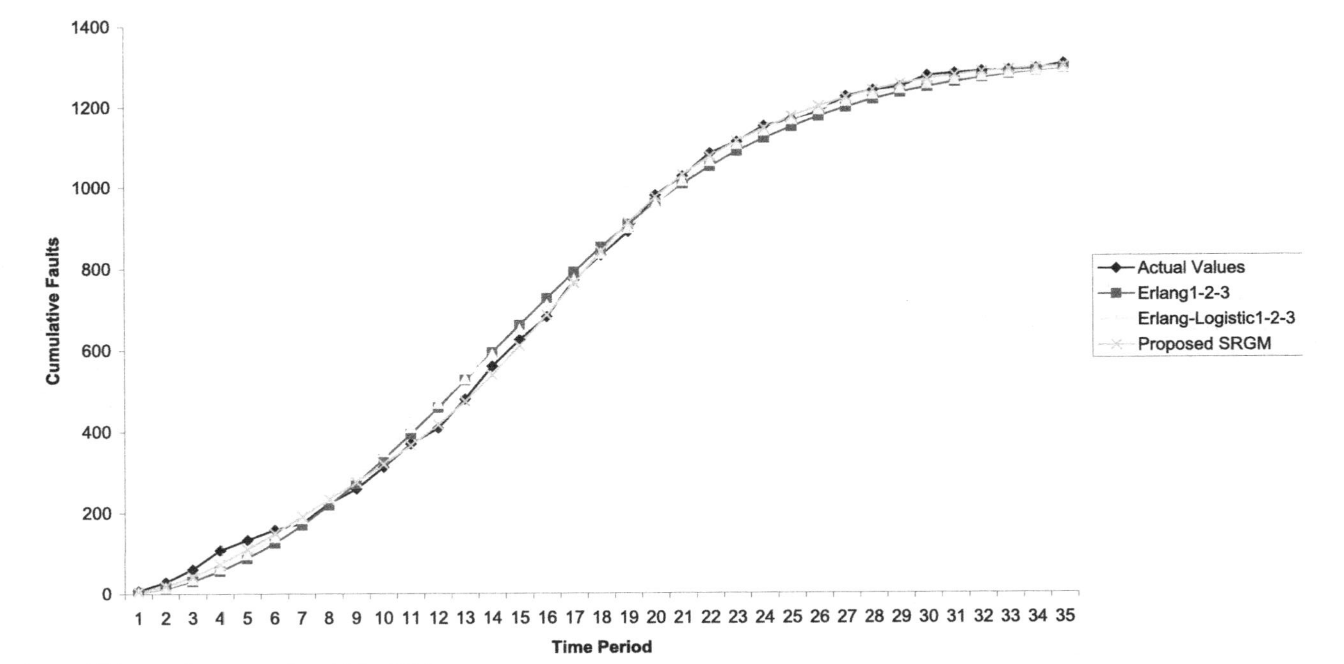

Goodness-of-fit curves for proposed model for DS-I.

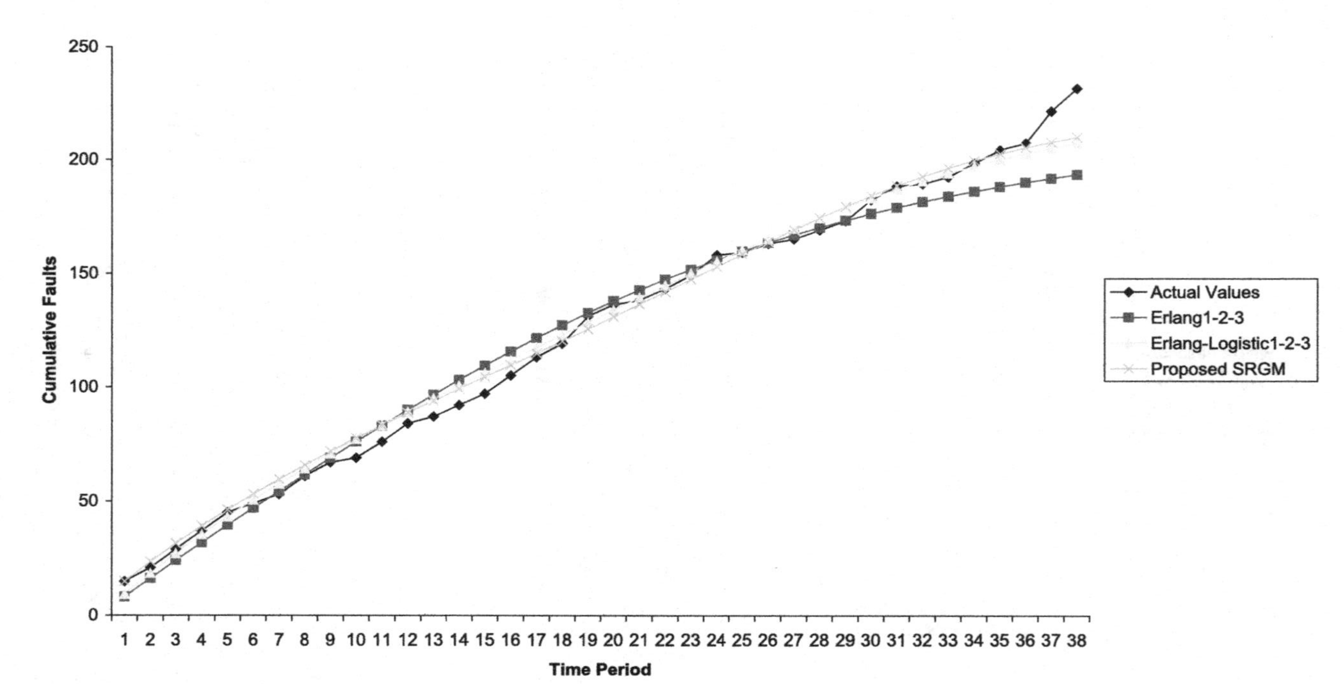

Goodness-of-fit curves for proposed model for DS-II.

6.6 Conclusion

In this chapter, the "software reliability growth model" (SRGM) to a distributed context is introduced. A software system, according to the paradigm, is made up of a small number of newly and reused built subsystems. The magnitude of the fault is represented by the period between subsequent fault and failure observation elimination in the suggested model. The flexible software reliability growth model [5] and the "generalized-Erlang model" [1995] are compared to the proposed model. The model considers a variety of defects (simple, hard, and complex). The proposed model was tested by applying it to a real-world software development project; it was tested and compared to well-documented NHPP paradigm. The proposed model results are pretty encouraging. We found that the proposed model reflects more exact results.

References

[1] M.R. Lyu, Handbook of Software Reliability Engineering, IEEE Computer Society Press, 1996.

[2] J.D. Musa, A. Iannini, K. Okumoto, Software Reliability: Measurement, Prediction, Applications, Professional Edition: Software Engineering Series, McGraw-Hill, New York, 1990.

[3] A. Geraci, L. McMonegal, P. Wilson, IEEE Standard Glossary of Software Engineering Terminology. IEEE Standards Board, September 28,1990.

[4] P.K. Kapur, A.K. Bardhan, O. Shatnawi, Why software reliability growth modelling should define errors of different severity, J. Indian Statist. Assoc. 40 (2) (2002) 119–142.

[5] P.K. Kapur, R.B. Garg, S. Kumar, Contributions to Hardware and Software Reliability, World Scientific, Singapore, 1999.

[6] R.S. Pressman, Software Engineering—A Practitioner's Approach, fourth ed., McGraw-Hill, 2010.

[7] T. Downs, A. Scott, Evaluating the performance of software reliability models, IEEE Trans. Reliab. 41 (4) (1992) 532–538.

[8] S. Bittanti, P. Bolzern, E. Pedrotti, R. Scattolini, A flexible modeling approach for software reliability growth, in: G. Goos, J. Harmanis (Eds.), Software Reliability Modelling and Identification, Springer Verlag, Berlin, 1998, pp. 101–140.

[9] A.L. Goel, K. Okumoto, Time dependent error detection rate model for software reliability and other performance measure, IEEE Trans. Reliab. R-28 (3) (1979) 206–211.

[10] P.K. Kapur, R.B. Garg, A software reliability growth model for an error removal phenomenon, Softw. Eng. J. 7 (1992) 291–294.

[11] M. Ohba, Software reliability analysis models, IBM J. Res. Dev. 28 (1984) 428–443.

[12] S. Yamada, S. Osaki, H. Narihisa, Software reliability growth models with two types of errors, RAIRO Oper. Res. 19 (1985) 87–104.

[13] P.K. Kapur, O. Shatnawi, V.S.S. Yadavalli, A software fault classification model, S. Afr. Comput. J. 33 (2004) 1–9.

[14] S. Yamada, M. Ohba, S. Osaki, S-shaped software reliability growth models and their applications, IEEE Trans. Reliab. R-33 (1984) 169–175.

[15] T.G. Woodcock, T.M. Khoshogoftaar, Software reliability model selection: a case study, in: Proceedings of the International Symposium on Software Reliability Engineering, 1991, pp. 183–191.

[16] P.K. Kapur, S. Younes, P.S. Grover, Software reliability growth model with errors of different severity, Comput. Sci. Inform. 25 (3) (1995) 51–65.
[17] S. Yamada, Y. Tamura, M. Kimura, A software reliability growth model for a distributed development environment, in: Electronics & Communication in Japan Part 3, vol. 83, 2003, pp. 1446–1453.
[18] S. Yamada, H. Ohtera, H. Narihisa, Software reliability growth models with testing-effort, IEEE Trans. Reliab. R-35 (1986) 19–23.
[19] M. Trachtenberg, A general theory of software reliability modeling, IEEE Trans. Reliab. 39 (1) (1990) 92–96.
[20] K. Pillai, V.S.S. Nair, A model for software development effort and cost estimation, IEEE Trans. Softw. Eng. 23 (8) (1997) 485–497.

Chapter 7

Design and performance analysis of MIMO PID controllers for a paper machine subsystem

Niharika Varshney, Parvesh Saini, and Ashutosh Dixit
Department of Electrical Engineering, Graphic Era Deemed to be University, Dehradun, Uttarakhand, India

7.1 Introduction

The important issue concerning the process industries is the control of the process. It is difficult to do the analysis of delayed processes and hence, it has been an interesting area of research for the last many decades. Most of the industrial processes are first-order plus dead time (FOPDT) types [1,2]. The dead time in the process is caused due to the measuring devices, controlling elements, controllers, etc. [3,4]. Any system has dead time when the propagation and transport phenomenon is modeled [5]. The consequence of delay or dead time in the process is that the design and analysis of such processes become difficult [5,6]. Hence, a robust and efficient controller is required to control such processes precisely and the effect of time delays, disturbances, etc., is suppressed. Process industries primarily use PID controllers due to their unpretentious and tough structure. PID controllers are easy to use in industries for process control [7]. Also, due to some unique features such as simple construction, robust performance, effective control of process parameters, a wide range of applications, quick response, and ability to eliminate steady-state error in the given time, PID controllers are the preferred choice for industrial applications [4–6,8–11]. In the past few decades, many PID tuning techniques have been developed. The controller design for dead time processes has been discussed and presented in Ref. [1], while Refs. [3, 12] presents the optimum way of determining settings for PID controllers. Astrom-Hagglund and coworkers [4,8,11,13] have discussed the PID controller tuning techniques. Comparative analysis of various PID tuning techniques for delayed MIMO systems has been

System Assurances. https://doi.org/10.1016/B978-0-323-90240-3.00007-2

presented in Refs. [9, 14] while Kadu et al. and others [2,5–7,10,15] have presented the PI/PID controller design for processes with delay. Lee et al. and others [16–18] have also presented PID tuning techniques for different processes. Ritala and Raunio have proposed a technique to enhance the control of basis weight using LQG technique [19], a review of drying techniques of paper is presented in Ref. [20], and a fuzzy logic-based analysis and control of moisture content of the paper have been presented and compared with conventional control strategies in Ref. [21].

However, in the current work, design and a comparative analysis of PID controllers have been presented. The PID controller has been tuned through various controller tuning algorithms. The system with a time delay taken into consideration is a paper machine headbox. A schematic arrangement of the paper machine headbox is shown in Fig. 7.1.

There are many subprocesses associated with a paper machine. Among all, the headbox is one of the most crucial subprocesses of the paper machine. Headbox aims to uniform distribution of the pulp onto the wire of the paper machine. The uniform spread of pulp on wire has to be ensured by efficient operation of the headbox. Headbox is a two input-two output system. The loops of headbox control are highly interactive. The important variables to be controlled as associated with the headbox are "dry weight (g/m^2)" and "ash content (%age)." The two manipulated variables of concern are thick stock valve position and filler valve position [23]. Since headbox is a highly interacting system, it has been a center of attraction for researchers for the last few years. Many control strategies have been developed for headbox control. A brief review of techniques developed for paper machine headbox has been presented in Ref. [24]. The next section presents the design aspects of paper machine headbox and PID controller tuning techniques.

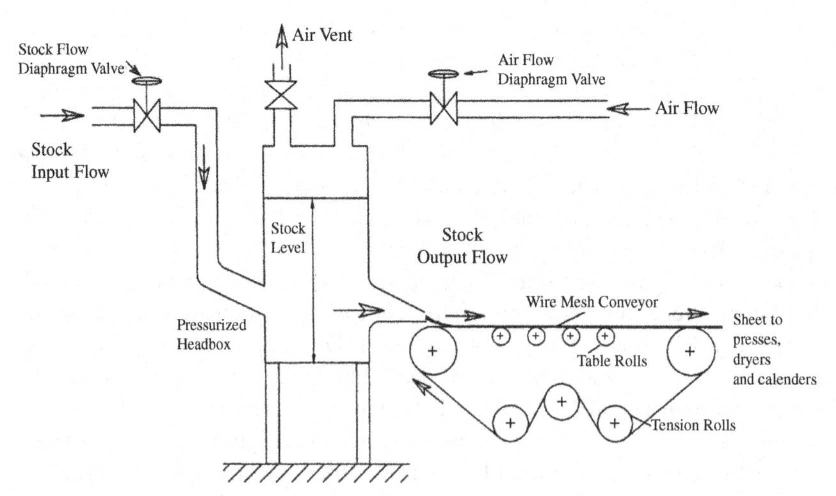

FIG. 7.1 Schematic arrangement of paper machine headbox [22].

7.2 Controller tuning

In this paper, a pressurized headbox has been taken into consideration. The mathematical model of the headbox is given in Eq. (7.1). In other words, a headbox is a multivariable system with two inputs [$y_1(s)$ and $y_2(s)$, which represent the controlled variables] and two outputs [$m_1(s)$ and $m_2(s)$ are the manipulated variables]. Since headbox model is highly interacting, the interaction among all loops is determined through relative gain array (RGA). Through RGA approach, SISO models are obtained with loop pairing between $y_1 - u_1$ and $y_2 - u_2$. The transfer function of the paper machine headbox is given below [23]

$$\begin{bmatrix} y_1(s) \\ y_2(s) \end{bmatrix} = \begin{bmatrix} \dfrac{0.214e^{-68s}}{125s+1} & \dfrac{-0.192e^{-68s}}{17s+1} \\ \dfrac{0.153e^{-68s}}{125s+1} & \dfrac{0.93e^{-68s}}{17s+1} \end{bmatrix} \begin{bmatrix} m_1(s) \\ m_2(s) \end{bmatrix} \tag{7.1}$$

The relative gain matrix is calculated which provides the extent of process influence in the multivariable process, by calculating the process interactions. It provides the relative gain between the input and output. The RGA matrix corresponding to the system is

$$\Lambda = \begin{bmatrix} 0.872 & 0.128 \\ 0.128 & 0.872 \end{bmatrix} \tag{7.2}$$

Obtained SISO system based on $y_1 - u_1$ and $y_2 - u_2$ loop pair recommendations are

$$G_1 = \frac{y_1(s)}{u_1(s)} = \frac{0.245e^{-68s}}{125s+1} \tag{7.3}$$

$$G_2 = \frac{y_2(s)}{u_2(s)} = \frac{1.066e^{-68s}}{17s+1} \tag{7.4}$$

Here, $G_1(s)$ and $G_2(s)$ represent dry weight (g/m^2) and ash content (%age) loops of headbox, respectively. The tuning techniques as mentioned in Table 7.1 are used to designing PID controller for the SISO loops as given in Eqs. (7.3), (7.4). The stability analysis of headbox loops as mentioned above is done based on the transient characteristics which are discussed in the next section.

In this work, some distinguished PID tuning techniques have been used to analyze the paper machine headbox as given by Eq. (7.1). The PID controller tuning parameters are calculated using Zeigler-Nichols, Tyreus-Luyben, Internal Model Control, Astrom-Hagglund, C-H-R set point regulation, and Cohen-coon tuning techniques [3,13,25–28]. The tuning techniques used for the given MIMO system for designing PID controllers are shown in Table 7.1.

TABLE 7.1 Tuning techniques [3,13,25–28].

Tuning techniques	k_c	τ_i	τ_d
Zeigler-Nichols	Ku/1.7	Pu/2	Pu/8
Tyreus-Luyben	Ku/2.2	2.2Pu	Pu/6.3
Internal model control Model type: $\frac{Ke^{\theta s}}{\tau s+1}$	$\frac{\tau+\frac{\theta}{2}}{K\left(\tau_c+\frac{\theta}{2}\right)}$	$\tau+\frac{\theta}{2}$	$\frac{\tau\theta}{2\tau+\theta}$
Astrom-Hagglund	$0.2+0.45*\frac{\tau}{L}$	$\left(\frac{0.4L+0.8\tau}{L+0.1\tau}\right)L$	$\frac{0.5L\tau}{0.3L+\tau}$
Cohen-coon	$\frac{0.35}{a}\left(1+\frac{0.18\tau}{1-\tau}\right)$	$\frac{2.5-2\tau}{1-0.39\tau}L$	$\frac{0.37-0.37\tau}{1-0.81\tau}L$
Set point regulation	$\frac{0.6}{a}$	T	$0.5L$

There are two methodologies to design the control systems for a multivariable system. In the first approach, the decoupling design is used. In this approach, decouplers are designed based on the loop pairing as obtained through a relative gain array (RGA). Through this approach, the MIMO system can be converted into a SISO system. Hence, scalar controllers are designed using this approach. This approach has been used to a large extent in process industries for the control of multivariable systems. The second approach is the MIMO approach wherein this approach, MIMO controllers are designed and implemented on the multivariable process, and it is not required to obtain decouplers. For a process of "n" states, n controllers are used. Each controller is implemented in such a way that it shall control its respective process variable. The MIMO controllers are matrix-based. If the process is "$n \times n$," then each controller's matrix will also be "$n \times n$." This approach is generally not considered for the higher-order processes (i.e., process with order greater than 2) because the reason is substantial mathematical calculations and analysis involved. For such processes, the decoupling technique is preferred. However, for a system of the lower order, MIMO technique can be used.

In the current work, the MIMO control approach has been used Hence, the process we have taken into consideration a 2×2 process (i.e., 2 inputs and 2 outputs). Hence, we need to design four controllers, one for each process element (i.e., g11, g12, g21, and g22). The MIMO controllers have been designed using the tuning equations for each technique as depicted in Table 7.1. The multivariable controllers that have been obtained for this work using the above-stated techniques are given by the following matrices:

1. Zeigler-Nichols technique (ZN)

$$K_p = \begin{bmatrix} 8.6878 & 1.7936 \\ -0.4778 & 0.6682 \end{bmatrix}; \; K_I = \begin{bmatrix} 0.0755 & 0.0156 \\ -0.0057 & 0.0080 \end{bmatrix} \text{ and } K_D = \begin{bmatrix} 249.9396 & 51.6004 \\ -9.9269 & 13.8847 \end{bmatrix} \tag{7.5}$$

2. Tyreus-Luyben technique (TL)

$$K_p = \begin{bmatrix} 6.5816 & 1.3588 \\ -0.3619 & 0.5062 \end{bmatrix}; \; K_I = \begin{bmatrix} 0.0130 & 0.0027 \\ -0.0010 & 0.0014 \end{bmatrix} \text{ and } K_D = \begin{bmatrix} 240.4421 & 49.6397 \\ -9.5497 & 13.3571 \end{bmatrix} \tag{7.6}$$

3. Cohen-Coon technique (CC)

$$K_p = \begin{bmatrix} 2.1444 & 0.4427 \\ -0.0474 & 0.0663 \end{bmatrix}; \; K_I = \begin{bmatrix} 0.0061 & 0.0013 \\ -0.0001 & 0.0002 \end{bmatrix} \text{ and } K_D = \begin{bmatrix} 66.7353 & 13.7776 \\ -1.4944 & 2.0902 \end{bmatrix} \tag{7.7}$$

4. Astrom-Hagglund technique (HA)

$$K_p = \begin{bmatrix} 0.5136 & 0.5136 \\ -0.1563 & 0.1563 \end{bmatrix}; \; K_I = \begin{bmatrix} 0.0048 & 0.0048 \\ -0.0039 & 0.0039 \end{bmatrix} \text{ and } K_D = \begin{bmatrix} 15.0125 & 15.0125 \\ -2.4148 & 2.4148 \end{bmatrix} \tag{7.8}$$

5. C-H-R set point regulation technique (CHR)

$$K_p = \begin{bmatrix} 4.4910 & 0.9272 \\ -0.1005 & 0.1405 \end{bmatrix}; \; K_I = \begin{bmatrix} 0.0359 & 0.0074 \\ -0.0059 & 0.0083 \end{bmatrix} \text{ and } K_D = \begin{bmatrix} 152.6953 & 31.5242 \\ -3.4164 & 4.7785 \end{bmatrix} \tag{7.9}$$

6. Internal model control (IMC)

$$K_p = \begin{bmatrix} 4.4910 & 0.9272 \\ -0.1005 & 0.1405 \end{bmatrix}; \; K_I = \begin{bmatrix} 0.0359 & 0.0074 \\ -0.0059 & 0.0083 \end{bmatrix} \text{ and } K_D = \begin{bmatrix} 152.6953 & 31.5242 \\ -3.4164 & 4.7785 \end{bmatrix} \tag{7.10}$$

The controller matrices as depicted from Eqs. (7.5)–(7.10) have been implemented on the headbox taken into consideration for this work. The results obtained have been discussed in the next section.

7.3 Result analysis

This section presents the results obtained by implementing the MIMO controllers design using six controller design techniques. To start with, open-loop time and frequency response of the headbox [as represented by Eq. (7.1)] has been presented in Table 7.2. The open-loop step response of the given headbox has been shown in Fig. 7.2 and open-loop bode plots are shown in Fig. 7.3.

From Table 7.2, it is observed that the phase margin of all four-process elements is infinity, which indicates that the given system does not have any gain crossover frequency. But the process transfer function has phase crossover frequency, which is also treated as the ultimate frequency. The ultimate frequency value has been used to obtain the controller for each process element through Z-N and T-L techniques. The ultimate gain and phase have also been given in Table 7.2. Fig. 7.2 depicts the open-loop step response of the MIMO headbox model. The first blue dot indicates the rise time and the second blue dot indicates the settling time. Out of the four responses (as depicted in Fig. 7.2), the first and fourth response is related to the main outputs (i.e., dry weight and ash content) while the second and third response acts as disturbances to both outputs. In other words, it is not wrong to mention that u1 acts as a disturbance to ash content and dry weight.

Once the MIMO controllers have been designed and implemented on the given headbox process, we can analyze the performance of the designed control systems on the basis of various aspects. The first aspect is to study the time response values of the headbox model using each tuning technique as discussed earlier. The next point is to discuss the frequency response values of the headbox. Time response analysis can be done through the time response characteristics (such as rise time, peak overshoot, settling time). Similarly, frequency response analysis can be done through the frequency response values (such as gain margin, phase margin, bandwidth, etc.). Also, the performance indices (i.e., integral errors—ISE, IAE, ITAE, and ITSE) have also been taken into consideration to have the proper analysis of the designed control systems.

Figs. 7.4 and 7.5 depict the closed-loop step response of dry weight and ash content, respectively. The respective time response (step response) values have been given in Tables 7.3 and 7.4, respectively.

Figs. 7.6 and 7.7 depict the comparison of the variation in the various time response values. This will give an insight into the technique that gives optimal results.

Similarly, the performance indices have been calculated for each tuning technique. The performance indices values as obtained through MATLAB have been depicted in Table 7.5 for dry weight and ash content.

TABLE 7.2 Open-loop time and frequency response values.

Process element	Rise time (s)	Peak overshoot (%age)	Settling time (s)	Gain margin (dB)	Phase margin (degrees)	Ultimate frequency (rad/s)	Ultimate gain (Ku) (dB)	Ultimate period (Pu) (s)
g11	274.66	0	557.03	24.4	Inf	0.0273	16.6168	230.1533
g12	37.35	0	134.51	15.8	Inf	0.0378	6.1910	166.2218
g21	274.66	0	557.03	27.3	Inf	0.0273	23.2418	230.1533
g22	37.35	0	134.51	2.13	Inf	0.0378	1.2781	166.2218

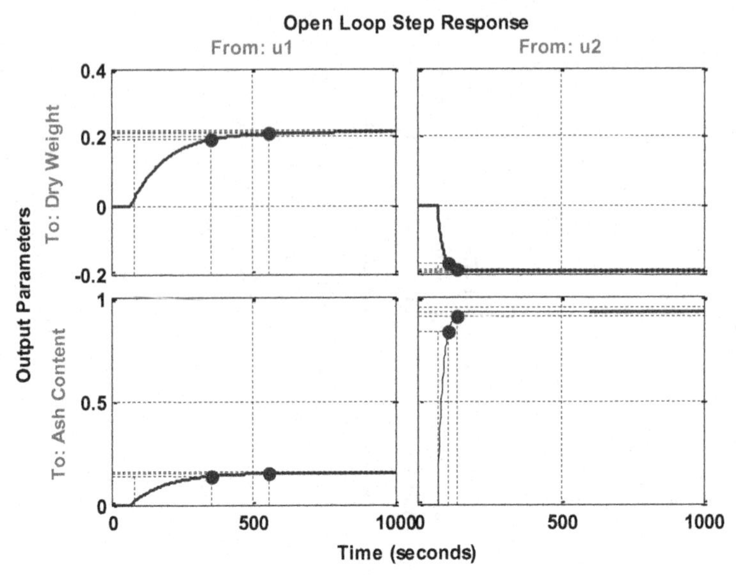

FIG. 7.2 Open-loop step response of the headbox.

Figs. 7.8 and 7.9 represent the variation in the performance indices (i.e., ISE and IAE) for dry weight and ash content, respectively. From the ISE and IAE values obtained and given in Table 7.5, and from Figs. 7.8 and 7.9, it is observed that Cohen-Coon technique gives higher ISE and IAE values. Ideally, these values must be low. However, Zeigler-Nichols tuning technique yields optimal ISE and IAE values (or minimum values) among all six tuning techniques that have been taken into consideration for this work.

The robustness of the controller tuning techniques has been assessed through the stability margin (i.e., gain margin and phase margin). Ideally, the gain margin and phase margin shall be infinite. Infinite stability margin indicates that the given system is all-time stable, meaning the system will remain stable for the infinite change in the system's gain and also will remain stable for the infinite shift in the phase of the system.

This is practically not feasible. Hence we can say that the gain margin and phase margin of a control system must be high. In the literature, the range of gain margin is from 8 to 10, while the phase margin is 45°–60°. But if we have the values of gain margin and phase margin greater than what is given in the literature, then it is acceptable. But large values of gain and phase margin would make the control system sluggish. Now, let us analyze the stability margins of the given headbox model through bode plots which are shown in Figs. 7.10 and 7.11 for dry weight and ash content, respectively. The statistical values of stability margin for both loops have been given in Table 7.6.

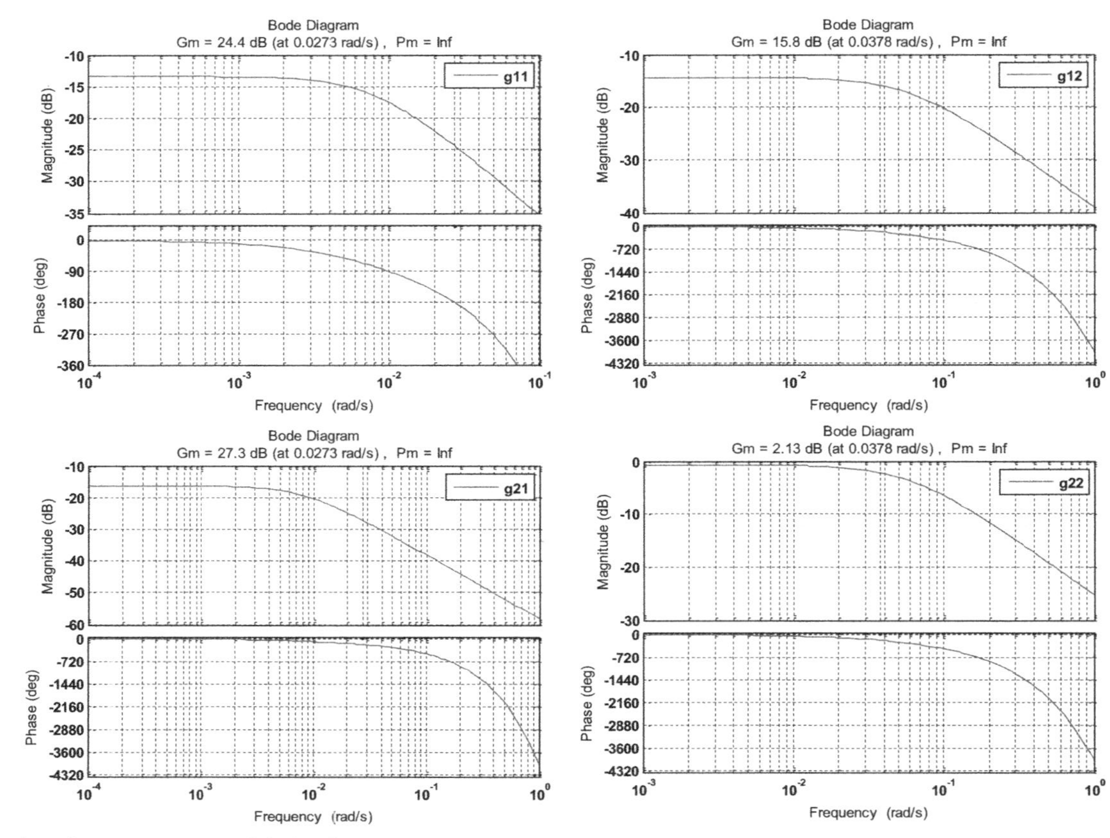

FIG. 7.3 Open-loop frequency response of the headbox.

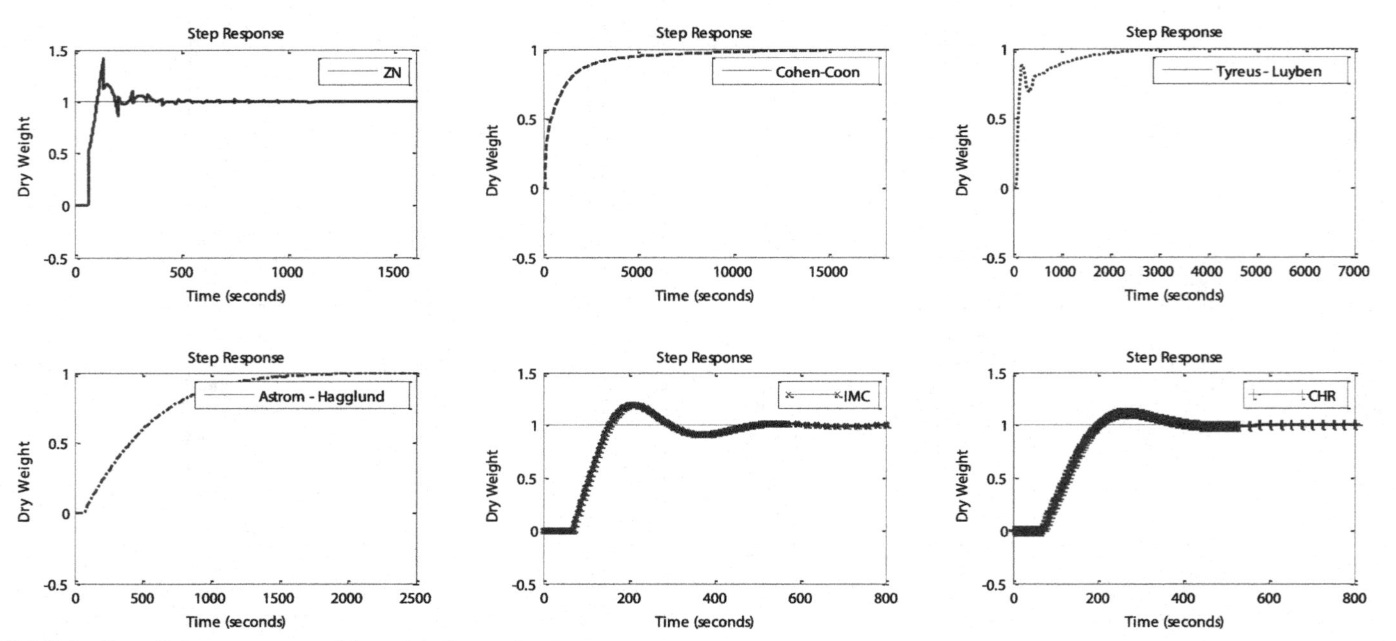

FIG. 7.4 Controlled step responses of dry weight ($y_1 - u_1$ loop).

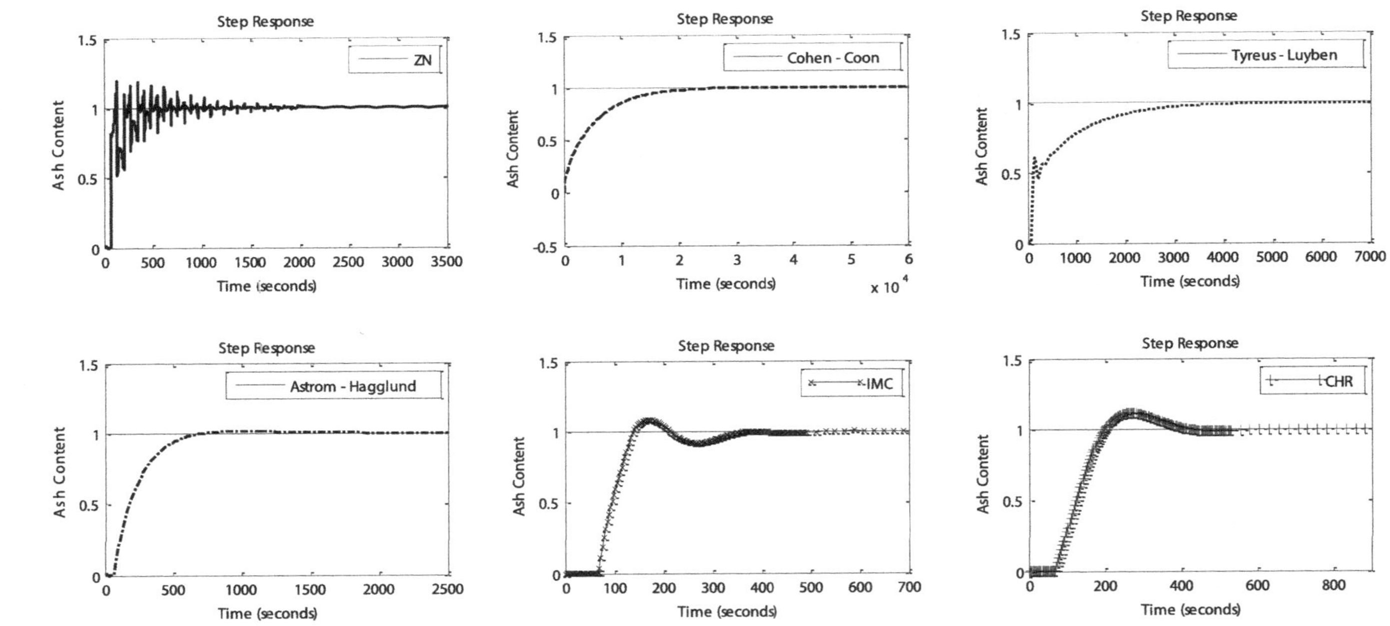

FIG. 7.5 Controlled step responses of ash content ($y_2 - u_2$ loop).

TABLE 7.3 Time response values of dry weight ($y_1 - u_1$ loop).

Controller technique	Rise time (s)	Peak overshoot (%age)	Settling time (s)
Zeigler-Nichols [3]	29.83	41.71	548.13
Cohen-Coon [27]	2703.4	0	10254
Tyreus-Luyben [25]	1008.4	0	2389.0
Astrom-Hagglund [13]	925.2539	0	1625.3
Internal model control [28]	67.67	19.30	467.14
C-H-R set point regulation [26]	98.35	11.65	386.78

TABLE 7.4 Time response values of ash content ($y_2 - u_2$ loop).

Controller technique	Rise time (s)	Peak overshoot (%age)	Settling time (s)
Zeigler-Nichols [3]	32.70	18.95	1567.7
Cohen-Coon [27]	11810	0	21113
Tyreus-Luyben [25]	1702.7	0	3347.1
Astrom-Hagglund [13]	350.65	1.44	600.27
Internal model control [28]	57.52	8.26	339.58
C-H-R set point regulation [26]	99.44	11.54	388.57

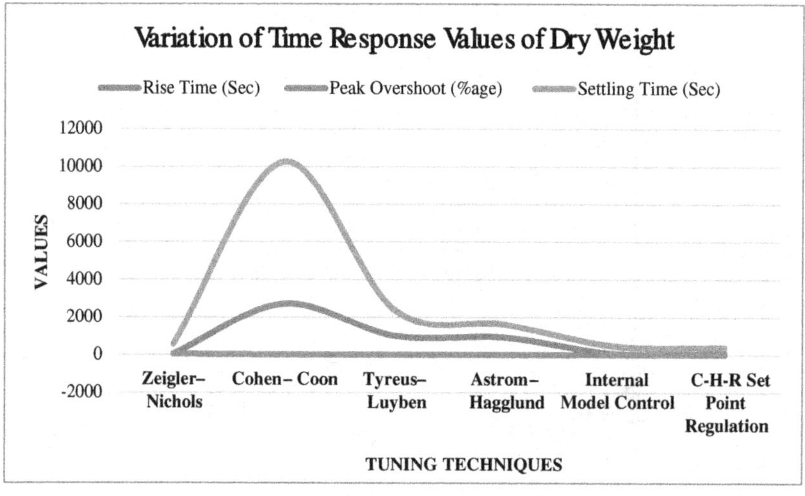

FIG. 7.6 Comparison of variation in the various time response parameters of dry weight ($y_1 - u_1$ loop).

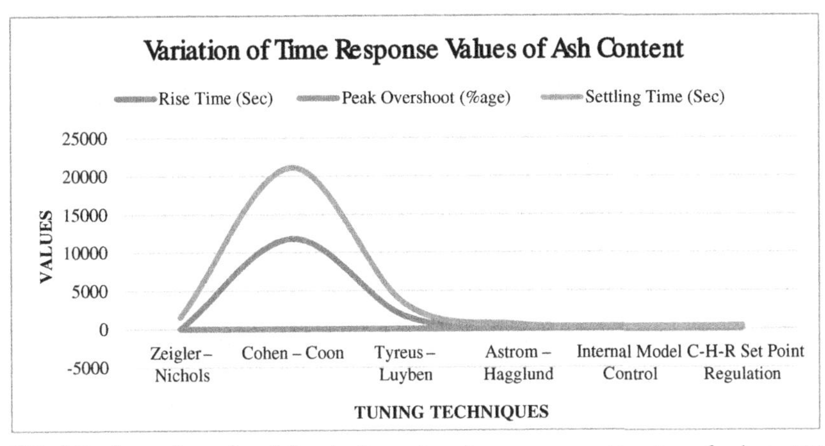

FIG. 7.7 Comparison of variation in the various time response parameters of ash content ($y_2 - u_2$ loop).

TABLE 7.5 Performance indices values.

	Dry weight ($y_1 - u_1$ loop)		Ash content ($y_2 - u_2$ loop)	
Controller technique	**ISE**	**IAE**	**ISE**	**IAE**
Zeigler-Nichols [3]	74.0167	100.2487	81.3387	139.1177
Cohen-Coon [27]	387.0008	1156.0	2009.5	3966.9
Tyreus-Luyben [25]	133.4515	360.0069	231.5999	630.0329
Astrom-Hagglund [13]	304.0312	511.2397	143.6789	229.7882
Internal model control [28]	97.2540	138.5241	86.3737	110.7473
C-H-R set point regulation [26]	107.4270	143.0699	107.1651	142.9789

From Table 7.6, it is observed that the gain margin for dry weight is highest for the Astrom-Hagglund technique and Zeigler-Nichols yields a lower value of gain margin. Similarly, for ash content, Cohen-Coon yields the highest and Zeigler-Nichols yields the lowest value of gain margin among all techniques. As far as the phase margin is concerned, Cohen-Coon gives the highest phase

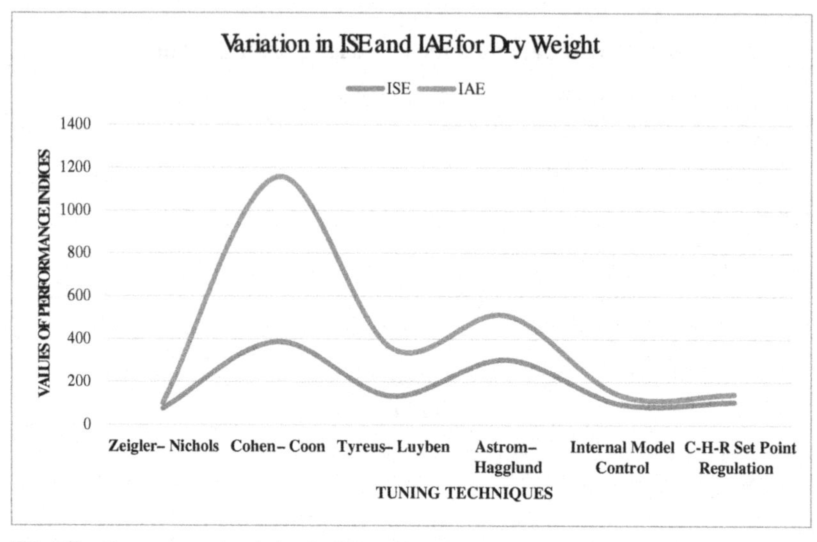

FIG. 7.8 Comparison of variation in ISE and IAE for dry weight ($y_1 - u_1$ loop).

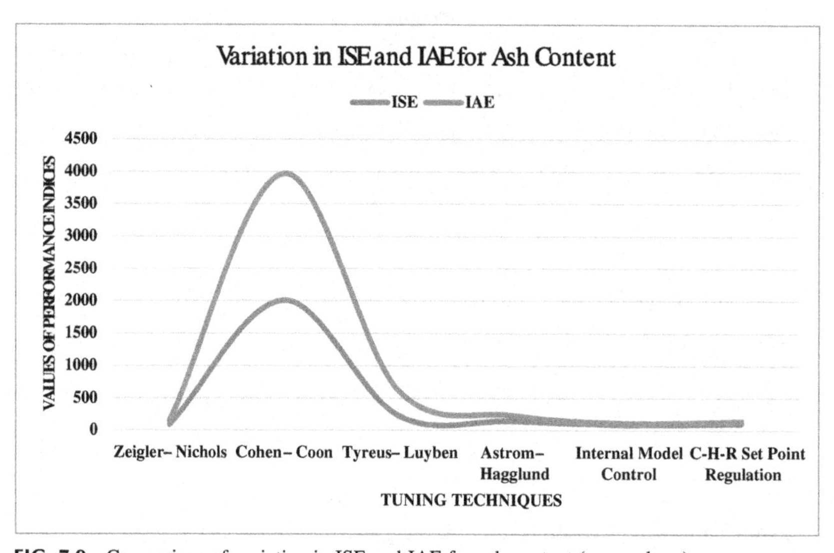

FIG. 7.9 Comparison of variation in ISE and IAE for ash content ($y_2 - u_2$ loop).

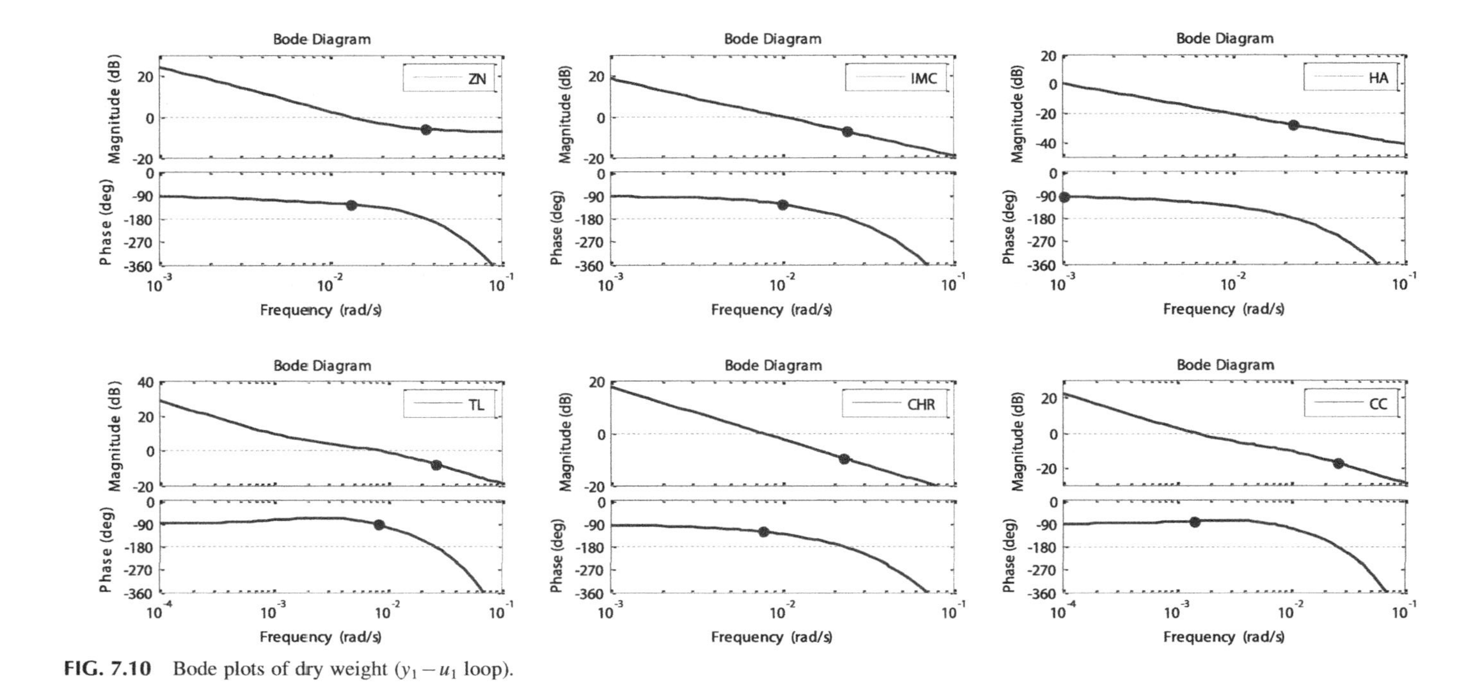

FIG. 7.10 Bode plots of dry weight ($y_1 - u_1$ loop).

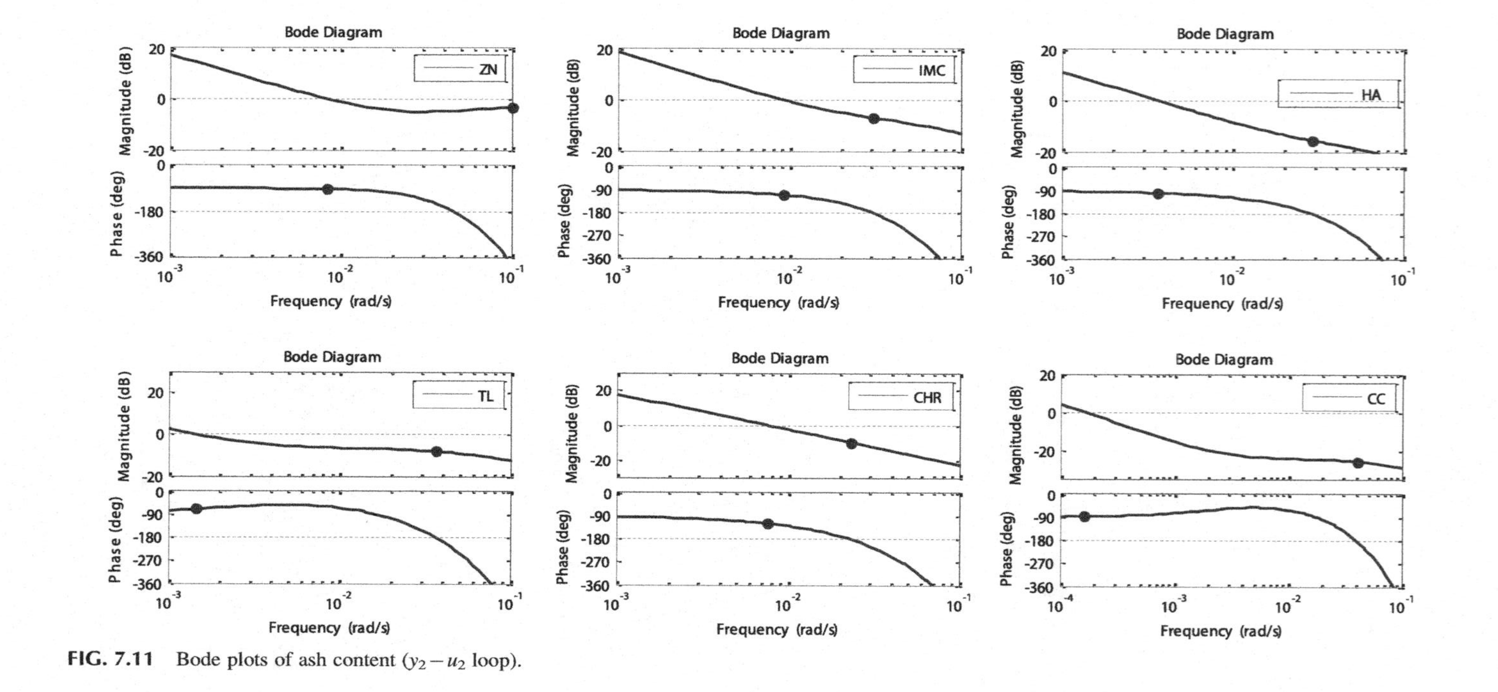

FIG. 7.11 Bode plots of ash content ($y_2 - u_2$ loop).

TABLE 7.6 Performance indices values of dry weight ($y_1 - u_1$ loop).

Controller technique	Dry weight ($y_1 - u_1$ loop)		Ash content ($y_2 - u_2$ loop)	
	Gain margin (dB)	Phase margin (degrees)	Gain margin (dB)	Phase margin (degrees)
Zeigler-Nichols [3]	5.76	54.2	2.39	87.7
Cohen-Coon [27]	17.3	101	25.7	92.7
Tyreus-Luyben [25]	7.73	87.9	7.97	111
Astrom-Hagglund [13]	27.9	85	15.5	80.4
Internal model control [28]	7.07	57.4	6.84	70.1
C-H-R set point regulation [26]	9.56	60	9.67	60.1

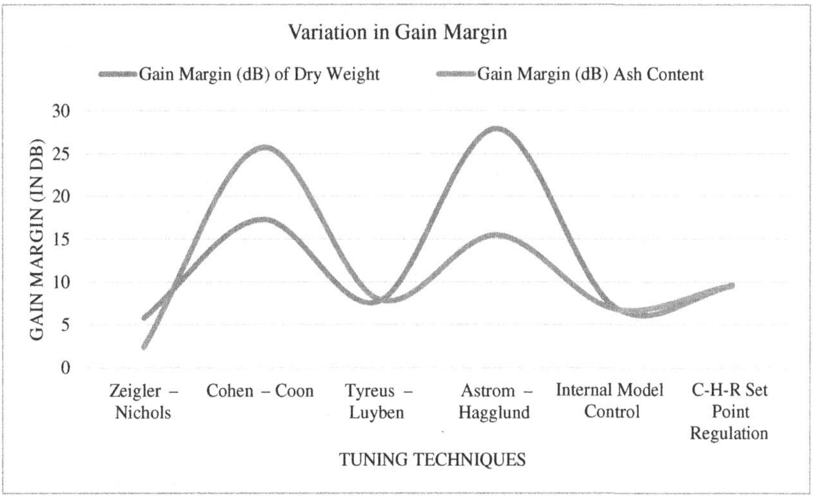

FIG. 7.12 Comparison of gain margin of both loops.

margin for dry weight and Tyreus-Luyben gives the highest phase margin for ash content. However, C-H-R set point regulation technique yields optimal and consistent gain and phase margin for both process loops. The variation in gain margin and phase margin for both process loops have been depicted in Figs. 7.12 and 7.13, respectively.

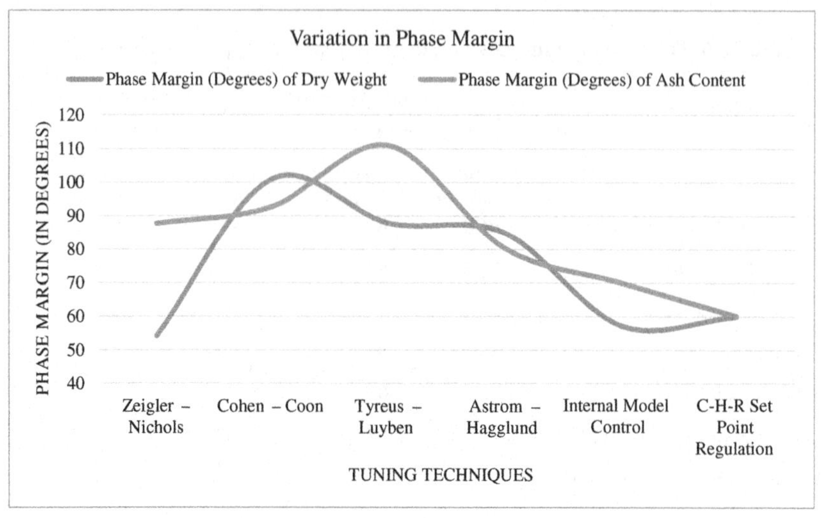

FIG. 7.13 Comparison of phase margin of both loops.

7.4 Conclusion

The PID controller design using seven tuning techniques has been successfully implemented on the paper machine headbox for controlling two important parameters. In this work, the novelty is that MIMO controllers have been designed using conventional techniques and implemented in the abovementioned process. For $G_1(s)$ and $G_2(s)$, controllers are designed, and time response characteristics are analyzed. It is revealed that for $G_1(s)$, set point tuning technique is better than the other ones. And it is further revealed that for $G_2(s)$, Cohen-Coon technique is better than the other ones.

References

[1] A. Haalman, Adjusting controllers for dead time processes, Control. Eng. 12 (1965) 71.

[2] C.B. Kadu, S.B. Lukare, S.B. Bhusal, Design of PI/PID controllers for FOPDT system, IJARCSSE 5 (2) (2015) 645–651.

[3] J.G. Ziegler, N.B. Nichols, Optimum settings for automatic controllers, ASME Trans. 64 (1942) 759.

[4] G.D. Pasgianos, K.G. Arvanitis, A.K. Boglou, Introduction to PID Controllers—Theory, Tuning and Application to Frontier Areas, InTech, 2012, pp. 51–74.

[5] P. Li, P. Wang, D. Xiuxia, An approach to optimal design of stabilizing PID controllers for time-delay systems, in: Control and Decision Conference, 2009. CCDC'09. Chinese, IEEE, 2009, pp. 3465–3470.

[6] M. Chidambaram, R. Padma Sree, A simple method of tuning PID controllers for integrator/dead time processe, Comput. Chem. Eng. (27) (2003) 211–215.

[7] M.S.M. Lee, PID controller design for integrating processes with time delay, Korean J. Chem. Eng. 25 (4) (2008) 637–645.
[8] Q.-G. Wang, T.-H. Lee, H.-W. Fung, B. Qiang, Y. Zhang, PID tuning for improved performance, IEEE Trans. Control Syst. Technol. 7 (4) (1999) 457–465.
[9] W. Tan, J. Liu, T. Chen, H.J. Morquez, Comparison of some well-known PID tuning formulas, Comput. Chem. Eng. 30 (2006) 1416–1423.
[10] M. Chidambaram, Design of PI controllers for integrator/dead time processes, Hung. J. Ind. Chem. 22 (1994) 37–43.
[11] A. O'Dwyer, Handbook of PI and PID Controller Tuning Rules, Imperial College Press, London, 2003.
[12] C.R. Madhuranthakam, A. Elkamel, H. Budman, Optimal tuning of PID controllers for FOPTD, SOPTD and SOPTD with lead processes, Chem. Eng. Process.: Process Intensif. 47 (2) (2008), 251–264.
[13] K.J. Astrom, T. Hagglund, PID Controllers: Theory, Design and Tuning, Instrument Society of America, 1995.
[14] P. Verma, P.K. Juneja, P. Saini, M. Chaturvedi, Comparative analysis of controllers designed for a MIMO system, in: 8th International Conference on Computing Communication and Networking Technologies 2017, IIT Delhi, 3–5 July, 2017.
[15] A. O'Dwyer, PI and PID controller tuning rules for time delay processes: a summary. Part 2: PID controller tuning rules, in: Proceedings of the Irish Signals and Systems Conference, National University of Ireland, Galway, June, 1999, pp. 339–346.
[16] Y. Lee, S. Park, M. Lee, C. Brosilow, PID controller tuning for desired closed-loop responses for SI/SO systems, AIChE J. 44 (1) (1998).
[17] Q.-G. Wang, T.-H. Lee, H.-W. Fung, Q. Bi, Y. Zhang, PID tuning for improved performance, IEEE Trans. Control Syst. Technol. 7 (4) (1999), 457–465.
[18] P. Saini, R. Kumar, N. Rajput, Cascade-PID control of a nonlinear chemical process, Nonlinear Stud. 23 (4) (2016) 561–568.
[19] P. Bajpai, Stock preparation, in: P. Bajpai (Ed.), Biermann's Handbook of Pulp and Paper, third ed., Elsevier, 2018, pp. 65–76 (Chapter 3).
[20] J.-P. Raunio, R. Ritala, Active scanner control on paper machines, J. Process Control 72 (2018) 74–90, https://doi.org/10.1016/j.jprocont.2018.09.012.
[21] S. Stenström, Drying of paper: a review 2000–2018, Dry. Technol. 38 (7) (2020) 825–845, https://doi.org/10.1080/07373937.2019.1596949.
[22] R. Whalley, M. Ebrahimi, Optimum control of a paper making machine headbox, Appl. Math. Model. 26 (6) (2002) 665–679.
[23] A.J. Isaksson, M. Hagberg, L.E. Jiinsson, Benchmarking for paper machine md-control: simulation results, Control. Eng. Pract. vol. 3 (10) (1995) 1491–1498.
[24] P. Saini, R. Kumar, P. Verma, A. Gupta, Control of the paper making subsystem—a brief review, Int. J. Control Theory Appl. 10 (18) (2017) 197–204.
[25] W.L. Luyben, Tuning proportional-integral-derivative controllers for integrator/deadtime processes, Ind. Eng. Chem. Res. 35 (10) (1996) 3480–3483.
[26] K.L. Chien, J.A. Hrons, J.B. Reswick, On the automatic control of generalized passive systems, Trans. Am. Soc. Mech. Eng. 74 (1972) 175–185.
[27] G.H. Cohen, G.A. Coon, Theoretical consideration of retarded control, Trans. ASME 75 (1953) 827–834.
[28] D.E. Rivera, M. Morari, S. Skogestad, Internal model control: PID controller design, Ind. Eng. Chem. Process. Des. Dev. 25 (1) (1986) 252–265.

Chapter 8

Network and security leveraging IoT and image processing: A quantum leap forward

Ajay Sudhir Bale[a], S. Saravana Kumar[b], S. Varun Yogi[a], Swetha Vura[a], R. Baby Chithra[c], N. Vinay[a], and P. Pravesh[a]
[a]*Department of ECE, SoET, CMR University, Bengaluru, India,* [b]*Department of CSE, SoET, CMR University, Bengaluru, India,* [c]*Department of ECE, New Horizon College of Engineering, Bengaluru, India*

8.1 Introduction

In the current era, the Internet of Things (IoT) plays a major role in technological development. IoT devices, as they are connected to the internet and data handling is world-wide, are often subject to several network privacy policies. Encryption and decryption time of images can be accomplished in a compressed sensing model, which is new and has lots of advantages in enhanced security in the transmission of data. The total time taken for computation is highly reduced—in the order of 413 ms for the storage of 3.13×10^6 amount of elements stowed for maximum size images [1]. A multiple number of virtual machines (VMs) in different configurations can hold good for data transfer overheads, wherein the machine focuses on a single workload at a time, which can be amalgamated through web applications [2]. IoT plays major role in solving the MitM attack which provides end to end security [3]. In practice, it is necessary to have an IoT capable of providing data communication with more vulnerability to provide good security and privacy [4]. DDoS (distributed denial of services) attack can cause severe issues in the social media and broadcasting business [5] and it can also consume the computing and communique resources within no time [6]. Privacy, prevention, and data aggregation are well organized in fog-based IoTs [7,8]. A distributed approach is used in the detection of attacks even though a smaller number of communication devices are linked in the fog node [9,10]. The study discusses security solutions to provide privacy to users using recent trends.

System Assurances. https://doi.org/10.1016/B978-0-323-90240-3.00008-4

8.2 Comparative study

8.2.1 Vehicular sensor networks

The idea of self-driving (SD) vehicles, which can be expected to be driven without human involvement, was suggested in 1956. A self-driving vehicle (SDV) should have the capacity to sense the environment around it with onboard circuitry, should know the correct process of vehicle driving, and also to control and decide the actions to be taken for safety of the vehicle in a timely manner. SD technology promises to enhance intelligent transportation systems (ITSs), providing a guarantee of better transportation and safety in journeys [11]. Vehicular sensor networks (VSNs) provide a network of sensing devices where the data of interest are gathered for accomplishing the above-mentioned features. This in turn will ensure the smooth run of traffic.These days, the vehicles have on-board improvements in circuitry consisting of various embedded sensors [12,12a]. The vehicles have the tendency to communicate among each other with the help of various protocols [11]. The communication between the vehicles ensures the smooth flow of data between them by considering the safe transmission of data. Remote vehicle health [3] has seen a great improvement due to these advances. The technologies are employed in providing safety to the user and ensuring smooth transit. These data can even be misused in injuring innocent users [11]. There is an expanding market for VSN for the betterment of modern society in terms of transportation. For example, there is a demand for SDVs, thus increasing the demand for higher quality and functionality networks with sensors, and [11] VSNs play a major role in monitoring of transport systems [13–15]. The applications are shown in Fig. 8.1.

The automotive world has progressed massively in recent years because of IoT advances [16]. The responsibility of such network is to gather real-time data, which can be best utilized in the smooth operation of the vehicle [17]. The various types of sensors that are used in vehicles are connected together to the control unit electronically. Controller area networks (CAN) are the desired wired technologies that are utilized in linking these sensors. On the other hand, the wired connections reduce flexibility. This happens due to the operating situations and the space for operation being limited. It will also affect the scalability of the system.

In VSNs, the vehicles are equipped with groups of sensors that are capable of recording data that describes the condition of the road and environmental scenarios. A remote cloud center receives the information for further processing and analyzing of the data [18–21]. However, uploading data in large quantities to the cloud data center requires wider bandwidth, which results in delayed communication. Fog computing [22] takes the lead in expand cloud computing competencies [23] around automobiles [24], as it is capable of grasping and handling the data reports produced locally by the vehicles and uploaded. Advantages like lower latency and location recognition are the outcome of these

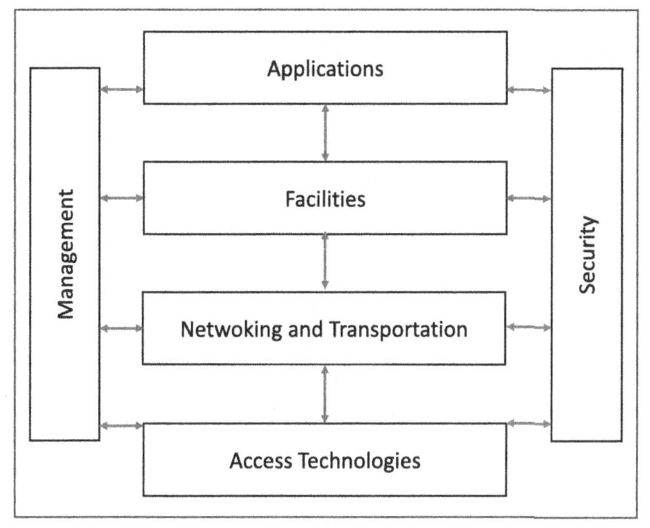

FIG. 8.1 Vehicle communique network points. *(Reprinted with copyright permission from F. Al-Turjman, J.P. Lemayian, Intelligence, security, and vehicular sensor networks in internet of things (IoT)-enabled smart-cities: an overview, Comput. Electr. Eng. 87 (2020) 106776, https://doi.org/10.1016/j.compeleceng.2020.106776.)*

aspects. Fog computing is in use to give out lower latency services in VSN, navigational services [25] and monitoring the surface conditions [26]. Typical fog-assisted vehicular sensor networks (F-VSNs) architecture comprises fog nodes, cloud center, trusted authority, and vehicles. The trusted authority is responsible for production of system parameters, and for recording the entities like vehicle details, fog nodes, and the cloud center. The centrally governed cloud center offers control through durable computational control and larger capacity for remote data. Although F-VSNs bring many advantages, there are still several issues concerned with data acquisition and data query. Specifically, vehicles produce sensory data in bulk, providing the conditions of the road and environmental conditions, uploaded to the cloud center to analyze and to process, but this increases the cost of communication. To address this issue, a process of converting multiple data into a single report by data aggregation technology has gained significant attention.

The data aggregation plans that currently exist [27–33] can neither predict the data reports number generated in respective road sections nor calculate the average data. To address this issue, the scheme [34] uses the Paillier cryptosystem and Chinese remainder theorem to measure the mean sensory data in the respective section; nevertheless, this leads to heavy computational and communication costs. The scheme idea of IoT [35,36] is a description of linking currently and formerly not linked networks to the current infrastructure of the internet.

MSIP-IoT-A is a potential framework for the internet, the main focal point of which is offering basic responses for the obstacles existing in the present day internet. The base structure is targeted on consecutively progressing MSIP-IoT-A for transparency, flexibility, and the huge complexity of the net. In order to attain these goals, MSIP-IoT-A removes the focus on certain structured targets, which also consists of continuous mobility as well as a host. However, this is not a method that can be entirely trusted, as it cannot be useful in fixing the data receipt plan or even changing the relative strength [35].

8.2.2 Biometrics

Biometrics is an arm of science concentrating on identification and affirmation based on an individual's behavior qualities. These qualities are unique and different for different people. Biometric identification systems (BIS) combine various complex operational and technological choices under different contexts. The systems will have similar authentication of technologies and tools, but biometric approaches merge with other authentication methodologies, which helps in enhancing various matters the user will have in response. BIS methods are preferred these days due to features that help with security parameters, help to understand when fraud technology is being used, and so on. Deep ANNs (artificial neural networks) are of great importance for successful biometrics systems [37,38]. IoT is a new technology where wireless technology concepts are starting to be used as intelligent systems. Basically, IoT works with the attributes of the environment. Privacy is considered to be the main part, which provides data security. Along with this, integrity is also an important aspect. Consequently, biometric technology is one of the most important ways of developing such systems. We also have to recognize an individual's identity on the basis of their physical traits with the assistance of data obtained from these systems.

IoT envisions the future of services infrastructure as a networking model where physical objects which are spatially distributed will be widely connected to create information networks, producing a smooth and state of the art system. Such a system requires many sensors, which sense changes, and actuators, used as output systems. RFID (radio frequency identification) tags are used in the monitoring of user data [39]. IoT services activate real world interactions with devices. Due to this, IoT is applicable much of the time: users benefit from smooth access to services by using IoT networks along with smart devices [40,41] as shown in Fig. 8.2. Then we have biometric securities using IoT. The two modalities of biometric are physical and behavioral [42].

8.2.2.1 First stage

In the first stage, the connection between devices and to the smart environment consists of many biometric sensors and smart objects, which collect the required

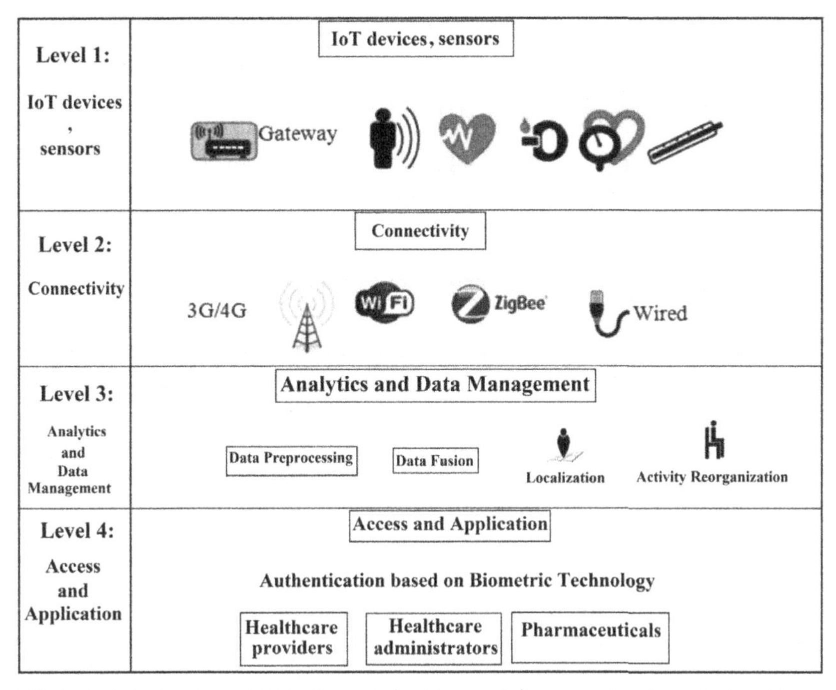

FIG. 8.2 Infrastructure of IoT. *(Reprinted with copyright permission from H. Hamidi, An approach to develop the smart health using internet of things and authentication based on biometric technology, Futur. Gener. Comput. Syst. 91 (2019) 434–449.)*

data from users [38]. This stage is able to collect all the concerned information from biometric sensing devices. The data obtained should be safeguarded from intruder attacks so that the data transferred to the weber via the communication level is safe. It is preferable to encrypt the IoT data before sending to avoid any intrusion that might collapse the system. Biometric data acquired by smart technologies can be thought of as a method for holding identity data more reliably.

8.2.2.2 Second stage

At the second stage, the data obtained from the biometric sensor network can be sent to data centers and weber. The information from smart watches and bracelets can be transmitted using wireless fidelity, fifth generation (5G), Bluetooth, radio frequency ID, ZigBee, and remote sensing satellites. The function of optical fiber in machine-to-machine transmission plays a major role in a communication system. For instance, when smartphones are linked to unstable internet, they exhibit security issues and show disturbances. On the contrary, connections with the weber may be attacked by insecurity. Implementation of security systems like encryption must be incorporated to avert attacks and the resulting leakage of data during transfer [38].

8.2.2.3 Third stage

Internet devices and smart objects are linked in a private weber within the communication system, including many short-range data centers. The public weber is related to centers with long-range data, where communication technologies are connected and data is interchanged at different levels of communication. With respect to biometric technology, the weber is of help in connecting the smart biometric machines with one another. Thus, the process of storing identity becomes quite easy anywhere. Generally, if the problems concerning private adytum increase and users do not have security related issues regarding individual information in the weber environment, then the users will not be deterred from placing their personal data in the weber space. To achieve secure processing of information, there ought to be a protected mechanism in the information storage centers to avoid influence and subversion, and to make sure the data is safe so that the information is confidential, and communication is protected. In this regard, cryptography has been used to provide protection at different levels of communication. This probability provides security against different types of imitation attacks [38].

8.2.2.4 Fourth stage

In this stage, the cloud based medical industry provides services with the use of APIs (application programming interfaces), which will help patients being taken care of at home. The intelligent system can analyze gathered data, which is extracted from the surface of the device.

The interaction between human and machine involves the use of biometrics at the application level. Door locks can also be operated with biometrics. The status of an elderly person can be monitored with the data available in the healthcare system using biometrics [43]. These systems can also be applied for car parking and have good applications in the field of smart grids [44]. Biometrics have massive application in the field of agriculture [45], reducing the burden on farmers.

However, as security loopholes are on the rise, new authentication tactics are needed to incorporate user's personal biometrics or data to secure intelligent systems [38,45].

8.2.3 Network and security

The internet of things (IoT) provides the prospect of improving our standard of living by creating a cardinal environment that can be insightful, flexible, and oriented to our requirements. The outbreaks in the network have to be masked or removed in order to utilize the technology benefits to the fullest.

Crypto tools like the Advanced Encryption Standard (AES) and Rivest-Shamir-Adleman (RSA) algorithms are employed to appraise IoT devices. AEC is comparatively quicker. As far as small devices are concerned, AES utilizes keys of 128 bits for one round, keys of 192 bits for 12 rounds, and keys of 256 bits for 14 rounds. The extra memory present in RSA is an added feature that is missing from AEC. Both AEC and RSA hold good for IoT in providing security against issues like viruses and denial of services [46]. Software attacks are due to influence at the interfaces of IoT. The outbreaks are mostly in the form of logic bombs, worms, Trojan horses, denial of services, and viruses. Data security and efficiency are improved using symmetric encryption algorithms [47].

An idea has been suggested that is an amalgam of Rivest Cipher (RC4) algorithm, Elliptic Curve Cryptography (ECC) algorithm, and one more algorithm called Secure Hash Algorithm (SHA-256), which safeguards sensitive statistics in smart irrigation systems based on IoT. The data integrity is amended, which secures against "man in the middle" (MitM) attacks [48]. An outline has been proposed for security-aware efficient distributed storage (SA-EDS) with the help of algorithms like Efficient Data Conflation (EDCon), Alternative Data Distribution (AD2), and Secure Efficient Data Distributions (SED2). Computation time involved is considerable but it safeguards the data from some of the leading threats of cloud.

The system has two main components, for example, data distributed storage process (D2SP) that protects data from the common threats and unanticipated manmade threats from the cloud side. Even though the cloud server is reliable, unforeseen human error also creates an insecure environment from the cloud side. One more component, called deterministic process (DP), is used in the system, which checks if the input data packets require an advanced security system or not [49].

8.2.4 Medical applications (healthcare)

IoT is being incorporated into various healthcare applications, such as patient monitoring, and classification and prediction of various diseases [50–52], which is associated with various network and data security issues. Elliptical Curve Cryptography (ECC) and Substitution Ceaser Cyphers (SCC) are used in order to maintain data security in medical sensor data. In order to augment the security of the system, an extra secret key is provided along with ECC. This provides lower computational cost with a strong algorithm, which provides an average correlation coefficient of 0.045. The time required to encrypt and decrypt the data is as little as 1.032 to 1.004 μs. The enactment of improved ECC is proven to be better in comparison with existing Rivest-Shamir-Adleman (RSA) and ECC algorithms.

The input data from IoT sensors are encrypted using Substitution Caesar Cipher (SCC), which is a plain substitution cipher, and then ciphered with the help of Caesar Cipher in the system mentioned in Ref. [53] and then directed to cloud server after encoded using improved ECC. The issues encountered in the IoT-based systems owing to the flexibility, partial computational competency, memory, and multiplicity are reviewed. The substantiation towards big data and cloud atmosphere is also analyzed [54].

The network security requirements in the case of healthcare are as follows: portability, traceability, accountability, and should be protected from malware and phishing attempts. In total the network must be reliable, scalable, and energy efficient. UT-Gate [55], which is a health monitoring system, possesses few of the abovementioned features. The energy consumption is much lower in IoTs if connected to fog nodes [56]. Fog is sandwiched between cloud and the other expedients to overcome some of the network security threats faced by IoT. Health-associated data of patients are taken from sensors, which are implanted in the body of the patient. The time-based health related data are continuously monitored and sent to staff through other devices and actuators. The model of fog computing is shown in Fig. 8.3.

6LoWPAN, Bluetooth, Wi-Fi, and ZigBee are used in the communication of transferring data from body sensors to the gateway. The edge/fog layer is designed by combining various gateways, which deals with a variety of combination of protocols. The fog layer converts protocols and makes them feasible for reducing the dimension of the data and also makes them compatible for filtering the data. The cloud layer forms the back end of the system and helps in the diagnosis of widespread diseases by implementing data analytics, warehousing, and broadcast. The smart gateways placed at the fog layer offer lower energy consumption, although complex algorithms are used in order to maintain proper data security systems [55].

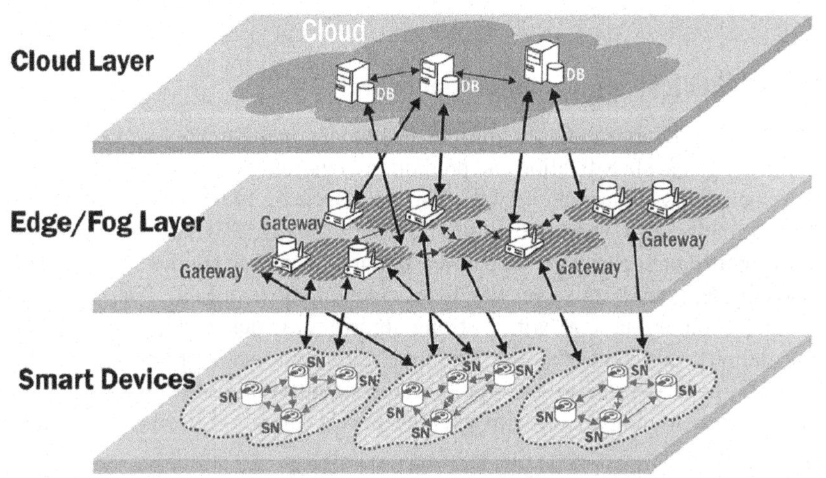

FIG. 8.3 Fog computing model. *(Reprinted with copyright permission from A.M. Rahmani, T.N. Gia, B. Negash, A. Anzanpour, I. Azimi, M. Jiang, P. Liljeberg, Exploiting smart e-Health gateways at the edge of healthcare internet-of-things: a fog computing approach, Futur. Gener. Comput. Syst. 78(Part 2) (2018) 641–658, ISSN 0167-739X, https://doi.org/10.1016/j.future.2017.02.014.)*

8.2.5 Protocols

8.2.5.1 DDoS attack and solution

Neural classifiers with KDD Cup data set and RBP Boost algorithm are convenient, and as accurate as 99.4% in the detection of a distributed denial of services (DDoS) attack [57]. REATO is used in sensing and alerting for DoS through middleware. A factual archetype has been designed in command to authorize the anticipated technique, by evaluating diverse related constraints and also based on the traffic in entropy [58–60]. IoT devices with less security are vulnerable to be an attack vector, which could develop into a wider and more expensive outbreak [61].

8.2.5.2 Man in the middle attack

The advisable methods involved in the uncovering of MitM threat are intrusion detection systems (IDS) as well as intrusion prevention systems (IPS). The fog nodes are debriefed by IDS using an AES possessing a key of 128 bit and block size [62]. Wi-Fi-enabled IoT devices are capable of differentiating legitimate access from unsecure access points [63]. As shown in Fig. 8.4, a software-enabled access point (SoftAP), known as a rogue access point, is generated by the intruder in order to steal data from the transmitting point. IoT devices must always be equipped with adequate features in order to distinguish the rogue access point.

Although, day by day, the usage of IoT devices increases [64,65], if the IoT devices are provided with adequate features then they are not so vulnerable to security threats [63].

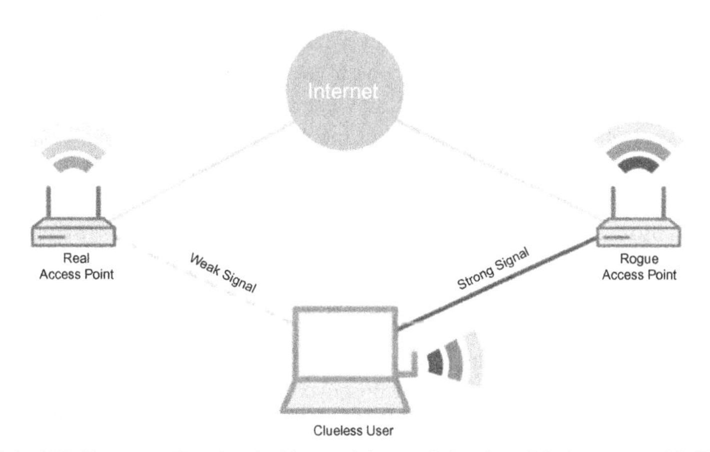

FIG. 8.4 DDoS systems. *(Reprinted with copyright permission from J.O. Agyemang, J.J. Kponyo, G.S. Klogo, J.O. Boateng, Lightweight rogue access point detection algorithm for WiFi-enabled internet of things (IoT) devices, Internet Things 11 (2020), 100200, ISSN 2542-6605, https://doi.org/10.1016/j.iot.2020.100200.)*

The transmitter from the IoT gathers data from sensors in order to reskill the data from the algorithm of channel access. A connective attack can be introduced to operate the data from the input to the transmitter. It has been proved that such attacks of various energy consumption and threat levels may cause loss in throughput. By introducing a defense mechanism to upsurge methodically is the way to attack and protect IoT networks with the help of deep learning [66]. Deep Q learning methods play a major role in the network security of healthcare systems in maintaining the reliability, security, and also the privacy of data [67].

8.2.6 Image processing:

In our fast-paced society, parents are always on the go and find it difficult to take care of their infants. Infants need continuous observation and instant care. A method is being proposed [68] where data is acquired using different sensors to monitor the infant. A moisture sensor is used to check whether the baby's cradle mattress is damp or not. Depending on a threshold value, warnings are sent to parents. The audio level of the area is monitored by a microphone sensor. When there is a loud noise, it alerts the USB microphone, which records the sound. The baby's movement is checked by a motion sensor and if the infant is out of the cradle, parents receive an alert. The facial expressions of the infant are recorded by a Raspberry Pi camera [69], which determines the emotion of the baby. Cry detection and its analysis [70] are also performed. Infant cries are classified [71] depending on analysis of time and frequency-based signal processing. The infant system detects cries, discerning the various actions like boredom, irritation, frustration, and hunger, as shown in Fig. 8.5.

The crucial information of the infant is examined by the system which calms the baby by rocking the cradle whenever required.

To curtail the Corona pandemic, one of the important steps is to wear a face mask. These can be detected by YOLO v3 and F-RCNN technologies. A methodology was proposed in Ref. [72] that incorporates bounding boxes to detect the number of people wearing a face mask. Data sets were generated manually and from the internet to accommodate labels for the boxes. The mask was recognized automatically using practical deep learning techniques. F-RCNN, as shown in Fig. 8.6, was observed to have greater accuracy when compared to YOLO v3 algorithm, but with a lower frame rate.

Autonomous vehicles incorporate various neural network models to detect humans, vehicles, lights, traffic sign boards, route, and objects to control the speed of the vehicle. The technology proposed in Ref. [73] uses a convolutional neural network (CNN) that connects different sensors to the IoT cloud platform. The lane is first detected using a Raspberry Pi camera with a 5 M pixel still resolution. Then the distance between an object and the vehicle is estimated by an ultrasonic sensor, which uses the concept of an echo signal. The resultant speed and direction are controlled by the motor drivers. The YOLO object detection

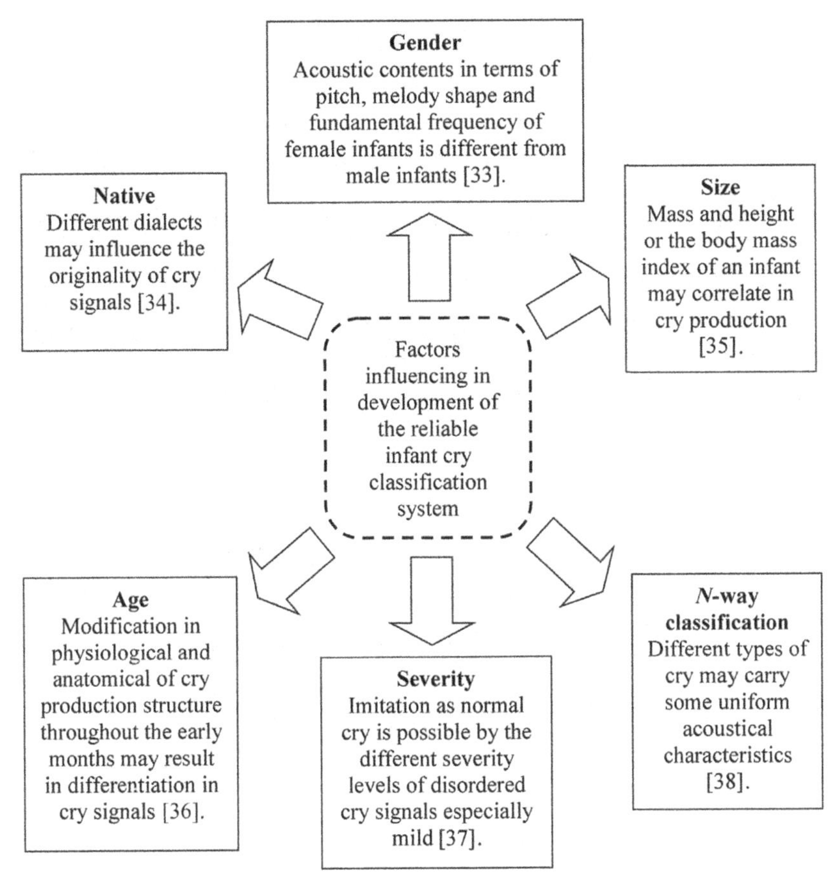

FIG. 8.5 Components used in determining the improvement of a reliable infant cry system. *(Reprinted with copyright permission from J. Saraswathy, et al., Time–frequency analysis in infant cry classification using quadratic time frequency distributions, Biocybern. Biomed. Eng. (2018) https://doi.org/10.1016/j.bbe.2018.05.002.)* Note: *Readers are requested to follow the corresponding references [33–38] from the original article.*

algorithm works on R-CNN models to extract the features in terms of bounding boxes when an object enters the cell of that matrix.

Canny edge detection [74] is used to find the image gradient, and smoothing of images is performed using Gaussian filters. The YOLO algorithm [75] reduces the time and space complexity and works well in a real-time situation.

A proposed cancer prediction system [76] studies blood cells to decide whether they are normal or cancerous in nature with the help of IoT. The results are stored in the cloud so that healthcare personnel can have information about the patient. The sensors implanted in the human body analyze differences in blood pressure and temperature. The features are extracted from the blood sample and classified using a DNN (deep neural network). The blood report is then

FIG. 8.6 F-RCNN model face mask detection. *(Reprinted with copyright permission from S. Singh, U. Ahuja, M. Kumar, et al., Face mask detection using YOLOv3 and faster R-CNN models: COVID-19 environment, Multimed Tools Appl. 80 (2021) 19753–19768, https://doi.org/10.1007/s11042-021-10711-8.)*

encrypted using Advanced Encryption Standard (AES) algorithm [77] so that the database of the patient remains secure. The calculations are based on bytes, where 16 bytes are divided into four sections and columns to form a grid. The data is encrypted and used as keys and the results are simulated using the CloudSim environment [78]. The security of the data is improved using this environment where only the healthcare team and the patient can access the information.

A new method called Multi-Context Fusion Network (MCFN) is employed in agricultural IoT [79] to detect crop diseases in the field. The images acquired by sensors extract features to classify diseases in the crop using ContextNet. Images of plants are captured using a CCD camera in fields, which are then noise filtered. The MCFN approach combines the temporal and spatial data with visual information to encode them. ResNet50 [80], used as the CNN backbone, attains high quality with improved accuracy. 50,000 crop disease samples with 74 categories have been used to include most of the crop diseases, as shown in Fig. 8.7.

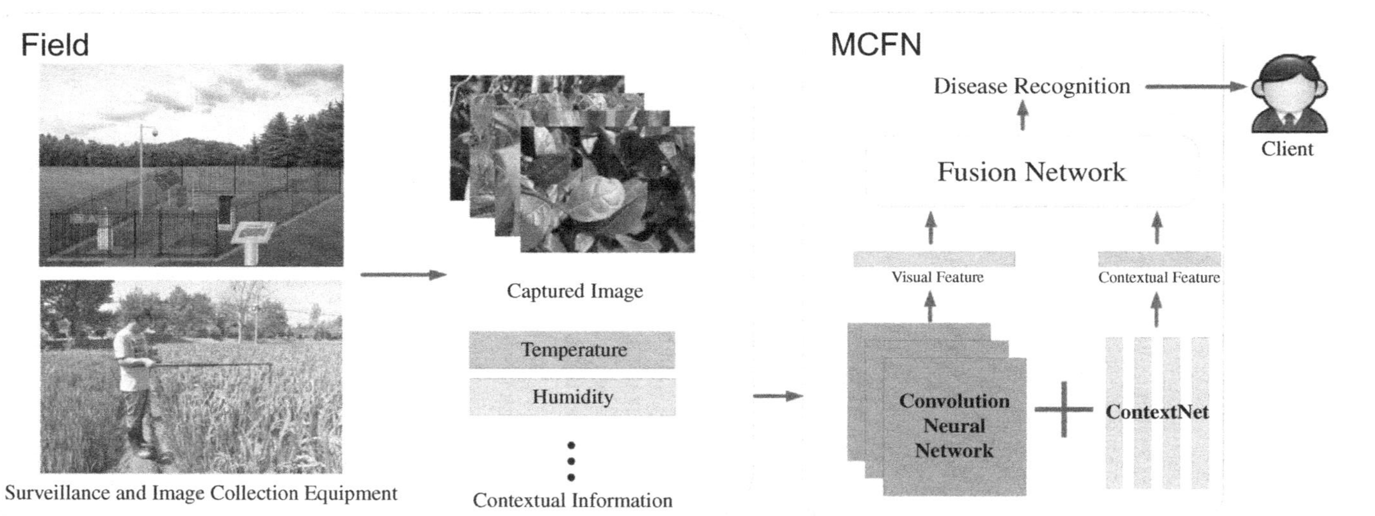

FIG. 8.7 System architecture. *(Reprinted with copyright permission from Y. Zhao, L. Liu, C. Xie, R. Wang, F. Wang, Y. Bu, S. Zhang, An effective automatic system deployed in agricultural internet of things using multi-context fusion network towards crop disease recognition in the wild, Appl. Soft Comput. 89 (2020) 106128, ISSN:1568-4946, https://doi org/10.1016/j.asoc.2020.106128.)*

8.3 Conclusion

This chapter overviews the applications of IoT in various fields along with their protocols and methodology in implementation. The critical findings of this review are summarized below:

- The VSN plays an important role in recent automated vehicle technology for monitoring transport system. Integration of IoT with VSNs can gather real time data leading to smooth operation of automated vehicles.
- The integration of advanced biometric identification systems, artificial neural networks, and IoT is currently enhancing the level of privacy and security in systems that rely on the uniqueness of people.
- In recent years, the application of IoT has been advanced in all fields, including medical, automobile, agriculture, etc., where the privacy of the data plays an important role.
- It was observed that the improved ECC algorithm gives better results when compared with Rivest-Shamir-Adleman (RSA) and traditional ECC algorithms.
- The healthcare network should be reliable, scalable, and energy efficient. These types of networks are discussed in this work. The UT-Gate is among the healthcare monitoring systems that have the required network security protocols. Fog computing model with cloud layer that are integrated in smart devices for healthcare are explained in this study.
- As image processing is advancing along with IoT technology, infant tracking and continuous monitoring is discussed in this chapter, which can become a major advantage for parents with busy schedules.
- Other major application of image processing include cancer prediction, Canny edge detection, automaton of vehicles, and agriculture are discussed, where CNNs, encryption algorithms, and Gaussian filters play a major role.
- Neural classifiers with KDD Cup data sets and RBP Boost algorithms give the best results for detection of DDoS attacks.
- In the network protection of healthcare systems, deep Q learning methods play a significant role.

As a future scope, there is a lot of demand for the energy optimization techniques in VSNs, wherein the consumption of power and the battery recharging time will be focused. The flow of the data since data acquisition from sensor nodes to the cloud and up to the end users with proper optimization techniques can provide a precise network security system with which an ultimate security frame work can be achieved for every node of IoT in the future. Metaheuristic optimization techniques can be employed in the choice of the best parameters of the method of time-frequency-based analysis of infant cry systems.

References

[1] L. Li, G. Wen, Z. Wang, Y. Yang, Efficient and secure image communication system based on compressed sensing for IoT monitoring applications, IEEE Trans. Multimedia 22 (1) (2020) 82–95, https://doi.org/10.1109/TMM.2019.2923111.

[2] A. Beloglazov, J. Abawajy, R. Buyya, Energy-aware resource allocation heuristics for efficient management of data centers for cloud computing, Futur. Gener. Comput. Syst. 28 (5) (2012) 755–768. ISSN 0167-739X https://doi.org/10.1016/j.future.2011.04.017.

[3] R.E. Navas, H. Le Bouder, N. Cuppens, F. Cuppens, G.Z. Papadopoulos, Demo: do not trust your neighbors! A small IoT platform illustrating a man-in-the-middle attack, in: N. Montavont, G. Papadopoulos (Eds.), Ad-hoc, Mobile, and Wireless Networks. ADHOC-NOW 2018, Lecture Notes in Computer Science, vol. 11104, Springer, Cham, 2018, https://doi.org/10.1007/978-3-030-00247-3_11.

[4] R.K. Pateriya, S. Sharma, The evolution of RFID security and privacy: a research survey, in: 2011 International Conference on Communication Systems and Network Technologies, Katra, India, 2011, pp. 115–119, https://doi.org/10.1109/CSNT.2011.31.

[5] C. Kolias, G. Kambourakis, A. Stavrou, J. Voas, DDoS in the IoT: Mirai and other botnets, Computer 50 (7) (2017) 80–84, https://doi.org/10.1109/mc.2017.201.

[6] C. Douligeris, A. Mitrokotsa, DDoS attacks and defense mechanisms: classification and state-of-the-art, Comput. Netw. 44 (5) (2004) 643–666. ISSN 1389-1286 https://doi.org/10.1016/j.comnet.2003.10.003.

[7] Z. Guan, Y. Zhang, L. Wu, J. Wu, J. Li, Y. Ma, J. Hu, APPA: an anonymous and privacy preserving data aggregation scheme for fog-enhanced IoT, J. Netw. Comput. Appl. 125 (2019) 82–92. ISSN 1084-8045 https://doi.org/10.1016/j.jnca.2018.09.019.

[8] S. Rathore, B.W. Kwon, J.H. Park, BlockSecIoTNet: blockchain-based decentralized security architecture for IoT network, J. Netw. Comput. Appl. 143 (2019) 167–177. ISSN 1084-8045 https://doi.org/10.1016/j.jnca.2019.06.019.

[9] S. Rathore, B. Wook Kwon, J.H. Park, BlockSecIoTNet: blockchain-based decentralized security architecture for IoT network, J. Netw. Comput. Appl. (2019), https://doi.org/10.1016/j.jnca.2019.06.019.

[10] K. Fan, H. Xu, L. Gao, H. Li, Y. Yang, Efficient and privacy preserving access control scheme for fog-enabled IoT, Futur. Gener. Comput. Syst. 99 (2019) 134–142. ISSN 0167-739X https://doi.org/10.1016/j.future.2019.04.003.

[11] S. Zhang, J. Chen, F. Lyu, N. Cheng, W. Shi, X. Shen, Vehicular communication networks in the automated driving era, IEEE Commun. Mag. 56 (9) (2018) 26–32, https://doi.org/10.1109/MCOM.2018.1701171.

[12] F. Al-Turjman, J.P. Lemayian, Intelligence, security, and vehicular sensor networks in internet of things (IoT)-enabled smart-cities: an overview, Comput. Electr. Eng. 87 (2020) 106776, https://doi.org/10.1016/j.compeleceng.2020.106776; (a) W.-W. Hu, F.-L. Chang, Y.-H. Zhang, L.-B. Chen, C.-T. Yu, W.-J. Chang, Design and implementation of a next-generation hybrid internet of vehicles communication system for driving safety, J. Commun. 13 (2018) 737–742, https://doi.org/10.12720/jcm.13.12.737-742.

[13] H.-W. Cheong, H. Lee, Requirements of AGV (automated guided vehicle) for SMEs (small and medium-sized enterprises), Procedia Comput. Sci. 139 (2018) 91–94, https://doi.org/10.1016/j.procs.2018.10.222.

[14] S.-C. Hu, Y.-C. Wang, C.-Y. Huang, Y.-C. Tseng, 2009 IEEE sensors—a vehicular wireless sensor network for CO2 monitoring, 2009, pp. 1498–1501 (IEEE 2009 IEEE Sensors—Christchurch, New Zealand (2009.10.25-2009.10.28)).

[15] I. Salhi, M.O. Cherif, S.M. Senouci, 2009 IEEE International Conference on Communications—A New Architecture for Data Collection in Vehicular Networks, 2009, pp. 1–6 (IEEE ICC 2009—2009 IEEE International Conference on Communications—Dresden, Germany (2009.06.14-2009.06.18)).

[16] M.A. Rahim, M.A. Rahman, M.M. Rahman, A.T. Asyhari, M.Z.A. Bhuiyan, D. Ramasamy, Evolution of IoT-enabled connectivity and applications in automotive industry: a review, Veh. Commun. 27 (2020) 100285.

[17] P. Hiep, Statistical method for performance analysis of WBAN in time-saturation, EURASIP J. Wirel. Commun. Netw. 2014 (1) (2014) 221.

[18] S. Tuohy, M. Glavin, C. Hughes, E. Jones, M. Trivedi, L. Kilmartin, Intra-vehicle networks: a review, IEEE Trans. Intell. Transp. Syst. 16 (2) (2015) 534–545.

[19] Y. Ming, X. Yu, Efficient privacy-preserving data sharing for fog-assisted vehicular sensor networks, Sensors 20 (2) (2020) 514.

[20] R. Yu, X. Huang, J. Kang, J. Ding, S. Maharjan, S. Gjessing, Y. Zhang, Cooperative resource management in cloud-enabled vehicular networks, IEEE Trans. Ind. Electron. 62 (12) (2015) 7938–7951.

[21] J. Ni, X. Lin, K. Zhang, X. Shen, Privacy-preserving real-time navigation system using vehicular crowdsourcing, in: 2016 IEEE 84th Vehicular Technology Conference (VTC-Fall), 2016.

[22] F. Bonomi, R. Milito, J. Zhu, S. Addepalli, Proceedings of the First Edition of the MCC Workshop on Mobile Cloud Computing—MCC '12—Fog Computing and Its Role in the Internet of Things, 2012, p. 13 (ACM Press the first edition of the MCC workshop—Helsinki, Finland (2012.08.17-2012.08.17)).

[23] M. Armbrust, A. Fox, R. Griffith, A.D. Joseph, R.H. Katz, A. Konwinski, G. Lee, D.A. Patterson, A. Rabkin, I. Stoica, et al., A view of cloud computing, Commun. ACM 53 (2010) 50–58.

[24] Y. Dai, D. Xu, S. Maharjan, Y. Zhang, Joint offloading and resource allocation in vehicular edge computing and networks, in: Proceedings of the IEEE Global Communications Conference, Abu Dhabi, UAE, 9–13, December 2018, pp. 1–7.

[25] J. Ni, K. Zhang, Y. Yu, X. Lin, X.S. Shen, Privacy-preserving smart parking navigation supporting efficient driving guidance retrieval, IEEE Trans. Veh. Technol. 67 (2018) 6504–6517.

[26] S. Basudan, X. Lin, K. Sankaranarayanan, A privacy-preserving vehicular crowdsensing based road surface condition monitoring system using fog computing, IEEE Internet Things J. 4 (2017) 772–782.

[27] G. Zhuo, Q. Jia, L. Guo, M. Li, P. Li, Privacy-preserving verifiable data aggregation and analysis for cloud-assisted mobile crowdsourcing, in: Proceedings of the 35th IEEE International Conference on Computer Communications, San Francisco, CA, USA, 10–14, April 2016, pp. 1–9.

[28] K. Rabieh, M.M.E.A. Mahmoud, M. Younis, Privacy-preserving route reporting schemes for traffic management systems, IEEE Trans. Veh. Technol. 66 (2017) 2703–2713.

[29] C. Xu, R. Lu, H. Wang, L. Zhu, C. Huang, PAVS: a new privacy-preserving data aggregation scheme for vehicle sensing systems, Sensors 17 (2017) 1–18.

[30] G. Sun, S. Sun, J. Sun, H. Yu, X. Du, M. Guizani, Security and privacy preservation in fog-based crowd sensing on the internet of vehicles, J. Netw. Comput. Appl. 134 (2019) 89–99.

[31] X. Lin, R. Lu, X. Shen, MDPA: multidimensional privacy-preserving aggregation scheme for wireless sensor networks, Wirel. Commun. Mob. Comput. 10 (2010) 843–856.

[32] R. Lu, K. Heung, A.H. Lashkari, A.A. Ghorbani, A light-weight privacy-preserving data aggregation scheme for fog computing-enhanced IoT, IEEE Access 5 (2017) 3302–3312.

[33] B. Wang, Z. Chang, Z. Zhou, T. Ristaniemi, Reliable and privacy-preserving task recomposition for crowdsensing in vehicular fog computing, in: Proceedings of the 87th Vehicular Technology Conference, Porto, Portugal, 3–6, June 2018, pp. 6–11.
[34] Q. Kong, R. Lu, M. Ma, H. Bao, A privacy-preserving sensory data sharing scheme in internet of vehicles, Futur. Gener. Comput. Syst. 92 (2019) 644–655.
[35] V. Vijaya Kumar, M. Devi, P. Vishnu Raja, P. Kanmani, V. Priya, S. Sudhakar, S. Krishnamoorthy, Design of peer-to-peer protocol with sensible and secure IoT communication for future internet architecture, Microprocess. Microsyst. 78 (2020) 103216.
[36] C. Karlof, N. Sastry, D. Wagner, TinySec: a link layer security architecture for wireless sensor networks, in: SenSys'04: Proceedings of the 2nd International Conference on Embedded Networked Sensor Systems, 2004.
[37] S. Dargan, M. Kumar, A comprehensive survey on the biometric recognition systems based on physiological and behavioral modalities, Expert Syst. Appl. 143 (2020) 113114.
[38] H. Hamidi, An approach to develop the smart health using internet of things and authentication based on biometric technology, Futur. Gener. Comput. Syst. 91 (2019) 434–449.
[39] Z. Yan, P. Zhang, A.V. Vasilakos, A survey on trust management for internet of things, J. Netw. Comput. Appl. 42 (2014) 120–134.
[40] P.K. Dhillon, S. Kalra, A lightweight biometrics based remote user authentication scheme for IoT services, J. Inf. Secur. Appl. 34 (2017) 255–270 (S2214212616301442–).
[41] M.S. Obaidat, I. Traore, I. Woungang, Biometric security and internet of things (IoT), in: Biometric-Based Physical and Cybersecurity Systems, 2019, pp. 477–509, https://doi.org/10.1007/978-3-319-98734-7 (Chapter 19).
[42] M.S. Obaidat, N. Boudriga, Security of e-Systems and Computer Networks, Cambridge University Press, 2007.
[43] M. Faundez-Zanuy, Biometric security technology, in: J.R. Rabuñal Dopico, J. Dorado, A. Pazos (Eds.), Encyclopedia of Artificial Intelligence, IGI Global, 2009, pp. 262–269.
[44] R. de Luis-Garcıa, C. Alberola-López, O. Aghzout, J. Ruiz-Alzola, Biometric identification systems, Signal Process. 83 (12) (2003) 2539–2557.
[45] M.S. Obaidat, S.P. Rana, T. Maitra, D. Giri, S. Dutta, Biometric security and internet of things (IoT), in: M. Obaidat, I. Traore, I. Woungang (Eds.) Biometric-based physical and cybersecurity systems. Springer, Cham., 2019, pp. 477–509. https://doi.org/10.1007/978-3-319-98734-7_19.
[46] E. Abinaya, K. Aishwarva, C.P.M. Lordwin, G. Kamatchi, I. Malarvizhi, A performance aware security framework to avoid software attacks on internet of things (IoT) based patient monitoring system Proceeding of 2018 IEEE International Conference on Current Trends toward Converging Technologies, Coimbatore, India.
[47] C. Ritika, S. Kuldeep, Efficiency and security of data with symmetric encryption algorithms, Int. J. Adv. Res. Comput. Sci. Softw. Eng. 2 (8) (2012) 1–6. ISSN: 2277 128X.
[48] S.K. Mousavi, A. Ghaffari, S. Besharat, H. Afshari, Improving the security of internet of things using cryptographic algorithms: a case of smart irrigation systems, J. Ambient. Intell. Humaniz. Comput. (2020), https://doi.org/10.1007/s12652-020-02303-5.
[49] Y. Li, K. Gai, L. Qiu, M. Qiu, H. Zhao, Intelligent cryptography approach for secure distributed big data storage in cloud computing, Inf. Sci. 387 (2017) 103–115. ISSN 0020-0255 https://doi.org/10.1016/j.ins.2016.09.005.
[50] C. Li, X. Hu, L. Zhang, The IoT-based heart disease monitoring system for pervasive healthcare service, Procedia Comput. Sci. 112 (2017) 2328–2334. ISSN 1877-0509 https://doi.org/10.1016/j.procs.2017.08.265.

[51] G. Manogaran, D. Lopez, C. Thota, M. Kaja Abbas, S. Pyne, R. Sundarasekar, Big data analytics in healthcare internet of things, in: Innovative Healthcare Systems for the 21st Century, Springer, Cham, Switzerland, 2017, pp. 263–284.

[52] L.M. Dang, M.J. Piran, D. Han, K. Min, H. Moon, A survey on internet of things and cloud computing for healthcare, Electronics 8 (7) (2019) 768.

[53] M.A. Khan, M.T. Quasim, N.S. Alghamdi, M.Y. Khan, A secure framework for authentication and encryption using improved ECC for IoT-based medical sensor data, IEEE Access (2020) 1, https://doi.org/10.1109/access.2020.2980739.

[54] M. Wazid, A.K. Das, R. Hussain, G. Succi, J.J.P.C. Rodrigues, Authentication in cloud-driven IoT-based big data environment: survey and outlook, J. Syst. Archit. 97 (2019) 185–196. ISSN 1383-7621 https://doi.org/10.1016/j.sysarc.2018.12.005.

[55] A.M. Rahmani, T.N. Gia, B. Negash, A. Anzanpour, I. Azimi, M. Jiang, P. Liljeberg, Exploiting smart e-Health gateways at the edge of healthcare internet-of-things: a fog computing approach, Futur. Gener. Comput. Syst. 78 (Part 2) (2018) 641–658. ISSN 0167-739X https://doi.org/10.1016/j.future.2017.02.014.

[56] R. Oma, S. Nakamura, D. Duolikun, T. Enokido, M. Takizawa, An energy-efficient model for fog computing in the internet of things (IoT), Internet Things 1–2 (2018) 14–26. ISSN 2542-6605 https://doi.org/10.1016/j.iot.2018.08.003.

[57] P.A. Raj Kumar, S. Selvakumar, Distributed denial of service attack detection using an ensemble of neural classifier, Comput. Commun. 34 (11) (2011) 1328–1341, https://doi.org/10.1016/j.comcom.2011.01.012.

[58] S. Sicari, A. Rizzardi, D. Miorandi, A. Coen-Porisini, REATO: REActing TO denial of service attacks in the internet of things, Comput. Netw. 137 (2018) 37–48, https://doi.org/10.1016/j.comnet.2018.03.020.

[59] R.T. Tiburski, L.A. Amaral, E. de Matos, D.F.G. de Azevedo, F. Hessel, The role of lightweight approaches towards the standardization of a security architecture for IoT middleware systems, IEEE Commun. Mag. 54 (12) (2016) 56–62, https://doi.org/10.1109/MCOM.2016.1600462CM.

[60] J. David, C. Thomas, DDoS attack detection using fast entropy approach on flow-based network traffic, Procedia Comput. Sci. 50 (2015) 30–36. ISSN 1877-0509 https://doi.org/10.1016/j.procs.2015.04.007.

[61] R. Syed, R.J. Orr, A. Cox, P. Ashokkumar, M.R. Rizvi, Identifying the attack surface for IoT network, Internet Things 9 (2020) 100162. ISSN 2542-6605 https://doi.org/10.1016/j.iot.2020.100162.

[62] F. Aliyu, T. Sheltami, E.M. Shakshuki, A detection and prevention technique for man in the middle attack in fog computing, Procedia Comput. Sci. 141 (2018) 24–31. ISSN 1877-0509 https://doi.org/10.1016/j.procs.2018.10.125.

[63] J.O. Agyemang, J.J. Kponyo, G.S. Klogo, J.O. Boateng, Lightweight rogue access point detection algorithm for WiFi-enabled internet of things(IoT) devices, Internet Things 11 (2020) 100200. ISSN 2542-6605 https://doi.org/10.1016/j.iot.2020.100200.

[64] D. Miorandi, S. Sicari, F. De Pellegrini, I. Chlamtac, Internet of things: vision, applications and research challenges, Ad Hoc Netw. 10 (7) (2012) 1497–1516. ISSN 1570-8705 https://doi.org/10.1016/j.adhoc.2012.02.016.

[65] E. Borgia, The internet of things vision: key features, applications and open issues, Comput. Commun. 54 (2014) 1–31. ISSN 0140-3664 https://doi.org/10.1016/j.comcom.2014.09.008.

[66] Y.E. Sagduyu, Y. Shi, T. Erpek, IoT network security from the perspective of adversarial deep learning, in: 2019 16th Annual IEEE International Conference on Sensing, Communication,

and Networking (SECON), Boston, MA, USA, 2019, pp. 1–9, https://doi.org/10.1109/SAHCN.2019.8824956.

[67] P. Mohamed Shakeel, S. Baskar, V.R. Sarma Dhulipala, et al., Maintaining security and privacy in health care system using learning based deep-Q-networks, J. Med. Syst. 42 (2018) 186, https://doi.org/10.1007/s10916-018-1045-z.

[68] C. Lobo, A. Chitrey, P. Gupta, Sarfaraj, A. Chaudhari, Infant care assistant using machine learning, audio processing, image processing and IoT sensor network, in: 2020 International Conference on Electronics and Sustainable Communication Systems (ICESC), Coimbatore, India, 2020, pp. 317–322, https://doi.org/10.1109/ICESC48915.2020.9155597.

[69] N.L. Pratap, K. Anuroop, P.N. Devi, A. Sandeep, S. Nalajala, IoT based smart cradle for baby monitoring system, in: 2021 6th International Conference on Inventive Computation Technologies (ICICT), Coimbatore, India, 2021, pp. 1298–1303, https://doi.org/10.1109/ICICT50816.2021.9358684.

[70] C. Ji, T.B. Mudiyanselage, Y. Gao, et al., A review of infant cry analysis and classification, J. Audio Speech Music Proc. 2021 (8) (2021), https://doi.org/10.1186/s13636-021-00197-5.

[71] J. Saraswathy, et al., Time–frequency analysis in infant cry classification using quadratic time frequency distributions, Biocybern. Biomed. Eng. (2018), https://doi.org/10.1016/j.bbe.2018.05.002.

[72] S. Singh, U. Ahuja, M. Kumar, et al., Face mask detection using YOLOv3 and faster R-CNN models: COVID-19 environment, Multimed. Tools Appl. 80 (2021) 19753–19768, https://doi.org/10.1007/s11042-021-10711-8.

[73] I. Ahmad, K. Pothuganti, Design & implementation of real time autonomous car by using image processing & IoT, in: 2020 Third International Conference on Smart Systems and Inventive Technology (ICSSIT), Tirunelveli, India, 2020, pp. 107–113, https://doi.org/10.1109/ICSSIT48917.2020.9214125.

[74] L.H. Gong, C. Tian, W.P. Zou, et al., Robust and imperceptible watermarking scheme based on Canny edge detection and SVD in the contourlet domain, Multimed. Tools Appl. 80 (2021) 439–461, https://doi.org/10.1007/s11042-020-09677-w.

[75] M.A.A. Al-qaness, A.A. Abbasi, H. Fan, et al., An improved YOLO-based road traffic monitoring system, Computing 103 (2021) 211–230, https://doi.org/10.1007/s00607-020-00869-8.

[76] M. Anuradha, T. Jayasankar, N.B. Prakash, M.Y. Sikkandar, G.R. Hemalakshmi, C. Bharatiraja, A.S.F. Britto, IoT enabled cancer prediction system to enhance the authentication and security using cloud computing, Microprocess. Microsyst. (2020), https://doi.org/10.1016/j.micpro.2020.103301.

[77] V. Savitha, N. Karthikeyan, S. Karthik, et al., A distributed key authentication and OKM-ANFIS scheme based breast cancer prediction system in the IoT environment, J. Ambient Intell. Human. Comput. 12 (2021) 1757–1769, https://doi.org/10.1007/s12652-020-02249-8.

[78] A. Sundas, S.N. Panda, An introduction of CloudSim simulation tool for modelling and scheduling, in: 2020 International Conference on Emerging Smart Computing and Informatics (ESCI), Pune, India, 2020, pp. 263–268, https://doi.org/10.1109/ESCI48226.2020.9167549.

[79] Y. Zhao, L. Liu, C. Xie, R. Wang, F. Wang, B. Yingqiao, S. Zhang, An effective automatic system deployed in agricultural internet of things using multi-context fusion network towards crop disease recognition in the wild, Appl. Soft Comput. 89 (2020) 106128. ISSN 1568-4946 https://doi.org/10.1016/j.asoc.2020.106128.

[80] S. Vemishetti, K. Swaraj, K. Meenakshi, P. Kora, A deep learning based crop disease classification using transfer learning, Mater. Today Proc. (2021), https://doi.org/10.1016/j.matpr.2020.10.846. ISSN 2214-7853.

Chapter 9

Modeling software patching process inculcating the impact of vulnerabilities discovered and disclosed

Deepti Aggrawal[a], **Jasmine Kaur**[b], **and Adarsh Anand**[b]
[a]*USME, DTU, East Delhi Campus, Delhi, India,* [b]*Department of Operational Research, Faculty of Mathematical Sciences, University of Delhi, Delhi, India*

9.1 Introduction

Software maintenance is important as it influences the perceived quality of the software. The after-sales support and customer service go a long way in building customer loyalty and retaining the customer for future products by the brand or organization. When the product is with the developers, i.e., during the development time, it can be carefully monitored but it becomes dynamic during the maintenance phase as numerous unpredictable forces come into play. The ever-evolving technology, varied platforms, varied uses, social media, consumer awareness, ease of information sharing, etc., all necessitate the product to be portable and robust.

Cyberattacks and cyberterrorism have become household terms. Data breaches are becoming a norm. Software security becomes imperative in such a situation. In 2019, 164.68 billion US records were exposed due to 1473 data breaches. Facebook's data breach compromised 267 million records while for the Google Cloud server it was 201 million records. Innumerable stats can be found which show the sad state of software security. Breach of personal information can lead to cyberbullying, monetary loss, identity theft, blackmailing, etc. The effects can vary from a deep psychological impact on an individual to political implications throughout the world with long-term repercussions. Hence, the software needs to be closely monitored and checked frequently for any loopholes that may cause it to be used maliciously or with ill intent. Any such loophole or fault in the software which can cause a security breach is termed as a vulnerability.

System Assurances. https://doi.org/10.1016/B978-0-323-90240-3.00009-6

Vulnerabilities have been classified into different categories such as SQL injection, buffer overflow, cross-site scripting, denial of service, memory corruption, file inclusion, directory traversal, etc., based on varied features such as the amount of damage they cause, the component they target, or their cause of creation. The vulnerabilities follow a typical life cycle. They are injected into the software due to bad programming practices, neglect, inexperience, etc. Vulnerabilities are then discovered with the help of various types of tools and testing techniques. The vulnerabilities may be discovered by white hat (or the ethical) hackers, black hat hackers, or the software vendor itself. The detected vulnerabilities are disclosed, which happens either as full disclosure or private disclosure. In full disclosure, the vulnerability is generally detected by black hat hackers who try to gain from the situation. They can release the vulnerability details on a public forum, sell it on Darknet, or release its exploit. An exploit is a tool or a set of code that helps to take advantage of the vulnerability. Such vulnerabilities are referred to as zero-day vulnerabilities as their fix is not available instantly. Another strategy of vulnerability disclosure, followed by white hat hackers and bug bounty hunters, is to reveal the details of the vulnerability to the respective vendor. This policy gives the vendor reasonable time to deal with the information and handle it before any real damage is done. The software vendor then works on creating a solution to fix the vulnerability via a patch. A patch is a piece of corrective code used to fix bugs and vulnerabilities in software. The timely implementation of a well-tested and correct patch helps to get rid of the vulnerability.

The details of discovered vulnerabilities, their severity, information regarding their exploits and patches, etc., are usually available publicly after disclosure. Many vulnerability databases are available, both government-owned and private, which maintain an updated record of the latest vulnerabilities. The National Vulnerability Database (NVD) contains Common Vulnerabilities and Exposures [1] and is maintained by MITRE Corporation. The CERT Division of Software Engineering Institute, Carnegie Mellon University is another such database [2]. WhiteSource Vulnerability Database contains details of many open-source software vulnerabilities. The number of vulnerabilities discovered each year keeps growing. According to WhiteSource Vulnerability Database, 6111 vulnerabilities were discovered in 2019 while it had risen to 9658 in 2020 [3].

The most widely accepted notation in the industry is the one given by the CVE. Each detected vulnerability is assigned a unique CVE number. The common vulnerability scoring system (CVSS) of the First.Org group assigns a severity score to each vulnerability [4]. The CVSS score is assigned based on base, temporal, and environmental scores. The base score considers only the properties specific to a particular vulnerability; the temporal score considers those factors that change with time due to external influences; the environmental score determines the impact of a particular vulnerability on a particular organization. NVD CVSS 3.0 version categorizes the vulnerabilities into five severity levels on a scale of 1–10. The base score of 9.0–10.0 is considered critical, 7.0–8.9 is considered high, 4.0–6.9 is considered medium, and 0.1–3.9 is considered low while 0.0 is none. This version considers only the base factors. This score helps in the prioritized handling of vulnerabilities.

The current chapter has determined the number of vulnerabilities detected, disclosed, and eventually patched. A mathematical model has been proposed to determine these numbers.

This chapter has been designed in the following way: Section 9.2 discusses the relevant literature; Section 9.3 discusses the notations used in the model; Section 9.4 presents the mathematical framework to simulate the vulnerability detection, disclosure, and patch policy; Section 9.5 presents an illustration of the discussed model, and Section 9.6 concludes the chapter. A list of references to the various sources used in the chapter has been supplemented.

9.2 Literature review

Vulnerability modeling has been widely studied. Different aspects of vulnerability such as their categorization, their exploitation, their remediation have been taken care of by different researchers. Various vulnerability detection techniques have been proposed. Alhazmi and Malaiya [5] proposed the Logarithmic Poisson Model. They have also discussed the other pioneering vulnerability discovery models such as the Anderson Thermodynamic Model [6], AML Model [7], and Rescorla Model [8]. Alhazmi et al. [9] have used vulnerability discovery models to predict the number of vulnerabilities not yet found in the software. Bhatt et al. [10] studied the vulnerability detection process in interdependent vulnerabilities. Kaur et al. [11] discussed a vulnerability correction model for dependent and independent vulnerabilities. Movahedi et al. [12] proposed a neural network model and compared it with existing vulnerability discovery models. Chen et al. [13] proposed particle swarm optimization for vulnerability detection. Hanif et al. [14] discussed the various machine learning approaches used in software vulnerability detection. Bhatt et al. [15] used machine learning to determine the exploitability of a vulnerability.

The vulnerability disclosure process has been studied by Arora and Telang [16] who identified the ideal approaches while disclosing a vulnerability. Böhme [17] compared vulnerability disclosure in different market types. Arora et al. [18] gave the optimal vulnerability disclosure policy and the related patch release behavior. Zhao et al. [19] studied the role of the white hat hacker's community in vulnerability discovery and disclosure. Tang et al. [20] used Big Data systems to understand the dependency between vulnerability discovery and disclosure. Ķinis [21] discussed the vulnerability disclosure policies practiced in Latvia. Böhme et al. [22] discussed vulnerability disclosure policy in the context of cryptocurrencies. According to the findings discussed by Shrestha et al. [23], numerous vulnerabilities had been discussed on social media platforms much earlier than they were officially released on databases like CVE.

The patch release behavior has also been studied widely in the literature. Jiang and Sarkar [24] discussed the economic benefits of patching service and gave a mathematical model for the same. Arora et al. [25] studied the impact of patches on software quality. Okamura et al. [26] discussed the optimal software patch release policy for a nonhomogenous vulnerability discovery process. Arora et al. [27] have studied the impact of competition on the patch release policy and the subsequent patching of vulnerabilities. Anand et al. [28] gave the optimal patch release policy using cost modeling. Anand et al. [29] studied the patch release policy in a multiupgraded software. Anand et al. [30] discussed an optimization problem to allocate resources for vulnerability removal using patches. Kaur et al. [31] studied the impact of an infected patch in a single version while Anand et al. [32] analyzed it in multiupgraded software.

The above-discussed works have been mostly dealt with one of the three processes, viz., vulnerability detection, vulnerability disclosure, and vulnerability patching. None of the abovementioned works has used a mathematical approach to discuss the above three categories simultaneously. Hence, the need for the current discussion.

9.3 Notations

The notations used in the model development are:

$\Omega(t)$ cumulative number of vulnerabilities detected till time t
$\upsilon(t)$ cumulative number of vulnerabilities disclosed till time t
$\Psi(t)$ cumulative number of vulnerabilities patched till time t
N number of vulnerabilities in the software
s_1 rate of vulnerability detection
s_2 rate of vulnerability disclosure
s_3 rate of patch release

9.4 Model development

As discussed earlier, software vulnerabilities follow a life cycle, wherein they are injected, discovered, disclosed, and removed. In this chapter, the latter three processes have been studied and a mathematical model to quantify the same has been discussed.

9.4.1 Vulnerability detection process

The faults/bugs in the software which go undetected during the development phase give rise to vulnerabilities in the operational phase. Postrelease testing by the software vendors leads to the detection of these vulnerabilities. Many a times the users report certain issues that they face during product usage which helps the software testing team to identify vulnerabilities in the software. Vendors sometimes also organize bug bounty programs where they pay people to find vulnerabilities in the software. White hat hackers are the major participants here. Black hat hackers are also involved in the vulnerability detection process, but their contribution is detrimental instead of progressive. Hence, for this study, their role has been overlooked. The vulnerabilities detected by the white hat hackers, the users, and the software vendors are the vulnerabilities discovered.

Using the exponential vulnerability discovery model proposed by Rescorla [8], the cumulative number of vulnerabilities detected during the product's useful life can be determined.

The number of **vulnerabilities detected** is represented by the following equations:

$$\frac{d}{dt}\Omega_1(t) = s_1(N - \Omega_1(t)) \tag{9.1}$$

Solving differential equation (9.1):

$$\Omega_1(t) = s_1(1 - \exp(-s_1 t)) \tag{9.2}$$

9.4.2 Vulnerability disclosure process

Private disclosure policy gives a safe period to the software vendors to deal with the discovered vulnerabilities in their own way. The number of vulnerabilities that will be disclosed to the public will be a subset of the number of vulnerabilities detected. All the discovered vulnerabilities may be disclosed. But it may also happen that some vulnerabilities are not important enough and hence they do not need to be dealt with and are not disclosed.

The number of **vulnerabilities disclosed** is represented by the following equations

$$\frac{d}{dt}\upsilon_2(t) = s_2(\Omega_1(t) - \upsilon_2(t)) \tag{9.3}$$

Solving differential equation (9.3), the cumulative number of vulnerabilities disclosed are

$$\upsilon_2(t) = N\left(1 - \left(\frac{s_1 \exp(-s_2 t) - s_2 \exp(-s_1 t)}{s_1 - s_2}\right)\right) \tag{9.4}$$

9.4.3 Vulnerability patching process

Software patching is considered the vulnerability correction process. Software patching helps to deal with vulnerabilities and other minor bugs in the software instantly. Security patches generally deal with a handful of vulnerabilities at a time. The quality of patches is also important as it determines whether the issue is permanently resolved or not. Often insufficiently tested or broken patches are released in the market which can further increment the issue instead of ending it.

The number of patches will again correspond to the number of vulnerabilities disclosed. Not all disclosed vulnerabilities are patched.

The cumulative number of **vulnerabilities patched** is represented by the following equations:

$$\frac{d}{dt}\psi_3(t) = s_3(v_2(t) - \psi_3(t)) \tag{9.5}$$

Solving differential equation (9.5):

$$\psi_3(t) = \left(\frac{(Ns_3)}{(s_1 - s_2)}\right)(F_1 + F_2 - F_3) \tag{9.6}$$

where

$$F_1 = \left(\frac{(s_1 - s_2)}{s_3}\right)(1 - \exp(-s_3 t)),\ F_2 = \left(\frac{s_2}{(s_3 - s_1)}\right)(\exp(-s_1 t) - \exp(-s_3 t)),$$
$$F_3 = \left(\frac{s_1}{(s_3 - s_2)}\right)(\exp(-s_2 t) - \exp(-s_3 t))$$

Thus, through the expressions obtained in Eqs. (9.2), (9.4), and (9.6) the number of vulnerabilities that are discovered, disclosed, and eventually patched can be determined. This modeling framework will help to allocate resources appropriately so that the optimal number of detected vulnerabilities are disclosed and patched promptly. This modeling further answers an important question of the optimal number of patches that should be released in a software's lifespan.

9.5 Model illustration

To show the working of the proposed model, it has been validated on the vulnerability discovery dataset of *Windows XP* from *2004 to 2015*. The data has been collected from the Microsoft Update Catalog [33]. Table 9.1 demonstrates the parameter estimate values obtained by solving the proposed model using SAS software [34].

Table 9.2 demonstrates the estimated values obtained by solving Eqs. (9.2), (9.4), and (9.6) using the parameters obtained in Table 9.1. Thus, the number of vulnerabilities discovered, disclosed, and patched have been determined using the proposed model.

Table 9.3 shows the percentage values of each group. Column 1 shows what percent of the discovered vulnerabilities were disclosed. Column 2 represents what percentage of the disclosed vulnerabilities were patched. To draw further attention to the number of discovered vulnerabilities that get eventually fixed, Column 3 shows what percent of the discovered vulnerabilities were patched.

TABLE 9.1 Estimated values of parameters.

Parameter	Estimated value
N	4620.59
s_1	0.07772
s_2	0.07876
s_3	0.07971

TABLE 9.2 Representation of estimated data.

Year	No. of vulnerabilities discovered	No. of vulnerabilities disclosed	No. of vulnerabilities patched
2004	350	14	0
2005	323	38	2
2006	299	59	6
2007	276	76	11
2008	255	90	16
2009	236	102	22
2010	218	112	29
2011	202	119	35
2012	186	125	42
2013	172	129	48
2014	159	132	54
2015	147	133	60

As can be seen in Table 9.3, a very large proportion of the discovered vulnerabilities are disclosed. But in contrast, a low percentage of it is patched. Similarly, the proportion of disclosed vulnerabilities that eventually get patched is also low. This points to the gap in the three processes. The ultimate goal is vulnerability fixation through the timely release of software patches. Thus, there is a need to optimally allocate available resources to maximize the number of discovered/disclosed vulnerabilities that get patched.

TABLE 9.3 Percentage of vulnerabilities dealt.

Year	% of Discovered Vulnerabilities that were Disclosed	% of Disclosed Vulnerabilities that were Patched	% of Discovered Vulnerabilities that were Patched
2004	4.00	0.00	0.00
2005	11.76	5.26	0.62
2006	19.73	10.17	2.01
2007	27.54	14.47	3.99
2008	35.29	17.78	6.27
2009	43.22	21.57	9.32
2010	51.38	25.89	13.30
2011	58.91	29.41	17.33
2012	67.20	33.60	22.58
2013	75.00	37.21	27.91
2014	83.02	40.91	33.96
2015	90.48	45.11	40.82

TABLE 9.4 Comparison criteria.

Parameter	Calculated value
SSE	611.4
MSE	76.43
RMSE	8.742
R^2	0.996
Adj. R^2	0.994

To check the performance of the model, various goodness-of-fit measures have been used. The results of the various comparison criteria have been discussed in Table 9.4. The low values of SSE, MSE, and RMSE show that the model was able to predict values very close to the original data. Similarly, the close to perfect values of R^2 and Adjusted R^2 show the high predictive capability of the model.

9.6 Conclusion

The operational phase of the software is very crucial as it determines the user's outlook towards the product as well as the firm. Hence, it needs to be carefully monitored for bugs and vulnerabilities. The vulnerabilities present in the software are hazardous and compromise the security of the software. Thus, the need to appropriately deal with vulnerabilities is imperative. In the current proposal, the vulnerability life cycle has been discussed with a major focus on vulnerabilities detection, disclosure, and patching. Mathematical modeling has been used to determine the number of each, i.e., vulnerabilities detected by various means, the number of vulnerabilities disclosed to the public, and the number of vulnerabilities patched. This mathematical modeling will help to determine the resource allocation to mitigate the damage caused by these vulnerabilities, thereby, ensuring higher software security and quality. In the current work, the model validation has been done on the dataset collected for vulnerability detection of closed source software. In future work, it would be interesting to explore how the modeling changes in the context of open-source software.

Acknowledgment

The second author thanks the Ministry of Minority Affairs (UGC) for providing financial assistance to carry out this research work under the Maulana Azad National Fellowship scheme (UGC Ref. No.: 190520033186, No. F. 82-27/2019 (SA-III)).

References

[1] Common Vulnerabilities and Exposures Details. https://cve.mitre.org/. Last accessed on 15 February 2021.

[2] CERT. https://www.sei.cmu.edu/about/divisions/cert/. Last accessed on 15 February 2021.

[3] WhiteSource Vulnerability Database. https://www.whitesourcesoftware.com/vulnerability-database/. Last accessed on 15 February 2021.

[4] National Vulnerability Database. https://nvd.nist.gov/vuln-metrics/cvss. Last accessed on 15 February 2021.

[5] O.H. Alhazmi, Y.K. Malaiya, Modeling the vulnerability discovery process, in: 16th IEEE International Symposium on Software Reliability Engineering (ISSRE'05), IEEE, 2005, November, p. 10.

[6] R. Anderson, Security in open versus closed systems—the dance of Boltzmann, Coase and Moore, 2002. Technical report, Cambridge University, England, pp. 1–15.

[7] O.H. Alhazmi, Y.K. Malaiya, Quantitative vulnerability assessment of systems software, in: Annual Reliability and Maintainability Symposium, 2005. Proceedings, IEEE, 2005, January, pp. 615–620.

[8] E. Rescorla, Is finding security holes a good idea? IEEE Security Privacy 3 (1) (2005) 14–19.

[9] O.H. Alhazmi, Y.K. Malaiya, I. Ray, Measuring, analyzing and predicting security vulnerabilities in software systems, Comput. Secur. 26 (3) (2007) 219–228.

[10] N. Bhatt, A. Anand, V.S.S. Yadavalli, V. Kumar, Modeling and characterizing software vulnerabilities, Int. J. Math. Eng. Manage. Sci. 2 (4) (2017) 288–299.

[11] J. Kaur, A. Anand, O. Singh, Modeling software vulnerability correction/fixation process incorporating time lag, Recent Adv. Software Reliab. Assur. (2019) 39–58.

[12] Y. Movahedi, M. Cukier, I. Gashi, Vulnerability prediction capability: a comparison between vulnerability discovery models and neural network models, Comput. Secur. 87 (2019) 101596.

[13] C. Chen, H. Xu, B. Cui, PSOFuzzer: a target-oriented software vulnerability detection technology based on particle swarm optimization, Appl. Sci. 11 (3) (2021) 1095.

[14] H. Hanif, M.H.N.M. Nasir, M.F. Ab Razak, A. Firdaus, N.B. Anuar, The rise of software vulnerability: taxonomy of software vulnerabilities detection and machine learning approaches, J. Network Comput. Appl. (2021) 103009.

[15] N. Bhatt, A. Anand, V.S.S. Yadavalli, Exploitability prediction of software vulnerabilities, Qual. Reliab. Eng. Int. 37 (2) (2021) 648–663.

[16] A. Arora, R. Telang, Economics of software vulnerability disclosure, IEEE Security Privacy 3 (1) (2005) 20–25.

[17] R. Böhme, A comparison of market approaches to software vulnerability disclosure, in: International Conference on Emerging Trends in Information and Communication Security, Springer, Berlin, Heidelberg, 2006, June, pp. 298–311.

[18] A. Arora, R. Telang, H. Xu, Optimal policy for software vulnerability disclosure, Manag. Sci. 54 (4) (2008) 642–656.

[19] M. Zhao, J. Grossklags, K. Chen, An exploratory study of white hat behaviors in a web vulnerability disclosure program, in: Proceedings of the 2014 ACM workshop on security information workers, pp. 51–58, 2014, November.

[20] M. Tang, M. Alazab, Y. Luo, Big data for cybersecurity: vulnerability disclosure trends and dependencies, IEEE Trans. Big Data 5 (3) (2017) 317–329.

[21] U. Ķinis, From responsible disclosure policy (RDP) towards state regulated responsible vulnerability disclosure procedure (hereinafter–RVDP): the Latvian approach, Comput. Law Security Rev. 34 (3) (2018) 508–522.

[22] R. Böhme, L. Eckey, T. Moore, N. Narula, T. Ruffing, A. Zohar, Responsible vulnerability disclosure in cryptocurrencies, Commun. ACM 63 (10) (2020) 62–71.

[23] P. Shrestha, A. Sathanur, S. Maharjan, E. Saldanha, D. Arendt, S. Volkova, Multiple social platforms reveal actionable signals for software vulnerability awareness: a study of GitHub, twitter and Reddit, Plos One 15 (3) (2020). e0230250.

[24] Z. Jiang, S. Sarkar, Optimal software release time with patching considered, in: Workshop on Information Technologies and Systems, Seattle, WA, USA, 2003, December.

[25] A. Arora, J.P. Caulkins, R. Telang, Research note—sell first, fix later: impact of patching on software quality, Manag. Sci. 52 (3) (2006) 465–471.

[26] H. Okamura, M. Tokuzane, T. Dohi, Optimal security patch release timing under non-homogeneous vulnerability-discovery processes, in: 2009 20th international symposium on software reliability engineering, IEEE, 2009, November, pp. 120–128.

[27] A. Arora, C. Forman, A. Nandkumar, R. Telang, Competition and patching of security vulnerabilities: an empirical analysis, Inf. Econ. Policy 22 (2) (2010) 164–177.

[28] A. Anand, M. Agarwal, Y. Tamura, S. Yamada, Economic impact of software patching and optimal release scheduling, Qual. Reliab. Eng. Int. 33 (1) (2017) 149–157.

[29] A. Anand, S. Das, D. Aggrawal, P.K. Kapur, Reliability analysis for upgraded software with updates, in: Quality, IT and Business Operations, Springer, Singapore, 2018, pp. 323–333.

[30] A. Anand, J. Kaur, A.A. Gokhale, M. Ram, Impact of available resources on software patch management, in: Systems Performance Modeling, vol. 4, Walter de Gruyter GmbH & Co KG, 2020, pp. 1–12, https://doi.org/10.1515/9783110619058-001.

[31] J. Kaur, A. Anand, O. Singh, V. Kumar, Measuring software reliability under the influence of an infected patch, Yugoslav J. Operat. Res. (2020). http://yujor.fon.bg.ac.rs/index.php/yujor/article/view/897/741/.

[32] A. Anand, J. Kaur, S. Inoue, Reliability modeling of multi-version software system incorporating the impact of infected patching, Int. J. Qual. Reliab. Manage. 37 (6/7) (2020) 1071–1085.

[33] Microsoft Update Catalog. https://www.catalog.update.microsoft.com/Home.aspx. Last accessed on 15 February 2021.

[34] SAS Institute Inc., SAS/ETS user's guide version 9.1, SAS Institute Inc., Cary, NC, 2004.

Chapter 10

Extension of software reliability growth models by several testing-time functions

Yuka Minamino[a], Shinji Inoue[b], and Shigeru Yamada[a]
[a]*Graduate School of Engineering, Tottori University, Tottori, Japan,* [b]*Faculty of Informatics, Kansai University, Osaka, Japan*

10.1 Introduction

10.1.1 Background for development of bivariate SRGMs

Most software reliability growth models (SRGMs) [1, 2] are developed based on the assumption that the software reliability growth is promoted by investing the testing time. However, significant testing effort is invested in the testing phase. Here, the testing effort represents test coverage, executed test cases, CPU time, execution time, and man-hour [2, 3]. It means that software reliability growth cannot be observed without substantial testing effort, not testing time. The existing SRGMs suggest that the number of detected faults increases unconditionally as the testing time elapses regardless of the amount of testing effort. Thus, the existing SRGMs cannot reflect the actual testing environment. They affect the accuracy of quantitative software reliability evaluation.

Various extension methods for the SRGMs have been introduced [4–10]. Among them, the simpler method is the extension method using the testing-time function. In this modeling approach, it is assumed that the testing time of the existing SRGMs is testing time as a software reliability growth factor. It is divided into the testing-time and testing-effort factors and expressed based on a specific testing-time function. The Cobb-Douglas-type testing-time function is a typical production function of economics used as a testing-time function [4, 5, 9, 11–14]. The Cobb-Douglas-type production function has a simple mathematical structure to handle as the characteristic; it is easy to apply as the testing-time function. Thus, bivariate SRGM based on the Cobb-Douglas-type testing-time function has been developed as previous research.

There are two main issues with the Cobb-Douglas-type testing-time function. First, the Cobb-Douglas-type testing-time function does not describe well

System Assurances. https://doi.org/10.1016/B978-0-323-90240-3.00010-2

the economic significance. Second, the elasticity of substitution between the software reliability growth factors (or production factors in economics) takes a value of 1. Here, the ease of substitution between the software reliability growth factors is called the *elasticity of substitution* in economics. Fig. 10.1 shows the conceptual figure for the elasticity of substitution. As shown in Fig. 10.1, when the testing-time factor decreases according to the arrows, the testing-effort factor increases later in the case of small elasticity of substitution. Thus, the substitution rate, which represents the incline, changes significantly. On the other hand, when the testing-time factor decreases according to the arrows, the testing-effort factor does not change significantly later in the case of large elasticity of substitution. Thus, the substitution rate, which represents the incline, does not change significantly.

The elasticity of substitution for the Cobb-Douglas-type production function, which takes a value of 1, means that the ratio of usage cost for each software reliability growth factor to the total software cost remains unchanged. It means that when the cost for using a software reliability growth factor rises, the share of software reliability growth factors does not change. Thus, it is desirable to apply more flexible testing-time functions since the Cobb-Douglas-type testing-time function assumes a strict software development environment.

To solve this issue, we apply the constant elasticity of substitution (CES)-type production function [12–14] as a new type of testing-time function. It has fewer constraints than the Cobb-Douglas-type production function because the elasticity of substitution can take a certain value. Besides, it is a generalized function of multiple production functions, including the Cobb-Douglas-type production function. As a new element, a substitution parameter is added to the parameters of the Cobb-Douglas-type production function. It enhanced the evaluation of the elasticity of substitution between the software reliability growth factors in more detail.

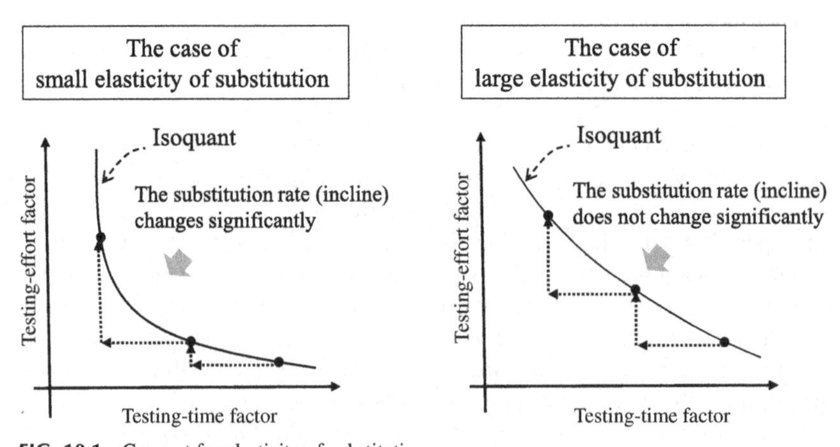

FIG. 10.1 Concept for elasticity of substitution.

The production factors assume that they can substitute with other production factors in economics. However, it gradually becomes difficult to continue the substitution of a certain production factor with another production factor. It means the marginal rate of substitution between the production factors gradually decreases, as shown in Fig. 10.1. The elasticity of substitution represents the strength of the diminishing degree.

The economic viewpoint can be applied to the software development environment. For example, if there is not the sufficient testing time in the testing phase, it is possible to maintain some software reliability growth using the testing effort, which compensates it. Besides, it is not easy to continue the substitution in the long term because the more testing time is substituted by the testing effort; the more testing effort required for the substitution increases. Thus, it is essential to clarify how much the software reliability growth factors can be substituted for each other to maintain software reliability growth.

10.1.2 Our proposed bivariate SRGMs

From the above background, we discussed the extension of the existing univariate SRGMs using two different testing-time functions. The software development environment was evaluated in terms of substitutability, which was an economic concept. We proposed the bivariate exponential (EXP), delayed S-shaped (DSS), and inflection S-shaped (ISS) SRGMs [2], and extended the software reliability assessment measures. Besides, the Cobb-Douglas- and CES-type testing-time functions were applied. We verified their effectiveness by comparing their goodness of fit, based on actual datasets. Furthermore, we proposed the bivariate Weibull-type SRGMs, which follow the Cobb-Douglas- and CES-type testing-time functions. We solved the optimization problems under budget constraints based on the proposed bivariate SRGMs to derive the maximum number of detectable faults. Finally, the sensitivity analysis was conducted, and the substitutability of the testing-time and testing-effort factors was evaluated.

10.2 Software reliability assessment using bivariate SRGMs

10.2.1 Existing nonhomogeneous Poisson process (NHPP) model

Now, let $\{N(t), t \geq 0\}$ denote a counting process representing the total number of faults detected up to the testing-time t. We formulate the probability mass function that m faults are detected up to the testing-time t as follows:

$$\begin{cases} \Pr\{N(t)=m\} = \dfrac{\{H(t)\}^m}{m!}\exp[-H(t)] \quad (m=0,1,2,\ldots), \\ H(t) = \displaystyle\int_0^t h(x)dx \quad (t \geq 0), \end{cases} \tag{10.1}$$

where $H(t)$ is the mean value function of NHPP. Besides, $H(t)$ represents the expected cumulative number of faults detected in the time-interval $(0, t]$. $h(t)$ is the intensity function, and it represents the software failure rate. When $\{N(t), t \geq 0\}$ follows the Poisson process with mean $H(t)$ as Eq. (10.1), the SRGM is called the *NHPP model*.

Most NHPP models assume that the expected number of detectable faults per unit time is proportional to the expected number of remaining faults at time t. Thus, it is formulated by a differential equation, as given in Eq. (10.2).

$$\frac{dH(t)}{dt} = b(t)[a - H(t)] \quad (b(t) > 0, t \geq 0), \tag{10.2}$$

where a represents the initial fault content, and $b(t)$ the fault-detection rate at time t. The fault-detection rate per unit time can be derived as

$$b(t) = \frac{\frac{d}{dt}H(t)}{[a - H(t)]}. \tag{10.3}$$

First, assuming that the initial conditions of Eq. (10.2) are $H(0) = 0$ and $b(t) = b(b > 0)$, the EXP SRGM is defined as

$$H(t) \equiv m(t) = a(1 - \exp[-bt]). \tag{10.4}$$

Suppose $b(t) = b$ and $a = m(t)$, then, we obtain the following differential equation:

$$\frac{dH(t)}{dt} = b[m(t) - H(t)]. \tag{10.5}$$

Solving Eq. (10.5) under Eq. (10.4), we obtain the DSS SRGM as follows:

$$H(t) \equiv M(t) = a(1 - (1 + bt)\exp[-bt]) \quad (a > 0, b > 0). \tag{10.6}$$

Finally, the fault-detection rate is assumed as follows:

$$b(t) \equiv b_I(t) = b\left\{ l + (1 - l)\frac{H(t)}{a} \right\} \quad (a > 0, b > 0, 0 < l \leq 1), \tag{10.7}$$

where l represents the inflection coefficient for the fault-detection ability. From Eqs. (10.2), (10.7), we obtain the ISS SRGM as follows:

$$H(t) \equiv I(t) = \frac{a(1 - \exp[-bt])}{(1 + c \cdot \exp[-bt])} \quad (c > 0), \tag{10.8}$$

where $c = (1 - l)/l$.

10.2.2 Bivariate NHPP models based on testing-time functions

This study assumes that the testing-time t in the existing SRGMs is the testing time as a software reliability growth factor. It consists of the testing-time factor

(s) and the testing-effort factor (u). However, it is assumed that the testing-time and testing-effort factors are independent. Accordingly, the NHPP is extended to bivariate NHPP, that is, $\{N(s, u), s \geq 0, u \geq 0\}$ indicates a counting process representing the total number of faults detected up to the testing-time s and the testing-effort u. Then, the bivariate NHPP is formulated as

$$\Pr\{N(s,u)=m\}=\frac{\{H(s,u)\}^m}{m!}\exp[-H(s,u)] \quad (m=0,1,2,\ldots), \tag{10.9}$$

where $H(s, u)$ represents the expected cumulative number of faults detected in the time-interval $(0, s]$ and testing-effort u, which is used by the testing-time s.

Next, the testing-time t in the existing models is represented using the following testing-time functions.

1. *Cobb-Douglas-type testing-time function*

$$t_{\mathrm{CD}}=s^{\alpha}u^{1-\alpha} \quad (0\leq\alpha\leq 1), \tag{10.10}$$

2. *CES-type testing-time function*

$$t_{\mathrm{CES}}=(\alpha s^{-\delta}+(1-\alpha)u^{-\delta})^{-\frac{1}{\delta}} \quad (0\leq\alpha\leq 1,\delta\geq -1), \tag{10.11}$$

where α represents the degree of influence for each factor on software reliability growth, and δ is the substitution parameter. The function form of the CES-type testing-time function (the CES production function) changes depending on the substitution parameter as follows:

- When $\delta = -1$, the CES-type testing-time function becomes a linear function.
- When $\delta \rightarrow 0$, it becomes the Cobb-Douglas-type function.
- When $\delta \rightarrow \infty$, it becomes the Leontief-type function.

The elasticity of substitution e, which represents the ease of substitution between software reliability growth factors, to maintain software reliability growth is given by

$$e=\frac{1}{1+\delta} \quad (\delta>-1). \tag{10.12}$$

As stated earlier, it is worth noting that the elasticity of substitution for the CES-type testing-time function can assume a certain value, and the elasticity of substitution for the Cobb-Douglas-type testing-time function always takes a value of 1.The bivariate SRGMs are constructed by introducing the Cobb-Douglas-type testing-time function to the existing EXP, DSS, and ISS SRGMs [11, 12, 14].

3. *CD-EXP SRGM*

$$H(s,u)\equiv m_{\mathrm{CD}}(s,u)=a(1-\exp[-bs^{\alpha}u^{1-\alpha}]). \tag{10.13}$$

4. *CD-DSS SRGM*

$$H(s,u) \equiv M_{\mathrm{CD}}(s,u) = a(1-(1+bs^{\alpha}u^{1-\alpha})\exp[-bs^{\alpha}u^{1-\alpha}]). \tag{10.14}$$

5. *CD-ISS SRGM*

$$H(s,u) \equiv I_{\mathrm{CD}}(s,u) = \frac{a(1-\exp[-bs^{\alpha}u^{1-\alpha}])}{(1+c\cdot\exp[-bs^{\alpha}u^{1-\alpha}])}. \tag{10.15}$$

Similarly, the bivariate SRGMs based on the CES-type testing-time function are constructed as follows:

6. *CES-EXP SRGM*

$$H(s,u) \equiv m_{\mathrm{CES}}(s,u) = a(1-\exp[-b(\alpha s^{-\delta}+(1-\alpha)u^{-\delta})^{-\frac{1}{\delta}}]). \tag{10.16}$$

7. *CES-DSS SRGM*

$$\begin{aligned} H(s,u) \equiv M_{\mathrm{CES}}(s,u) &= a(1-(1+b(\alpha s^{-\delta}+(1-\alpha)u^{-\delta})^{-\frac{1}{\delta}}) \\ &\quad \times \exp[-b(\alpha s^{-\delta}+(1-\alpha)u^{-\delta})^{-\frac{1}{\delta}}]). \end{aligned} \tag{10.17}$$

8. *CES-ISS SRGM*

$$H(s,u) \equiv I_{\mathrm{CES}}(s,u) = \frac{a(1-\exp[-b(\alpha s^{-\delta}+(1-\alpha)u^{-\delta})^{-\frac{1}{\delta}}])}{(1+c\cdot\exp[-b(\alpha s^{-\delta}+(1-\alpha)u^{-\delta})^{-\frac{1}{\delta}}])}. \tag{10.18}$$

10.2.3 Parameter estimation

We estimate the parameters of the proposed bivariate SRGMs using the maximum-likelihood method. Supposed we have observed $(s_k, u_k, y_k)(k=0,1,2,\ldots,K)$, where s_k represents the testing time, u_k the testing effort, and y_k the total number of detected faults. Then, the log-likelihood function is as follows:

$$\begin{aligned} \ln L(\boldsymbol{\theta}) &= \sum_{k=1}^{K}(y_k-y_{k-1})\ln[H(s_k,u_k;\boldsymbol{\theta})-H(s_{k-1},u_{k-1};\boldsymbol{\theta})] \\ &\quad -H(s_K,u_K;\boldsymbol{\theta})-\sum_{k=1}^{K}\ln[(y_k-y_{k-1})!]. \end{aligned} \tag{10.19}$$

The parameters are estimated by solving the simultaneous log-likelihood equation with respect to the parameters as

$$\frac{\partial \ln L(\boldsymbol{\theta})}{\partial \boldsymbol{\theta}} = 0, \tag{10.20}$$

where $\boldsymbol{\theta}$ is a set of the parameters in each proposed bivariate SRGM.

10.2.4 Bivariate software reliability assessment measures

We extend the software reliability assessment measures based on the existing SRGMs to bivariate assessment measures. First, software reliability function [2] represents the probability for which a software failure does not occur in the time-interval $(s, s + x)(s \geq 0, x \geq 0)$ given that the testing or operation has gone up to time s. It is worth nothing that the testing-effort value has been attained up to u by testing termination time s. Thus, the software reliability function, which follows the bivariate SRGMs, is derived as follows [6, 12, 14]:

$$R(x|s,u) = \exp[-\{H(s+x,u) - H(s,u)\}]. \tag{10.21}$$

The expected number of remaining faults by arbitrary testing-time s and testing-effort u is defined as follows [6, 12, 14]:

$$\begin{aligned} M(s,u) \equiv \mathrm{E}[\bar{N}(s,u)] &= \mathrm{E}[N(\infty,\infty) - N(s,u)] \\ &= H(\infty,\infty) - H(s,u) \\ &= a - H(s,u). \end{aligned} \tag{10.22}$$

10.2.5 Numerical examples

We apply the following datasets (DS) collected in the actual testing phase [15–18]. The results of calculating and drawing have been obtained using R, which is a statistical analysis free software.

DS1: $(s_k, u_k, y_k)(k = 1,2,\ldots,19; s_{19} = 19, u_{19} = 47.65, y_{19} = 328)$
DS2: $(s_k, u_k, y_k)(k = 1,2,\ldots,21; s_{21} = 21, u_{21} = 8736, y_{21} = 43)$
DS3: $(s_k, u_k, y_k)(k = 1,2,\ldots,20; s_{20} = 20, u_{20} = 10{,}000, y_{20} = 100)$
DS4: $(s_k, u_k, y_k)(k = 1,2,\ldots,19; s_{19} = 19, u_{19} = 10{,}272, y_{19} = 120)$
DS5: $(s_k, u_k, y_k)(k = 1,2,\ldots,12; s_{12} = 12, u_{12} = 5053, y_{12} = 61)$
DS6: $(s_k, u_k, y_k)(k = 1,2,\ldots,19; s_{19} = 19, u_{19} = 11{,}305, y_{19} = 42)$

In the above datasets, s_k is the calendar time as the testing-time factor, u_k is the CPU time/execution time as the testing-effort factor, and y_k is the total number of faults detected during $[0, s_k]$, $[0, u_k]$. DS1 and DS2 are the database application program test data and the telecommunication system test data [16–18]. DS3 and DS6 are software failure data including four major releases of software products at Tandem Computers [15]. Besides, we use the mean squared errors (MSE) as the assessment criteria of the goodness-of-fit comparison. The MSE is defined as

$$\text{MSE} = \frac{1}{K}\sum_{k=1}^{K}(y_k - \hat{y}_k(s_k, u_k))^2, \tag{10.23}$$

where K is the total number of data pairs (s_k, u_k, y_k).

Table 10.1 shows the result of parameter estimation and the MSE for DS1 as an example. From Table 10.1, the goodness of fit of the proposed bivariate SRGMs improves compared to the existing SRGMs. Focusing on the bivariate SRGMs based on the CES-type testing-time function, and the estimated $\hat{\alpha}$ is above 0.5. Thus, we obtain that the effect of the testing-effort factor on software reliability growth is larger than the effect of the testing-time factor. Besides, we obtain the estimated $\hat{\alpha}$ of CES-DSS SRGM is very large. The result of parameter estimation for the CES-ISS SRGM is the most appropriate from the result of MSE.

Table 10.2 presents the result of goodness-of-fit comparison for all datasets. It can be seen that the bivariate SRGMs based on the CES-type testing-time function are more suitable than the existing SRGMs and bivariate SRGMs based on the Cobb-Douglas-type testing-time function for all datasets. As an example, the result of the software reliability assessment based on the CES-ISS SRGM of DS2 is as follows: The software reliability has been estimated to be $\hat{R}(1.0|21.0, 8736) = 0.41$, and the expected number of remaining faults has been estimated to be $\hat{M}(21.0, 8736) \approx 5$.

Table 10.3 shows the estimated elasticity of substitution for all datasets. It is worth nothing that the bold indicates that the goodness of fit for each data is the highest model. We adopt the results of the elasticity of substitution, which are bold. From Table 10.3, the testing-time and testing-effort factors are not easy to substitute for DS1 because the estimated value is relatively small. The software development environment from DS3 to DS5 is relatively easy to substitute between the testing-time and testing-effort factors.

TABLE 10.1 Result of parameter estimation of each model (DS1).

	$\hat{a}$	$\hat{b}$	$\hat{l}$	$\hat{\alpha}$	$\hat{\delta}$	MSE
EXP	513.23	0.053	–	–	–	222.00
CES-EXP	550.17	0.048	–	0.63	55.37	193.35
DSS	359.93	0.21	–	–	–	188.93
CES-DSS	359.90	0.21	–	0.92	68.05	188.89
ISS	355.09	0.21	0.22	–	–	96.65
CES-ISS	355.07	0.21	0.22	0.63	31.51	96.64

TABLE 10.2 Result of goodness-of-fit comparison using MSE.

	EXP	CD-EXP	CES-EXP	DSS	CD-DSS	CES-DSS	ISS	CD-ISS	CES-ISS
DS1	220.00	206.24	193.35	188.93	204.79	188.89	96.65	101.86	96.64
DS2	7.14	7.13	7.01	3.42	3.42	3.41	1.77	1.96	1.77
DS3	20.17	16.36	11.67	28.38	28.32	27.51	11.74	14.42	11.74
DS4	31.20	57.93	31.13	14.04	13.86	13.79	6.55	31.96	5.17
DS5	27.20	6.41	6.24	10.93	6.60	5.70	2.30	2.50	2.16
DS6	6.01	2.79	2.62	1.09	6.28	1.09	0.95	1.06	0.95

TABLE 10.3 Estimated elasticity of substitution.

	CES-EXP	CES-DSS	CES-ISS
DS1	0.018	0.014	**0.031**
DS2	0.27	0.17	**0.22**
DS3	**5**	0.086	0.15
DS4	0.11	0.072	**0.76**
DS5	1.41	2.04	**0.75**
DS6	0.85	0.27	**0.46**

10.3 Bivariate Weibull-type SRGMs and their application

In this chapter, we discuss the extension of bivariate SRGMs and their application to software development management.

10.3.1 Bivariate Weibull-type SRGMs based on testing-time functions

The univariate Weibull-type SRGM [7, 9, 13, 19] is defined as follows:

$$H(t) \equiv \gamma(t) = \left(\frac{t}{\rho}\right)^{\beta} \quad (0 < \beta < 1, \rho > 0), \tag{10.24}$$

where ρ represents the scale parameter, and β is a parameter, representing the degree of influence to software reliability growth. Note that Eq. (10.24) becomes $\gamma(0) = 0$ and $\gamma(\infty) = \infty$.

Applying the Cobb-Douglas-type testing-time function (10.10), the bivariate Weibull-type SRGM (CD-Weibull-type SRGM) [7, 9, 13, 19] is defined as

1. *CD-Weibull-type SRGM*

$$H(s,u) \equiv \gamma_{\mathrm{CD}}(s,u) = \left(\frac{s^{\alpha}u^{1-\alpha}}{\rho}\right)^{\beta}. \tag{10.25}$$

Similarly, applying the CES-type testing-time function in Eq. (10.11), the bivariate Weibull-type SRGM (CES-Weibull-type SRGM) [13] is defined as

2. *CES-Weibull-type SRGM*

$$H(s,u) \equiv \gamma_{\mathrm{CES}}(s,u) = \left[\frac{(\alpha s^{-\delta} + (1-\alpha)u^{-\delta})^{-\frac{1}{\delta}}}{\rho}\right]^{\beta}. \tag{10.26}$$

When Eq. (10.26) is formulated with $c = \rho^{-\beta}$, we obtain the following functional form:

$$\begin{aligned} H(s,u) \equiv \gamma_{\mathrm{CES}}(s,u) &= \left[\frac{(\alpha s^{-\delta} + (1-\alpha)u^{-\delta})^{-\frac{1}{\delta}}}{\rho}\right]^{\beta} \\ &= \rho^{-\beta}[\alpha s^{-\delta} + (1-\alpha)u^{-\delta}]^{-\frac{\beta}{\delta}} \\ &= c[\alpha s^{-\delta} + (1-\alpha)u^{-\delta}]^{-\frac{\beta}{\delta}}. \end{aligned} \tag{10.27}$$

Eq. (10.27) is equivalent to the CES-type production function considered as *economies of scale* in economics.

The economies of scale are a concept regarding whether the production efficiency is improved by increasing production factors. Besides, the economies of scale in Eq. (10.27) means the degree to which the fault-detection efficiency is promoted when the scale of investing the testing-time and testing-effort factors increases. β is the software reliability growth parameter, indicating the cost effectiveness per unit investing of software reliability growth factors. The cost effectiveness can be evaluated based on the behavior of β as follows:

$\beta > 1$: As investing in the software reliability growth factors increases, the amount of increase for the expected number of detected faults increases gradually, that is, it is more cost effective.

$\beta = 1$: Even if investing in the software reliability growth factors increases, the amount of increase for the expected number of detected faults is constant, that is, the cost effectiveness is constant.

$\beta < 1$: Even if investing in the software reliability growth factors increases, the amount of increase for the expected number of detected faults decreases gradually, that is, it is less cost effective.

10.3.2 Parameter estimation

Suppose we have (s_k, u_k, y_k) $(k = 0, 1, 2, \ldots, K)$, where s_k is the testing time, u_k is the amount of testing effort, y_k is the total number of faults detected during $[0, s_k]$, $[0, u_k]$, and K is the total number of the dataset.

Taking the natural logarithm to Eqs. (10.25), (10.27), the following regression equations are derived, respectively:

$$\ln \gamma_{\mathrm{CD}}(s,u) = -\beta \log \rho + \alpha\beta \log s + (1-\alpha)\beta \log u, \tag{10.28}$$

$$\begin{aligned} \ln \gamma_{\mathrm{CES}}(s,u) &= \ln c - \frac{\beta}{\delta} \ln(\alpha s^{-\delta} + (1-\alpha)u^{-\delta}) \\ &\cong \ln c + \alpha\beta \ln s + (1-\alpha)\beta \ln u - \frac{1}{2}\alpha\beta\delta(1-\alpha)[\ln s - \ln u]^2. \end{aligned} \tag{10.29}$$

It is worth nothing that Eq. (10.29) is obtained by conducting a quadratic Taylor expansion for second order since the CES-type testing-time function is non-linear. From Eq. (10.28), the following multiple regression equation defined by rewriting the coefficients is as follows:

$$\ln \gamma_{CD}(s,u) \equiv Y_k = a_0 + a_1 K_k + a_2 L_k + \varepsilon_k, \tag{10.30}$$

$$\begin{cases} Y_k = \ln y_k, \\ K_k = \ln s_k, \\ L_k = \ln u_k, \\ a_0 = -\beta \ln \rho, \\ a_1 = \alpha\beta, \\ a_2 = (1-\alpha)\beta, \end{cases} \tag{10.31}$$

where ε_k is the error term. From Eq. (10.29), the following multiple regression equation defined by rewriting the coefficients is as follows:

$$\ln \gamma_{CES}(s,u) \equiv Y_k = a_0 + a_1 K_k + a_2 L_k + a_3 [K_k - L_k]^2 + \varepsilon_k, \tag{10.32}$$

$$\begin{cases} Y_k = \ln y_k, \\ K_k = \ln s_k, \\ L_k = \ln u_k, \\ a_0 = \ln c, \\ a_1 = \alpha\beta, \\ a_2 = (1-\alpha)\beta, \\ a_3 = -\frac{1}{2}\delta\beta\alpha. \end{cases} \tag{10.33}$$

From Eqs. (10.30), (10.33), the following sum of squares of errors is formulated, respectively:

$$S_{CD}(a_0,a_1,a_2) = \sum_{k=1}^{K} \varepsilon_k^2 = \sum_{k=1}^{K} \{Y_k - (a_0 + a_1 K_k + a_2 L_k)\}^2, \tag{10.34}$$

$$S_{CES}(a_0,a_1,a_2,a_3) = \sum_{k=1}^{K} \varepsilon_k^2 = \sum_{k=1}^{K} \{Y_k - (a_0 + a_1 K_k + a_2 L_k + a_3 [K_k - L_k]^2)\}^2. \tag{10.35}$$

Solving the simultaneous equation as shown in Eq. (10.36), the following parameters of the bivariate CD-Weibull-type SRGM are estimated:

$$\frac{\partial S_{CD}}{\partial a_0} = \frac{\partial S_{CD}}{\partial a_1} = \frac{\partial S_{CD}}{\partial a_2} = 0. \tag{10.36}$$

$$\begin{cases} \hat{\alpha} = \dfrac{\hat{a}_1}{\hat{a}_1 + \hat{a}_2}, \\ \hat{\beta} = \hat{a}_1 + \hat{a}_2, \\ \hat{\rho} = \exp[-\dfrac{\hat{a}_0}{\hat{a}_1 + \hat{a}_2}]. \end{cases} \tag{10.37}$$

Solving the simultaneous equation as shown in Eq. (10.38), the following parameters of the bivariate CES-Weibull-type SRGM are estimated:

$$\frac{\partial S_{\text{CES}}}{\partial a_0} = \frac{\partial S_{\text{CES}}}{\partial a_1} = \frac{\partial S_{\text{CES}}}{\partial a_2} = \frac{\partial S_{\text{CES}}}{\partial a_3} = 0. \tag{10.38}$$

$$\begin{cases} \hat{\alpha} = \dfrac{\hat{a}_1}{\hat{a}_1 + \hat{a}_2}, \\ \hat{\beta} = \hat{a}_1 + \hat{a}_2, \\ \hat{c} = \exp[\hat{a}_0], \\ \hat{\delta} = -\dfrac{2\hat{a}_3}{\hat{a}_1 - \dfrac{\hat{a}_1{}^2}{\hat{a}_1 + \hat{a}_2}}. \end{cases} \tag{10.39}$$

10.3.3 Bivariate Weibull-type SRGMs under budget constraints

As an application problem for software development management, this study estimates the number of detectable faults under budget constraints. This study applies the bivariate Weibull-type SRGMs based on the Cobb-Douglas- and CES-type testing-time functions, and formulates their optimization problems.

The following optimization problem based on the bivariate CD-Weibull-type SRGM is formulated:

$$\begin{aligned} \max \ \gamma_{\text{CD}}(s,u) &= \left(\frac{s^{\alpha}u^{1-\alpha}}{\rho}\right)^{\beta} \\ &\text{s.t.} \ \ p_s s + p_u u \le I, s \ge 0, u \ge 0, \end{aligned} \tag{10.40}$$

where p_s and p_u represent the cost of testing-time and testing-effort factors, respectively. I represents the budget for the testing phase. From Eq. (10.40), the following optimal software reliability growth factors (s^*, u^*) are obtained using the Kuhn-Tucker condition.

$$(s^*, u^*) = \left(\frac{\alpha I}{p_s}, \frac{(1-\alpha)I}{p_u}\right). \tag{10.41}$$

Substituting Eq. (10.41) into Eq. (10.25), the bivariate CD-Weibull-type SRGM subject to budget constraint [7, 13, 19] is derived as follows:

1. *CD-Weibull-type SRGM subject to budget constraint*

$$H(s^*,u^*)=\left\{\frac{\left(\frac{\alpha I}{p_s}\right)^{\alpha}\left(\frac{(1-\alpha)I}{p_u}\right)^{1-\alpha}}{\rho}\right\}^{\beta}. \tag{10.42}$$

Similarly, the following optimization problem based on the bivariate CES-Weibull-type SRGM is formulated:

$$\begin{aligned}\max\ \gamma_{\text{CES}}(s,u)\ &=c(\alpha s^{-\delta}+(1-\alpha)u^{-\delta})^{-\frac{\beta}{\delta}}\\ &\text{s.t.}\ \ p_s s+p_u u\leq I, s\geq 0, u\geq 0.\end{aligned} \tag{10.43}$$

From Eq. (10.43), we obtain the following optimal software reliability growth factors (s^*, u^*) using the Kuhn-Tucker condition:

$$(s^*,u^*)=\left(\frac{I}{p_s\left[\left(\frac{p_u}{p_s}\right)^{\frac{\delta}{\delta-1}}\left(\frac{\alpha}{1-\alpha}\right)^{\frac{1}{\delta-1}}+1\right]},\frac{I}{p_u\left[\left(\frac{p_s}{p_u}\right)^{\frac{\delta}{\delta-1}}\left(\frac{1-\alpha}{\alpha}\right)^{\frac{1}{\delta-1}}+1\right]}\right). \tag{10.44}$$

Substituting Eq. (10.44) into Eq. (10.27), the bivariate CES-Weibull-type SRGM subject to budget constraint [13] is derived as follows:

2. *CES-Weibull-type SRGM subject to budget constraint*

$$H(s^*,u^*)=c\left[\alpha\left(\frac{I}{p_s\left[\left(\frac{p_u}{p_s}\right)^{\frac{\delta}{\delta-1}}\left(\frac{\alpha}{1-\alpha}\right)^{\frac{1}{\delta-1}}+1\right]}\right)^{-\delta}\right.$$

$$\left.+(1-\alpha)\left(\frac{I}{p_u\left[\left(\frac{p_s}{p_u}\right)^{\frac{\delta}{\delta-1}}\left(\frac{1-\alpha}{\alpha}\right)^{\frac{1}{\delta-1}}+1\right]}\right)^{-\delta}\right]^{-\frac{\beta}{\delta}}. \tag{10.45}$$

10.3.4 Numerical examples

We use the following actual datasets [7, 20] to illustrate numerical examples. The result of calculating has been obtained using R.

DS7: $(s_k, u_k, y_k)(k=1,2,\ldots,24; s_{24}=24, u_{24}=90.946, y_{24}=296)$,

DS8: $(s_k, u_k, y_k)(k=1,2,\ldots,21; s_{21}=21, u_{21}=91.981, y_{21}=212)$,

where s_k is the calendar time measured based on weeks as the testing-time factor, u_k is the test coverage (%) as the testing-effort factor, and y_k is the total number of faults detected during $[0, s_k]$, $[0, u_k]$. These datasets have been corrected in the testing phase for embedded software, which has been developed by software development engineers with about 7–10 years of development experience. The development scale is 1.972×10^5 (lines of code [LOC]) for DS7 and 1.630×10^5 (LOC) for DS8. Here, it is necessary to adjust the unit of the testing-time and detected fault count data as with the test coverage. Thus, they are adjusted as in Eqs. (10.46), (10.47).

$$\text{The execution rate of the specified testing period}: s \equiv \frac{s_k}{s_K}, \tag{10.46}$$

$$\text{The expected fault}-\text{detection rate}: y \equiv \frac{y_k}{y_K}, \tag{10.47}$$

where K represents the total number of datasets (s_k, u_k, y_k).

Tables 10.4 and 10.5 show the estimated parameters for the CD-Weibull- and CES-Weibull-type SRGMs, respectively. The estimated elasticity of substitution ê in Table 10.5 is derived using the estimated substitution parameter

TABLE 10.4 Estimated parameters for bivariate CD-Weibull-type SRGM [13].

Data	$\hat{a}_0$	$\hat{a}_1$	$\hat{a}_2$	$\hat{\alpha}$	$\hat{\beta}$	$\hat{\rho}$
DS7	0.089	0.17	0.79	0.18	0.96	0.91
DS8	0.051	0.28	0.67	0.29	0.95	0.95

TABLE 10.5 Estimated parameters for bivariate CES-Weibull-type SRGM [13].

Data	$\hat{a}_0$	$\hat{a}_1$	$\hat{a}_2$	$\hat{a}_3$	ĉ	$\hat{\alpha}$	$\hat{\beta}$	$\hat{\delta}$	ê
DS7	0.09	0.18	0.78	0.055	1.095	0.18	0.96	−0.76	4.17
DS8	0.058	0.40	0.56	−0.066	1.06	0.42	0.96	0.56	0.64

$\hat{\delta}$. From Table 10.5, we observe that the software development environment for DS7 is easier to substitute between software reliability growth factors than the software development environment for DS8, relatively, because the elasticity of substitution for DS7 is larger than the other one. Table 10.6 shows the result of goodness-of-fit comparison for the existing univariate Weibull-type SRGM and bivariate Weibull-type SRGMs using MSE. From each value of MSE, we obtain that the bivariate Weibull-type SRGMs have high goodness of fit. The bivariate CES-Weibull-type SRGM has the best performance.

Tables 10.7 and 10.8 show the result of sensitivity analysis for the CD-Weibull-type SRGM under budget constraint for DS7. The p_s, p_u, and I are given parameters, and we assume that I and p_s or p_u are constant in each table. In the case of increasing the cost for the testing-effort factor p_u

TABLE 10.6 Result of goodness-of-fit comparison using MSE [13].

Data	Univariate Weibull	CD-Weibull	CES-Weibull
DS7	1.15	0.54×10^{-3}	**0.27×10^{-4}**
DS8	1.18	0.41×10^{-2}	**0.61×10^{-4}**

TABLE 10.7 Result of sensitivity analysis (DS7, CD-Weibull-type SRGM under budget constraint, p_s and I: constant) [13].

p_s	p_u	I	s^*	u^*	$H(s^*, u^*)$	Faults
2	6	5	0.440	0.687	0.708	209.647
2	7	5	0.440	0.589	0.627	185.661
2	8	5	0.440	0.515	0.565	167.114
2	9	5	0.440	0.458	0.515	152.298
2	10	5	0.440	0.412	0.474	140.161
2	11	5	0.440	0.375	0.439	130.017
2	12	5	0.440	0.343	0.410	121.399
2	13	5	0.440	0.317	0.385	113.977
2	14	5	0.440	0.294	0.363	107.510
2	15	5	0.440	0.275	0.344	101.820

TABLE 10.8 Result of sensitivity analysis (DS7, CD-Weibull-type SRGM under budget constraint, p_u and I: constant) [13].

p_s	p_u	I	s^*	u^*	$H(s^*, u^*)$	Faults
6	5	5	0.147	0.824	0.680	201.215
7	5	5	0.126	0.824	0.662	196.066
8	5	5	0.110	0.824	0.648	191.712
9	5	5	0.098	0.824	0.635	187.952
10	5	5	0.088	0.824	0.624	184.651
11	5	5	0.080	0.824	0.614	181.715
12	5	5	0.073	0.824	0.605	179.075
13	5	5	0.068	0.824	0.597	176.681
14	5	5	0.063	0.824	0.590	174.492
15	5	5	0.059	0.824	0.583	172.480

(Table 10.7), the amount of the testing-effort factor decreases gradually. Note that the testing-time factor does not change although the testing-effort factor changes since the testing-time factor is unaffected from the cost for the testing-effort factor p_u (Eq. 10.41). Besides, the higher the cost for the testing-effort factor, the greater the number of detectable faults decreases. Thus, we obtain that the higher the cost for the testing-effort factor, the more the cost effective becomes small. It is worth nothing that software reliability growth is not maintained by the effective substitution since the number of detectable faults decreases. Besides, the amount of change for the total expected number of detectable faults gradually decreases since $\hat{\beta} = 0.96$.

Next, we focus on Table 10.8 in the case of increasing the cost for the testing-time factor p_s. It is seen that the testing-time factor decreases gradually. The testing-effort factor does not change although the testing-time factor changes since the testing-effort factor is unaffected from the cost for the testing-time factor p_s (Eq. 10.41). We obtain that when the cost for the testing-time factor increases, the number of detectable faults decreases. Thus, the higher the cost for the testing-time factor, the more the cost effective becomes small. Besides, software reliability growth is not maintained and the amount of change for the total expected number of detectable faults gradually decreases as with Table 10.7.

Tables 10.9 and 10.10 show the result of sensitivity analysis for the CES-Weibull-type SRGM under budget constraint for DS7. Similar to

TABLE 10.9 Result of sensitivity analysis (DS7, CES-Weibull-type SRGM under budget constraint, p_s and I: constant) [13].

p_s	p_u	I	s^*	u^*	$H(s^*, u^*)$	Faults
2	6	5	0.527	0.658	1.700	503.192
2	7	5	0.500	0.571	1.919	568.058
2	8	5	0.478	0.506	2.131	630.915
2	9	5	0.458	0.454	2.338	692.038
2	10	5	0.442	0.412	2.539	751.638
2	11	5	0.427	0.377	2.736	809.881
2	12	5	0.414	0.348	2.929	866.901
2	13	5	0.402	0.323	3.118	922.808
2	14	5	0.391	0.301	3.303	977.695
2	15	5	0.381	0.282	3.485	1031.641

TABLE 10.10 Result of sensitivity analysis (DS7, CES-Weibull-type SRGM under budget constraint, p_u and I: constant) [13].

p_s	p_u	I	s^*	u^*	$H(s^*, u^*)$	Faults
6	5	5	0.264	0.683	1.793	530.781
7	5	5	0.237	0.668	1.846	546.341
8	5	5	0.215	0.655	1.893	560.400
9	5	5	0.198	0.644	1.937	573.282
10	5	5	0.183	0.633	1.977	585.213
11	5	5	0.171	0.624	2.015	596.356
12	5	5	0.161	0.615	2.050	606.835
13	5	5	0.151	0.606	2.084	616.744
14	5	5	0.143	0.599	2.115	626.158
15	5	5	0.136	0.592	2.146	635.138

Tables 10.9 and 10.10, p_s, p_u, and I are given parameters, and I and p_s or p_u are constant in each table. From Table 10.9, the higher the cost of testing-effort factor p_u, the greater the number of detectable faults increases. We note that the amount of change for the detectable faults decreases since $\hat{\beta} = 0.96$. Thus,

the higher the cost for the testing-effort factor, the more the cost effective becomes small. However, software reliability growth is maintained by the effective substitution because the number of detectable faults increases.

Finally, we consider Table 10.10 in the case of increasing the cost for the testing-time factor p_s. Table 10.10 shows that the higher the cost of testing-time factor p_s, the greater the number of detectable faults increases. Besides, the amount of change for the detectable faults is smaller than the former shown in Table 10.9, and it decreases gradually. Thus, we obtain that the higher the cost for the testing-time factor, the more the cost effective becomes small. However, software reliability growth is maintained by the effective substitution as with Table 10.9.

10.4 Conclusions

We presented the extension of the existing SRGMs using two different testing-time functions. First, the bivariate EXP, DSS, and ISS SRGMs based on the Cobb-Douglas- and CES-type testing-time functions were developed. Besides, we extended software reliability assessment measures using their models and showed numerical examples. We developed bivariate Weibull-type SRGMs based on the Cobb-Douglas- and CES-type testing-time functions, based on the same extension method. As a further extension, the bivariate Weibull-type SRGMs under budget constraints have also been discussed in terms of software development management.

It is worth noting that the evaluation method for software development management with economic significance has been proposed. We introduced the substitution parameter as a new parameter and estimated the elasticity of substitution although with relative evaluation. The CES-type testing-time function can change depending on the substitution parameter. Thus, it is possible to express the relationship between software reliability growth factors using appropriate functional forms according to the actual datasets. For data of testing-effort factor, we used the CPU time/execution time and test coverage data. There are many other types of testing-effort factors, such as executed test cases and man-hour. For future studies, we will apply other kinds of data related to the testing-effort factor.

There are several problems for software development management discussed based on the SRGMs, such as *optimal software release problem* [2, 11, 14] and *optimal testing-resource allocation problem* [2]. In the future study, the proposed bivariate SRGMs will be applied to these problems and contribute to the improvement of software development management technology.

Acknowledgments

This research was financially supported by JSPS KAKENHI, Grant No. 20K14983, and Technology of Japan and the Telecommunications Advancement Foundation in Japan.

References

[1] H. Pham, Software Reliability, Springer, Singapore, 2000.
[2] S. Yamada, Software Reliability Modeling—Fundamentals and Applications, Springer, Tokyo, 2013.
[3] A. Tickoo, P.K. Kapur, A.K. Shrivastava, S.K. Khatri, Testing effort based modeling to determine optimal release and patching time of software, Int. J. Syst. Assur. Eng. Manag. 7 (4) (2016) 427–434.
[4] P.K. Kapur, A.G. Aggarwal, G. Kaur, Simultaneous allocation of testing time and resources for a modular software, Int. J. Syst. Assur. Eng. Manag. 1 (4) (2010) 351–361.
[5] P.K. Kapur, H. Pham, A.G. Aggarwal, G. Kaur, Two-dimensional multi-release software reliability modeling and optimal release planning, IEEE Trans. Reliab. 61 (3) (2012) 758–768.
[6] S. Inoue, K. Fukuma, S. Yamada, Two-dimensional change-point modeling for software reliability assessment, Int. J. Reliab. Qual. Saf. Eng. 17 (6) (2010) 531–542.
[7] S. Inoue, K. Hotta, S. Yamada, On estimation of number of detectable software faults under budget constraint, Int. J. Math. Eng. Sci. 2 (3) (2017) 135–139.
[8] S. Inoue, S. Yamada, Two-dimensional software reliability assessment with testing-coverage, in: Proceedings of 2nd IEEE International Conference on Secure System Integration and Reliability Improvement, 2008, pp. 150–157.
[9] S. Inoue, S. Yamada, A bivariate Weibull-type software reliability growth model and its goodness-of-fit evaluation, J. Inf. Process. Soc. Jpn. 49 (8) (2008) 2851–2861 (Japanese).
[10] T. Ishii, T. Dohi, Testing-effort depending software reliability models based on two-dimensional NHPPs, J. Inf. Process. Soc. Jpn. J89-D (8) (2006) 1684–1694 (Japanese).
[11] S. Sakaguchi, Y. Minamino, S. Inoue, S. Yamada, Bivariate software reliability models depending on a Cobb-Douglas type testing-time function and their applications, in: Proceedings of the 24th ISSAT International Conference on Reliability and Quality in Design, 2018, pp. 611–615.
[12] Y. Minamino, S. Inoue, S. Yamada, Two-dimensional software reliability growth modeling based on a CES type time function, in: Proceedings of 2017 Infocom Technologies and Unmanned Systems (ICTUS'2017), 2017, pp. 120–125.
[13] Y. Minamino, S. Inoue, S. Yamada, Bivariate software reliability growth models under budget constraint for development management, Int. J. Math. Eng. Manag. Sci. 5 (1) (2020) 56–65.
[14] Y. Minamino, S. Sakaguchi, S. Inoue, S. Yamada, Two-dimensional NHPP models based on several testing-time functions and their applications, Int. J. Reliab. Qual. Saf. Eng. 26 (4) (2019). 1950018-1–1950018-14.
[15] A. Wood, Predicting software reliability, IEEE Comput. Mag. 11 (1996) 69–77.
[16] H. Pham, A generalized fault-detection software reliability model subject to random operating environments, Vietnam J. Comput. Sci. 3 (3) (2016) 145–150.
[17] M. Ohba, Software reliability analysis models, IBM J. Res. Dev. 28 (4) (1984) 428–443.
[18] M. Zhu, H. Pham, A software reliability model with time-dependent fault detection and fault removal, Vietnam J. Comput. Sci. 3 (2) (2016) 71–79.
[19] K. Hotta, S. Inoue, S. Yamada, Estimation of detectable number of software faults under budget constraint, in: Proceedings of the 22nd ISSAT International Conference on Reliability and Quality in Design, 2016, pp. 172–175.
[20] T. Fujiwara, S. Yamada, C0 coverage-measure and testing-domain metrics based on a software reliability growth model, Int. J. Reliab. Qual. Saf. Eng. 9 (4) (2002) 329–340.

Chapter 11

A semi-Markov model of a system working under uncertainty

R.K. Bhardwaj[a], Purnima Sonker[a], and Ravinder Singh[b]
[a]*Department of Statistics, Punjabi University Patiala, Patiala, Punjab, India,* [b]*Department of Statistics, Central University of Haryana, Narnaul, Haryana, India*

11.1 Introduction

In this chapter, we have developed a stochastic model for a standby system using semi-Markov processes and the regenerative point technique of probability theory. The system consists of two units, similar in all respects. It starts operation with one unit in operation and the other in cold-standby modes. The standby needs to be switched into operation on failure of the operating unit. A switching mechanism, called a switch, performs all such switching. A repairman, known as a server, is present in the system to accomplish all restorative tasks. The server is assumed to fail during a working period but not in the idle time. Posttreatment, it rejoins the system. The switch is also random, i.e., it may fail. If the switch and the unit fail at the same time, then priority is given to the repair of the switch. The different random durations of time (i.e., time to failure of server and that of the switch and the unit, time to repair or treatment, etc.) are independently distributed. The system performance is evaluated and expressions are derived for some system performance indices such as MTSF, availability, busy period, expected number of visits, etc. To solve the expressions, the Laplace transform method is used. A simulation study is also given to show the significance of theoretical results obtained in this chapter.

The stimulus for the study is the prevailing uncertainties in the operating environment of systems. Nowadays, technological changes have their impressions all around us, whether it is a tech enabled cell phone, a multirole war machine, a manufacturing plant, a power generation system, or a life-support system. A common feature of all such systems is the nonpredictable or uncertain nature of the working environment, which puts challenges on the system

System Assurances. https://doi.org/10.1016/B978-0-323-90240-3.00011-4

engineers and planners to develop reliable and cost-effective systems. In fact, the existing uncertainty in the working environment of a system can severely affect the economic health as well as reliability of such systems. Managing the accurate performance of the system is a tedious task. It is always advisable to develop a model before the actual system setup. There are two broad categories of approaches available to study the performance of systems with uncertainties: deterministic and probabilistic. Over the deterministic approaches, the semi-Markovian processes, which are the generalization of the Markovian processes, are much more capable in handling these uncertainties. So the effective way to develop a system model is the use of probabilistic methods. Therefore, stochastic modeling techniques are gaining more popularity.

The standby system models have been widely discussed in the literature. In a cold-standby system, the failure rate of a standby unit is zero or negligible in comparison to the operative unit. A two dissimilar-unit cold-standby system with the option of inspection, to decide about minor or major repair, is studied by Mokaddis et al. [1]. A two-component cold-standby system working under shocks is debated by Wu [2]. An alternative approach is used by Fathizadeh and Khorshidian [3] for reliability analysis of standby systems on the basis of matrix renewal function. The reliability of a parallel multicomponent and single cold-standby system is analyzed using a copula-based technique by Yongjin et al. [4]. A standby repairable system with the rest of the server between repairs is proposed by Aggarwal and Malik [5]. The semi-Markov processes are gaining popularity (see Refs. [6–8]). The pioneer work in this area was initiated by Levy [9], Smith [10], and Pyke [11]. The different issues of server failure [12,13], switch failure [14], and standby failures [15] are also discussed in the literature.

In this chapter, a probabilistic model is developed to study the effect of instability, of service facility and that of the switching mechanism of the cold standby, on the performance of the system.

11.2 Notations

Most of the notations used in Ref. [13] are utilized here. Some further notations are as follows:

Sh	the switching mechanism/switch is good
Sv	the server is good
p/q	probability that the switch is operative/failed
ShF_{ur}/ShF_{UR}	switch is under repair/under repair continuously from a previous state after failure
ShF_{wr}/ShF_{WR}	the failed switch waiting for repair/waiting for repair continuously from a previous state
SvF_{ut}/SvF_{UT}	the failed server is under treatment/under treatment continuously from a previous state
$z(t)/Z(t)$	pdf/cdf of failure time of the unit

***r*(*t*)/*R*(*t*)**	pdf/cdf of failure time of the server
***f*(*t*)/*F*(*t*)**	pdf/cdf of repair time of the failed unit
***h*(*t*)/*H*(*t*)**	pdf/cdf of repair time of the failed switch
***s*(*t*)/*S*(*t*)**	pdf/cdf of the treatment time of the server

11.3 Development of system model

11.3.1 States of system model

The following are the possible transition states of the system model (Fig. 11.1).

Regenerative states:

$$S_0 = (\mathrm{O, Cs, Sh, SG}),\quad S_1 = (\mathrm{O}, \mathrm{F}_{ur}, \mathrm{SG}),\quad S_2 = (\mathrm{F}_{wr}, Cs_w, \mathrm{ShF}_{ur})$$
$$S_3 = (\mathrm{O}, \mathrm{F}_{wr}, \mathrm{SvF}_{ut})$$

Nonregenerative states:

$$S_4 = (\mathrm{F}_{UR}, \mathrm{F}_{wr}),\quad S_5 = (\mathrm{F}_{wr}, \mathrm{F}_{WR}, \mathrm{SvF}_{ut}),\quad S_6 = (\mathrm{F}_{ur}, \mathrm{F}_{WR})$$

$$S_7 = (\mathrm{F}_{WR}, \mathrm{Cs}_w, \mathrm{ShF}_{wr}, \mathrm{SvF}_{ut}),\quad S_8 = (\mathrm{F}_{WR}, \mathrm{Cs}_w, \mathrm{ShF}_{ur}),$$

$$S_9 = (\mathrm{F}_{wr}, \mathrm{F}_{WR}, \mathrm{SvF}_{UT})$$

11.3.2 Transition probability

Simple probabilistic considerations yield the following expressions for the non-zero element:

$$p_{ij} = Q_{ij}(\infty) = \int_0^\infty q_{ij}(t)dt = \widetilde{Q}_{ij}(0) \tag{11.1}$$

We obtain

$$dQ_{0,1}(t) = pz(t)dt,\quad dQ_{0,2}(t) = qz(t)dt,$$

$$dQ_{1,0}(t) = \mathrm{f(t)}\overline{\mathrm{Z}}(\mathrm{t})\overline{\mathrm{R}}(\mathrm{t})\mathrm{dt},\quad dQ_{1,4}(t) = z(\mathrm{t})\overline{\mathrm{F}}(\mathrm{t})\overline{\mathrm{R}}(\mathrm{t})\mathrm{dt},$$

$$dQ_{1,3}(t) = r(\mathrm{t})\overline{\mathrm{Z}}(\mathrm{t})\overline{\mathrm{F}}(\mathrm{t})\mathrm{dt},\quad dQ_{2,1}(t) = h(t)\overline{\mathrm{R}}(\mathrm{t})dt,$$

$$dQ_{2,7}(t) = r(t)\overline{\mathrm{H}}(\mathrm{t})dt,\quad dQ_{3,1}(t) = s(t)\overline{\mathrm{Z}}(\mathrm{t})dt,$$

$$dQ_{3,9}(t) = z(t)\overline{\mathrm{S}}(\mathrm{t})dt,\quad dQ_{4,1}(t) = f(t)\overline{\mathrm{R}}(\mathrm{t})dt,$$

$$dQ_{4,5}(t) = r(t)\overline{\mathrm{F}}(\mathrm{t})dt,\quad dQ_{5,6}(t) = s(t)dt,$$

$$dQ_{6,1}(t) = f(t)\overline{\mathrm{R}}(\mathrm{t})dt,\quad dQ_{6,5}(t) = r(t)\overline{\mathrm{F}}(\mathrm{t})dt,$$

$$dQ_{7,8}(t) = s(t)dt,\quad dQ_{8,1}(t) = h(t)dt,$$

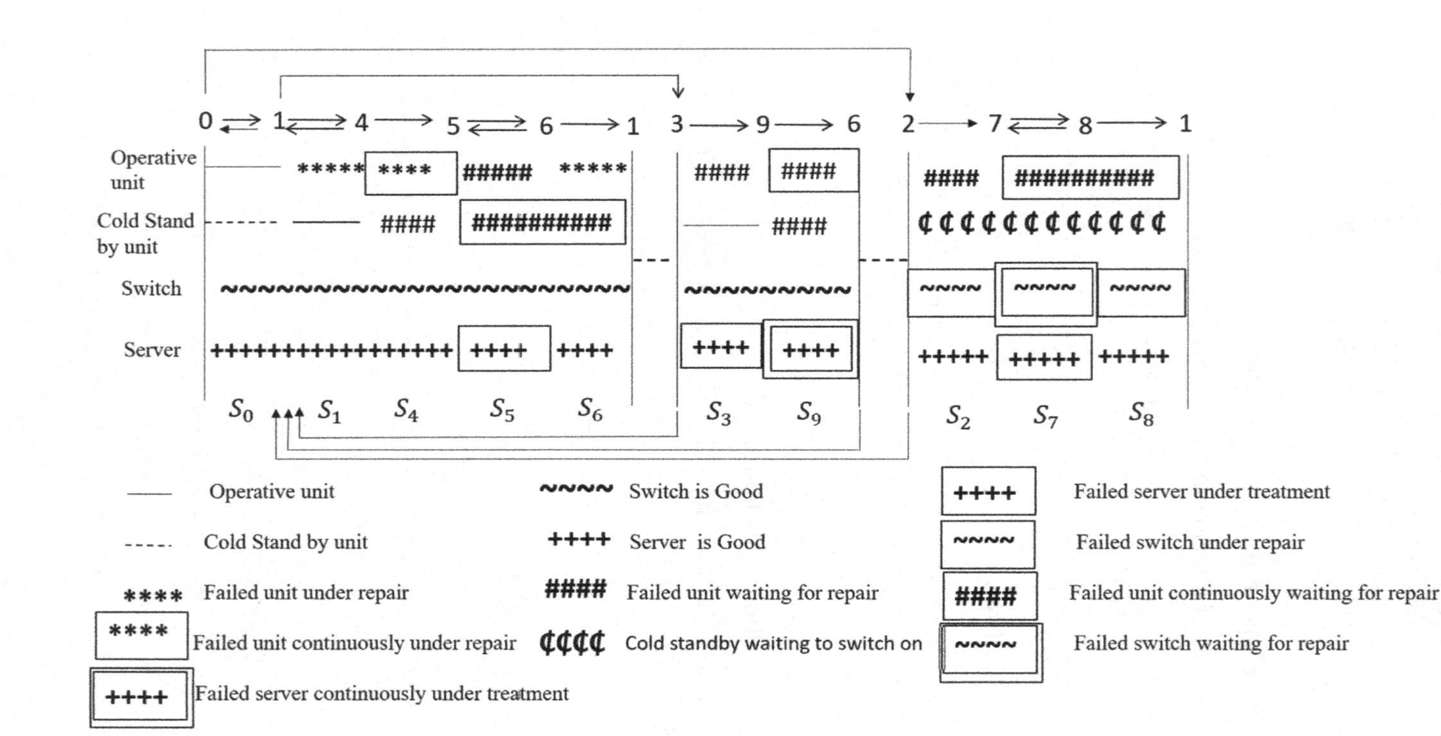

FIG. 11.1 State transition diagram.

$$dQ_{9,6}(t)=s(t)dt,\ \ dQ_{11.4}(t)=\mathrm{d}[\mathrm{Q}_{14}(t)[s]\mathrm{Q}_{41}(t)],$$

$$dQ_{11.4,(5,6)^n}(t)=\mathrm{d}[\mathrm{Q}_{14}(t)[s]\mathrm{Q}_{45}(t)[s]\mathrm{Q}_{56}(t)[s]\mathrm{Q}_{61}(t)],$$

$$dQ_{21.78}(t)=\mathrm{d}[\mathrm{Q}_{27}(t)[s]\mathrm{Q}_{78}(t)[s]\mathrm{Q}_{81}(t)],$$

$$dQ_{31.96}(t)=\mathrm{d}[\mathrm{Q}_{39}(t)[s]\mathrm{Q}_{96}(t)[s]\mathrm{Q}_{61}(t)],$$

$$dQ_{31.9,(6,5)^n}(t)=\mathrm{d}[\mathrm{Q}_{39}(t)[s]\mathrm{Q}_{96}(t)[s]\mathrm{Q}_{96}(t)[s]\mathrm{Q}_{65}(t)[s]\mathrm{Q}_{56}(t)[s]\mathrm{Q}_{61}(t)]$$

Taking the Laplace transformation of equations, we obtain

$$p_{01}=\int_0^\infty pz(t)dt,\ \ p_{02}=\int_0^\infty qz(t)dt,\ \ p_{10}=\int_0^\infty \mathrm{f(t)}\overline{\mathrm{Z}}(\mathrm{t})\overline{\mathrm{R}}(\mathrm{t})dt,$$

$$p_{13}=\int_0^\infty r(\mathrm{t})\overline{\mathrm{Z}}(\mathrm{t})\overline{\mathrm{F}}(\mathrm{t})dt,\ \ p_{14}=\int_0^\infty \mathrm{z(t)}\overline{\mathrm{F}}(\mathrm{t})\overline{\mathrm{R}}(\mathrm{t})dt,\ \ p_{21}=\int_0^\infty h(t)\overline{\mathrm{R}}(\mathrm{t})dt,$$

$$p_{27}=\int_0^\infty r(t)\overline{\mathrm{H}}(\mathrm{t})dt,\ \ p_{31}=\int_0^\infty \mathrm{s(t)}\overline{\mathrm{Z}}(\mathrm{t})dt,\ \ p_{39}=\int_0^\infty z(t)\overline{\mathrm{S}}(\mathrm{t})dt,$$

$$p_{41}=\int_0^\infty f(t)\overline{\mathrm{R}}(\mathrm{t})dt,\ \ p_{45}=\int_0^\infty r(t)\overline{\mathrm{F}}(\mathrm{t})dt,\ \ p_{56}=\int_0^\infty \mathrm{s(t)}dt,$$

$$p_{61}=\int_0^\infty f(t)\overline{\mathrm{R}}(\mathrm{t})dt,\ \ p_{65}=\int_0^\infty f(t)\overline{\mathrm{R}}(\mathrm{t})dt,\ \ p_{78}=\int_0^\infty \mathrm{s(t)}dt,$$

$$p_{81}=\int_0^\infty \mathrm{h(t)}dt,\ \ p_{96}=\int_0^\infty \mathrm{s(t)}dt,\ \ p_{11.4}=p_{14}[c]\mathrm{p}_{41},$$

$$p_{11.4,(5,6)^n}=p_{14}[c]\mathrm{p}_{45}[c]\mathrm{p}_{56}[c]\mathrm{p}_{61},\ \ p_{21.78}=p_{27}[c]\mathrm{p}_{78}[c]\mathrm{p}_{81},\ \ p_{31.96}=p_{39}[c]p_{96}[c]\mathrm{p}_{61},$$

$$p_{31.9,(6,5)^n}=p_{39}[c]\mathrm{p}_{96}[c]\mathrm{p}_{65}[c]p_{56}[c]\mathrm{p}_{61}$$

For these transition probabilities, it can be verified that

$$p_{01}+p_{02}=p_{10}+p_{13}+p_{14}=p_{21}+p_{27}=p_{31}+p_{39}=p_{41}+p_{45}=p_{56}=p_{61}+p_{65}$$
$$=p_{78}=p_{81}=p_{96}=p_{10}+p_{13}+p_{11.4}+p_{11.4,(5,6)^n}=p_{21}+p_{21.78}$$
$$=p_{31}+p_{31.96}+p_{31.9,(6,5)^n}=1$$

11.3.3 Mean sojourn times

The mean Sojourn times μ_i in state S_i are given by

$$\mu_i = E(t) = \int_0^\infty P(T > t)dt$$

$$\mu_0 = \int_0^\infty \overline{Z}(t)dt, \quad \mu_1 = \int_0^\infty \overline{F}(t)\overline{Z}(t)\overline{R}(t)dt, \tag{11.2}$$

$$\mu_2 = \int_0^\infty H(t)\overline{R}(t)dt, \quad \mu_3 = \int_0^\infty \overline{S}(t)\overline{Z}(t)dt$$

11.4 Performance measures

11.4.1 Reliability and mean time-to-system failure

Let $\phi_i(t)$ be the cdf of the first passage time from the regenerative state S_i to a failed state. Regarding the failed state as absorbing state, we have the following recursive relations for $\phi_i(t)$:

$$\phi_i(t) = \sum_j Q_{i,j}(t)[s]\phi_j(t) + \sum_k Q_{i,k}(t), \quad i = 0,1,3 \tag{11.3}$$

Taking LST of Eq. (11.3) and solving for $\widetilde{\phi}_0(s)$, we have

$$R^*(s) = \frac{1 - \widetilde{\phi}_0(s)}{s} \tag{11.4}$$

The reliability R(t) can be obtained by taking the inverse Laplace transition of Eq. (11.4) and MTSF is given by

$$MTSF(t) = \lim_{s \to 0} R^*(s) = \lim_{s \to 0} \frac{1 - \widetilde{\phi}_0(s)}{s} = \frac{N_1}{D_1} \tag{11.5}$$

where

$$N_1 = \mu_0(1 - p_{13}p_{31}) + p_{01}(\mu_1 + p_{13}\mu_3)$$
$$D_1 = 1 - p_{13}p_{31} - p_{10}p_{01}$$

11.4.2 Steady-state availability

$M_i(t)$ is the probability that the system initially in state $S_i \in E$ is up at time t without visiting any other regenerative state, we have

$$M_0 = \int_0^\infty \overline{Z}(t)dt, \quad M_1 = \int_0^\infty \overline{F}(t)\overline{Z}(t)\overline{R}(t)dt, \quad M_3 = \int_0^\infty \overline{S}(t)\overline{Z}(t)dt$$

Let $A_i(t)$ be the probability that the system is in up-state at an instant 't' given that the system entered the regenerative state S_i at $t=0$. The recursive relations for $A_i(t)$ are as follows:

$$A_i(t) = M_i(t) + \sum_j q_{i,j}^{(n)}(t)[c]A_j(t), \quad i=0,1,2,3 \tag{11.6}$$

where S_j is any successive regenerative state to which the regenerative state S_i can transit through n transitions.

Taking LT of Eq. (11.6) and solving, the steady-state availability is given by

$$A_0 = \lim_{s\to 0} sA_0^*(s) = \frac{N_2}{D_2} \tag{11.7}$$

where

$$N_2 = \mu_0 p_{10} + \mu_1 + \mu_3 p_{13}$$
$$D_2 = \mu_0 p_{10} + \mu'_1 + \mu'_2 p_{10} p_{02} + p_{13}\mu'_3$$

11.4.3 Busy period analysis for the server

Let $B_i(t)$ be the probability that the server is busy in repair of the unit or switch at an instant t given that the system entered the regenerative state S_i at $t=0$. The recursive relations for $B_i(t)$ are as follows:

$$B_i(t) = W_i(t) + \sum_j q_{i,j}^{(n)}(t)[c]B_j(t), \quad i=0,1,2,3 \tag{11.8}$$

where S_j is any successive regenerative state to which the regenerative state S_i can transit through n transitions. $W_i(t)$ is the probability that the server is busy in state S_i due to repair of the unit up to time 't' without making any transition to any other regenerative state or returning to the same via one or more nonregenerative states

$$\begin{aligned}
W_1(t) &= \overline{F}(t)\overline{Z}(t)\overline{R}(t) + \left(z(t)\overline{F}(t)\overline{R}(t)(c)1\right)\overline{F}(t)\\
&\quad + \left(z(t)\overline{F}(t)\overline{R}(t)(c)r(t)\overline{F}(t)(c)1\right)\overline{S}(t)\\
&\quad + \left(z(t)\overline{F}(t)\overline{R}(t)(c)r(t)\overline{F}(t)(c)s(t)(c)1\right)\overline{F}(t)\\
W_2(t) &= \overline{H}(t)\overline{R}(t) + \left(r(t)\overline{H}(t)(c)1\right)\overline{S}(t) + \left(r(t)\overline{H}(t)(c)s(t)(c)1\right)\overline{H}(t)
\end{aligned}$$

Using LT of Eq. (11.8) and solving for $B_0^*(s)$, we have

$$B_0 = \lim_{s\to 0} sB_0^*(s) = \frac{N_3}{D_3} \tag{11.9}$$

where

$$N_3 = W_1{}^*(0) + p_{02}p_{10}W_1{}^*(0)$$
$$D_3 = \mu_0 p_{10} + \mu'_1 + \mu'_2 p_{10} p_{02} + p_{13}\mu'_3$$

11.4.4 Expected number of treatment given to the server

Let $T_i(t)$ be the expected number of treatments given to the server in (0, t] given that the system entered the regenerative state S_i at t = 0. The recursive relations for $T_i(t)$ are as follows:

$$\mathrm{T_j(t)} = \sum_{\mathrm{j}} \mathrm{Q_{i,j}(t)[s]}\left[\varphi_{\mathrm{j}} + \mathrm{T_i(t)}\right], \quad i=0,1,2,3 \tag{11.10}$$

where

$$\varphi_{\mathrm{j}} = \begin{cases} 1 & \text{if the server performs the task in state} \, \mathrm{S_j}. \\ 0 & \text{otherwise} \end{cases}$$

Using LT of Eq. (11.10) and solving for $\widetilde{T}_0(s) = (s)$, we obtain

$$T_0 = \lim_{s \to 0} s\widetilde{T}_0(s) = \frac{N_4}{D_4} \tag{11.11}$$

where

$$N_4 = p_{31}p_{13}$$
$$D_4 = \mu_0 p_{10} + \mu'_1 + \mu'_2 p_{10} p_{02} + p_{13}\mu'_3$$

11.4.5 Expected number of repairs given to the switch

Let $U_i(t)$ be the expected number of repairs given to the switch in (0, t] given that the system entered the regenerative state S_i at t = 0. The recursive relations for $U_i(t)$ are as follows:

$$\mathrm{U_i(t)} = \sum_{\mathrm{j}} \mathrm{Q_{i,j}(t)[s]}\left[\varphi_{\mathrm{j}} + \mathrm{U_j(t)}\right], \quad i=0,1,2,3 \tag{11.12}$$

where

$$\varphi_{\mathrm{j}} = \begin{cases} 1 & \text{if the server performs the task in state} \, \mathrm{S_j}. \\ 0 & \text{otherwise} \end{cases}$$

Using LT of Eq. (11.12) and solving for $\widetilde{U}_0(s)$, we obtain

$$U_0 = \lim_{s \to 0} s\widetilde{U}_0(s) = \frac{N_5}{D_5} \tag{11.13}$$

where

$$N_5 = p_{10}p_{02}$$
$$D_5 = \mu_0 p_{10} + \mu'_1 + \mu'_2 p_{10} p_{02} + p_{13}\mu'_3$$

11.4.6 Expected number of repairs given to the unit

Let $O_i(t)$ be the expected number of repairs given to the unit in (0, t] given that the system entered the regenerative state S_i at $t = 0$. The recursive relations for $O_i(t)$ are as follows:

$$\mathrm{O_i(t)} = \sum_{\mathrm{j}} \mathrm{Q_{i,j}(t)[s]}\left[\varphi_{\mathrm{j}} + \mathrm{O_j(\tau)}\right], \quad i = 0,1,2,3 \tag{11.14}$$

where

$$\varphi_{\mathrm{j}} = \begin{cases} 1 & \text{if the server performs the task in state} \mathrm{S_j}. \\ 0 & \text{otherwise} \end{cases}$$

Using LT of Eq. (11.14) and solving for $\widetilde{O}_0(s)$, we obtain

$$O_0 = \lim_{s \to 0} s\widetilde{O}_0(s) = \frac{N_6}{D_6} \tag{11.15}$$

where

$$N_6 = 1 - p_{31}p_{13}$$
$$D_6 = \mu_0 p_{10} + \mu'_1 + \mu'_2 p_{10} p_{02} + p_{13}\mu'_3$$

11.4.7 Expected number of server visits

Let $N_i(t)$ be the expected number of visits by the server in (0, t] given that the system entered the regenerative state S_i at t = 0. The recursive relations for $N_i(t)$ are as follows:

$$\mathrm{N_i(t)} = \sum_{\mathrm{j}} \mathrm{Q_{i,j}(t)[s]}\left[\varphi_{\mathrm{j}} + \mathrm{N_j(t)}\right], \quad i = 0,1,2,3 \tag{11.16}$$

where

$$\varphi_{\mathrm{j}} = \begin{cases} 1 & \text{if the server performs the task in state} \mathrm{S_j}. \\ 0 & \text{otherwise} \end{cases}$$

Using LT of Eq. (11.16) and solving for $\widetilde{N}_0$ (s), we obtain

$$N_0 = \lim_{s \to 0} s\widetilde{N}_0(s) = \frac{N_7}{D_7} \tag{11.17}$$

where

$$N_7 = p_{10} + p_{31}p_{13}$$
$$D_7 = \mu_0 p_{10} + \mu'_1 + \mu'_2 p_{10} p_{02} + p_{13}\mu'_3$$

11.4.8 Profit

The profit incurred to the system in time (0, t] is calculated by.

Profit = Total revenue – Total expenditures

$$P(t) = C_0 A_0(t) - \sum_{j=1}^{5} C_j Y_j(t)$$

where

$$Y_j(t) = \begin{cases} B_0; & \text{for} j = 1 \\ T_0; & \text{for} j = 2 \\ U_0; & \text{for} j = 3 \\ O_0; & \text{for} j = 4 \\ N_0; & \text{for} j = 5 \end{cases} \tag{11.18}$$

C_0 = Revenue per unit up time of the system.
C_1 = Cost per unit time for which server is busy.
C_2 = Cost per server treatment.
C_3 = Cost per repair of the switch.
C_4 = Cost per repair of the unit.
C_5 = Cost per server visit.
A_0, B_0, T_0, U_0, O_0, and N_0 are already defined.

11.5 Simulation study

In this section, we present a simulation study to highlight the significance of the theoretical results obtained. For that let us suppose that all the random variables follow exponential distributions with different density functions, i.e.,

$$\begin{aligned} z(t) &= \lambda e^{-\lambda x} \\ f(t) &= \alpha e^{-\alpha x} \\ r(t) &= \mu e^{-\mu x} \\ s(t) &= \beta e^{-\beta x} \\ h(t) &= \gamma e^{-\gamma x} \end{aligned}$$

Furthermore, we assign set values to the parameters, i.e., $\alpha = 0.5$, $\beta = 0.7$, $\gamma = 0.3$, $\mu = 0.2$, $p = 0.4$, $q = 0.6$. Fig. 11.2 presents the behavior of MTSF; similarly, Fig. 11.3 availability, and Fig. 11.4 presents profit analysis of the system model.

The performance of the system is established through simulations. The results are displayed graphically using Figs. 11.2–11.4. Fig. 11.2 graphically illustrates the trends of mean time to the system failure against the unit failure rate for different values of other parameters. It reveals that MTSF gradually decreases as the failure rate of the unit increases. The decrease in the value

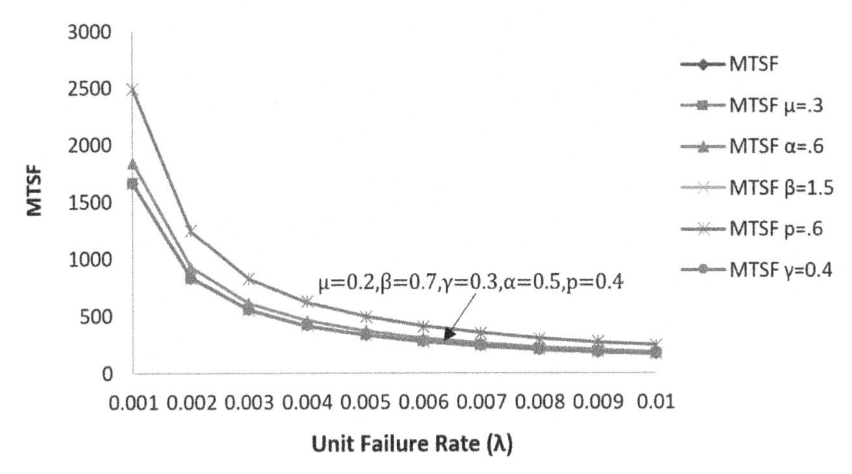

FIG. 11.2 MTSF behavior against failure rate of the unit.

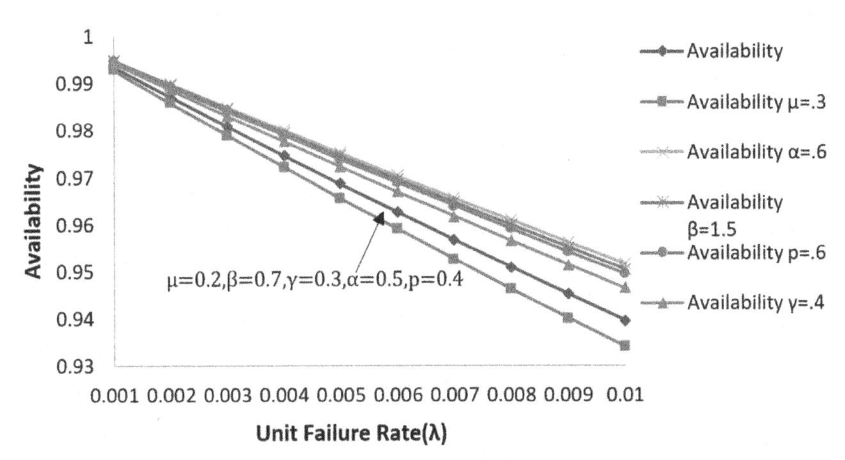

FIG. 11.3 Availability behavior against failure rate of the unit.

of MTSF is slow with p increasing from $p=0.4$ to $p=0.6$. This indicates that except for the value of p, the other parameters have the same effect on the MTSF.

Fig. 11.3 shows the behavior of the system availability graphically against the failure rate of the unit for different values of other parameters. The system availability trend remained strictly decreasing for increasing unit failure rates. The availability exhibits higher values for higher values of α. Availability shows almost the same results for higher values of p and β. This concludes that high availability of the system can be achieved with a working switch and operating server.

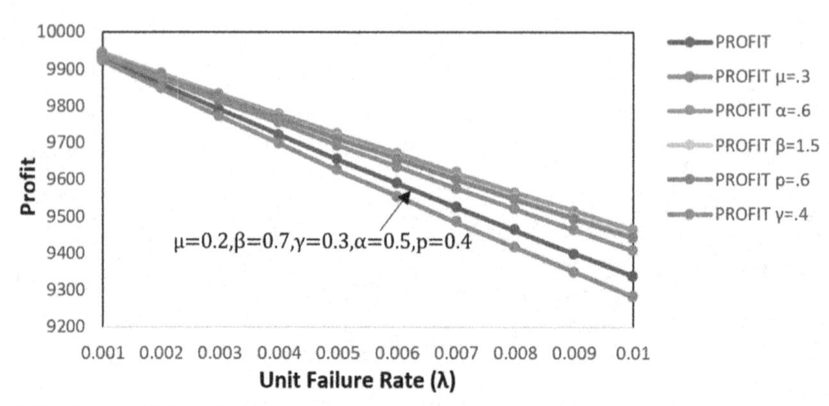

FIG. 11.4 Profit behavior against failure rate of the unit.

The trend in Fig. 11.4 shows the effect of different parameters on the system profit. The profit trends are similar to that of the availability. The profit shows declining trends as the unit failure rate rises but the trends revert with higher values of the repair and treatment rates.

11.6 Concluding remarks

The advancement of science and technology on the one side provides us the high-end systems to use but simultaneously, on the other side, it enhances their intrinsic complexities. These complexities indeed give way to unpredictable or uncertain situations. Therefore, stochastic processes, specifically the semi-Markov processes, provide us an efficient tool to develop the alternative conceptual models of such systems. The study presented in this chapter advocates the use of semi-Markov processes for investigating the performance of the systems working under uncertainties. The graphical trends of various performance measures have underlined the practical importance of the switching device and the service facility in a cold-standby system. To keep a cold-standby system reliable and profitable, the states of both the switching device and the service facility must be taken care of.

Acknowledgment

The authors are grateful to the reviewers for their valuable comments and suggestions.

References

[1] G.S. Mokaddis, M.L. Tawfek, S.A.M. Elhssia, Cost analysis of a two dissimilar-unit cold standby redundant system subject to inspection and two types of repair, Microelectron. Reliab. 37 (2) (Feb. 1997) 335–340, https://doi.org/10.1016/0026-2714(95)00216-2.

[2] Q. Wu, Reliability analysis of a cold standby system attacked by shocks, Appl. Math Comput. 218 (23) (2012) 11654–11673, https://doi.org/10.1016/j.amc.2012.05.051.
[3] M. Fathizadeh, K. Khorshidian, An alternative approach to reliability analysis of cold standby systems, Commun. Stat. Theory Methods 45 (21) (Nov. 2016) 6471–6480, https://doi.org/10.1080/03610926.2014.944660.
[4] Z. Yongjin, S. Youchao, L. Longbiao, Z. Ming, Copula-based reliability analysis for a parallel system with a cold standby, Commun. Stat. Theory Methods 47 (3) (Feb. 2018) 562–582, https://doi.org/10.1080/03610926.2017.1309432.
[5] C. Aggarwal, S.C. Malik, A standby repairable system with rest of server between repairs, J. Stat. Manag. Syst. (May 2020) 1–12, https://doi.org/10.1080/09720510.2020.1737382.
[6] N. Limnios, Reliability measures of semi-Markov systems with general state space, Methodol. Comput. Appl. Probab. 14 (4) (Jan. 2012) 895–917, https://doi.org/10.1007/s11009-011-9211-5.
[7] F. Grabski, Semi-Markov Processes: Applications in System Reliability and Maintenance, Elsevier Inc., 2014.
[8] Y. Zhang, J.D. Fricker, Multi-state semi-Markov modeling of recurrent events: estimating driver waiting time at semi-controlled crosswalks, Anal. Methods Accid. Res. 28 (Dec. 2020) 100131, https://doi.org/10.1016/j.amar.2020.100131.
[9] P. Levy, Processus semi-Markoviens, in: International Congress of Mathematicians-III, 1954, pp. 416–426.
[10] W.L. Smith, Regenerative stochastic processes, in: Proceedings of the Royal Society of London. Series A, Mathematical and Physical Sciences, vol. 232, Royal Society, 1955, pp. 6–31, https://doi.org/10.2307/99680.
[11] R. Pyke, Markov renewal processes: definitions and preliminary properties, Ann. Math. Stat. 32 (4) (1961) 1231–1242, https://doi.org/10.1214/aoms/1177704863.
[12] T. Guo, J. Cao, Reliability analysis of a two-unit redundant system with a replaceable repair facility, Microelectron. Reliab. 32 (9) (Sep. 1992) 1237–1240, https://doi.org/10.1016/0026-2714(92)90646-3.
[13] R.K. Bhardwaj, R. Singh, Stochastic model of a cold-stand by system with waiting for Arrival & Treatment of server, Am. J. Oper. Res. 6 (2016) 334–342, https://doi.org/10.4236/ajor.2016.64031.
[14] C. Shekhar, A. Kumar, S. Varshney, Load sharing redundant repairable systems with switching and reboot delay, Reliab. Eng. Syst. Saf. 193 (Jan. 2020) 106656, https://doi.org/10.1016/j.ress.2019.106656.
[15] K. Kaur, R.K. Bhardwaj, A standby system model under maximum redundancy time and gamma repairs, Int. J. Eng. Sci. Technol. 10 (3) (2018) 123–131, https://doi.org/10.21817/ijest/2018/v10i1/181003100.

Chapter 12

Design and evaluation of parallel-series IRM system

Sridhar Akiri[a], P. Sasikala[b], Pavan Kumar Subbara[b], and VSS Yadavalli[c]
[a]Department of Mathematics, G.I.S, GITAM (Deemed to be University), Visakhapatnam, Andhra Pradesh, India, [b]Department of Mathematics, G.S.S, GITAM (Deemed to be University), Bangalore, Karnataka, India, [c]Department of Industrial and Systems Engineering, Pretoria University, Pretoria, South Africa

12.1 Introduction

Reliability engineering has evolved widely since World War II with a crucial furnishing by the defense personnel, thus resulting in the development of more reliable services/products. Today, it has acquired a lot of significance among rehearsing engineers and researchers. By mixing reliability ideas in all periods of the product life cycle from the initial stages of proposal to later manufacture, it has become straightforward to develop compelling and cost-effective systems that perform better.

The efficiency of the system can be increased, depending on the resource restriction, which can lead to the determination of the ideal number of unuseful factors for each phase, only when the efficiency of every factor is identified. On the other hand, it can also be maximized depending on the source limitation to regulate the reliabilities of those factors in the system, only when the efficiency of every factor is identified.

An integrated efficiency model allows to calculate the number of factors, factor efficiencies, stage efficiencies, and the system efficiency where this issue takes into account both the undetermined factor efficiencies and the total number of factors in every stage for the given price restriction for increasing the system's efficiency. Till now in the history of integrated efficiency models are idealized by the only cost restriction and there is actual truth between worth and efficiency.

The current orientation of these pieces is to idealize a group of integrated efficiency models for unuseful systems with load and size with extra restrictions, among the common price restriction. We are going to negotiate the impact

System Assurances. https://doi.org/10.1016/B978-0-323-90240-3.00012-6

of load and size (apart from worth) as restrictions in idealizing the unuseful systems based on considering the designed mathematical conceptual function. Although the worth is directly related to increasing the system's efficiency, the analysis of this work is an attempt to describe the concealed fact that the effect of load and size as factors in idealizing the efficiency of an unuseful system gives a great start in the main domain of a careful study of a subject.

The integrated efficiency models for unuseful systems with worth, load, and size as restrictions for the mathematical function are considered for the proposed work. The author attempts to establish the naked truth that these models are handy with high application value, particularly in the case of IRM for redundant systems with a parallel-series configuration, perhaps the best suited whenever the worth of the system is very low.

For a given parallel-series system, the IRM alludes to determine the number of factors (z_j), factor efficiencies (r_j), phase efficiencies (Rp_j), and the system efficiency (R_{sys}). For determining the above arguments, the Lagrangean method is followed because of the quality of easy to understand and coherent. The IRM is developed using IRM concepts and solved by using the Lagrangean method, and an integer solution is derived by applying the dynamic programming approach.

12.1.1 Lagrangean way to understand equality constraints

Think about the problem carefully

$$\text{Minimize}\, z = f(z)$$

$$\text{Subject to}\, h(z) = 0$$

$$\text{where}\ z = (z_1, z_2, z_3, \ldots z_n)$$

$$h = (h_1, h_2, h_3, \ldots h_m)^T$$

The functions $f(z)$ and $h(z)$, $i = 1, 2, 3, \ldots m$ are assumed twice continuously differentiable and consider $P(z,\lambda) = f(z) - \lambda\, h(z)$.

The function 'P' is known as the Lagrangean function and the argument λ is known as the Lagrangean multiplier, which is used as a constant.

The conditions $\frac{\delta L}{\delta \lambda} = 0$ and $\frac{\delta L}{\delta z} = 0$ yield the similar requisite boundaries given above; thus, to generate the necessary conditions, we can use Lagrangean function straightaway. This means that the optimization of $f(z)$ subject to $g(z) = 0$ is the same as the optimization of the Lagrangean function $L(z,\lambda)$. Given the immovable condition (z_0, λ_0) for the method of the Lagrangean function $L(z,\lambda)$, HB, a bordered Hessian matrix, is evaluated at (z_0, λ_0). If z_0 is biggest, each polynomial $|\Delta| = 0$ must have a negative real root (n–m); otherwise, if it is positive, z_0 0 is lowest.

12.1.2 Lagrangean way to understand unequal factors

Consider a problem

$$\text{Maximize}\, z = f(z)$$

$$\text{Subject to}\, h(z_i) \leq 0$$

where $i = 1, 2, 3, \ldots m$ and

Nonnegativity restrictions $z_i \geq 0$.

The general notion behind expanding the Lagrangean technique is that if the unfactored optimum of $f(z)$ does not match all of the factors' conditions, the optimal factor should occur at the working solution space's boundary condition. This means that one or more of the 'm' components must be satisfied using equations. The steps in this procedure are outlined below:

Step 1:. Find the answer to Max $z = f(z)$. There is nothing to do if the resulting optimum meets all of the factors, because all of the factors are useless. Otherwise, set $k = 1$ and move to step 2.

Step 2:. Using the Lagrangean technique, operate any 'k' factors (i.e., convert them to equalities and idealize $f(z)$ with regard to the 'k' working factors. Exit the problem if the final solution is feasible in terms of the remaining elements and treat it as a local optimum. Otherwise, use a different set of 'k' factors and repeat the process. If no acceptable solution can be found after considering all sets of working factors 'k' at a time, move on to step 3.

Step 3:. If $k = m$, exit the question; it shows that there is no achievable solution. Otherwise, replace 'k' with $k + 1$ and go and repeat step 2.

Even when the problem is well behaved, the approach outlined above does not ensure the global ideal. This is an important fact that is sometimes overlooked when presenting the process outlined above (possess a unique optimum). Another important point is the implicit misconception that, for $p < q$, the optimum of $f(z)$ subject to 'p' equality constraints is always better than its optimum subject to 'q' equality constraints. Unfortunately, this is true, in general, only if the 'q' constraints from a subset of the constraints, 'p'.

The quantity of redundancies is treated as a real number in the Lagrangean multiplier Method. After obtaining a real-number solution, the integer solution is found using the branch-and-bound technique or dynamic programming approach. There are two types of efficiency evaluation and optimization approaches available: exact methods and repetitive methods. The exact method gives the solution analytically and produces a more detailed and exact solution. The Lagrangean multiplier method is used along with the Khun-Tucker conditions and examples for the exact methods are dynamic programming.

The Lagrangean multiplier method for the proposed mathematical function in the present thesis is carried out by using the MATLAB software and the results are presented.

12.2 Lagrangean procedure for formulation of the problem function: $c_j = b_j e^{\left[\frac{a_j}{(1-r_j)}\right]}$

The quantity of redundancies, which are integers, the factor efficiencies, which are real values, or a combination of both, are the determining parameters in most efficiency optimization situations. This method employs a differentiation of the determined parameters, all of which must be continuous. Consider there are 'n' statistically independent phases in series with z_j components in each that are statistically unrelated in the current study to create the integrated efficiency model based on the given mathematical function.

12.2.1 Model analysis

System's efficiency to the provided worth function

$$R_{sys} = \prod_{j=1}^{n} Rp_j \tag{12.1}$$

The worth coefficient of each unit in phase j is derived in the following relationship between worth and efficiency,

$$c_j = b_j e^{\left[\frac{a_j}{(1-r_j)}\right]} \tag{12.2}$$

Since the worth factor is linear in z_j,

$$\sum_{j=1}^{n} w_j \,.\, z_j \leq W_0 \tag{12.3a}$$

Similarly, the load and size factors are also linear in z_j,

$$\sum_{j=1}^{n} l_j \,.\, z_j \leq L_0 \tag{12.3b}$$

$$\sum_{j=1}^{n} s_j \,.\, z_j \leq S_0 \tag{12.3c}$$

Substituting (12.2) in (12.3),

$$\sum_{j=1}^{n} \left[b_j e^{\left[\frac{a_j}{(1-r_j)}\right]} \right] .\, z_j - W_0 \leq 0 \tag{12.4a}$$

$$\sum_{j=1}^{n}\left[q_j\, e^{\left[\frac{p_j}{(1-r_j)}\right]}\right].z_j - L_0 \leq 0 \tag{12.4b}$$

$$\sum_{j=1}^{n}\left[v_j\, e^{\left[\frac{u_j}{(1-r_j)}\right]}\right].z_j - S_0 \leq 0 \tag{12.4c}$$

Consider the problem

$$\textit{Maximize}\, R_{sys}(t) = 1 - \prod_{i=1}^{m}\left[1 - \prod_{j=1}^{n} R_{ij}\right] \tag{12.5}$$

Subject to the constraints

$$\sum_{j=1}^{n}\left[b_j\, e^{\left[\frac{a_j}{(1-r_j)}\right]}\right].z_j - W_0 \leq 0 \tag{12.6a}$$

$$\sum_{j=1}^{n}\left[q_j\, e^{\left[\frac{p_j}{(1-r_j)}\right]}\right].z_j - L_0 \leq 0 \tag{12.6b}$$

$$\sum_{j=1}^{n}\left[v_j\, e^{\left[\frac{u_j}{(1-r_j)}\right]}\right].z_j - S_0 \leq 0 \tag{12.6c}$$

Nonnegativity restrictions: $z_j \geq 0$

Through the relationship, the changed equations are

$$z_j = \frac{\ln(Rp_j)}{\ln(r_j)} \tag{12.7}$$

$$\text{Maximize}\, R_{sys}(t) = 1 - \prod_{i=1}^{n}\left[1 - \prod_{j=1}^{m} R_{ij}\right] \tag{12.8}$$

Subject to the constraints

$$\sum_{j=1}^{n}\left[b_j\, e^{\left[\frac{a_j}{(1-r_j)}\right]}.\frac{\ln(1-Rp_j)}{\ln(1-r_j)}\right]-W_0\leq 0 \tag{12.9a}$$

$$\sum_{j=1}^{n}\left[q_j\, e^{\left[\frac{p_j}{(1-r_j)}\right]}.\frac{\ln(1-Rp_j)}{\ln(1-r_j)}\right]-L_0\leq 0 \tag{12.9b}$$

$$\sum_{j=1}^{n}\left[v_j\, e^{\left[\frac{u_j}{(1-r_j)}\right]}.\frac{\ln(1-Rp_j)}{\ln(1-r_j)}\right]-S_0\leq 0 \tag{12.9c}$$

Positivity restrictions: $z_j \geq 0$

A Lagrangean function is defined by

$$F = R_{sys} + \lambda_1\left[\sum_{j=1}^{m}\left[b_j.e^{\left[\frac{a_j}{(1-r_j)}\right]}.\frac{\ln(1-Rp_j)}{\ln(1-r_j)}\right]-W_0\right]$$

$$+\lambda_2\left[\sum_{j=1}^{m}\left[q_j.e^{\left[\frac{p_j}{(1-r_j)}\right]}.\frac{\ln(1-Rp_j)}{\ln(1-r_j)}\right]-L_0\right]$$

$$+\lambda_3\left[\sum_{j=1}^{m}\left[v_j.e^{\left[\frac{u_j}{(1-r_j)}\right]}.\frac{\ln(1-Rp_j)}{\ln(1-r_j)}\right]-S_0\right]$$

$$= 0 \tag{12.10}$$

The ideal point can be gained by taking the Lagrangean function and differentiating it by Rp_j, r_j, λ_1, λ_2, and λ_3,

$$\frac{\partial L}{\partial r_j}=\lambda_1\left[\left[\left\{\sum_{j=1}^{m} ln\left(1-Rp_j\right)\left[b_j.e^{\left[\frac{a_j}{(1-r_j)}\right]}.\frac{a_j}{(1-r_j)}\frac{a_j}{(1-r_j)^2}\frac{1}{\ln(1-r_j)}\right]\right.\right.\right.$$

$$\left.\left.+\frac{b_j.e^{\left[\frac{a_j}{(1-r_j)}\right]}}{ln\ (1-r_j)^2}.\frac{1}{\ln(1-r_j)}\right\}\right]$$

$$+\lambda_2\left[\left[\left\{\sum_{j=1}^{m} ln\left(1-Rp_j\right)\left[q_j.e^{\left[\frac{p_j}{(1-r_j)}\right]}.\frac{p_j}{(1-r_j)}\frac{p_j}{(1-r_j)^2}\frac{1}{\ln(1-r_j)}\right]\right.\right.\right.$$

$$\left.\left.+\frac{q_j.e^{\left[\frac{p_j}{(1-r_j)}\right]}}{ln\ (1-r_j)^2}.\frac{1}{\ln(1-r_j)}\right\}\right]$$

$$+\lambda_3\left[\left[\left\{\sum_{j=1}^{m} ln\left(1-Rp_j\right)\left[v_j.e^{\left[\frac{u_j}{(1-r_j)}\right]}.\frac{u_j}{(1-r_j)}\frac{u_j}{(1-r_j)^2}\frac{1}{ln\ (1-r_j)}\right]\right.\right.\right.$$

$$\left.\left.+\frac{v_j.e^{\left[\frac{u_j}{(1-r_j)}\right]}}{ln\ (1-r_j)^2}.\frac{1}{ln\ (1-r_j)}\right\}\right]=0 \tag{12.11}$$

$$\frac{\partial L}{\partial Rp_j}=1+\lambda_1\left[\left\{\sum_{j=1}^{n}\frac{b_j.e^{\left[\frac{a_j}{(1-r_j)}\right]}}{ln\ (1-r_j)}.\frac{-1}{\ln(1-Rp_j)}\right\}\right]$$

$$+\lambda_2\left[\sum_{j=1}^{n}\left\{\frac{q_j.e^{\left[\frac{p_j}{(1-r_j)}\right]}}{(1-r_j)}\cdot\frac{-1}{\ln(1-Rp_j)}\right\}\right]$$

$$+\lambda_3\left[\sum_{j=1}^{n}\left\{\frac{v_j.e^{\left[\frac{u_j}{(1-r_j)}\right]}}{(1-r_j)}\cdot\frac{-1}{\ln(1-Rp_j)}\right\}\right]$$

$$=0 \tag{12.12}$$

$$\frac{\partial L}{\partial \lambda_1}=\left[b_j.e^{\left[\frac{a_j}{(1-r_j)}\right]}\frac{\ln(1-Rp_j)}{ln(1-r_j)}\right]-W_0=0 \tag{12.13}$$

$$\frac{\partial L}{\partial \lambda_2}=\left[q_j.e^{\left[\frac{p_j}{(1-r_j)}\right]}\frac{\ln(1-Rp_j)}{ln(1-r_j)}\right]-L_0=0 \tag{12.14}$$

$$\frac{\partial L}{\partial \lambda_3}=\left[v_j.e^{\left[\frac{u_j}{(1-r_j)}\right]}\frac{\ln(1-Rp_j)}{ln(1-r_j)}\right]-S_0=0 \tag{12.15}$$

Eqs. (12.11)–(12.15) can be rewritten after simplification as

$$\frac{\partial F}{\partial r_j}=\left[\ln(1-Rp_j)\left\{a_j.e^{\frac{b_j}{(1-r_j)}\cdot(b_j\ ln(1-r_j))+(1-r_j)}\right\}\right]$$

$$+\left[\ln(1-Rp_j)\left\{p_j.e^{\frac{q_j}{(1-r_j)}\cdot(q_j\ ln(1-r_j))+(1-r_j)}\right\}\right]$$

$$+\left[\ln(1-Rp_j)\left\{u_j.e^{\frac{v_j}{(1-r_j)}\cdot(v_j\ ln(1-r_j))+(1-r_j)}\right\}\right]$$

$$=0 \tag{12.16}$$

$$\frac{\partial F}{\partial Rp_j} = 1 - \lambda_1 \left[\sum_{j=1}^{n} \frac{b_j.e^{\left[\frac{a_j}{(1-r_j)}\right]}}{\ln(1-r_j)\ln(1-Rp_j)} \right] - \lambda_2 \left[\sum_{j=1}^{n} \frac{q_j.e^{\left[\frac{p_j}{(1-r_j)}\right]}}{\ln(1-r_j)\ln(1-Rp_j)} \right] - \lambda_3 \left[\sum_{j=1}^{n} \frac{v_j.e^{\left[\frac{u_j}{(1-r_j)}\right]}}{\ln(1-r_j)\ln(1-Rp_j)} \right] = 0 \tag{12.17}$$

$$\frac{\partial F}{\partial \lambda_1} = \left[\sum_{j=1}^{m} b_j.e^{\left[\frac{a_j}{(1-r_j)}\right]} \frac{\ln(1-Rp_j)}{\ln(1-r_j)} \right] - W_0 = 0 \tag{12.18}$$

$$\frac{\partial F}{\partial \lambda_2} = \left[\sum_{j=1}^{m} q_j.e^{\left[\frac{p_j}{(1-r_j)}\right]} \frac{\ln(1-Rp_j)}{\ln(1-r_j)} \right] - L_0 = 0 \tag{12.19}$$

$$\frac{\partial F}{\partial \lambda_1} = \left[\sum_{j=1}^{m} v_j.e^{\left[\frac{u_j}{(1-r_j)}\right]} \frac{\ln(1-Rp_j)}{\ln(1-r_j)} \right] - S_0 = 0 \tag{12.20}$$

12.3 Case problem

Examine the above instance of a mechanical system with three phases for which the factor efficiency is defined by the equation: $c_j = b_j\, e^{\left[\frac{a_j}{(1-r_j)}\right]}$.

For determining the optimum factor efficiency (r_j), stage efficiency (Rp_j), the number of factors in each phase (z_j), and the system efficiency (R_{sys}) the system worth Rs. 5000, load of the system 7450 kg, and system of size 5000 cm^3 are required.

The important information for constants are given in detail in the following table

	Worth constants		Load constants		Size constants	
Phase	b_j	a_j	q_j	p_j	v_j	u_j
1	100	0.7	100	0.9	100	0.7
2	120	0.8	150	0.8	120	0.8
3	90	0.5	100	0.7	90	0.5

The following tables indicate the factor efficiencies, phase efficiencies, number of factors in each stage, and system efficiency.

(A) Worth constraint details

The efficiency design related to worth is depicted in the following table:

Stage	r_j	Rp_j	z_j	w_j	$w_j. z_j$
01	0.4697	0.8626	3.2	374	1197
02	0.4481	0.9205	4.3	512	2201
03	0.6000	0.9900	5.1	314	1602
Total worth					5000

(B) Load constraint details

The equivalent results related to load are depicted in the following table:

Stage	r_j	Rp_j	z_j	l_j	$l_j. z_j$
01	0.4697	0.8626	3.2	548	1755
02	0.4481	0.9205	4.3	641	2757
03	0.6000	0.9900	5.0	576	2938
Total load					7450

(C) Size constraint details

The equivalent results related to size are depicted in the following table:

Stage	r_j	Rp_j	z_j	s_j	$s_j. z_j$
01	0.4697	0.8626	3.2	374	1197
02	0.4481	0.9205	4.3	512	2201
03	0.6000	0.9900	5.0	314	2602
Total size					5000

System efficiency = R_{sys} = 0.9998.

12.3.1 Efficiency design with z_j rounding off

The efficiency design is extracted by properly considering the values of z_j to be integers (by rounding off the value of z_j to the nearest integer) and the admissible results in relation to worth, load, and size are conferred in the given tables, furthermore stating the information by calculating the variation due to worth, load, size, and system efficiency (before and after rounding off z_j to the nearest integer).

(A) Efficiency design relating to worth with rounding off

Phase	r_j	Rp_j	z_j	w_j	$w_j.\ z_j$
01	0.4697	0.9209	4	374	1496
02	0.4481	0.9488	5	512	2560
03	0.6000	0.9897	5	314	1570
Total worth					5626
System efficiency (R_{sys})					0.999

Difference in total worth: 20.46%
Difference in system efficiency: 00.10%
The difference in efficiency design because of rounding off z_j values for the case problem is shown in the table with respect to load.

(B) Efficiency design relating to load with rounding off

Phase	r_j	Rp_j	z_j	l_j	$l_j.\ z_j$
01	0.4697	0.9209	4	548	2192
02	0.4481	0.9487	5	641	3205
03	0.6000	0.9897	5	576	2880
Total load					8277
System efficiency (R_{sys})					0.9999

Difference in total load: 20.52%
Difference in system efficiency: 00.10%
The difference in efficiency design because of rounding off z_j values for the case problem is depicted in the below table with respect to size.

(C) Efficiency design relating to size with rounding off

Phase	r_j	Rp_j	z_j	s_j	$s_j.\ z_j$
01	0.4697	0.9209	4	374	1496
02	0.4481	0.9487	5	512	2560
03	0.6000	0.9897	5	314	1570
Total size					5626
System efficiency (R_{sys})					0.9999

Difference in total load: 20.46%
Difference in system efficiency: 00.10%

Using the Lagrangean method which has definitely certain impediments as in the quantity of components needed at each stage, (z_j) shall be specified in real numbers which may be practically infeasible for implementation. The commonly used method of rounding off the value of (z_j) leads to the variations in worth, load, and size, thereby affecting the system reliability, which has a telling effect on the efficiency design of the model. This defect could be regarded for which the author proposes a substitute empirical implementation to get an integer solution by way of applying the dynamic programming method by getting the solutions which are derived from the Lagrangean method as parameters for the proposed dynamic programming method.

12.4 Dynamic programming approach

12.4.1 Introduction

The Lagrangean multiplier method allows an explanation to arrive at an ideal design more quickly than complex conclusions. Of course, this is done at the expense of investigating the number of components in each stage (z_j) as an actual possibility. As a result of rounding off z_j, the overall worth of each phase and the total worth of the system are all influenced by the values of phase efficiencies (Rp_j) and system efficiencies (R_{sys}). In the case issues, the changes in efficiency design caused by rounding off z_j are depicted. To overcome this disadvantage, use the dynamic programming approach.

The factor efficiency is used as one of the parameters in the dynamic programming approach to control the quantity of components in each phase, phase efficiencies, and system efficiency. However, the main disadvantage of the dynamic programming method is that it cannot be utilized to construct an integrated efficiency model in a straightforward manner, that is, without employing the factor efficiencies' parameters. As a result, the factor efficiencies from the previous approach, namely, the Lagrangean method, may be employed as parameters in the dynamic programming method to determine phase efficiencies, system efficiencies, stage worth, and system worth. The obligatory code is widely applicable and commonly used in programming in *C* language.

12.4.2 Dynamic way to approach programming

When a problem is considered with reference to its parameters varying over time, an approach, which includes the time element, has to be used for achieving a solution. In this scenario, the dynamic programming technique will be used. Dynamic programming is a mathematical technique developed by **R. Bellman and G.B. Dantzig**. Their contributions to these techniques were first published in 1950, and later **R. Bellman** published the advances in this approach in 1957.

In the beginning, this technique was termed as stochastic linear programming or linear programming technique dealing with uncertainty. Afterward, dynamic programming was developed to handle a large range of problems, including allocation, inventory, replacement, etc.

Dynamic programming is a mathematical approach that approaches the optimization of multistage decision questions. A multistage decision question could be split up into a sequential number of stages and steps, which can be executed in one or more than one way. The result of every step is a decision and the decisions for all the stages represent a decision policy. With each decision is associated some return in the form of worth or benefits. The objective in dynamic programming is to select a decision policy, so as to optimize the returns.

Some important concepts in dynamic programming are:

(A) **Phase:** A phase denotes a part of the problem as a whole, for which a conclusion can be made. At every phase there are numerous substitutes, and the finest choice of those is called the phase decisions, which may or may not be optimal for the phase but further advances to obtain the minimum decision policy.
(B) **State:** The constraint of the decision process at a phase is known as its state. The parameters that state the conditions of the decision process, i.e., describe the position of the system at a particular stage, are called state variables. The quantity of state variables should be as minute as is logically possible; otherwise, the decision process is more sophisticated.
(C) **Principle of optimality: R. Bellman's (1957)** principle of optimality states: "An optimal policy (A sequence of decisions) has the property that whatever the initial state and decisions are, the remaining decisions must constitute an optimal policy with regard to state resulting from the first decision." This process hints that a false decision taken at a phase will not prevent from taking minimum decisions for the remaining phases. The time factor can be considered here in two ways, i.e., the present

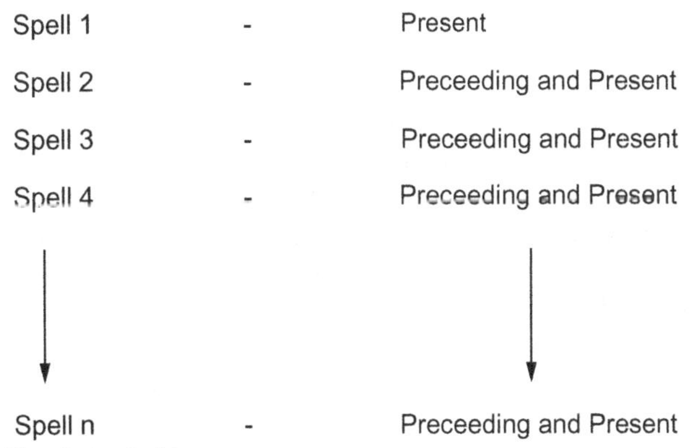

FIG. 12.1 Time factor-decision processes.

and its immediately preceding period. The decision processes in dynamic programming can be represented as shown in Fig. 12.1.

This approach is based on the mathematical interpretation of recursion, which is established in continued fractions. This decision can have a finite or an infinite number of stages. On the basis of the above, the inherent characteristics of dynamic programming are as follows:

1. Multistage processes are involved. In general, these stages refer to the time intervals, but in some special cases it may be treated as stages also.
2. At each stage, the problem can be described by a small number of variables. The original problem with a large number of variables is broken into a number of subtasks or phases which are interrelated and involve a lesser number of variables compared to other tasks in mathematical programming.
3. The effect of a decision at any phase is to change this set of variables into a similar set of variables for the next stage.

12.4.3 Basic features of the dynamic programming approach

1. Declarations about an issue are usually idealized at different stages rather than instantaneously in dynamic programming problems. This means that if dynamic programming is to be used to solve a problem, it must be divided into 'N' subproblems.
2. Dynamic programming approaches problems by allowing decisions to be made at every stage. At every phase, the state reflects and/or determines the set of all desirable possibilities.
3. A return function is associated with each decision at each phase, and it assesses the chance made at each decision in relation to the contribution that the decision is capable of contributing to the overall goal (Maximum or Minimum). Each decision made at each stage is connected with a return function, which evaluates the chance taken at each decision in terms of the contribution.
4. A quantitative relationship, known as a transition function, connects each phase '$\boldsymbol{n}$' of the whole decision process to its neighboring phases in some way. Depending on the nature of the situation, this function can be used to represent discrete or continuous numbers.
5. Given the current state, an optimal policy for the next phases in terms of a feasible input state is independent of the policy used in previous phases.
6. The method of determination always starts by finding the best optimal policy for each potential input state at the time.

To connect the optimal policy at stage '$\boldsymbol{n}$' to the $(\boldsymbol{n}-\mathbf{1})$ stage that follows, a recursive relationship is always used. This connection is established through

$$f_n{}^*(S_n) = opt_{d_n}\{r_n(S_n, d_n) + f_{n-1}{}^*(S_n{}^*d_n)\}$$

Any mathematical relationship between S_n and d_n is represented by the symbol. Addition, subtraction, multiplication, and root operations are all included in this category.

Using this recursive relation, the determination method shifts from one phase to the next, identifying an optimal policy in each phase until the final phase's optimal policy is discovered. The *'N'* factor decision vector can be retrieved by tracing back through the *'N'* phase transition function once the *'N'* phase optimal policy has been determined.

12.4.4 Classification of dynamic programming problems

The problems that could be handled by dynamic programming are branched as shown in Fig. 12.2. In a few problems, uncertainties are not involved. The probabilistic problems, on the other hand, generally involve acts of nature, and each conclusion has a probability associated with it. Similarly, problems can be divided into subcategories based on the number of declarations to be made or the number of phases (either limited or endless).

12.4.5 Computational procedure in dynamic programming

In general, depending on whether the system is discrete or continuous, the solution of a recursive equation can necessitate two sorts of computations. In the first scenario, each phase is followed by a tabular computational strategy. Each table's number of rows represents the number of possible state values, whereas the number of columns represents the number of decisions.

In the case of a continuous system, the best judgments are made at each phase utilizing traditional techniques such as differentiation, etc.

12.4.6 Computation in the forward and backward directions

If the recursive equations in a dynamic programming problem are solved in the manner given below, the problem will be solved.

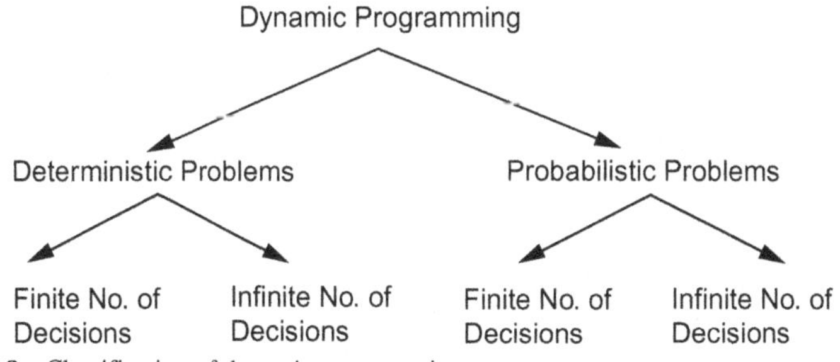

FIG. 12.2 Classification of dynamic programming.

$$f_1 \rightarrow f_2 \rightarrow, \ldots \rightarrow f_n$$

i.e., first find f_1, then f_2, etc., where f_1 denotes the best return for stage 1 and all preceding stages, the calculation is referred to as a forward computing technique. Recursive equations, on the other hand, can be constructed in a variety of ways and solved in any sequence.

$$f_n \rightarrow f_{n-1} \rightarrow, \ldots \rightarrow f_1$$

i.e., first f_n, then f_{n-1}, etc. Such type of computation is termed as "backward computational procedure." In forward computation, stage 1 is to be solved first and then stage 2 and so on, and in the backward computation the nth (Last) stage is to be solved first and then $(n - 1)^{th}$, etc. In practice, backward computation is more convenient.

12.4.7 Development of an optimal decision policy using dynamic programming

The important characteristics of a multistage system are as follows:

1. There are exactly 'N' points at which a decision must be made.
2. If started at stage 1, then nothing affects an optimal decision except the knowledge of the state of the system at stage 1 and the choice of the decision variable.
3. Stage 2 only affects the decisions at stage 1; the choice at stage 2 is governed only by the state of the system at stage 2 and the restrictions on the decision variable, etc., to stage n.

To begin, assume that for each potential state at phase 1, the optimal policy (the set of decisions that would lead to an optimal value of our returned function) is known. True, whatever happens in phases 2, 3, ..., N has no effect on the optimal decision taken at phase 1 corresponding to a particular state S_1 (since they precede phase 1). Assume that phase 1 and phase 2 are linked (Fig. 12.3).

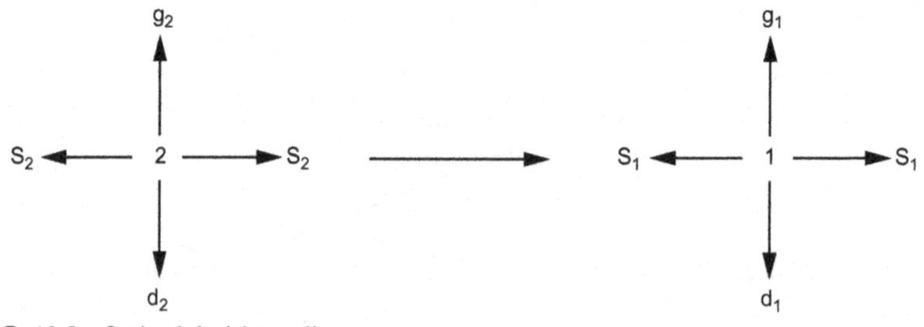

FIG. 12.3 Optimal decision policy.

12.5 The integrated efficiency model by using the dynamic programming method for the function $c_j = b_j e^{\left[\frac{a_j}{(1-r_j)}\right]}$

12.5.1 Dynamic programming solution

By inheriting the technique outlined in Section 12.3, integrated efficiency models for unuseful systems with numerous components for the mathematical function under discussion can be created. The required program is written in the powerful C language, with inputs derived from the Lagrangean method, and the number of factors and stage efficiency are determined using dynamic programming methods. The number of factors was previously a real number, but it has now been converted to an integer. The end outcome was achieved in two stages, each with a high efficiency.

(A) Stage 1 of the dynamic programming table

Number of factors	Phase reliability
z_j	Rp_j
01	0.4697
02	0.7187
03	0.8508
04	0.9209
05	0.9580
06	0.9777

(B) Dynamic programming table—Stage 2

Number of factors	Phase reliability							
z_j	Rp_j							
04	0.3812	0.4997	0.3906					
05	0.4126	0.5916	0.5978	0.4237				
06	0.4292	0.6403	0.7070	0.6484	0.4456			
07	0.4381	0.6661	0.7660	0.7675	0.6818	0.4564		
08	0.4428	0.6798	0.7968	0.8480	0.8071	0.6903	0.6983	0.4623
09	0.4452	0.6871	0.8132	0.8643	0.8736	0.8267	0.7074	0.4656
10	0.4465	0.6910	0.8219	0.8820	0.9487	0.8948	0.8375	0.7124

(C) Dynamic programming table—Stage 3

Number of factors	Phase reliability							
z_j	Rp_j							
09	0.5089	0.6447	0.6575	0.5798	0.4897	0.3249	0.2100	–
10	0.5241	0.7123	0.7137	0.6850	0.5858	0.4972	0.3260	0.2102
11	0.5692	0.7738	0.7816	0.7464	0.7521	0.5948	0.4988	0.3263
12	0.5584	0.7969	0.8124	0.8220	0.7521	0.7034	0.5967	0.4983
13	0.5859	0.7818	0.8822	0.8473	0.8310	0.7636	0.7057	0.5973
14	0.5772	0.8202	0.8656	0.9202	**0.8561**	0.8437	0.7661	0.7065

(D) Worth related reliability design

The maximum system efficiency, calculated using dynamic programming tables, is 0.999 with a total value of Rs. 5696, and the optimal values are listed below.

Phase	r_j	Rp_j	z_j	w_j	$w_j z_j$
01	0.4697	0.9209	4	374.34	1498
02	0.4481	0.9487	5	511.33	2557
03	0.6000	0.9897	5	328.09	1641
Total worth					5696
System efficiency (R_{sys})					0.999

(E) Load related reliability design

The maximum system reliability is 0.999 with a total load of 8430, according to the dynamic programming tables, and the optimal values are indicated below:

Phase	r_j	Rp_j	z_j	l_j	$l_j z_j$
01	0.4697	0.9209	4	545.83	2184
02	0.4481	0.9487	5	639.17	3196
03	0.6000	0.9897	5	609.98	3050
Total load					8430
System efficiency (R_{sys})					0.999

(F) Size related reliability design

The maximum system reliability is 0.999 with a total size of 5696, according to the dynamic programming tables, and the optimal values are indicated below.

Stage	r_j	Rp_j	z_j	s_j	s_jz_j
01	0.4697	0.9209	4	374.34	1498
02	0.4481	0.9487	5	511.33	2557
03	0.6000	0.9897	5	328.09	1641
Total size					5696
System efficiency (R_{sys})					0.999

12.6 Conclusions

An integrated reliability model for a parallel-series unuseful system with multiple factors idealizing the system efficiency is given in this thesis. The IRM refers to the determination of the number of factors (z_j), factor efficiencies (r_j), phase efficiency (Rp_j), and system efficiency (R_{sys}) and these are calculated using the Lagrangean multiplier method where the values are found to be in reals, and to have practical applicability. The dynamic way of programming method is applied to derive an integer solution by treating the inputs from the Lagrangean method.

The model is presented as a problem that infers the traditional analogy that system efficiency grows as the system's worth increases. When the value of the number of factors (z_j) is rounded off for want of an integer solution, a customary increase in worth is found but the approach of rounding off the values leads to variation in worth, load, and size, and to overcome this the dynamic way of programming method is applied. The results of the dynamic programming wherein the values regarding worth, load, and size are more or less approximately adhering to the concerned values in the case problem [when (z_j) is rounded off] but the dynamic programming approach is a proven scientific method and the advantage being in overcoming the variations due to the constraints. The IRM model thus established is very handy, particularly for practical applications when the reliability engineer would like to establish a parallel-series configuration IRM with redundancy. The suggested model is particularly useful for the reliability design engineer to generate quality and efficient goods where in a situation when the worth of the system is very low.

The writers also propose to establish similar IRMs with redundancy with a novel concept of restriction on the minimum and maximum values of factor reliabilities while idealizing the system reliability using any of the available Heuristic procedures in future studies.

Further reading

[1] K.K. Aggarwal, J.S. Gupta, On minimizing the expense of reliable systems, IEEE Trans. Reliab. R-24 (3) (1975).

[2] K.B. Mishra, Reliability Optimization of series—parallel system, IEEE Trans. Reliab. R-21 (4) (1972).

[3] K.B. Mishra, A method of solving redundancy optimization problems, IEEE Trans. Reliab. R-20 (3) (1971).

[4] L.T. Fan, T. Wang, Optimization of system reliability, IEEE Trans. Reliab. R-16 (1997).

[5] K.K. Aggarwal, K.B. Mishra, T.S. Gupta, Reliability evaluation: a comparative study of different techniques, Microelectron. Reliab. 14 (1975) 49–56.

[6] A. Mettas, Reliability allocation and optimization for complex systems, in: Proceedings Annual Reliability and Maintainability Symposium, Los Angeles, California, USA, 2000.

[7] W. Kuo, V. Rajendra Prasad, An annotated overview of system reliability optimization, IEEE Trans. Reliab. 49 (2) (2000) 176–187.

[8] Ushakov-Levitin-Lisnianski, Multi state system reliability: from theory to practice, in: Proceedings of the 3rd International Conference on Mathematical Methods in Reliability, Trondheim, Norway, 2002, pp. 635–638.

[9] V. Kumar, Optimized a set of integrated reliability models for redundant systems by appling Lagrangean and dynamic programming approaches, 2004. Ph.D. thesis, Sri Krishnadevaraya University, India.

[10] G. Sankaraiah, B.D. Sarma, C. Umashankar, V.S.S. Yadavalli, Design, modelling and optimizing of an integrated reliability redundant system, South African J. Ind. Eng. 22 (2011) 100–106.

[11] A. Sridhar, S. Pavankumar, Y. Raghunatha Reddy, G. Sankaraiah, C. Umashankar, The k out of n redundant IRM optimization with multiple constraints, Int. J. Res. Sci. Adv. Technol. 2 (2013) 1–6.

[12] P. Sasikala, A. Sridhar, S. Pavankumar, C. Umashankar, Optimization of IRM—parallel-series redundant system, Int. J. Eng. Res. Technol. (IJERT) 2 (2) (2013).

[13] S. Pavankumar, A. Sridhar, P. Sasikala, Optimization of integrated redundant reliability coherent systems—Heuristic method, J. Crit. Rev. 7 (05) (2020) 2480–2491.

[14] P. Sasikala, A. Sridhar, S. Pavankumar, Optimization of IRM series—parallel and parallel—series redundant system, J. Crit. Rev. 7 (17) (2020) 1801–1811.

[15] Sridhar Akiri, Sasikala, P, Pavankumar, S, (2021) Systems Reliability Engineering Modeling and Performance Improvement, De Gruyter, ISBN: 9783110617375, https://doi.org/10.1515/9783110617375.

Chapter 13

Modeling and availability assessment of smart building automation systems with multigoal maintenance

Yuriy Ponochovnyi[a], Vyacheslav Kharchenko[b], and Olga Morozova[b]
[a]*Department of Information Systems and Technologies, Poltava State Agrarian University, Poltava, Ukraine,* [b]*Department of Computer Systems, Networks and Cybersecurity, National Aerospace University "KhAI", Kharkiv, Ukraine*

13.1 Introduction

13.1.1 Motivation

The rapid development of virtualization technologies and the cloud computing environment creation have led to the development and implementation of new versions of IT-systems architectures, for which it is necessary to take into account, evaluate, and provide indicators of computer systems and Web Services dependability. These include various systems, in particular smart building automation system (BAS).

Automation of houses is performed for efficient and safe maintenance of all technical systems of the residential or industrial building. Modern devices and control modules control (block, regulate) working of all equipment used for the automation of houses. Software control is carried out for preset conditions using the centralized system [1].

The evolution of BAS is interrelated with the development of the concept of the Internet of Things (IoT) [2], [3] and includes the following stages:

- "Building Automation" (lighting, climate and music equipment are connected to the data network, begin in 2010);
- "Connected Building" (cloud technologies and advanced capabilities over the Internet, begin in 2017);
- "Adaptive Building" (data analysis, event forecasting, adequate reactions, begin in 2021);

System Assurances. https://doi.org/10.1016/B978-0-323-90240-3.00013-8

- "Intellectual Building" (artificial intelligence, virtual reality, digital consciousness, predict begin in 2025).

The set of "smart building" subsystems that perform information and control functions is considered as building automation system [3]. The architecture of the BAS [4] includes controller levels (on microprocessor or programmable logic device), databases, and communications (wired and/or wireless).

An important component of information and control systems (ICS) (these include BAS) quality is dependability. Under dependability, in the general case, the complex property is the ability of the system to provide the necessary services that can be justifiably trusted [5,6]. There are a few taxonomic models of dependability for different applications [5–8]. The dependability combines groups of primary properties (reliability, availability, maintainability, high confidence, functional safety, survivability, integrity, confidentiality) and secondary properties (for example, information and cybersecurity, authenticity).

The complex nature of external factors in relation to the system is the rationale for choosing the concept of the dependability. In this case, the property of ICS and BAS dependability will be considered as the ability of the system to perform specified functions (provide the required services), maintaining over time the level of its defined indicators within specified limits and operating conditions, as well as changing requirements, environmental parameters, and the manifestation of unspecified failures.

The indicators of the BAS are determined by the properties (including the properties of the dependability) of its hardware and software components [9]. The factors that affect the dependability of the system [10,11] are divided into internal (reliability) and external (cybersecurity) factors. The failures and errors of the programmed logic automated means as well as the manifestation of production faults are considered as internal system factors. External factors include vandalism and attacks on software vulnerabilities (hidden intervention, cyberattacks).

13.1.2 Work-related analysis

The concept of dependability as complex property joining reliability, availability, security, and safety is a chronological and consistent development of the proposed [5]. The problem of integrating complex property of dependability definition has been revealed in Refs. [12–15], the issues of dependability web-services modeling in Refs. [8,16], and researched system dependability models of critical infrastructures revealed in Refs. [7,17]. The models of software components of dependability systems were developed in Refs. [18,19].

Different mathematical apparatus are used to assess the dependability indicators such as Monte Carlo method [18], stochastic reward nets Petri model based on package SHARPE [14,20], and models of multistate systems [21]. Issues in ensuring the reliability and cybersecurity of smart buildings were considered in Refs. [22–24] and others.

The analysis of research results provides an opportunity to identify specific problems that need to be addressed in terms of development and application of mathematical apparatus for dependability systems with different goals and strategies for maintenance and updates.

13.1.3 Goals and contribution

The goal of this study is to assess the availability functions (as an indicator of dependability) of maintained BAS in terms of their evolution as a result of occurrence of hardware and software failures and unspecified software failures (as attacks on vulnerabilities). To achieve this goal the following tasks are performed: development and analysis the principles of multigoal maintenance concept (Section 13.2), development of multifragment models for assessing the BAS availability (Section 13.3), study of simulation results for interval values of partial input parameters (Section 13.4), and conclusions and suggestions for further research (Section 13.5).

An approach to research is based on three main ideas and steps. First, we suggest principles of multigoal maintenance, in particular, two-goal maintenance of BAS and similar systems considering reliability- and cybersecurity-oriented maintenance activities performed separately (only for the reliability assurance or for cybersecurity assurance) or jointly (for reliability and security assurance).

Second idea is based on general model such as structural graph for describing system behavior in case of different failures caused by physical (hardware) and design (software) faults and attacks on vulnerabilities, and implementation of different maintenance strategies (separate or joint maintenance for the assurance of reliability and/or cybersecurity).

The next step is developing and researching Markov's models according to general model and maintenance strategies.

Hence, the contributions of the research are the development of a set of maintenance strategies; the models of BAS availability considering different failures and strategies; and recommendations concerning the choice of parameters of system, its components, and maintenance strategies.

13.2 Concept of multigoal maintenance

To ensure the normalized indicators of smart building BAS availability, it is proposed to use the concept of dependable ICS with multigoal maintenance [25]. It is obtained by developing the paradigm of von Neumann as follows: to build dependable (reliable and secure) systems with not dependable enough components and multigoal maintenance strategies in changing the requirements and the environment parameters. The proposed concept is based on the following conditions:

(1) The development of von Neumann's concept of "building reliable systems from unreliable components" [26] in the direction of ensuring the dependability [7,27], i.e., building dependable ICS from insufficiently reliable and secure/safe components.
(2) The intended use of ICS in changing conditions, changing environmental parameters, and the occurrence of unspecified failures due to physical and design faults and vulnerabilities prevents or significantly complicates the direct application of the von Neumann paradigm for nonmaintenance ICS components and requires updating and restoring the system, and/or giving it the properties of resilience [15,20].
(3) The use of multigoal maintenance with the use of strategies combined on various indicators, in particular, the goal of the maintenance (confirmation of requirements, changes in parameters, changes in functions, prevention of changes in parameters), maintenance processes (verification, update, patching), properties that are maintained through maintenance (reliability, functional safety, cybersecurity).

13.2.1 Principles of multigoal maintenance

The proposed conception is based on the concept of multigoal maintenance strategy, the formation of such strategies, and their choice to ensure dependability. In the multiple aspects the maintenance strategy will be defined as a set of identified components in the form of

$$\mathrm{StrMaint} = \{\{\mathrm{GM_i}\}, \{\mathrm{TM}_i\}, \{\mathrm{ProcM}_i\}, \{\mathrm{PropM}_i\}, \{\mathrm{ParM}_i\}\},$$

where GM is the goal of maintenance, TM is the type of maintenance, ProcM is the process of maintenance, PropM is the properties for which maintenance is performed, and ParM is the parameters of maintenance.

For coverage of the conceptual principles proposed in the work, it was decided to limit by two procedures, namely, patching and proof testing.

Patching (ProcM2) involves the replacement of blocks/modules or sections of program code with detected faults or vulnerabilities. In some cases, corrective actions are taken to disable defective blocks or code sections. At the same time the system degrades but retains a partially operational state.

The key point is to take corrective actions after the fault is detected or manifested as a failure or error.

Proof testing (ProcM4) aims to detect hidden failures and vulnerabilities. Such testing is characterized by parameter D (the probability of detecting hidden failures/vulnerabilities), the value of which may vary and must be justified.

The issues of complexing these maintenance procedures are interrelated with the domain of ICS use (building automation systems) and models of availability assessment. These are discussed in the following sections.

13.2.2 Assessment of availability

The function of ICS availability with multigoal maintenance has the following nature of changes: at the first stage the system availability is reduced to a minimum, then it asymptotically tends to a constant value.

Thus, in the further analysis of the results it is necessary to take into account three indicators:

- the minimum value of the availability function $A_M$min;
- the value of the availability function in the constant mode $A_M$const;
- time interval of transition of the availability function to the constant mode $T_M$const.

In the case of using the proposed models to compare service strategies, additional result indicators were used (Fig. 13.1):

- reduction of the system availability level with maintenance at the initial stage of operation relative to the coefficient of similar system availability of hardware and software system without maintenance $-\Delta Ai$;
- gain in the system availability with maintenance relatively similar in terms of hardware and software system without maintenance $+\Delta Ai$;
- time, after which there is a gain in the system availability with maintenance Ti_{up}.

To calculate these additional result indicators for a group of models with multigoal maintenance requires development and study of at least two additional supporting models.

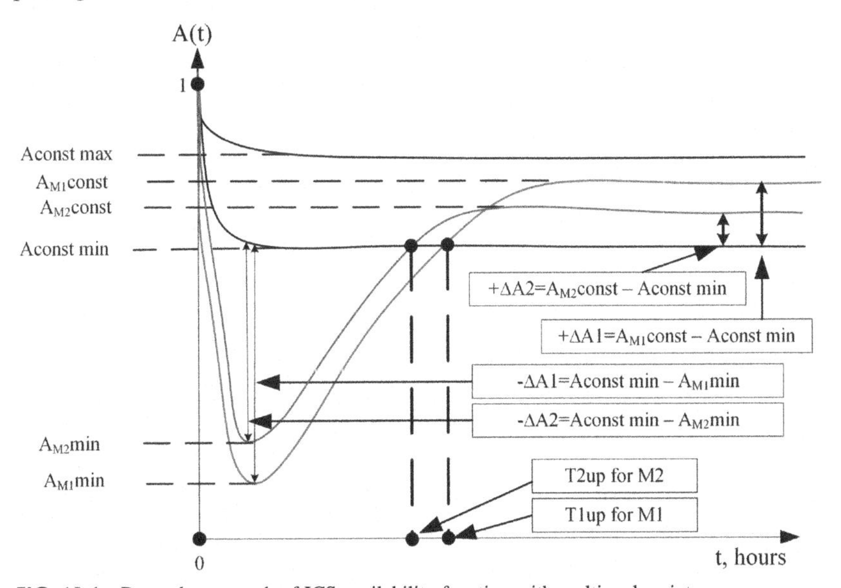

FIG. 13.1 Dependence graph of ICS availability function with multigoal maintenance.

13.3 Development of models

13.3.1 General model

The model of multigoal maintenance states in the context of patching and proof testing procedures is shown in Fig. 13.2. The oriented state graph of this model contains S_0 a fragment that includes operational states, S_1 a state that simulates patching, and S_2 a state that simulates the conduct of proof testing. Given the specifics of application domains, this model can be modified in the following scenarios:

(1) multifragment model, which describes only the patching procedures, after which the parameters of the ICS change (according to this scenario, the tests after the installation of patches are modeled not by a separate state S_1, but are taken into account when weighing the transition $S_1 \rightarrow S_0$*);
(2) one-fragment model, which describes only the procedures of proof testing, because their implementation does not change the parameters of the ICS;
(3) multifragment model, which includes both procedures (patching and proof testing), and these procedures are carried out sequentially: after installing the patch is performed proof testing of ICS components;
(4) multifragment model, which includes patching and proof testing procedures that are performed independently of each other.

This section will specify the options for building models in the fourth and second scenarios.

Model $M_{DEP}2.2^{(i)}$ describes BAS in the event of hardware and software failures caused by physical faults (Fp), design faults (Fd), and vulnerabilities of components (Fai). It summarizes the BAS models developed by studies in Refs. [10,11,28] and expands them with space of hardware faults for database server architecture components and wireless modules.

13.3.2 Availability model considering reliability and cybersecurity

Marked graph of the model $M_{DEP}2.2^{(1)}$ is shown in Fig. 13.3.

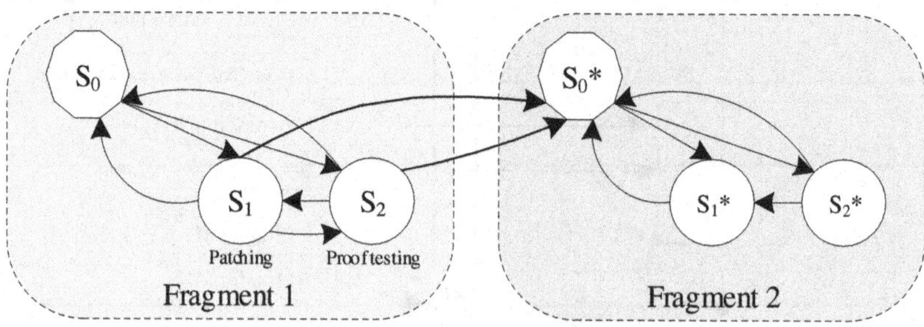

FIG. 13.2 Generalized model of patching and proof testing states of dependability ICS in the form of a two-fragment graph.

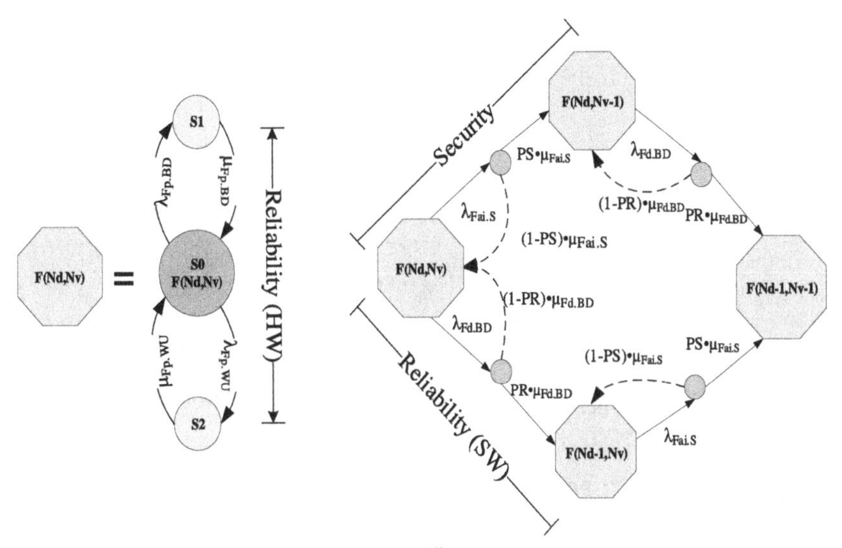

FIG. 13.3 Marked graph of the model $M_{DEP}2.2^{(1)}$ with the elimination of one SW fault and one vulnerability.

Dependability model $M_{DEP}2.2^{(1)}$ describes the operation of smart building BAS with a three-tier architecture:

(1) the level of sensors with controllers (sensor level);
(2) the level of wireless communications (wireless unit level);
(3) the level of data servers (BD Level).

BAS operates in the conditions of physical fault manifestation of components of the second and third levels (physical faults of components of the first level practically do not affect the system dependability as a whole), design faults of software components of the third level, and attacks on software components of the first level. On the marked graph these types of failures are illustrated by transitions with intensities $\lambda_{\text{Fp.BD}}$, $\lambda_{\text{Fp.WU}}$, $\lambda_{\text{Fd.BD}}$, $\lambda_{\text{Fai.S}}$.

The assumptions of the simplest flow of failures and recoveries that change the state of the system are accepted when developing the model. These assumptions are substantiated by studies in Refs. [12,14] and confirmed by the data of failures of real systems given in Refs. [11]. After the onset of hardware failure, the repair of the corresponding component is performed (in the graph of Fig. 13.3 it is shown by transitions with intensities $\mu_{\text{Fp.BD}}$, $\mu_{\text{Fp.WU}}$).

After the manifestation of the design fault, the system with the probability PR stops working until its complete elimination, or in the system with the probability (1-PR) the BAS returns to the previous working state due to the restart of the program. The same mechanism simulates the actions after an attack on the vulnerability of the component of the first level of the BAS architecture: with the probability PS the vulnerability will be eliminated, with the probability

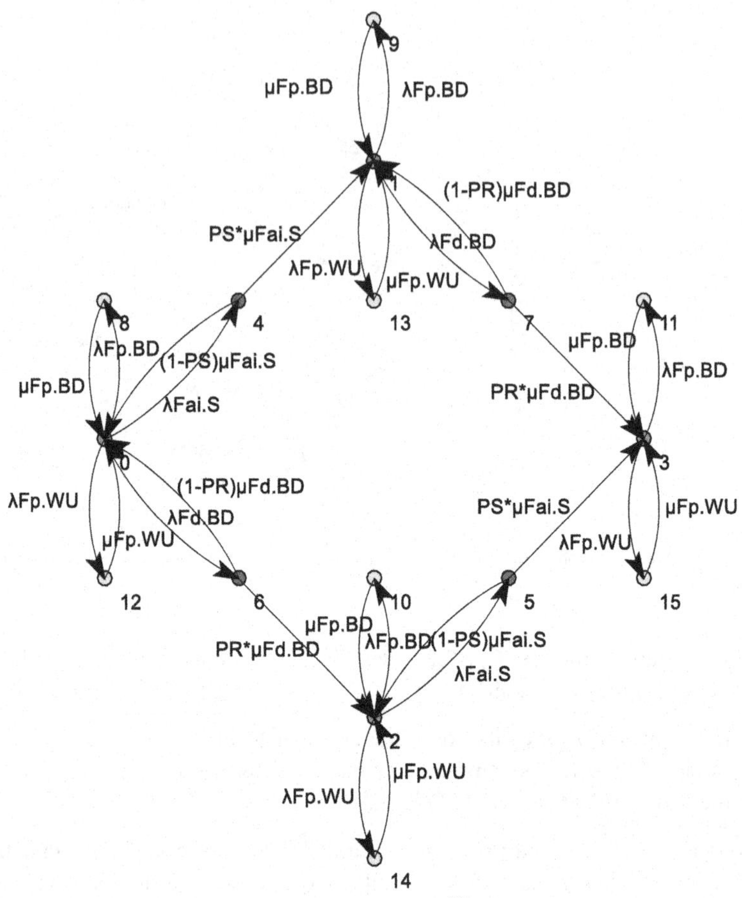

FIG. 13.4 The marked directed graph of model $M_{DEP}2.2^{(1)}$ with numbering of states, constructed by means of function grPlot_marker.

(1-PS) the BAS will continue to function with vulnerability after system restart. The system stops functioning to completely eliminate the software design faults and vulnerabilities of the first level architecture components with PR = 1 and PS = 1. Thus, model $M_{DEP}2.2^{(1)}$ describes the procedures for patching the software code of the BAS architecture components. But with PR = 0 and PS = 0 the system operates only in the conditions of software restart without elimination of design faults and vulnerabilities of components. The marked directed graph of states and transitions (Fig. 13.4) includes end-to-end numbering of states, constructed using the modified function grPlot_marker [28].

Kolmogorov's system of differential equations, built on the graph of the model $M_{DEP}2.2^{(1)}$, are expressed by

$$
\begin{cases}
dP_0(t)/dt = -(\lambda_{\text{Fp.BD}} + \lambda_{\text{Fp.WU}} + \lambda_{\text{Fd.BD}} + \lambda_{\text{Fai.S}})P_0(t) + \mu_{\text{Fp.BD}}P_8(t) + \mu_{\text{Fp.WU}}P_{12}(t) \\
\quad + (1-PS)\mu_{\text{Fai.S}}P_4(t) + (1-PR)\mu_{\text{Fd.BD}}P_6(t), \\
dP_1(t)/dt = -(\lambda_{\text{Fp.BD}} + \lambda_{\text{Fp.WU}} + \lambda_{\text{Fd.BD}})P_1(t) + \mu_{\text{Fp.BD}}P_9(t) + \mu_{\text{Fp.WU}}P_{13}(t) \\
\quad + (1-PR)\mu_{\text{Fd.BD}}P_7(t) + PS\mu_{\text{Fai.S}}P_4(t), \\
dP_2(t)/dt = -(\lambda_{\text{Fp.BD}} + \lambda_{\text{Fp.WU}} + \lambda_{\text{Fai.S}})P_2(t) + \mu_{\text{Fp.BD}}P_{10}(t) + \mu_{\text{Fp.WU}}P_{14}(t) \\
\quad + (1-PS)\mu_{\text{Fai.S}}P_5(t) + PR\mu_{\text{Fd.BD}}P_6(t), \\
dP_3(t)/dt = -(\lambda_{\text{Fp.BD}} + \lambda_{\text{Fp.WU}})P_3(t) + \mu_{\text{Fp.BD}}P_{11}(t) + \mu_{\text{Fp.WU}}P_{15}(t) \\
\quad + PS\mu_{\text{Fai.S}}P_5(t) + PR\mu_{\text{Fd.BD}}P_7(t), \\
dP_4(t)/dt = -\mu_{\text{Fai.S}}P_4(t) + \lambda_{\text{Fai.S}}P_0(t), \\
dP_5(t)/dt = -\mu_{\text{Fai.S}}P_5(t) + \lambda_{\text{Fai.S}}P_2(t), \\
dP_6(t)/dt = -\mu_{\text{Fd.BD}}P_6(t) + \lambda_{\text{Fd.BD}}P_0(t), \\
dP_7(t)/dt = -\mu_{\text{Fd.BD}}P_7(t) + \lambda_{\text{Fd.BD}}P_1(t), \\
dP_{8..11}(t)/dt = -\mu_{\text{Fp.BD}}P_{8..11}(t) + \lambda_{\text{Fp.BD}}P_{0..3}(t), \\
dP_{12..15}(t)/dt = -\mu_{\text{Fp.WU}}P_{12..15}(t) + \lambda_{\text{Fp.WU}}P_{0..3}(t), \\
\sum_{i=0}^{20} P_i(t) = 1;
\end{cases}
$$

$$
P_0(0) = 1; \quad \forall i \in [1\ldots15] \Rightarrow P_i(0) = 0. \tag{13.1}
$$

The standby function is defined as

$$
A(t) = \sum_{i=0}^{(\text{Nd}+1)\cdot(\text{Nv}+1)-1} P_i(t). \tag{13.2}
$$

13.3.3 Availability model considering joint maintenance strategy

Availability model $M_{DEP}2.2^{(2)}$ expands the previous model with additional procedures of the general proof maintenance during which both detection and elimination of software faults of database servers and vulnerabilities of sensors of the first level of architecture of ICS are possible. The marked graph of the model is shown in Fig. 13.5. It is a generalization of the MBAS2.2 availability model proposed in Ref. [11] and takes into account hardware failures caused by physical faults.

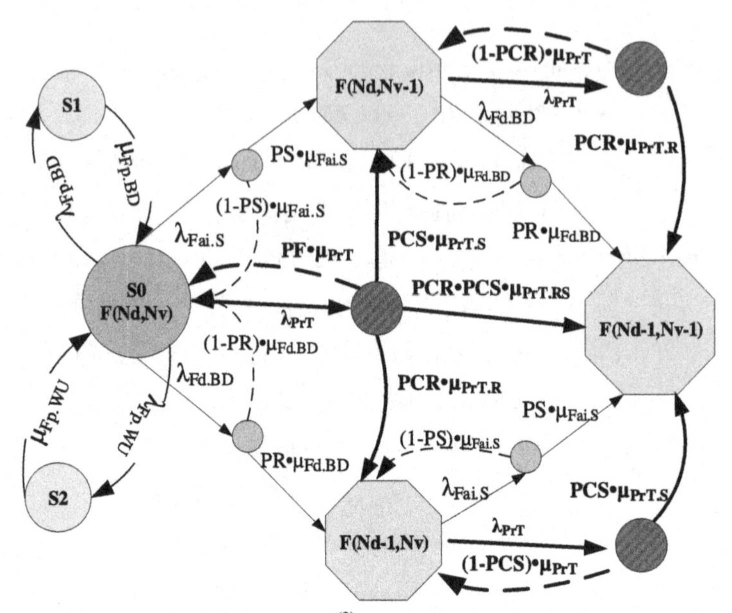

FIG. 13.5 Marked graph of model $M_{DEP}2.2^{(2)}$ with elimination of one software design faults and one vulnerability at two procedures of the general proof testing.

The condition describing the procedure of general proof testing is shown by a shaded brown circle. The transition to the maintenance state is carried out with the intensity λ_{PrT}. The PCR and PCS parameters correspond to the probabilities of detecting a software fault or vulnerability during maintenance. The detection of software faults and vulnerabilities is carried out with the probability PCR*PCS, and detection of both the reasons for refusal is carried out with the probability PF, moreover.

$$\text{PCS*PCR} + \text{PCS} + \text{PCR} + \text{PF} = 1.$$

The following transitions are possible from the state of proof testing:

(a) transition up with intensity PCS*$\mu_{\text{PrT.S}}$ ($\mu_{\text{PrT.S}}$ is the value inverse of the average time of detection and elimination of vulnerabilities [12]),

$$\mu_{\text{PrT.S}} = 1/\left(T_{\text{det.v}} + T_{\text{rem.v}}\right);$$

(b) transition down with the intensity PCR*$\mu_{\text{PrT.R}}$ ($\mu_{\text{PrT.R}}$ is the value inverse of the average time of detection and elimination of software design faults of database server [29]),

$$\mu_{\text{PrT.R}} = 1/(T\,\text{det.D} + T\text{rem.D});$$

(c) transition to the right with intensity PCS*PCR* $\mu_{\mathrm{PrT.RS}}$,

$$\mu_{\mathrm{PrT.RS}} = \frac{\mu_{\mathrm{PrT.R}} \cdot \mu_{\mathrm{PrT.S}}}{\mu_{\mathrm{PrT.R}} + \mu_{\mathrm{PrT.S}}} \tag{13.3}$$

(d) transition to the left with intensity PF*μ_{PrT}, where μ_{PrT} is the value inverse of the average time of the general proof testing procedure, $\mu_{\mathrm{PrT}} = 1/T_{\mathrm{PrT}}$.

After detection and elimination of all design faults, the proof testing procedure is aimed at finding and eliminating only vulnerabilities (provided that there are unresolved vulnerabilities), i.e., it is separate. Therefore, it has only two transitions weighted by the intensities PCR*$\mu_{\mathrm{PrT.R}}$ and (1–PCR)*μ_{PrT}.

Similarly, for the case where all vulnerabilities have been eliminated, the proof testing procedure has transitions weighted by the intensities PCS*$\mu_{\mathrm{PrT.S}}$ and (1–PCS)*μ_{PrT}.

After identifying and eliminating all design faults and vulnerabilities, the proof testing procedure has one transition with intensity μ_{PrT}.

13.3.4 Availability model considering separate maintenance strategy

Model $M_{DEP}2.2^{(3)}$ is a generalization of the availability model MBAS3.2 proposed in Ref. [11] and takes into account hardware failures caused by physical faults. It extends $M_{DEP}2.2^{(1)}$ model with additional separate proof maintenance procedures, during which it is possible to detect and eliminate only the software faults of the database servers (reliability maintenance) or only the vulnerabilities of the first level sensors of the ICS architecture (cybersecurity maintenance). The marked graph of the model is shown in Fig. 13.6.

13.4 Research of models

The values of the input parameters of model $M_{DEP}2.2$ are substantiated by studies in Refs. [11,12,29] and are given in Table 13.1.

Because models $M_{DEP}2.2^{(2)}$ and $M_{DEP}2.2^{(3)}$ describe two different maintenance strategies, each of which has certain parameters (intensity of conduction, average duration, probability of detecting faults, and vulnerabilities during maintenance), it is necessary to consider them separately and then compare.

Fig. 13.7 shows the results of models $M_{DEP}2.2$ (1, 2, and 3) for the basic values of the input parameters of Table 13.1. The following values of partial result indicators of dependability are received:

- the time when model $M_{DEP}2.2^{(2)}$ begins to give a gain in availability relative to model $M_{DEP}2.2^{(1)}$ Tup1.1 = 963.4 h;

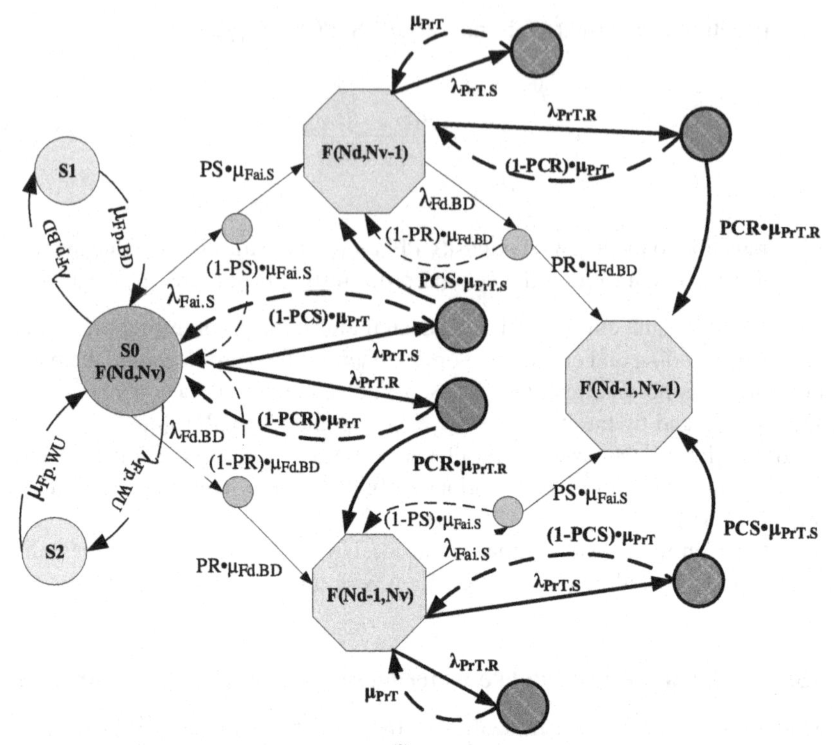

FIG. 13.6 Marked graph of model $M_{DEP}2.2^{(3)}$ with elimination of one software design fault and one vulnerability.

- the time when model $M_{DEP}2.2^{(3)}$ begins to give a gain in availability relative to model $M_{DEP}2.2^{(1)}$ Tup1.2 = 338.2 h;
- the time when models $M_{DEP}2.2^{(2)}$ and $M_{DEP}2.2^{(3)}$ begin to provide a requirement for the availability level of 0.999 Tup2 ≈ 1600 h;
- the largest decrease in the availability level of model $M_{DEP}2.2^{(2)}$ relative to model $M_{DEP}2.2^{(1)}$ $\Delta A = 6e-3$;
- the largest decrease in the availability level of model $M_{DEP}2.2^{(3)}$ relative to model $M_{DEP}2.2^{(1)}$ $\Delta A = 5{,}5e-3$.

Fig. 13.8 shows the results of the evaluation of the performance indicators obtained using a model with general proof testing $M_{\mathrm{DEP}}2.2^{(2)}$.

The influence of the values of the input parameter λ_{PrT} on the dynamics of change of partial dependability indicators was investigated.

A linear dependence was found for the partial Amin index, which decreases by 10.52% with a fourfold increase in λ_{PrT}. The nonlinear dependences were found for partial indicators Tup1 and Tup2, at the same time, with an increase in λ_{PrT} to 0.01 1/h and the time Tup2 for which the service system will take precedence over the maintenance system will be the same as the time Tup1

TABLE 13.1 The values of the input parameters of the model $M_{DEP}2.2$.

No.	Parameter	Matlab-variable	The value
1.	The hardware failure rate of the database server component	la_fp_bd	4e−5 (1/h)
2.	The intensity of hardware recovery of the database server component	mu_fp_bd	1.5 (1/h)
3.	The hardware failure rate of the wireless communications	la_fp_wu	2e−5 (1/h)
4.	The intensity of hardware recovery of the wireless communications	mu_fp_wu	0.66 (1/h)
5.	Intensity of manifestation of software design faults of the database server	la_fd_bd	[5e−4 4.5e−4] (1/h)
6.	Intensity of attack on sensors vulnerability	la_fai_s	[3e−3 3.5e−3] (1/h)
7.	Intensity of recovery after manifestation of software design faults of the database server	mu_fd_bd	[0.5 0.4] (1/h)
8.	Intensity of recovery after an attack on sensor vulnerabilities	mu_fai_s	[0.45 0.34] (1/h)
9.	The probability of eliminating the fault of the database server software during recovery	PR	0.87
10.	The probability of eliminating the vulnerability of sensors during recovery	PS	0.68
11.	The number of faults in the database server software in the system	Nd	2
12.	The number of sensor vulnerabilities in the system	Nv	2
13.	Intensity of BAS proof testing for faults and vulnerabilities detection (general maintenance)	la_prt	4e−3 (1/h)
14.	Intensity of BAS proof testing for vulnerability detection (separate maintenance for vulnerabilities)	la_prt_s	5e−3 (1/h)
15.	Intensity of BAS proof testing for fault detection (separate maintenance by faults)	la_prt_r	1e−4 (1/h)
16.	The inverse of the average duration of general service activities	mu_prt	0.4 (1/h)

Continued

TABLE 13.1 The values of the input parameters of the model $M_{DEP}2.2$—cont'd

No.	Parameter	Matlab-variable	The value
17.	Intensity of detection and elimination of sensors vulnerability	mu_prt_s	1 (1/h)
18.	The intensity of detection and elimination of the database server fault	mu_prt_r	1.5 (1/h)
19.	The probability of detecting the vulnerability of sensors during proof testing in the model $M_{DEP}2.2^{(3)}$	PCS	0.9
20.	The probability of detecting a software fault of the database server during proof testing in the model $M_{DEP}2.2^{(3)}$	PCR	0.97
21.	The probability of detecting the vulnerability of sensors in general proof testing in the model $M_{DEP}2.2^{(2)}$	PCS	0.4409
22.	The probability of detecting a software fault of the database server during the general proof testing in the model $M_{DEP}2.2^{(2)}$	PCR	0.388
23.	Predicted number of vulnerability prevention in the model $M_{DEP}2.2^{(3)}$	Nvp	2
24.	The projected number of software design faults prevention in the model $M_{DEP}2.2^{(3)}$	Ndp	2
25.	The projected number of general prevention in the model $M_{DEP}2.2^{(2)}$	Np	4

(time for which a general maintenance system will take precedence over a patching system).

The results of model $M_{DEP}2.2^{(3)}$ research are presented in the form of a graphical dependence of the availability function on the system operation time and partial indicators Tup1, Tup2, Tup3, and Amin in Fig. 13.9.

In the previous study of model $M_{DEP}2.2^{(2)}$, linear dependences of the partial dependability index Amin on the input parameters λ_{PrTr} and λ_{PrTs} were revealed. At the same time, the graph Amin(λ_{PrTs}) has a larger angle of inclination relative

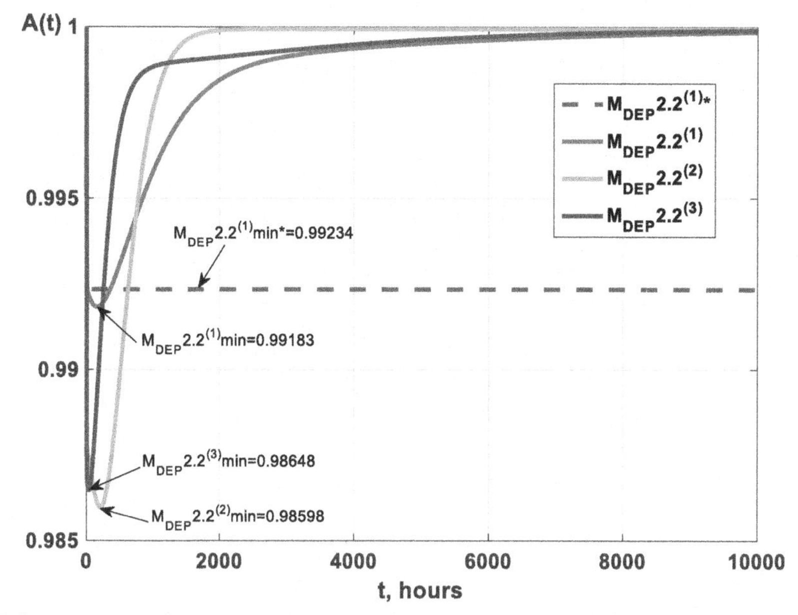

FIG. 13.7 Assessment results of the availability by models $M_{DEP}2.2$ under the baseline values of the parameters ($M_{DEP}2.2^{(1)}$*—model without patching procedures with PR = PS = 0).

to the horizontal axis. The dynamics of partial dependability indicators Tup1 and Tup2 change in the model with separate maintenance will differ significantly from the model with general maintenance (Figs. 13.8B and 13.9B).

As a result of the assessment of the dependability indicators for software patcherization procedures, an increase in accuracy by 0.00636 (Fig. 13.7, models $M_{\mathrm{DEP}}2.2^{(2)}$ and $M_{\mathrm{DEP}}2.2^{(3)}$) relative to model $M_{\mathrm{DEP}}2.2^{(1)}$ was obtained. The results of the assessment of the dependability indicators Amin, Tup1, and Tup2 allowed to determine the impact of service parameters and their values, which meet the requirements for the level of the dependability, in particular to increase the partial Amin by 10.5% (Fig. 13.8, model $M_{\mathrm{DEP}}2.2^{(2)}$ relative to the model $M_{\mathrm{DEP}}2.2^{(1)}$).

13.5 Conclusions

The multifragment models of dependability BAS in the event of failures caused by hardware physical faults, software design faults, and attacks on component vulnerabilities have been developed. Two strategies of maintenance have been analyzed: separate strategy for reliability and cybersecurity maintenance and assurance (RMA and CMA), and joint strategy for their maintenance and assurance (JMA). The function of BAS availability is used as a main indicator depending on component failure, recovery, and maintenance rates. Three models are developed and researched: the model of BAS without proactive testing and models of system with RMA-CMA and JMA maintenance strategies.

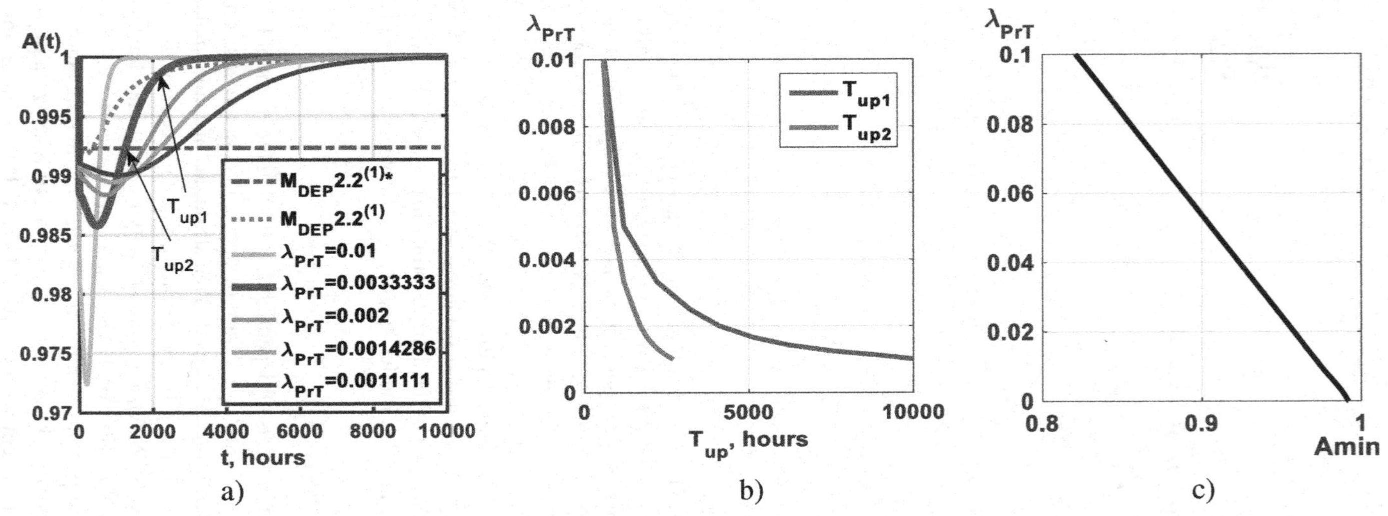

FIG. 13.8 Graphical dependences of availability function change on system operation time (A), partial dependability indicators Tup1 and Tup2 (B), and Amin (C) with different values λ_{PrT}, obtained using model $M_{DEP}2.2^{(2)}$.

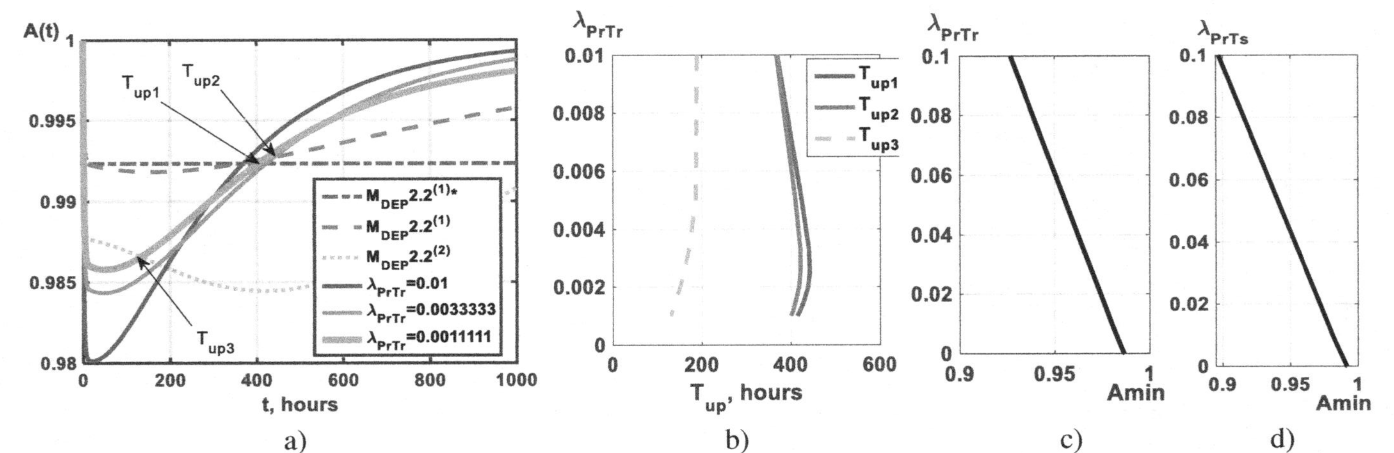

FIG. 13.9 Graphical dependences of availability function change on system operation time (A), partial dependability indicators Tup1, Tup2, and Tup3 (B), Amin with different values λ_{PrTr} (C), and λ_{PrTs} (D) obtained using the model $M_{DEP}2.2^{(3)}$.

Studies of $M_{DEP}2.2^{(i)}$ models have shown that patching procedures in conjunction with separate maintenance and cybersecurity procedures provide a more effective level of BAS availability in comparison with carrying out procedures with general maintenance. The increase in efficiency is to gain an increase in availability at the initial stage of operation by $1.5e-3$ with approximately the same time interval of bringing the system to the planned level of availability by 0.999 in 1600 h.

Further using the obtained results, development and research of multifragment models of smart building ICS should be carried out to create a single service to automate the process of assessing the BAS dependability, integration of developments with known packages for decision-making at all stages of ICS life cycle and their components that operate in changing requirements, environmental parameters, and unspecified failures including failures caused by attacks on vulnerabilities of IoT communication protocols [30].

References

[1] M. Froufe, C. Chinelli, A. Guedes, A. Haddad, A. Hammad, C. Soares, Smart buildings: systems and drivers, Buildings 10 (2020) 153, https://doi.org/10.3390/buildings10090153.

[2] D. Nicol, W. Sanders, K. Trivedi, Model-based evaluation: from dependability to security, IEEE Trans. Dependable Secure Comput. 1 (2004) 48–65, https://doi.org/10.1109/tdsc.2004.11.

[3] M. Alam, M. Reaz, M. Ali, A review of smart homes—past, present, and future, IEEE Trans. Syst. Man Cybern. Part C Appl. Rev. 42 (2012) 1190–1203, https://doi.org/10.1109/tsmcc.2012.2189204.

[4] S. Budijono, J. Andrianto, M.A.N. Noor, Design and implementation of modular home security system with short messaging system, EPJ Web Conf. 68 (2014), https://doi.org/10.1051/epjconf/20146800025, 00025.

[5] A. Avizienis, J. Laprie, B. Randell, C. Landwehr, Basic concepts and taxonomy of dependable and secure computing, IEEE Trans. Dependable Secure Comput. 1 (2004) 11–33, https://doi.org/10.1109/tdsc.2004.2.

[6] International Electrotechnical Commission, IEC 60050–192. International Electrotechnical Vocabulary—Part 192: Dependability, 2015. https://webstore.iec.ch/publication/21886.

[7] O. Drozd, V. Kharchenko, A. Rucinski, T. Kochanski, R. Garbos, D. Maevsky, Development of models in resilient computing, in: 10th International Conference on Dependable Systems, Services and Technologies (DESSERT), 2019, pp. 1–6, https://doi.org/10.1109/DESSERT.2019.8770035.

[8] A. Gorbenko, V. Kharchenko, P. Popov, A. Romanovsky, Dependable composite web services with components upgraded online, in: Architecting Dependable Systems III, 2005, pp. 92–121, https://doi.org/10.1007/11556169_5.

[9] International Organization for Standardization, ISO 16484-1:2010. Building Automation and Control Systems (BACS)—Part 1: Project Specification And Implementation, 2010. https://www.iso.org/standard/37300.html.

[10] V. Kharchenko, Y. Ponochovnyi, A. Abdulmunem, A. Andrashov, Availability models and maintenance strategies for smart building automation systems considering attacks on component vulnerabilities, in: Advances in Dependability Engineering of Complex Systems, 2017, pp. 186–195, https://doi.org/10.1007/978-3-319-59415-6_18.

[11] V. Kharchenko, Y. Ponochovnyi, A.-S.M.Q. Abdulmunem, A. Ivasiuk, O. Ivanchenko, Model of information and control systems in smart buildings with separate maintenance by reliability and security, in: 14th Int. Conf. on ICT in Education, Research and Industrial Applications. Integration, Harmonization and Knowledge Transfer, Kyiv, Ukraine, May 14–17, 2018, pp. 583–595.

[12] A. Gorbenko, V. Kharchenko, O. Tarasyuk, Y. Chen, A. Romanovsky, The threat of uncertainty in service-oriented architecture, in: SERENE 2008, RISE/EFTS Joint International Workshop on Software Engineering for REsilient SystEms, Newcastle Upon Tyne, UK, November 17–19, 2008, https://doi.org/10.1145/1479772.1479781.

[13] M. Bhuiyan, S. Kuo, D. Lyons, Z. Shao, Dependability in cyber-physical systems and applications, ACM Trans. Cyber-Phys. Syst. 3 (2019) 1–4, https://doi.org/10.1145/3271432.

[14] K. Trivedi, A. Bobbio, Reliability and Availability Engineering, Modeling, Analysis, and Applications, Cambridge University Press, 2017, https://doi.org/10.1017/9781316163047.

[15] S. Lysenko, V. Kharchenko, K. Bobrovnikova, R. Shchuka, Computer systems resilience in the presence of cyber threats: taxonomy and ontology, Radioelectron. Comput. Syst. 1 (2020) 17–28, https://doi.org/10.32620/reks.2020.1.02.

[16] A. Gorbenko, A. Romanovsky, O. Tarasyuk, Fault tolerant internet computing: benchmarking and modelling trade-offs between availability, latency and consistency, J. Netw. Comput. Appl. 146 (2019) 102412, https://doi.org/10.1016/j.jnca.2019.102412.

[17] A. Rucinski, I. Kovalev, M. Drozd, O. Drozd, V. Antoniuk, Y. Sulima, Development of computer system components in critical applications: problems, their origins and solutions, Herald Adv. Inf. Technol. 3 (2020) 252–262, https://doi.org/10.15276/hait.04.2020.4.

[18] E. Zio, The Monte Carlo Simulation Method for System Reliability and Risk Analysis, Springer Series in Reliability Engineering, 2013, https://doi.org/10.1007/978-1-4471-4588-2.

[19] R. Pietrantuono, P. Popov, S. Russo, Reliability assessment of service-based software under operational profile uncertainty, Reliab. Eng. Syst. Saf. 204 (2020) 107193, https://doi.org/10.1016/j.ress.2020.107193.

[20] K. Trivedi, D. Kim, R. Ghosh, Resilience in computer systems and networks, in: Proceedings of the 2009 International Conference on Computer-Aided Design—ICCAD'09, 2009, https://doi.org/10.1145/1687399.1687415.

[21] E. Zaitseva, V. Levashenko, J. Rabcan, M. Kvassay, P. Rusnak, Reliability evaluation of multi-state system based on incompletely specified data and structure function, in: 2019 10th IEEE International Conference on Intelligent Data Acquisition and Advanced Computing Systems: Technology and Applications (IDAACS), 2019, https://doi.org/10.1109/idaacs.2019.8924454.

[22] J. Batalla, F. Gonciarz, Deployment of smart home management system at the edge: mechanisms and protocols, Neural Comput. Applic. 31 (2018) 1301–1315, https://doi.org/10.1007/s00521-018-3545-7.

[23] K. Milanovic, Smart home security: how safe is your data? [opinion], IEEE Technol. Soc. Mag. 39 (2020) 26–29, https://doi.org/10.1109/mts.2020.2967490.

[24] S. Zheng, N. Apthorpe, M. Chetty, N. Feamster, User perceptions of smart home IoT privacy, Proc. ACM Hum.-Comput. Interact. 2 (2018) 1–20, https://doi.org/10.1145/3274469.

[25] Y. Ponochovnyi, V. Kharchenko, Dependability assurance methodology of information and control systems using multipurpose service strategies, Radioelectron. Comput. Syst. 3 (2020) 43–58, https://doi.org/10.32620/reks.2020.3.05.

[26] J. Neumann, C.E. Shannon, J. McCarthy, Probabilistic logics and the synthesis of reliable organisms from unreliable components, in: Automata Studies, Princeton Univ. Press, 1956, pp. 43–98.

[27] J. Henke, Dependable software for undependable hardware, in: 7th IEEE International Symposium on Industrial Embedded Systems (SIES'12), 2012, p. 1, https://doi.org/10.1109/SIES.2012.6356614.
[28] S. Iglin, grPlot—A Function for Drawing Graphs and Digraphs Using MATLAB, 2020. http://www.iglin.epizy.com/All/GrMatlab/grPlot.html.
[29] M. Ge, H. Kim, D. Kim, Evaluating security and availability of multiple redundancy designs when applying security patches, in: 2017 47th Annual IEEE/IFIP International Conference on Dependable Systems and Networks Workshops (DSN-W), 2017, https://doi.org/10.1109/dsn-w.2017.37.
[30] M. Kolisnyk, Vulnerability analysis and method of selection of communication protocols for information transfer in Internet of Things systems, Radioelectron. Comput. Syst. 1 (2021) 133–149, https://doi.org/10.32620/reks.2021.1.12.

Chapter 14

A study of bitcoin and Ethereum blockchains in the context of client types, transactions, and underlying network architecture

Rohaila Naaz and Ashendra Kumar Saxena
Teerthanker Mahaveer University, Moradabad, Uttar Pradesh, India

Blockchain is a growing technology possessing various budding platforms. Bitcoin and Ethereum are widely accepted and growing constituents of blockchain. The major difference between Bitcoin and Ethereum is that while Bitcoin is specifically a cryptocurrency application with a limited set of data types, transactions types, and size of data storage, Ethereum is Turing complete general purpose blockchain: in Ethereum you do not need to worry about underlying peer-to-peer networks, infrastructure details, and consensus mechanisms; it abstracts all these details and provide a secure and deterministic environment to programmers for building various types of applications. This chapter constitutes a comparative study of Bitcoin and Ethereum blockchains, specifically in their "Clients Types," "Transactions," and "Networks." it starts from explaining Bitcoin reference client, also known as satoshi client, and Bitcoin core client specifications. Information about Bitcoin network status is obtained by Bitcoin core client for, e.g., version, protocol version, and wallet version, etc., while Ethereum clients are software that implements Ethereum specifications and make communication with other peers in the Ethereum network.There are two types of clients in Ethereum: full node and remote client, such as parity and geth. Bitcoin uses public key cryptography, public key which is unique derived from private key is used to receive Bitcoins and private key is used to sign transactions to spend those Bitcoins. In Bitcoin it is also required to cryptographically secure random number generator for creating private keys. Ethereum transactions are also cryptographically signed messages, but in contrast with Bitcoin, transactions can only trigger a change of

System Assurances. https://doi.org/10.1016/B978-0-323-90240-3.00014-X

state of singleton blockchain or cause a smart contract to execute. After this, the chapter discusses wallets in both the platforms. In Bitcoin, wallets are databases that contain private keys and are of two types: deterministic, also known as seeded wallets, and nondeterministic wallets; while in Ethereum wallets exist at two levels: at application level, wallets serve as an interface to the Ethereum-based DApps, also known as interface browsers, and wallets that are able to communicate with other smart contracts. Network architecture used in Bitcoin and Ethereum is also described. Firstly, it comprises of peer-to-peer network with various types of nodes and their roles in network. As Bitcoin is a cryptocurrency-based platform, it provides an SPV (simplified payment verification) scheme, while in Ethereum, the basic architecture is based on a web of DApps also synonyms with web3, a decentralized web comprised of smart contracts and P2P technologies.

14.1 Blockchain: An insight

Blockchain is a term known by almost any computer science professional but when it comes to explaining it to a layperson, it must be explained in a simple manner, or would require a number of assumptions and constraints to explain it. For example, when we talk about its definition, it is "a number of blocks connected cryptographically makes a chain that is blockchain," but it is more than that: Blockchain is a revolutionary idea if applied ethically. There must be Governance Regulatory control and standards for blockchain-based applications especially in cryptocurrency.

In today's world, artificial intelligence is widely used in almost everything. Artificial intelligence tries to create an artificial human on which one can rely for getting things done with accuracy so that we can trust machines. We are targeting to achieve this accuracy from the artificially created machines so that we can create a trustless environment for getting things done with both accuracy and transparency, and for this Blockchain can be a solution.

Cryptocurrency is one of the major applications of blockchain technology. Apart from this, it can be applied in various fields of human life where trustless (transparent) digital environment is needed. In order to understand more about blockchain technology, this chapter discusses two major platforms of blockchain—Bitcoin and Ethereum.

14.2 Blockchain architectures

The term blockchain was first described back in 1991. It is a combination of different technologies, such as cryptography, decentralized storage, consensus to maintain networks, and timestamping of documents. Timestamping was introduced in 1991 by Haber and Stornetta, while consensus was first introduced in 1937 by Archie Blake. The technique was adapted and reinvented by Satoshi Nakamoto [1]. In 2008, Nakamoto created the first cryptocurrency, the blockchain-based project called Bitcoin.

Blockchain is applicable mainly on applications where decentralization, accountability, and security have been of primary concern, like FinTech, supply chain, and identity management, to provide a certain level of transparency. This technology increases cost efficiency and speed efficiency, which are also major constraints and challenges in existing technology, so blockchain can provide a promising solution. For the past two years, blockchain has evolved in the right direction and is now adaptable and proving to have convincing capacities among stakeholders and organizations.

Let's learn the key insights to easily understand how blockchain technology works. Fig. 14.1 shows how a traditional legacy system works with one central server while in a peer-to-peer network all nodes can communicate as both client and server. Maybe this will encourage you to think more about building your own blockchain solution.

In 1991, Stuart Haber and W. Scott Stornetta [2] introduced Timestamping of documents and files communicated over a network. Instead of securing a medium for communicating those files, they provided this timestamping even when timestamping service collision would occur. Later, along with Bayer, Haber and Stornetta [2] used the Merkle tree to make the model efficient. There is famous saying of Stornetta that when he designed the timestamping storage concept, his major concern was how to make a network without centralization of a single authority—so if he adds a certificate authorization node in a network, which is still required to be authenticated by itself and so on, so the number of participants in a network becomes over populous, then network itself becomes the authorizing authority by a majority wins concept. That's why Satoshi Nakamoto cites Haber work three times out of total eight citations in his paper. When decentralization came in to existence in 1979, through David Chaum proposing a decentralized network system known as MIX Networks [3], it became widely acceptable.

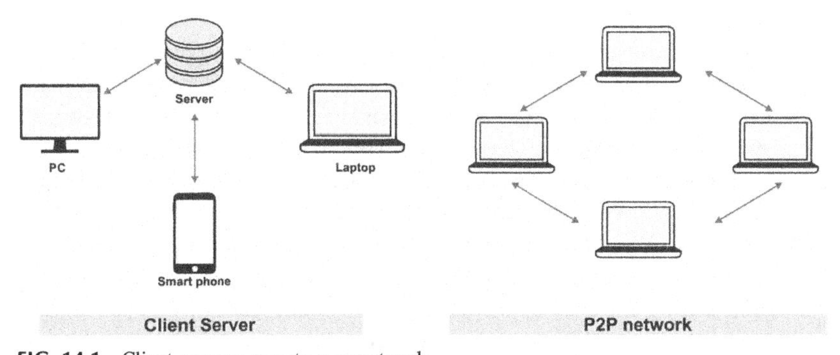

FIG. 14.1 Client sever vs peer-to-peer network.

In 2008, Satoshi Nakamoto applied proof of concept on this over populous network with cryptographically linking blocks containing time stamped transactions in sequential order without the need of any single centralized certification authority, known as Blockchain. The intention behind developing this technology in the form of cryptocurrency is to remove mediacy between people and their money by bodies such as banks and governments, as in the existing system. However, the actual implementation does not producing this intention, as proved by research conducted by IIT Kanpur India on Bitcoin—we are not discussing this concept here as it is beyond the scope of this chapter.

An abstract view of blockchain is shown in Fig. 14.2.

Although blockchain architecture, as shown in Fig. 14.2, mainly depends on the application in which it is going to be used, it is not a rigid architecture and there may be some simpler means available for that particular application that do not require all layers of the generalized architecture.

There are mainly four layers in any blockchain application in the context of developing applications, as described below.

14.2.1 Infrastructure layer

The infrastructure layer consists of basic building blocks that create the infrastructure of any application.

(i) **Nodes.** The participating nodes in the blockchain network include light nodes and full nodes. Light nodes require less space in terms of storage as they store only the header of the block in Blockchain—they are also called SPV (simple payment verification nodes)—while a full node contains all Blockchain history and all information related to blocks.

 A node is highly available nodes and also called Bootstrap node. Every node is having address which is hardcoded in the software. This address is recognizable in the blockchain network and nodes can communicate with each other using this address. These nodes are called blockchain application developer nodes.

(ii) **Storage**. Storage is another key component of deploying blockchain infrastructure, where data is cryptographically stored in a decentralized manner. This decentralization provides smaller and more efficient storage space and less maintenance of the network to store data. This is in contrast to a traditional storage network where data is centrally stored and managed and has a single point of failure challenge.

(iii) **Network.** According to application, there are different infrastructures. Every application needs different networks as well as different blockchain platforms. Considering this, there are two main types of blockchain network: permissionless and permissioned.

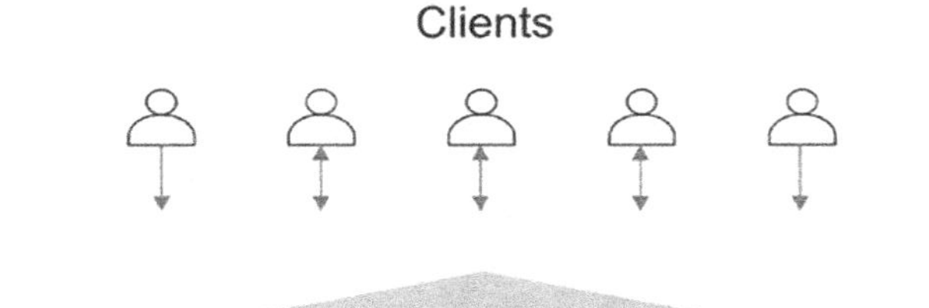

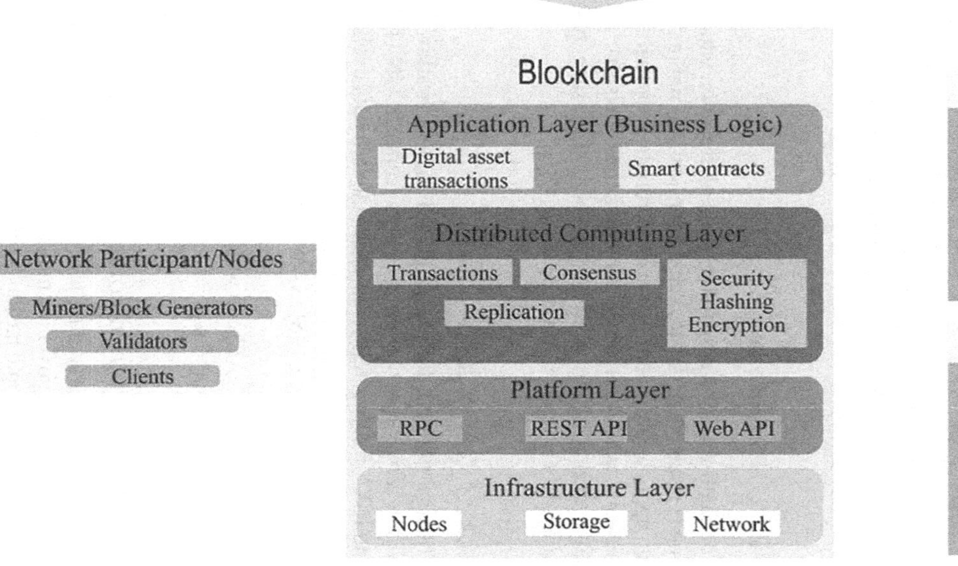

FIG. 14.2 Generalized blockchain architecture. *(Source: www.mdpi.com.)*

In permissionless blockchain, every participant can add into the network with pseudonymity. Examples of this blockchain application include Bitcoin and Ethereum.

In permissioned blockchain, there is a central authority that issues authentication certificate to the user to be part of the network. Again, it is divided in to two:

(1) Public permissioned blockchain, which is type of private blockchain with certain privileges and restrictions: first, a user is be added into network after authentication and verification, now this user has been given only privileges to read transactions data in blockchain means it can see all transaction occurs into blockchain and validate them but can not initiate a transaction or do mining.

(2) Private permissioned blockchain, which comes with more restrictions: it cannot perform any validation, transaction initiation, or mining without authentication and verification.

14.2.2 Platform layer

The platform layer is specific to the platform on which the blockchain application is implemented. Blockchain applications run in decentralized networks. To run applications over it we need to run the subprograms on a shared distributed network, which runs on over blockchain network nodes as local procedural calls, executing via a request-response message passing system. These are called RPCs (remote procedure calls).

14.2.2.1 JSON-RPC

This is a protocol for exchanging data, which allows a client to communicate with a server by combining two concepts: JSON (JavaScript object notation) and RPC. JSON refers to an open standard data exchange format to be communicated with remote procedure calls. Bitcoin, Ethereum and other platforms use different RPCs, e.g., JSON-RPC, Web3.js for JavaScript, Ethers.js for JavaScript, Web3j for Java, Ethereum for Microsoft. NET / C#, and Web3.py for Python, etc. REST API and web API are also used here for developing and communicating applications over the internet.

14.2.2.2 Distributed computing layer

The distributed computing layer considers designing and deploying of different consensus running over the network, how encryption is applied over the communication and data, and hashing techniques which are used in encrypting data. Creation of blockchain is also taken care of in this layer. However, this architecture is not rigid in context that all blockchain applications will require all technicalities of blockchain, but in general these are the main components discussed in the distributed computing layer.

14.2.2.3 Transactions

A transaction is a basic operation which is going to happen in a blockchain network, it is atomic in nature such as if you want to develop an asset management Application in which one transaction is like a user Alice sell her Rolls Royce for 4000 $ to another user on blockchain network then it will be a transaction.

14.2.2.4 Consensus

Consensus is a mechanism in order to achieve a single state of blockchain among all participating nodes over a network. This technique aims to achieve a fault-tolerant, consistent system. There are various consensus mechanism that can be applied specifically for different blockchain use cases; some examples are proof of work (PoW), which is used in the two popular blockchain platforms, Bitcoin and Ethereum, and others such as proof of stake, proof of elapsed time, proof of burn, proof of luck, proof of capacity, etc.

14.2.2.5 Replication

As blockchain works on a distributed network, working in a decentralized manner to store the same state of data over all blockchains will require consistent data replication. All users in a blockchain network can access data from any storage. This consistent state can be managed by updating the global state of blockchain data in almost real time. There are the following methods for updating and storing this data:

(1) Synchronous—which means that if data is changed over one node in a blockchain network, then it must be updated immediately on all other storage devices.
(2) Asynchronous—which means that data will be updated after a certain period of time using some checkpoints.

Storage of data in this distributed system also has the following methods for storage. All data is stored in storage disks using some partitioning technique, for which, two techniques are used:

(1) Vertical partitioning—different fields of the same data are stored in different storage devices over the network.
(2) Horizontal partitioning—in which data of one record is saved on one storage device and another record is saved on another storage device [4].

14.2.2.6 Security

Security is inherent in blockchain—it uses decentralization, cryptography, and consensus to make it efficient in a trustless environment. Security works at different levels in blockchain this may be at infrastructure level or platform level, but predominantly it is required at the distributed computing level. In blockchain, transactions are grouped together in order to make a block and this block gets validated by using consensus between participating nodes over the blockchain network; the final validated node is added in to the longest chain and this chain's distributed ledger is updated by all other nodes in the blockchain network to make a final global state of blockchain. These blocks are cryptographically linked with each other to provide immutability and are tamper proof in nature, but still vulnerable to cyber-attacks and threats. Some examples of these threats and attacks are stolen keys, especially in cryptocurrency applications, code exploitation, etc.

14.2.2.7 Application layer business logic

Bitcoin, invented in 2008 by Satoshi Nakamoto, was the first real application of blockchain technology; however, the capabilities and scope of blockchain technology are not fully utilized in it as it is mainly designed for cryptocurrency applications. Vitalik Butarin, a former Bitcoin developer, expert, and writer found these limitations in Bitcoin and introduced Ethereum, with a special feature known as smart contract, which has the power to deploy business logic in terms of executable code to develop different real use applications using blockchain technology [5]. Some real use case business applications that can deploy blockchain technology are:

(1) Smart contracts
(2) Internet of things (IoT)
(3) Money transfer
(4) Personal identity security
(5) Logistics
(6) Digital media

Apart from this, additional features in blockchain architecture like accuracy and applicability of consensus according to applications, security from different attacks, and interoperability of blockchain applications are major concerns for blockchain deployment.

14.3 Blockchain adequacy

Over the past two to three years, evolution of blockchain has mainly occurred in the area of architecture, which is highly dependent on certain key characteristics that would be needed in blockchain applications deployment before actual implementation. For any use case or problem solution, there are certain checks to see if blockchain is the best solution, or what other alternatives are available for solving that particular use case or application [6]. Fig. 14.3 shows some prior checks for feasibility of applying blockchain solution to particular problem such as in Fig. 14.3 Quick checks should be done before starting developing any application on blockchiain, if an application requires tracking of transactions,

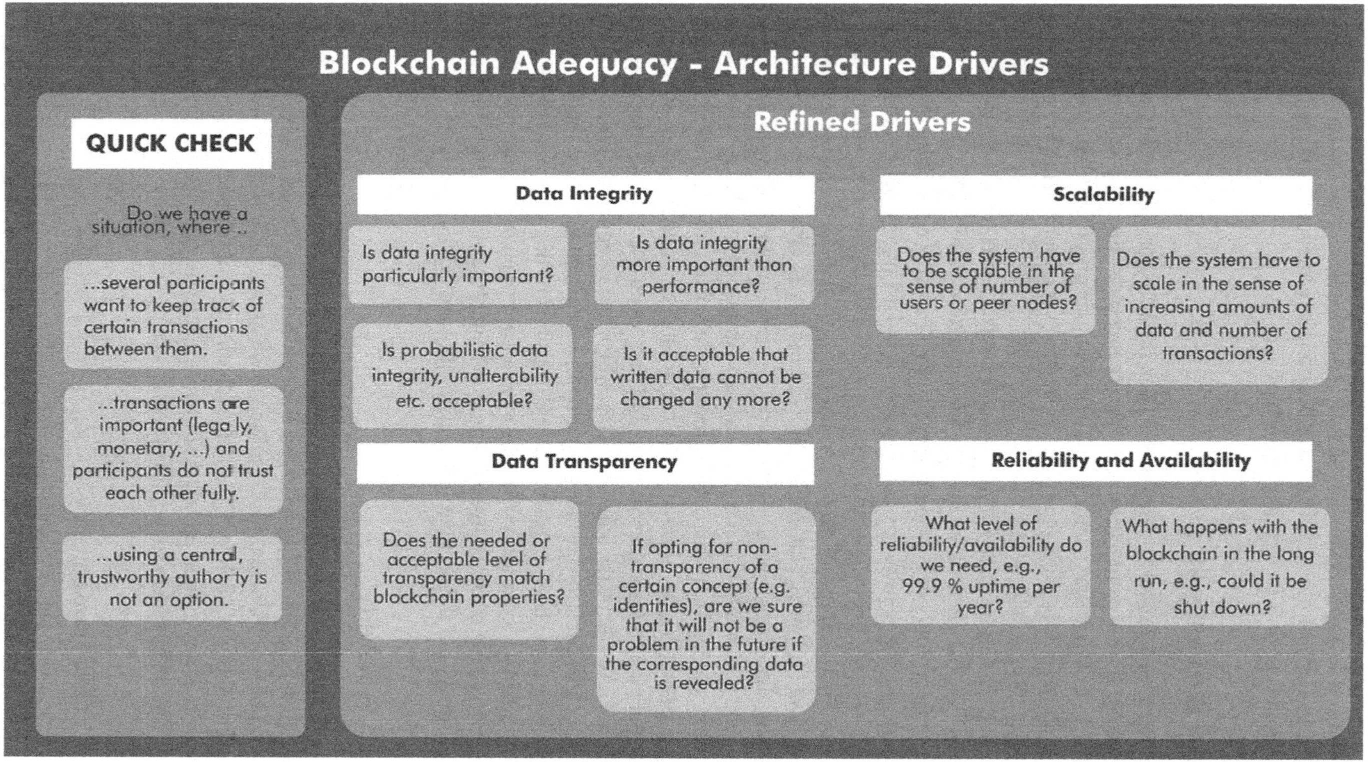

Blockchain Adequacy - Architecture Drivers

QUICK CHECK

Do we have a situation, where ..

...several participants want to keep track of certain transactions between them.

...transactions are important (legally, monetary, ...) and participants do not trust each other fully.

...using a central, trustworthy authority is not an option.

Refined Drivers

Data Integrity

Is data integrity particularly important?

Is data integrity more important than performance?

Is probabilistic data integrity, unalterability etc. acceptable?

Is it acceptable that written data cannot be changed any more?

Scalability

Does the system have to be scalable in the sense of number of users or peer nodes?

Does the system have to scale in the sense of increasing amounts of data and number of transactions?

Data Transparency

Does the needed or acceptable level of transparency match blockchain properties?

If opting for non-transparency of a certain concept (e.g. identities), are we sure that it will not be a problem in the future if the corresponding data is revealed?

Reliability and Availability

What level of reliability/availability do we need, e.g., 99.9 % uptime per year?

What happens with the blockchain in the long run, e.g., could it be shut down?

FIG. 14.3 Blockchain adequacy drivers. *(Source: www.iese.fraunhofer.de.)*

or immutability of transaction due to lack of trust or there is no chance of central authorization in application. The above prior checks should be referenced to ensure adequacy of blockchain in Application development. Adequacy drivers are classified in four categories, as follows.

14.3.1 Data integrity

Data integrity ensures that data has not been modified or seen by any unauthorized means. If your application's primary concern is this, then you are on the right track, as blockchain solutions provide tamper proofing of data and would make it almost impossible, or at least cause much cost, for an attacker to change or modify previously stored blockchain data [7]. In real world applications where data has to be added in real time and also needs to be not modified in future or not lost, then blockchain is an adequate solution. This scenario must also have flexibility between data integrity and transactions speed as blockchain technology speed is lower as compared to standard traditional databases.

Data integrity in blockchain works in a probabilistic manner, which means it does not provide full proof surety on data integrity or an "impossible to modify" concept. Data integrity and impossibility to modify is greatly dependent on the consensus algorithm used in order to remove central authorization. Modification of data is computationally very hard and impossible to break with existing computing capability; it can only be broken with quantum computing. Furthermore, ledger retroactivity can be possible in blockchain technology, so again if your application does require a strict full sure evidence on data integrity then blockchain may not be the solution. Also, when data integrity is a concern, some applications, especially business logic applications, may require modifiability as a must for some legal or governance constraints of that particular region. So, for that integrity, we must be clear in whether our objective is that once data is written it must never be changed in any way in the future; if so, then blockchain is good for it, and if not, then blockchain may not be the solution.

14.3.2 Scalability

Adding peers or nodes is not a major concern for blockchain. Nodes can be easily added to a distributed blockchain network because it does not require centralized structure for scalability. However, when transaction speed and storage capacity of nodes is a concern, then blockchain may not be an adequate solution, as blockchain throughput and data storage capacity is not so scalable, and also depends on other blockchain components.

14.3.3 Data transparency

Blockchain's first application, Bitcoin, is a permissionless blockchain where anyone can be a participating node and the complete blockchain is visible to all nodes. However, Bitcoin provides pseudonymity, because transactions, participants, and data are to be recognized only by their public keys not by IP address or any KYC (know your customer), but other metadata or certain sequences of account numbers can be monitored or analyzed in public blockchain [8]. Hence, it is called pseudonymous identity blockchain. This transparency sometimes is not required in all real life use cases and applications, so there are some adjustments needed for transparency of data stored in particular blockchains. This pseudonymity can be achieved at higher level with some explicit implementations in blockchain.

If nontransparency is a primary concern for a blockchain application, then you must ensure that in future, even if any participant performs some illegal activity on this blockchain, the identity of that participant can't be revealed. This is a major drawback in cryptocurrencies, which can be used for illegal activities like drug smuggling and dark web activities [9]. Furthermore, if one can trace a public key that belongs to a specific person, then all the transactions made by using that public key can be traced and all the transactions done in bitcoin using that particular private key could be found. In addition, it may be that, right now, with existing computational capacity, it is not possible to decrypt data, but in future, it may be possible with quantum computing, and information could be revealed. A third major concern is the level of transparency required, which determines which type of blockchain is used.

The fourth major concern is legal and governance issues related to protection of personal data of users, as some do individuals not want their data to be stored permanently. This requires additional compliance to make it more customized. GDPR, IT Act 200 is still at the development stage and needs to form standard governance use of blockchain in different applications.

14.3.4 Reliability and availability

Reliability and availability are high in blockchain applications. Suppose we want availability of data of 99.9%, if we are using a permissionless blockchain, then we need not be worried—large blockchain networks have good fault tolerance, which means that if high availability in one node fails, then other nodes will have backup, so availability is not compromised [10]. If we can achieve availability without investing too much in blockchain solutions, then we are moving in the right direction.

For a public blockchain, we have to consider what happens with the blockchain in the longer run. For example, it is possible that nobody will run Bitcoin nodes in a few years, since Bitcoin may be replaced by a different cryptocurrency. Then, applications building on the respective blockchain might become unavailable and data stored in the blockchain will be lost unless it is backed up in some way.

Another consideration is where we want to make a trade-off between transparency and reliability. For example, in order to grow a network in its initial phase the network might be permissionless, with pseudonymity, meaning anyone can participate, but this can cause reliability problems if the number of malicious users increases so much that they can make a 51% attack; and in this situation, availability would also be at stake, because a small number of faulty nodes crash can cause a whole system failure [11].

14.3.5 Blockchain architecture design space guidelines

Designing the architecture of blockchain-based applications comes with a large number of degrees of freedom. There is no standard architecture and, depending on the concrete requirements and blockchain technology chosen, an architect has to come up with many individual decisions. Thus, we chose to give the architecting guidance in the form of questions to be answered and decided on by the architects and developers of blockchain-based applications.

14.4 Bitcoin architecture

Bitcoin was the first real application of blockchain. It was first introduced in 2008 in a white paper named "Bitcoin: A peer-to-peer Electronic cash system" by an initialy anonymous Satoshi Nakamoto [12]. From then on, Bitcoin has evolved in many ways, including changing block size capacity, which led to a new blockchain known as Bitcoin Cash, changing functionalities, developing side chains, etc. Bitcoin code is open source, so anyone can change it, but in order for a change to be effective one needs to get network consensus, which is really hard to achieve [13].

Fig. 14.4 shows working scenario for bitcoin. It is consisting of Bitcoin nodes which includes merchants, Bitcoin Exchanges, Bitcoin Miners, and Bitcoin users.

Bitcoin is a public blockchain network, which allows anyone to be part of the network. After joining the network, the user is given a user wallet in which private key and public key pairs are stored. When a transaction is done by this user, the transaction is signed with the user's private key and verified by receiving nodes and all other nodes with the user's public key, so asymmetric cryptography is used here.

14.4.1 Bitcoin components

Bitcoin architecture has mainly the following components:

- Wallets
- Network nodes
- Consensus
- Miner
- Transactions

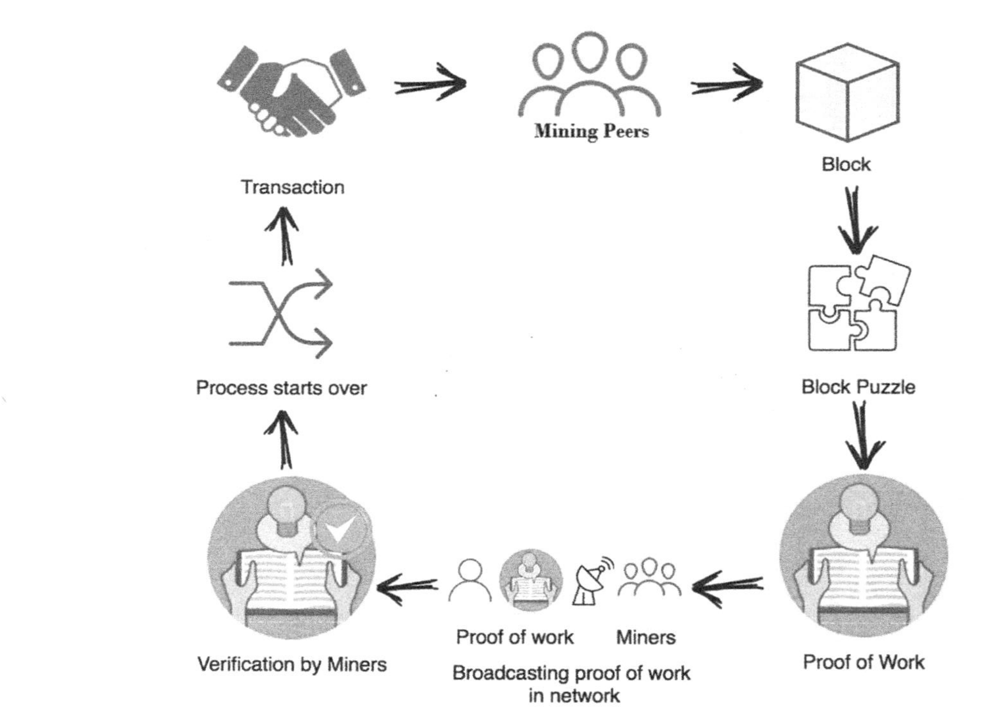

FIG. 14.4 Bitcoin components. *(Source: upgrad.com.)*

14.4.1.1 Bitcoin wallets

A Bitcoin wallet shows ownership of a particular account address to the user and contains a public key, private key, and digital signature, which is unique for that particular wallet.

One main advantage of this wallet usage is that the wallets are not stored on the public blockchain network; instead they are stored in the user's machine, as the database of this wallet is created at the user's end, which makes the blockchain network more secure and efficient. As earlier mentioned, the wallet contains private keys and public keys. Private keys are generated using a random number. This random number is then used to generate a public key using elliptic curve cryptography by using ECC point multiplication method. From this public key, a cryptographic hash is generated, which is not reversible to generate a bitcoin address [12].

Fig. 14.5 shows the procedure of converting a public key, which has been generated from a private key to a Bitcoin address.

In origin, Bitcoin address are alphanumeric characters. Base 58 encoding is used; this encoding scheme doesn't allow zero (0), O (capital o), I (capital i),

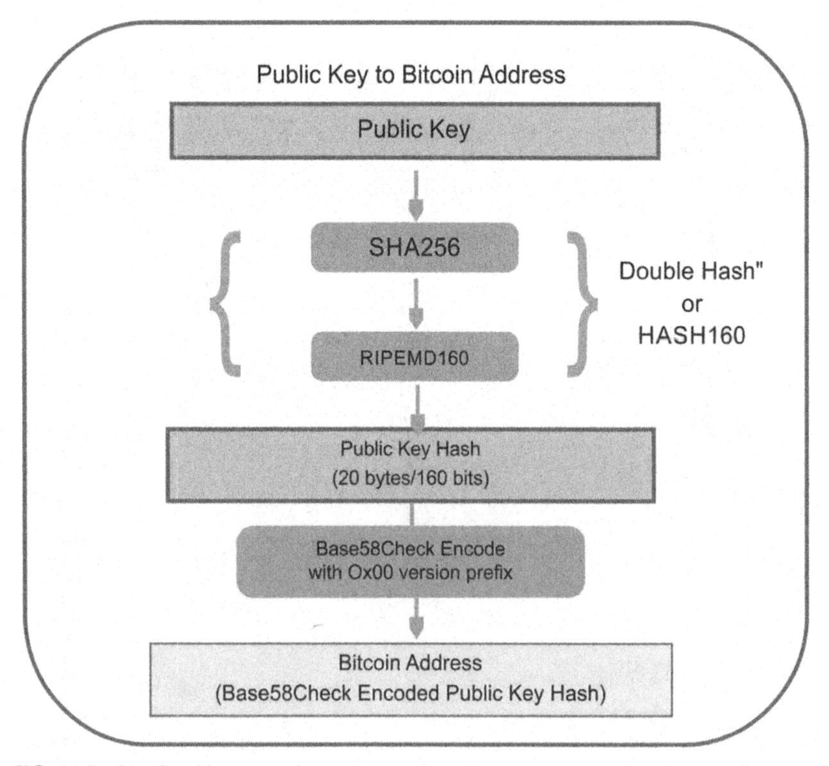

FIG. 14.5 Bitcoin address mapping.

l (lower l), and +,/ signs. Bitcoin address are formed using the following group of characters and numbers:

123456789ABCDEFGHJKLMNPQRSTUVWXYZabcdefghijkmnopqrst uvwxyz

Nowadays, there are only three different address formats used by Bitcoin:

(1) **P2PKH** (Pay to Pubkey Hash): 1KwBRs6CioGM2pFdzQsxyrSZ9ynJQ-r7Amd, addresses start with a 1.
(2) **P2SH** (Pay to script hash): 3DnW8JGpPViEZdpqat8qky1zc26EKbXnmM, addresses start with 3.
(3) **Bech32** (SegWit): bc1qngw83fg8dz0k749cg7k3emc7v98wy0c74dlrkd, addresses start with bc1.

Fig. 14.6 shows steps to generate Bitcoin wallet address using Open-source libraries which are based on algorithmic hash functions SHA 256, RIPEMD 160 are used in double hash computation in order to generate bitcoin wallet address [12].

0: Private Key:
a4f228d49910e8ecb53ba6f23f33fbfd2bad442e902ea20b8cf89c473237bf9f
0: Private Key Base58:
C6t2iJ7AXA2X1KQVnqw3r7NKtKaERdLnaGmbDZNqxXjk
1: public key:
03564213318d739994e4d9785bf40eac4edbfa21f0546040ce7e6859778dfc e5d4
2: SHA-256 public key:
482c77b119e47024d00b38a256a3a83cbc716ebb4d684a0d30b8ea1af12d4 2d9
3: RIPEMD-160 hashing on the result of SHA-256 public key:
0c2c910a661178ef63e276dd0e239883b862f58c
4: RIPEMD-160 hash with version byte:
000c2c910a661178ef63e276dd0e239883b862f58c
5–6: 2 * SHA-256 hash for RIPEMD-160 hash with version byte:
c3c0439f33dc4cf4d66d3dd37900fc12597938a64817306b542a75b922321 3e0
7: CheckSum: *c3c0439f*
8: 25 Byte Binary Bitcoin Address:
000c2c910a661178ef63e276dd0e239883b862f58cc3c0439f
9: Bitcoin Address: *127NVqnjf8gB9BFAW2dnQeM6wqmy1gbGtv*

FIG. 14.6 Bitcoin address generation using SHA 256.

Nondeterministic wallet

Wallets of the type we have generated above are nondeterministic, in which we have randomly chosen private keys. These private keys are to be used by Bitcoin core clients, so in order to use them, one has to store or back up of these private keys somewhere; this storage of private keys in a wallet makes this wallet a JBOK (just a bunch of keys) [14], so nondeterministic wallets are not so feasible, because backup and storage have some simplicity and security challenges [15].

Deterministic wallet

In deterministic wallets, key generation is derived from one root key, also known as a seed key. Seeded wallets contain a chain of keys that are basically generated from this seed key, so that if one has to create a new key it uses a one way hash from the previous private key, and so on. These private keys are linked in sequence with each other and can be recreated to backtrack from the seed key—so the seed key has great importance as all the keys are generated from it, and therefore you only need to backup the seed key at creation and it will be enough to recreate all the private keys generated from it.

The seed is also required for wallet additional functionalities, such as if you want to import or export the wallet, you only need to migrate with the seed key, which in turn recreates the sequence of all private keys generated from it, making migration easy.

A wallet application that implements deterministic wallets with mnemonic code will show the user a sequence of 12–24 words when first creating a wallet. That sequence of words is the wallet backup and can be used to recover and recreate all the keys in the same or any compatible wallet application.

Bitcoin client types

Bitcoin client has various types, depending on how you download and use the bitcoin platform. It is mainly a piece of software installed on the end user's machine, which contains the facility of creating private key generation and security, and performing transactions on behalf of this private key without using any identity, thus making it pseudonymous. In addition, the transactions related to that particular account are visible to all the nodes in the Bitcoin network, which provides information about the state and network of transactions.

Bitcoin client and Bitcoin wallets are nowadays used interchangeably. Previously, Bitcoin client was of two types:

(1) Full client
(2) Light client or thin client

Full client

also known as "Bitcoin core" or "Satoshi client." This full client contains and implements all components of Bitcoin including all history of transactions and blocks in the Bitcoin network, starting from the genesis block. Generally, it takes two days to update with all the blockchain history and require100 Gb of storage. This full client implements all aspects of the bitcoin system, including wallets, a transaction verification engine with a full copy of the entire transaction ledger (blockchain), and a full network node in the peer-to-peer bitcoin network.

When you go to http://bitcoin.org/en/choose-your-wallet and select "Bitcoin Core" to download the reference client, you can choose the compatible operating system of your machine and download the executable installer. You will see types of available wallets for bitcoin, as shown in Fig. 14.7.

Mobile wallets

Mobile wallets are software or apps that can be easily used on cellular phones; this is mainly for people who want to use Bitcoins on a daily basis. Hot wallets, which are wallets that can easily connect to the internet through smartphones or personal computers, come under this category. If your phone and wallet are password protected, you can always download your wallet onto another phone, restore it, and then change your passphrase.

In a similar vein to mobile wallets, web wallets can be used to easily access your funds on the move, and your private keys are stored on a server that is online and controlled by a third party. If not properly secured, however, people can still gain access to your private keys, and take over your funds.

Desktop wallets

In desktop wallets you can store your private keys on your hard drive or SSD, which is more secure than mobile or web wallets.

Hardware wallets

This is the most secure and protected form of wallet, as the private key is stored in a hardware device, which can be directly communicate with your end machine. Some hardware wallets also come with additional features, like screens which show your private key phrase, balances, accounts address with which you have transacted, etc.

14.4.2 Bitcoin network nodes

Bitcoin nodes are synonymous with any participating entity in the Bitcoin blockchain network. This may be a computer, a smartphone, or a browser-based wallet running over the top of blockchain. In order to be a part of the Bitcoin network, it must be connected to it using some address anonymously, once user is connected to the network, it can be a miner or just verifier of the transaction, to be a miner you need to require some additional hardware and

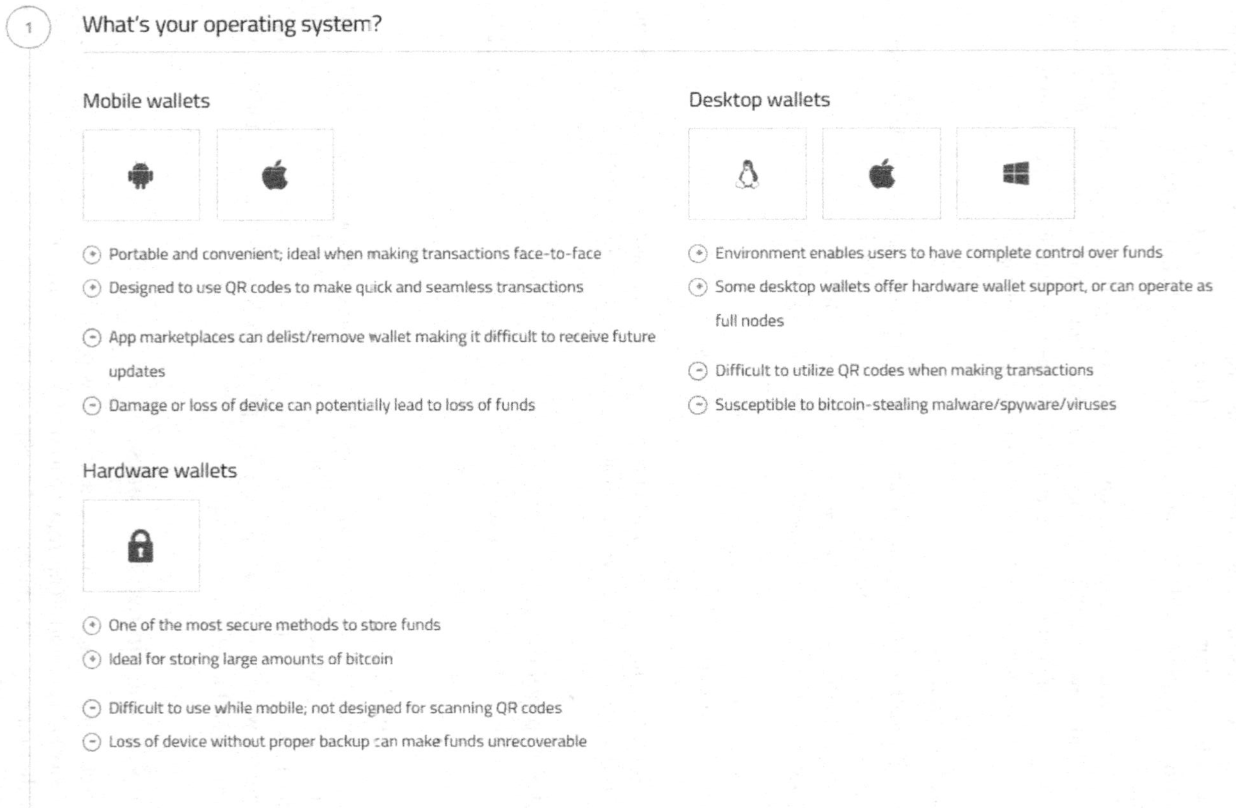

1

What's your operating system?

Mobile wallets

- (+) Portable and convenient; ideal when making transactions face-to-face
- (+) Designed to use QR codes to make quick and seamless transactions
- (−) App marketplaces can delist/remove wallet making it difficult to receive future updates
- (−) Damage or loss of device can potentially lead to loss of funds

Desktop wallets

- (+) Environment enables users to have complete control over funds
- (+) Some desktop wallets offer hardware wallet support, or can operate as full nodes
- (−) Difficult to utilize QR codes when making transactions
- (−) Susceptible to bitcoin-stealing malware/spyware/viruses

Hardware wallets

- (+) One of the most secure methods to store funds
- (+) Ideal for storing large amounts of bitcoin
- (−) Difficult to use while mobile; not designed for scanning QR codes
- (−) Loss of device without proper backup can make funds unrecoverable

FIG. 14.7 Types of wallets in Bitcoin. *(Source Bitcoin.com.)*

computing resources but to be verifier you can connect simple with your general computers [16].

The main idea behind the working of the Bitcoin network is its decentralized nature. The Bitcoin network is evolving with great speed and, as the Bitcoin network is expanding, the value of Bitcoin money is also increasing.

This decentralization of the network provides advantages of efficient and secure data communication, which can work in a trustless environment, as centralization is not an issue here, all the vulnerabilities relating to centralized servers being removed [17]. Fault tolerance consensus, cryptographically linked data, and validation with digital signatures make it usable in scenarios where data hacking or leakage is a major concern.

There are two types of nodes in Bitcoin:

(1) **Light node**—nodes which contains only the root hash of a block not all block transactions.
(2) **Full Node**—nodes which actually store all the blockchain, history starting from the genesis block.

14.4.2.1 Functionalities of full node

Every blockchain network is made up of nodes that are connected in a peer-to-peer network. While every node is equal in capabilities, their responsibility of being a part of the network may be different: they may be full node, reference client, i.e., bitcoin client, SPV, solo miner, etc. Fig. 14.8 shows four key components of any node. If a node has all four components, then it is treated as a full node. The components are: routing, the blockchain database, mining, and wallet services.

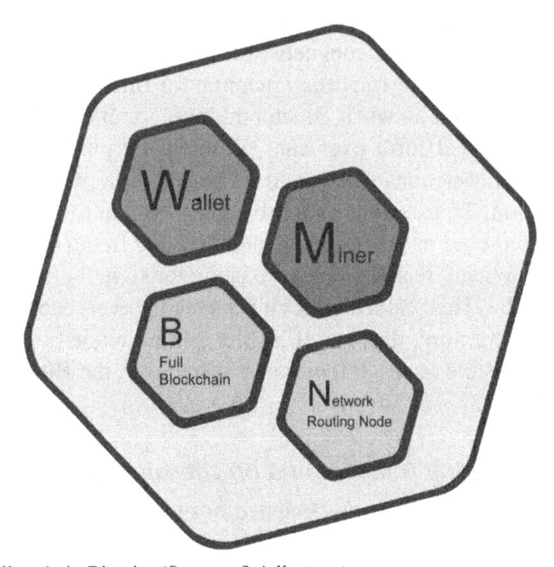

FIG. 14.8 Full node in Bitcoin. *(Source: Orielly.com.)*

14.4.2.2 *Thin client*

Thin client contains only block headers, not all the transaction history of the blockchain on the end user's machine. It stores only block header and request for required information such as transaction details, a particular block, etc. It stores the state roots in block header and, according to that, can verify and validate the transactions or blocks. It is mainly useful for smartphones and low capacity computers, which cannot store the huge amounts of data that are in GBs required for storing the entire history of blockchain (Fig. 14.9).

Note: Generally "Bitcoin clients" and "Bitcoin nodes" are used Synonymously.

14.4.3 Bitcoin consensus

14.4.3.1 *Broadcasting transactions to the bitcoin network*

Bitcoin transaction main concept started here as Bitcoin is first application of Blockchain and Blockchain provides transfer of data or transfer of any value in trustless environment, bitcoin network exactly provide this, when Bitcoin transaction is become valid with authenticated signatures then it reaches to the bitcoin network now it propagated through the bitcoin nodes, it does not matter here if the nodes are trusted or not until and unless they are propagating the transaction, also nodes does not need to verify the source of transaction as it does not contain any confidential data and is getting signed by the authenticated owner, also it do not have private information like credentials or other account related information like in credit card in case of fiat currency, bitcoin transactions are publicly broadcasted through any medium for propagation, it does not require any encrypted network, also the main purpose of bitcoin transaction is that it reaches to all nodes in bitcoin network, nodes are also not worry about the identity of the sender. The transporting medium for Bitcoin transactions can be insecure networks, such as wi-fi, Bluetooth, NFC, or Barcodes. Transactions can be sent in encoded form over chat or on public platforms such as like smileys, Bitcoin transactions that would be posted on a public platform. [18] In Communication, Transactions can also be sent as text message over skype, can be transmitted over packet radio or satellite relay frequency by doing this, Bitcoin can solve many real challenges existing today in the context of secure transmission media. This feature makes it possible for everyone to crate Bitcoin transactions. Furthermore, the size of a Bitcoin transaction is 300–400 bytes, which make it possible to reach thousands of nodes in the Bitcoin network in a feasible time [12].

14.4.3.2 *Propagating transactions on the bitcoin network*

The Bitcoin network is inherently designed in such a way that propagation of only valid transactions are possible. Once a transaction is validated and reaches the network, then it is its originator's responsibility to send it to three or four

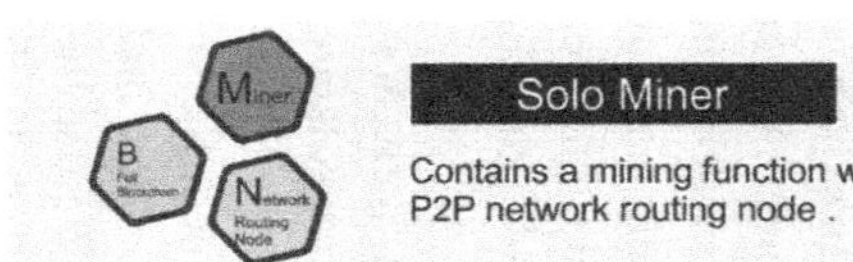
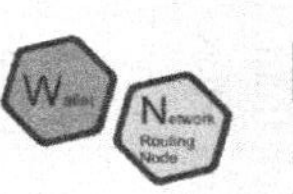
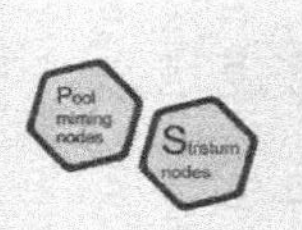
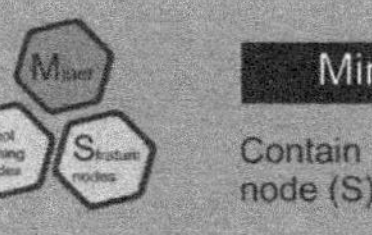
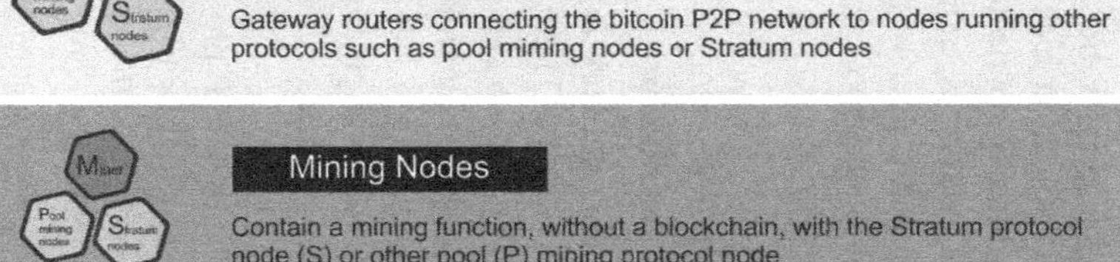
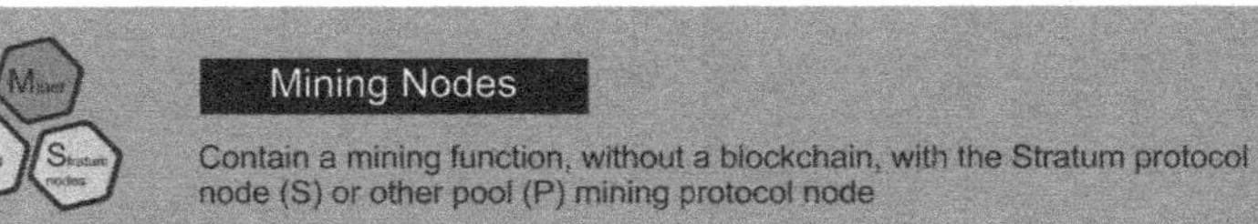
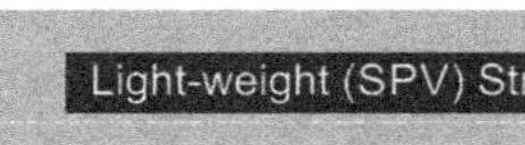

FIG. 14.9 Types of light nodes in Bitcoin with their components. *(Source: Orielly.com.)*

neighbors. If this transaction is validated by these nodes, then synchronously, validation messages reach the originator; if these nodes reject validation, then a rejection message synchronously reaches the originator. The size of transactions and blocks small enough that within seconds, transactions are propagated and reach to all nodes of the network in a ripple through a gossip protocol; these protocols are also designed in a way that they are able to combat certain attacks, like spamming, denial of service, or other types of attacks. Every node in the network individually validates the transaction independently so that a malicious transaction could not propagate further, because every node is connected to every other node in a peer-to-peer network manner, and it would be identified by nodes whether a transaction is valid or malicious, based on the number of messages of validation or rejection received from its peers [19]. Every node in this network creates a loosely connected mesh topology and all nodes carry equal roles in the network. Messages, transactions, and blocks are propagated in the overall network through this peer-to-peer connection, and so on. This propagation of transactions is done exponentially until all nodes receive the same message across the network.

14.4.3.3 Bitcoin network

Peer-to-peer network architecture

Bitcoin's P2P (peer-to-peer) network is a loosely connected mesh topology, where each peer is treated as equal. There is a difference between existing client server architecture in a way that no controller, centralized storage, or centralized decision-making is happening in this network; instead, it is a decentralized network architecture designed in a way that the network itself maintains its integrity on data and decision through consensus. It is mainly designed for digital cash systems, where this digital cash value is maintained and controlled in a decentralized manner.

The Bitcoin network runs a P2P protocol. Other than that, there are other protocols such as Stratum, which is used for mining, these protocols are connected to the bitcoin network through a router gateway and then the Bitcoin network is extended to run these protocols. Stratum servers connect the stratum mining nodes to the P2P Bitcoin network via the Stratum protocol and then bridge the Stratum protocol to the Bitcoin P2P 139 protocol. We use the term "extended bitcoin network" to refer to the overall network that includes the Bitcoin P2P protocol, pool mining protocols, the Stratum protocol, and any other related protocols connecting the components of the Bitcoin system.

Network discovery

Network discovery starts with an originating node. First, in order to discover a network, it must find a new neighbor node—geographical location does not matter here, any node which can be found from the originating node can be its first peer and will connect with it. The node is selected at random as Bitcoin

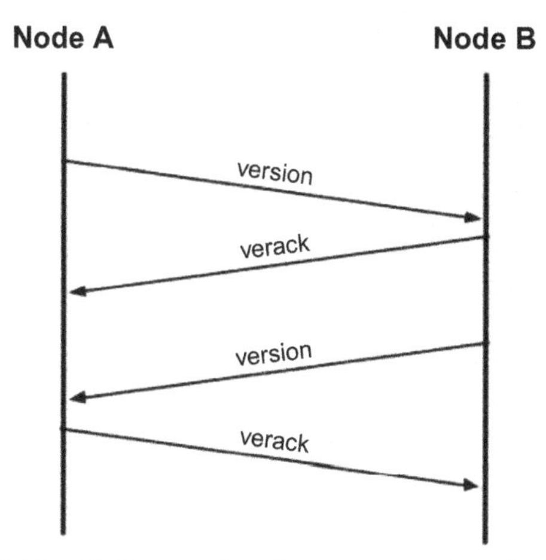

FIG. 14.10 The initial handshake protocol used to connect to peer node.

is not geographically confined. To connect with this peer, a protocol known as a handshake protocol establishes a TCP (transmission control protocol) connection, generally with port 8333. After connecting through this port, the node will start a handshake by transmitting a version message, which gives information regarding protocol-version, local services supported by the node, current IP address of the remote node as seen from this node, IP address of the local node, a subversion, which shows the type of software running on this node, and the block height of this node's blockchain. In Fig. 14.10 "Version" shows Version of protocol used by node which started connection and Verack shows Acknowledged version of Node with its own version, no communication will start until both peers have exchanged their versions.

14.4.4 Bitcoin miners

The mining process is carried out by some specified nodes in the network that are dedicated for mining—these nodes are called miners. After getting created and validated by miner, "Block" has been added in to Blockchain and in return of this incentive reward has been given to the node who created that "block", in bitcoin reward is currently 12.5 bitcoin, there is some additional incentive also for miners knowns as Transaction fees which is given by nodes who created that transaction.

A new block is mined every 10 min. Transactions that are added in a validated block are considered as confirmed transactions.

The process of getting incentivizes and creating blocks is known as mining because, after solving a complex mathematical puzzle, the block is generated and produces Bitcoin as a reward. The amount of Bitcoin reward decreases after every four years; currently it is 12.5 BTC, and will be 6.5 BTC in 2022.

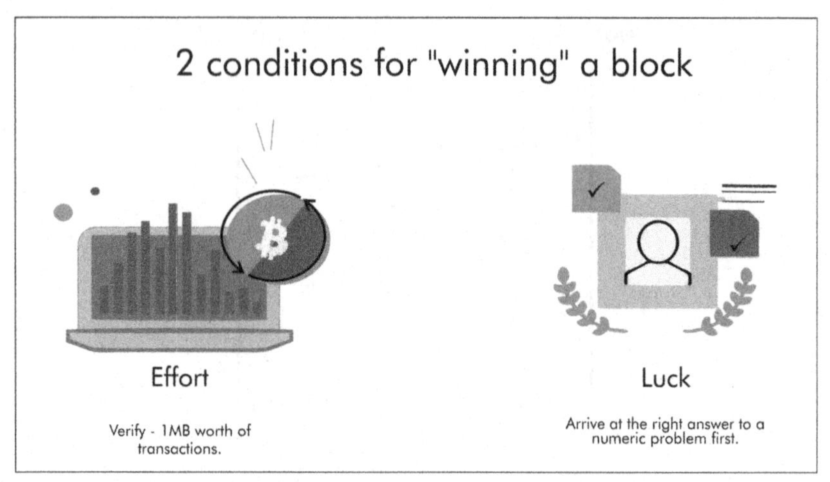

FIG. 14.11 Bitcoin mining process.

14.4.4.1 Bitcoin mining process

Fig. 14.11 shows that in order to win a block which is to be added into the main blockchain, one has to verify 1 MB worth of transactions in Bitcoin (Bitcoin block size is 1 MB), and also solve a complex mathematical puzzle in order to find a nonce (number only used once) value. Nonce is a random number that varies with time so that difficulty level also varies in Bitcoin.

14.4.5 Bitcoin transactions

Everything than can be changed or transfer in bitcoin is through transactions; these transactions are actually transferring values, storing value, creating value, or updating value in the global blockchain state in Bitcoin. The transaction is a data structure possessing a certain format with specific fields. When this creation, updating, or transfer of value is validated through a network in a blockchain and added to a block, then the group of these transactions in the form of the block is added to the blockchain. Every transaction that executed on Blockchain is maintained in public ledger known as "Bitcoin Global double entry Ledger."

14.4.5.1 Transaction lifecycle

In Bitcoin, when a transaction is created by a participant in a network, that participant may be a light node. One or more signatures are required to initiate this transaction. After signing, this transaction is propagated through the network via a gossip protocol until it reaches all the nodes in the network. After this validation by the participating node, miners verify the transaction and add it into a block they are creating. This block is then propagated to all participating nodes and, if that block is validated by participants through consensus, it will be added into the blockchain after certain confirmation of subsequent blocks. In the case of Bitcoin, a block is confirmed after six more blocks are added in to it. When a block is added in the longest sequence of blockchain, this transfer of value can be transferred to another recipient through creating a new transaction, and the transaction cycle starts again.

14.4.5.2 Creating transactions

Transaction creation is generally done by a participant in the network with an authorized account, but this is not only the way: a different participant who might not have an authorized account or who is not an authorized signer can create a transaction online or offline [20]. For example, in a real life scenario where an account department can prepare payable checks on behalf of the CEO and then have them signed by the CEO, the same can happen in Bitcoin, where an account department employee can create bitcoin transactions to which the CEO then applies their own digital signature to make them valid. Here, the source of funds in the Bitcoin transaction is a specific previous transaction rather than an account. The transaction is then signed by the source fund account owner or owners. After this signing of the transaction by authenticated owners, it gets validated and can be executed to transfer the funds. This valid transaction reaches the Bitcoin network and every participant in the network when it reaches a miner for verification and is added into the miner's block.

14.4.5.3 Transaction structure

In Bitcoin, a transaction is generally a data structure that has a source address, an input, and an output. Its input and output do not require identities—or its identities cannot be defined from the account address—as the input and output information is integrated with confidential information known only to the owner, those who are authorized to know that information, or those who maliciously know it [21] (Figs. 14.12 and 14.13).

Bytes	Name	Data Type	Description
4	version	int32_t	Transaction version number (note, this is signed); currently version 1 or 2. Programs creating transactions using newer consensus rules may use higher version numbers. Version 2 means that BIP 68 applies.
Varies	tx_in count	compactSize uint	Number of inputs in this transaction.
Varies	tx_in	txIn	Transaction inputs. See description of txIn below.
Varies	tx_out count	compactSize uint	Number of outputs in this transaction.
Varies	tx_out	txOut	Transaction outputs. See description of txOut below.
4	lock_time	uint32_t	A time (Unix epoch time) or block number. See the locktime parsing rules.

FIG. 14.12 Bitcoin transaction structure.

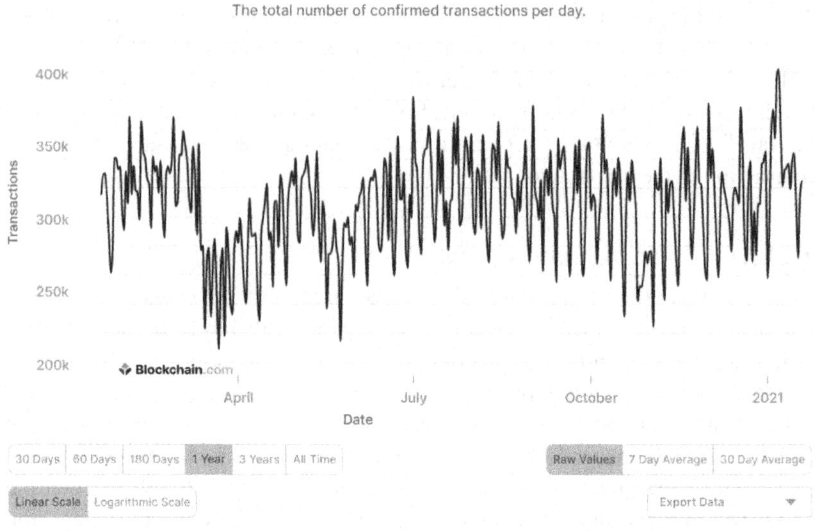

FIG. 14.13 The number of daily confirmed transactions highlights the value of the Bitcoin network as a way to securely transfer funds without employing a third party. *(Source: Blockchain.com as on 21 Jan 2021.)*

14.4.5.4 Summary

This section discussed the introduction of blockchain and where blockchain is used. Adequacy standards for blockchain architecture has also been discussed—it is not necessary that all standards are required by every blockchain application, as there may be some other simpler alternatives for a particular application. Bitcoin blockchain is explained in terms of its architecture and components and it is found that we are having limited customized features in Bitcoin. Components of Bitcoins such as Wallets, Size of Metadata we integrate in Bitcoin Transaction, a computationally complex consensus known as Proof of Work, its mathematical formation is discussed in above section.

14.5 Introduction of Ethereum

Ethereum and Bitcoin are similar in certain basic ways, such as Ethereum being open-source public blockchain the same as Bitcoin, a decentralized peer-to-peer network, all nodes being connected with each other, proof of work consensus algorithm being used for synchronization of state updates that are to be done to maintain global ledger state, and also, blocks are cryptographically linked with each other to provide immutability; Ethereum also has its own cryptocurrency, known as Ether.

The difference between Bitcoin and Ethereum is the capability in Ethereum of executing smart contracts—executable codes written in general purpose programming language like solidity—in terms of transactions. Unlike Bitcoin [16], which has a very limited scripting language, Ethereum is able to execute programs that are Turing complete, which means they must be deterministic in nature, and can execute code of arbitrary and unbounded complexity.

14.6 Ethereum's evolution

Ethereum was the creation of Vitalik Butarin's mind: he wanted to use blockchain technology for a broader aspect of implementation and use, so he introduced Ethereum in 2015. After this, evolution occurred in four stages, and every change required a hard forks initiator: in Ethereum, hard forks are used to change functionality to existing blockchain when there is no compatibility of these changes to previously existing nodes. The four stages were as follows:

(1) Frontier
(2) Homestead
(3) Metropolis
(4) Serenity

Furthermore, some hard forks occurred between these stages, such as Ice Age, DAO, Tangerine Whistle, Spurious Dragon, Byzantium, and Constantinople. These hard forks are used to change something in existing blockchain or used in reference of any event [22]. Details of these intermediate hard forks are beyond the scope of this chapter.

Development stages are shown on the timeline in Table 14.1, which is "dated" by block number.

TABLE 14.1 Ethereum's four stages of development.

Block no	Hard fork name	Duration	Functionality
Block #0	Frontier	Lasting from July 30, 2015, to March 2016	Start of Ethereum
Block #1,150,000	Homestead	Launched in March 2016	A hard fork to introduce an exponential difficulty increase, to motivate a transition to PoS when ready.
Block #4,370,000	Metropolis Byzantium	September 2017	Very small change in it from previous fork, it includes requirements for implementing Zero Knowledge Proofs" (e.g., zk-snarks)
	Serenity (Ethereum 2.0)	January 12, 2020	Serenity is the last phase and will move Ethereum from proof of work to proof of stake.

Apart from the main development stages given in Table 14.1, the following intermediate hard forks were introduced in Ethereum:

(1) Block #1,192,000 DAO—A hard fork that reimbursed victims of the hacked DAO contract and caused Ethereum and Ethereum Classic to split into two competing systems.
(2) Block #200,000 Ice Age—A hard fork to introduce an exponential difficulty increase, to motivate a transition to PoS when ready.
(3) Block #2,463,000 Tangerine Whistle—A hard fork to change the gas calculation for certain I/O heavy operations and to clear the accumulated state from a denial of service (DoS) attack that exploited the low gas cost of those operations.
(4) Block #2,675,000 Spurious Dragon—A hard fork to address more DoS attack vectors, and another state clearing. Also, a replay attack protection mechanism.

14.7 Architecture of Ethereum

14.7.1 Ethereum virtual machine

Ethereum virtual machine (EVM) is a platform for developers to develop decentralized applications over Ethereum blockchain network. Objects created by programmers are safe from modification because an underlying Ethereum blockchain network exists, which provides functionalities of blockchain to the developer for their program or DApp. Fig. 14.14 explains the architecture of EVM.

EVM architecture consists of all state objects which are written in solidity code, Ethereum Compiler then converted that code into fixed binary format

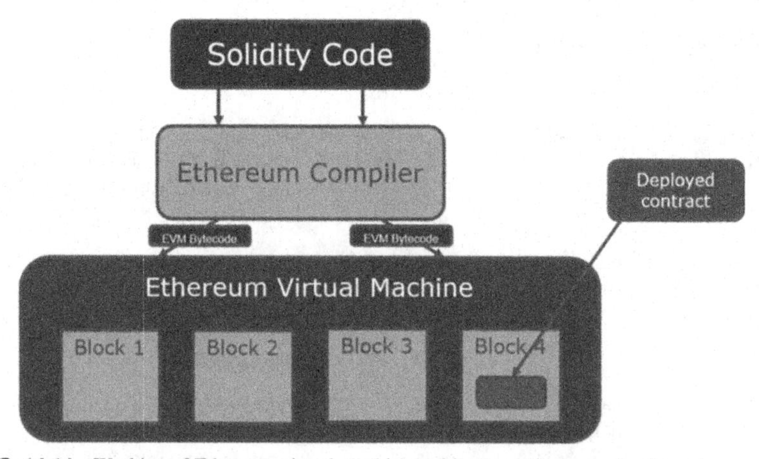

FIG. 14.14 Working of Ethereum virtual machine architecture. *(Source: edureka.co.)*

known as EVM byte code, this byte code then run on EVM which contains all smart contract deployed in Blocks.

14.7.1.1 Example of smart contract in solidity

Ether and gas

Ether is an inherent cryptocurrency used in Ethereum to pay for transactions over the network [23]. Ethers are used to buy "gas" for computation of smart contract codes over EVM.

Ether is a metric unit and has a range of denominations, which helps to accurately pay for transactions and gas. The smallest denomination or base unit is called a Wei. The denominations, along with their specific names, are as follows:

Units	Wei value	Wei
Wei	1 Wei	1
Ether	1e18 Wei	1,000,000,000,000,000,000

Gas provides a constraint on overusing the network, as gas is used to pay for computations performed on the Ethereum network; so if any smart contract requires computations greater than the gas available, it would be halted or stopped. It also puts constraints on use of indefinite loop execution in smart contracts. Operation computations have a fixed gas requirement in Ethereum, and maximum available gas for executing smart contracts is known as the gas limit.

Gas price is the cost of gas in terms of tokens like Ether and its other denominations. To stabilize the value of gas, the gas price is a floating value such that if the cost of tokens or currency fluctuates, the gas price changes to keep the same real value.

Gas fee is effectively the amount of gas needed to be paid to run a particular transaction or program (called a contract).

Hence, if someone tries to run a piece of code that runs forever, the contract will eventually exceed its gas limit and the entire transaction that invoked the contract will be rolled back to its previous state [24].

14.8 Ethereum components

Ethereum architecture consists of the following components:

(1) Ethereum accounts
(2) EVM
(3) Ethereum networks
(4) Ether and gas
(5) Mining

14.8.1 Ethereum accounts

Ethereum accounts are of two types:

(1) External accounts
(2) Contract accounts

Ethereum blockchain is also known as single world state computer, All Ethereum accounts in this network are referred as state objects of this world state computer where each state represent a state of Ethereum network, these states are well defined and includes account balances. smart contract accounts are defined by the memory storage and balances [24].

These accounts are generally controlled with the help of public key cryptography algorithms like RSA (Rivest-Shamir-Adleman). The main purpose of external accounts is to serve as a medium for users to interact with the Ethereum blockchain, while contract accounts are used to contain smart contract codes, and these contracts are initiated by externally owned accounts. These contracts are written in general-purpose programming language, which possesses Turing completeness features such as solidity, serpent, etc.

Smart contracts are stored in a specific format known as bytecode that can run only on Ethereum Virtual Machine [26].

14.8.2 Mining in Ethereum

Like bitcoin Ethereum also uses an incentive model and proof of work consensus. Fig. 14.15 shows the mining process in Ethereum.

Fig. 14.16 is an example of a one-use case for a crowd funding DApp, showing its smart contract excerpt in solidy. It takes a decentralized methodology to address the problem, as explained in the figure.

14.8.3 Ethereum clients

Client verifies all transactions and creates a block, and maintains security and integrity of the Ethereum network. There are various networks provided by the Ethereum network, such as Testnet, Mainnet, Rinkeby, and Ropsten. If we want to simulate any DApp before deploying on to Mainnet, Testnet can be used, as Testnet provides all the Ethers and transfer of Ethers, which have no value in the real world but can be used for simulation of application.

Ethereum has several interoperable implementations of the client software, the most prominent of which are Go-Ethereum (Geth) and Parity.

14.8.3.1 Geth

Geth (Go-Ethereum) is a command line interface for running Ethereum nodes implemented in Go language. Using Geth, you can join an Ethereum network, transfer Ether between accounts or even mine Ethers.

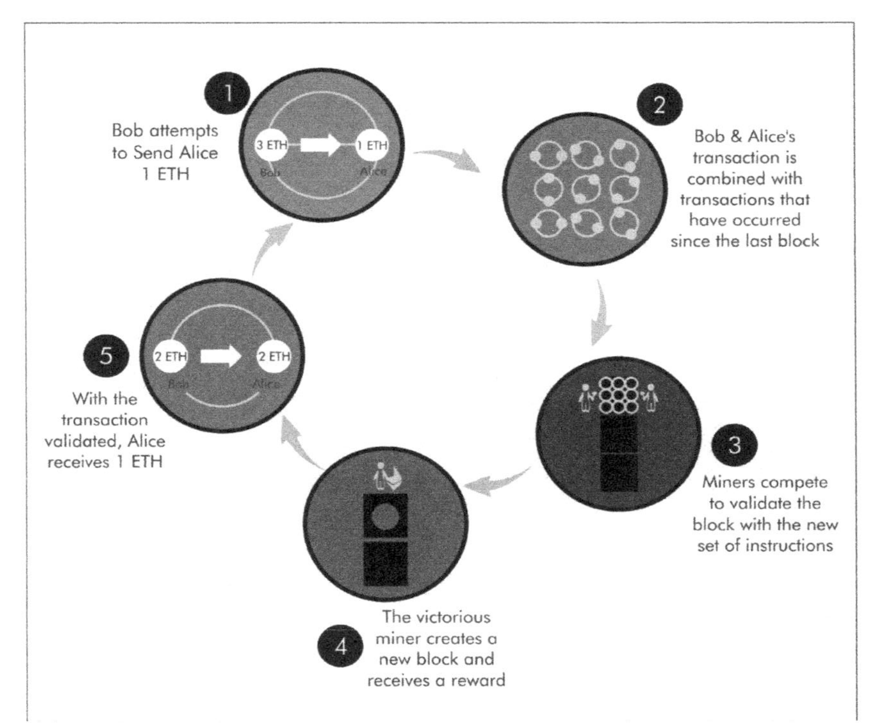

FIG. 14.15 Proof of work mining in Ethereum is called Ethash. *(Source: Edureka.com.)*

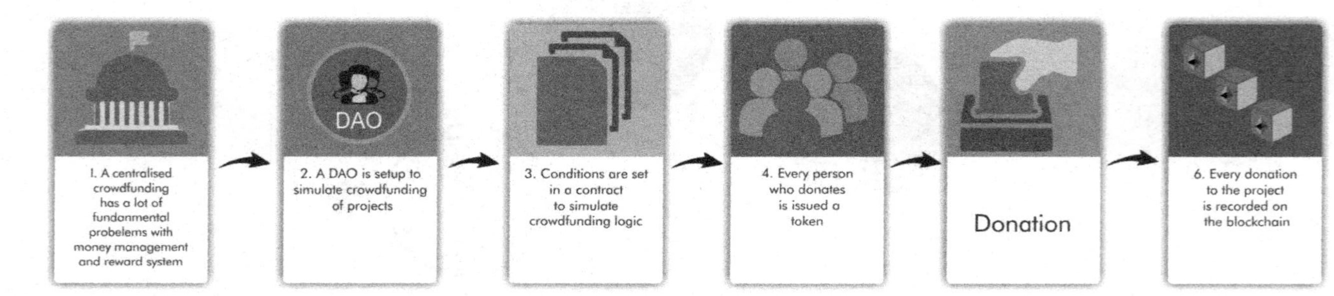

FIG. 14.16 Decentralized use case for crowd funding DApp. *(Source: Edureka.com.)*

14.8.3.2 Parity

Parity Ethereum is an open-source software solution that allows an individual to run a node on the public Ethereum network, or any other blockchain network that uses the Ethereum protocol. The software constitutes an alternative to the Geth Ethereum client [27].

14.8.3.3 Ethereum clients (nodes)

Full node

Advantages and disadvantages Choosing to run a full node helps with the operation of the networks you connect it to, but also incurs some mild to moderate costs for you. Ethereum full node contains all transactions from all Ethereum networks, while Ethereum light client retrieves the blockchain state from nodes operating as a light protocol server [28]. Light client and full client both have advantages and disadvantages, as follows.

Advantages

- Supports the resilience and censorship resistance of Ethereum-based networks
- Authoritatively validates all transactions
- Can interact with any contract on the public blockchain without an intermediary
- Can directly deploy contracts into the public blockchain without an intermediary
- Can query (read-only) the blockchain status (accounts, contracts, etc.) offline
- Can query the blockchain without letting a third party know the information you're reading

Disadvantages

- Requires significant and growing hardware and bandwidth resources
- May require several days to fully sync when first started
- Must be maintained, upgraded, and kept online to remain synced

14.8.3.4 Software requirements for building and running a client (node)

This section covers Parity and Geth client software. It also assumes you are using a Unix-like command-line environment. The examples show the commands and output as they appear on an Ubuntu GNU/Linux operating system running the bash shell (command-line execution environment). Typically, every blockchain will have its own version of Geth, while Parity provides support for multiple Ethereum-based blockchains (Ethereum, Ethereum Classic, Ellaism, Expanse, Musicoin) with the same client download.

14.8.4 Ethereum transaction

Ethereum transactions are the same as Bitcoin with some prespecified structure and initiated by an externally owned account (EAO). However, a smart contract integrated in the transaction can either be initiated by an EAO or in response to a trigger initiated by the EAO to another smart contract, which in turn may invoke another smart contract contained in a different account> After this invocation of smart contract or transaction, the output of smart contract or transaction will be saved on the blockchain network. These transaction and smart contract executions cause a single atomic state change in the global state of the Ethereum network. It must be noted that smart contracts can't run on their own: a transaction is always needed to execute smart contracts on EVM.

14.8.4.1 Structure of Ethereum transaction

Each client and application that receives a serialized transaction will store it in memory using its own internal data structure, perhaps embellished with metadata that doesn't exist in the network serialized transaction itself. The network serialization is the only standard form of a transaction. A transaction is a serialized binary message that contains the following data:

- **Nonce**—A sequence number, issued by the originating EOA, used to prevent message replay
- **Gas price**—The price of gas (in Wei) the originator is willing to pay
- **Gas limit**—The maximum amount of gas the originator is willing to buy for this transaction
- **Recipient**—The destination Ethereum address
- **Value**—The amount of Ether to send to the destination
- **Data**—The variable-length binary data payload
- **v, r, s**—The three components of an ECDSA (elliptic curve digital signature algorithm) digital signature of the originating EOA

The transaction nonce

Unlike Bitcoin, there are two nonces in Ethereum: block nonce and transaction nonce. Block nonce is used in proof of work in order to cause a miner to confirm a block in a blockchain [29]. Bitcoin is cryptocurrency based on blockchain, where transactions are done to transfer currency from sender to receiver with UTXO (unspent transaction output)-based protocol, which means the remaining amount after transmitting the intended amount to the receiver will also be transfer in the form of UTXO as a different transaction to another address. Blockchain traces all UTXOs and their rightful owners. In practical terms, the nonce is an up-to-date count of the number of confirmed (i.e., on-chain) transactions that have originated from an account.

Transaction gas

Gas is used for computation in world state computers on the Ethereum Network. While Ether is a reward, gas is used for investment or spending; however, it is also virtual currency but it is used to determine the amount of resources required for a transaction execution or to control resources for transaction execution by limiting it by gas limit [30].

Transaction recipient

The recipient of a transaction is specified in the "to" field. This contains a 20-byte Ethereum address. The address can be an EOA or a contract address. For example, Transaction Recipient | 107 www.EBooksWorld.ir Ethereum requires no further validation of this field. Any 20-byte value is considered valid. If the 20-byte value corresponds to an address without a corresponding private key, or without a corresponding contract, the transaction is still valid. Ethereum has no way of knowing whether an address was correctly derived from a public key (and therefore from a private key) in existence.

Transaction value and data

The main "payload" of a transaction is contained in two fields: value and data. Transactions can have both value and data, only value, only data, or neither value nor data. All four combinations are valid. A transaction with only value is a payment. A transaction with only data is an invocation. A transaction with both value and data is both a payment and an invocation. A transaction with neither value nor data is just a waste of gas, but it is still possible [31].

There are transaction which are initiated by both types of account in Ethereum, either by EAO or by contract account. When you are transmitting only value then it is called a payment transaction. Payment transactions may have a destination account that is an EAO or a contract account. If the destination account is an EAO, Ethereum will record that value and make a state change and will send that value to the destination account's balance. If this account is used for the first time, then it will be added to the client's internal representation of the state and the internal balance of the client is initialized with this payment. If the destination account is not an EAO and a contract account, then Ethereum will execute the smart contract contained in this transaction and called the function in it. If there is no data in your transaction, then EVM will call a fallback function, and if that function in a smart contract is payable, it will execute according to the conditions written in it. If there is no fallback function in the transaction's smart contract, then the effect of the transaction will be to increase the balance of the contract like payment to a wallet. In addition, a contract can reject incoming payments by coding some conditions in a function; if the function terminates successfully without any rejection then the contract's state is updated and the contract's Ether balance is increased.

Special transaction

In contract creation, one special case that we should mention is a transaction that creates a new contract on the blockchain, deploying it for future use. Contract creation transactions are sent to a special destination address called the zero address; the "to" field in a contract registration transaction contains the address 0×0. This address represents neither an EOA (there is no corresponding private-public key pair) nor a contract. It can never spend ether or initiate a transaction. It is only used as a destination, with the special meaning "create this contract."

14.8.5 Ethereum networks

Ethereum networks are defined by the Ethereum Yellow Paper, authored by Vitalik Buterin and Nick Szabo, which defines formal specifications about Ethereum networks, and they may be interoperable or may not. Among these networks, the main platform of Ethereum defines two types of networks: Mainnet and Testnet. In Mainnet you can deploy your network directly on the Ethereum network, while in Testnet you can test your application on real Ethereum platforms, but Ether and gas would have no real value. Ethereum developers at the beginning are recommended to use Ethereum Testnet for deploying applications [32]. There are different implementations of Ethereum protocols, such as:

- Parity (written in Rust)
- Geth (stands for Go Ethereum and written in Go installation of Geth shown in Fig. 14.17)

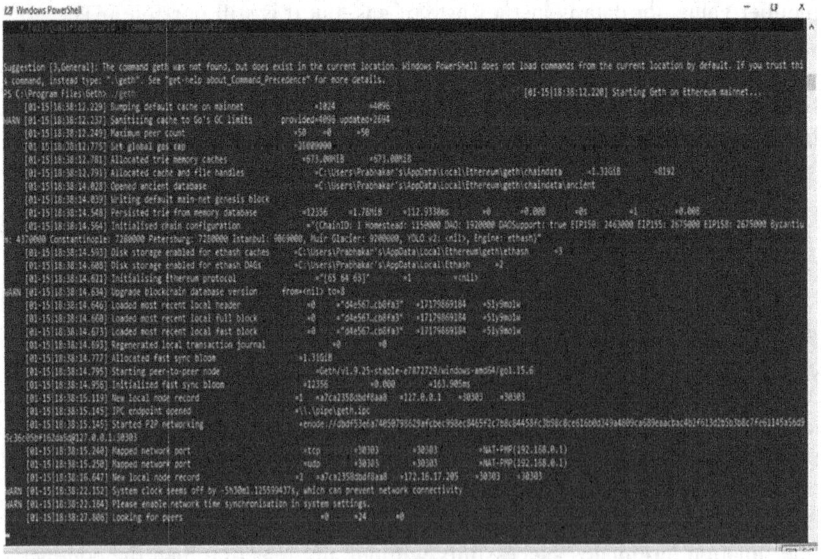

FIG. 14.17 Installation of Geth on Windows.

- Pyethereum (written in Python)
- Mantis (written in Scala)
- Harmony (written in Java)

14.8.5.1 Public Testnet advantages and disadvantages

Whether or not you choose to run a full node, you will probably want to run a public Testnet node. Let's look at some of the advantages and disadvantages of using a public Testnet.

Advantages

- A Testnet node needs to sync and store much less data—about 10 GB, depending on the network (as of April 2018)
- A Testnet node can sync fully in a few hours
- Deploying contracts or making transactions requires test Ether, which has no value and can be acquired for free from several "faucets"
- Testnets are public blockchains with many other users and contracts, running "live"

Disadvantages

- You can't use "real" money on a Testnet; it runs on test Ether. Consequently, you can'*t*-test security against real adversaries, as there is nothing at stake
- There are some aspects of a public blockchain that you cannot-test realistically on a Testnet; for example, transaction fees, although necessary to send transactions, are not a consideration on a Testnet, since gas is free, and the Testnets do not experience network congestion like the public Mainnet sometimes does

14.8.5.2 Local Blockchain simulation advantages and disadvantages

For many testing purposes, the best option is to launch a single-instance private blockchain. Ganache (formerly named testrpc) is one of the most popular local blockchain simulations that you can interact with, without any other participants. It shares many of the advantages and disadvantages of the public Testnet, but also has some differences.

Advantages

- No syncing and almost no data on disk; you mine the first block yourself
- No need to obtain test Ether; you "award" yourself mining rewards that you can use for testing
- No other users, just you
- No other contracts, just the ones you deploy after you launch it

Disadvantages

- Having no other users means that it doesn't behave the same as a public blockchain: there's no competition for transaction space or sequencing of transactions
- No miners other than you means that mining is more predictable; therefore, you can't-test some scenarios that occur on a public blockchain
- Having no other contracts means you have to deploy everything that you want to test, including dependencies and contract libraries
- You can't recreate some of the public contracts and their addresses to test some scenarios (e.g., the DAO contract)

14.8.5.3 Running an Ethereum client

If you have the time and resources, you should attempt to run a full node, even if only to learn more about the process. In this section we cover how to download, compile, and run the Ethereum clients Parity and Geth. This requires some familiarity with using the command-line interface on your operating system. It's worth installing these clients, whether you choose to run them as full nodes, as Testnet nodes, or as clients to a local private blockchain.

14.8.5.4 Hardware requirements for a full node

Before we get started, you should ensure you have a computer with sufficient resources to run an Ethereum full node. You will need at least 80 GB of disk space to store a full copy of the Ethereum blockchain. If you also want to run a full node on the Ethereum Testnet, you will need at least an additional 15 GB. Downloading 80 GB of blockchain data can take a long time, so it's recommended that you work on a fast internet connection. Syncing the Ethereum blockchain is very input/output (I/O) intensive. It is best to have a solid-state drive (SSD). If you have a mechanical hard disk drive (HDD), you will need at least 8 GB of RAM to use as a cache. Otherwise, you may discover that your system is too slow to keep up and sync fully.

14.8.5.5 Summary

This section discussed Ethereum blockchain in the context of its architecture, which relies on Ethereum Virtual Machine. This blockchain maintains the global state of the Ethereum network dynamically and is Turing complete; it also has an added feature that is smart contract, which provides Ethereum's ability to be used in wider applications beyond cryptocurrencies. Ethereum mining is also based on proof of work, but in Ethereum 2.0 it is heading towards a more robust blockchain network that uses proof of stake consensus in subsequent phases. The section also discussed Ethereum networks such as Mainnet and Testnet, and how they are used. In Ethereum, one can also simulate its application over different Testnets provided by Ethereum, which provides dummy Ethers for transactions.

14.9 Conclusion

We have seen various technical features of both Bitcoin and Ethereum blockchain, which are two major platforms of blockchain applications, and conclude that while Bitcoin provides a detailed concept of blockchain technology, Ethereum offers a more sophisticated application of it in different applications; it also removes some major limitation of Bitcoin, such as use of object oriented programming, which helps in writing smart contracts, maintenance of world state ledger, and use of Ethereum Virtual Machine for more interoperability of Blockchain applications.

References

[1] N. Szabo, Formalizing and securing relationships on public networks, First Monday 2 (1997) 9 (1997).
[2] S. Haber, W.S. Stornetta, How to time-stamp a digital document, J. Crypt. 3 (2) (1991) 99–111.
[3] D. Chaum, BUntraceable electronic mail, return addresses, and digital pseudonyms, Commun. ACM 24 (2) (1981) 84–88.
[4] Blockchain Technology, Cryptocurrency and other Applications by Sandeep Kumar Shukla Mohan Dhawan Venkatesan Subramanian, October 24 2019 (initial draft).
[5] G. Wood, Ethereum: A Secure Decentralised Generalised Transaction Ledger, Ethereum Project Yellow Paper, 2014.
[6] Azure, Microsoft Azure: Blockchain as a Service, 2016. https://azure.microsoft.com/enus/solutions/blockchain.
[7] R.C. Merkle, Secrecy, Authentication, and Public Key Systems, Ph.D. Dissertation, Stanford University, Stanford, CA, USA, 1979. AAI8001972.
[8] S. Meng, L. Luo, P. Sun, Y. Gao, Reliability service assurance in public clouds based on blockchain, in: 2020 IEEE 20th International Conference on Software Quality, Reliability and Security Companion (QRS-C), Macau, China, 2020, pp. 688–689, https://doi.org/10.1109/QRS-C51114.2020.00122.
[9] H. Desai, M. Kantarcioglu, L. Kagal, A hybrid blockchain architecture for privacy-enabled and accountable auctions, in: 2019 IEEE International Conference on Blockchain (Blockchain), Atlanta, GA, USA, 2019, pp. 34–43, https://doi.org/10.1109/Blockchain.2019.00014.
[10] T.H. Kim, J. Lampkins, SSP: self-sovereign privacy for internet of things using blockchain and MPC, in: 2019 IEEE International Conference on Blockchain (Blockchain), Atlanta, GA, USA, 2019, pp. 411–418, https://doi.org/10.1109/Blockchain.2019.00063.
[11] L. Ismail, H. Materwala, A review of Blockchain architecture and consensus protocols: use cases, challenges, and solutions, Symmetry 11 (2019) 1198, https://doi.org/10.3390/sym11101198.
[12] S. Nakamoto, Bitcoin: A Peer-to-Peer Electronic Cashsystem, 2008.
[13] A. Suresh, A.R. Nair, A. Lal, M.S. Kumaran, G. Sarath, A hybrid proof based consensus algorithm for permission less blockchain, in: 2020 Second International Conference on Inventive

Research in Computing Applications (ICIRCA), Coimbatore, India, 2020, pp. 707–713, https://doi.org/10.1109/ICIRCA48905.2020.9183109.

[14] M. Conti, E. Sandeep Kumar, C. Lal, S. Ruj, A survey on security and privacy issues of bitcoin, IEEE Commun. Surv. Tutorials 20 (4) (2018) 3416–3452. Fourth quarter https://doi.org/10.1109/COMST.2018.2842460.

[15] P.K. Kaushal, A. Bagga, R. Sobti, Evolution of bitcoin and security risk in bitcoin wallets, in: 2017 International Conference on Computer, Communications and Electronics (Comptelix), Jaipur, 2017, pp. 172–177, https://doi.org/10.1109/COMPTELIX.2017.8003959.

[16] C. Gupta, A. Mahajan, Evaluation of proof-of-work consensus algorithm for blockchain networks, in: 2020 11th International Conference on Computing, Communication and Networking Technologies (ICCCNT), Kharagpur, India, 2020, pp. 1–7, https://doi.org/10.1109/ICCCNT49239.2020.9225676.

[17] J. Kim, S. Kang, H. Ahn, C. Keum, C. Lee, Poster: architecture reconstruction and evaluation of blockchain open-source platform, in: 2018 IEEE/ACM 40th International Conference on Software Engineering: Companion (ICSE-Companion), Gothenburg, Sweden, 2018, pp. 185–186.

[18] T. Robinson, Bitcoin is not Anonymous. [online], 2015, Available: http://www.respublica.org.uk/disraeli-room-post/2015/03/24/bitcoin-isnot-anonymous/.

[19] A. Judmayer, N. Stifter, K. Krombholz, E. Weippl, E. Bertino, R. Sandhu, Blocks and Chains: Introduction to Bitcoin, Cryptocurrencies, and Their Consensus Mechanisms, Morgan & Claypool, 2017, https://doi.org/10.2200/S00773ED1V01Y201704SPT020.

[20] Y. Wu, A. Luo, D. Xu, Forensic analysis of bitcoin transactions, in: 2019 IEEE International Conference on Intelligence and Security Informatics (ISI), Shenzhen, China, 2019, pp. 167–169, https://doi.org/10.1109/ISI.2019.8823498.

[21] S. Hong, H. Kim, Analysis of bitcoin exchange using relationship of transactions and addresses, in: 2019 21st International Conference on Advanced Communication Technology (ICACT), PyeongChang, Korea (South), 2019, pp. 67–70, https://doi.org/10.23919/ICACT.2019.8701992.

[22] Q. Bai, C. Zhang, Y. Xu, X. Chen, X. Wang, Poster: evolution of ethereum: a temporal graph perspective, in: 2020 IFIP Networking Conference (Networking), Paris, France, 2020, pp. 652–654.

[23] Xie, C. Zhang, L. Wei, Y. Niu, F. Wang, J. Liu, A privacy-preserving ethereum lightweight client using PIR, in: 2019 IEEE/CIC International Conference on Communications in China (ICCC), Changchun, China, 2019, pp. 1006–1011, https://doi.org/10.1109/ICCChina.2019.8855900.

[24] edureka.com.

[25] Y. Chen, Z. Rong, Evolution of the external owned account trading network on ethereum, in: 2020 Chinese Automation Congress (CAC), Shanghai, China, 2020, pp. 3369–3373, https://doi.org/10.1109/CAC51589.2020.9327434.

[26] Ethereum.org.

[27] www.parity.io.

[28] S. Paavolainen, C. Carr, Security properties of light clients on the Ethereum Blockchain, IEEE Access 8 (2020) 124339–124358, https://doi.org/10.1109/ACCESS.2020.3006113.

[29] G.A. Pierro, H. Rocha, The influence factors on ethereum transaction fees, in: 2019 IEEE/ACM 2nd International Workshop on Emerging Trends in Software Engineering for Blockchain (WETSEB), Montreal, QC, Canada, 2019, pp. 24–31, https://doi.org/10.1109/WETSEB.2019.00010.

[30] M.S. Bhargavi, S.M. Katti, M. Shilpa, V.P. Kulkarni, S. Prasad, Transactional data analytics for inferring behavioural traits in ethereum blockchain network, in: 2020 IEEE 16th

International Conference on Intelligent Computer Communication and Processing (ICCP), Cluj-Napoca, Romania, 2020, pp. 485–490, https://doi.org/10.1109/ICCP51029.2020.9266176.

[31] Andreas M. Antonopoulos, Gavin Wood, Mastering Ethereum Building Smart Contracts and Dapps, Orielly, 2018.

[32] M. Dabbagh, M. Kakavand, M. Tahir, A. Amphawan, Performance analysis of blockchain platforms: empirical evaluation of hyperledger fabric and ethereum, in: 2020 IEEE 2nd International Conference on Artificial Intelligence in Engineering and Technology (IICAIET), Kota Kinabalu, Malaysia, 2020, pp. 1–6, https://doi.org/10.1109/IICAIET49801.2020.9257811.

Further reading

D. Ronand, A. Shamir, Quantitative analysis of the full bitcoin transaction graph, in: Financial Cryptography and Data Security (Lecture Notes in Computer Science), Springer, Berlin, Germany, Apr. 2013, pp. 6–24. [Online]. Available: http://link.springer.com/chapter/10.1007/978-3-64239884-1_2.

Coanalytic—Blockchain Intelligence, accessed on Mar. 15, 2016. [Online]. Available http://coinalytics.co/.

K. Delmolino, M. Arnett, A. Kosba, A. Miller, E. Shi, Step by step towards creating a safe smart contract: lessons and insights from a cryptocurrency lab, in: International Conference on Financial Cryptography and Data Security, Springer, 2016, pp. 79–94.

K. Bhargavan, A. Delignat-Lavaud, C. Fournet, A. Gollamudi, G. Gonthier, N. Kobeissi, N. Kulatova, A. Rastogi, T. Sibut-Pinote, N. Swamy, et al., Formal verification of smart contracts: short paper, in: Proceedings of the 2016 ACM Workshop on Programming Languages and Analysis for Security, ACM, 2016, pp. 91–96.

L. Luu, D.-H. Chu, H. Olickel, P. Saxena, A. Hobor, Making smart contracts smarter, in: Proceedings of the 2016 ACM SIGSAC Conference on Computer and Communications Security, CCS '16, ACM, 2016, pp. 254–269.

B. Marino, A. Juels, Setting standards for altering and undoing smart contracts, in: International Symposium on Rules and Rule Markup Languages for the Semantic Web, Springer, 2016, pp. 151–166.

https://blockchainhub.net.

The DAO smart contract. http://etherscan.io/address/0xbb9bc244d798123fde783fcc1c72d3bb8c189413#cod.

An Overview of Blockchain based Smart Contract, Available from: [accessed Mar 03 2021]. https://www.researchgate.net/publication/332606108_An_Overview_of_Blockchain_based_Smart_Contract.

Chapter 15

High assurance software architecture and design

Muhammad Ehsan Rana[a] and Omar S. Saleh[b]
[a]*Asia Pacific University of Technology & Innovation (APU), Technology Park Malaysia, Kuala Lumpur, Malaysia,* [b]*Studies, Planning and Follow-Up Directorate, Ministry of Higher Education and Scientific Research, Baghdad, Iraq*

15.1 Introduction

When a software system is constructed, it needs to be carefully architected and designed to get benefited from the inherent features of software quality. In addition to hardware, which is usually taken as the primary source for improving performance, design and architecture have a high impact on the performance of a software application. Software design impacts various aspects of software quality including reusability, maintainability, flexibility, reliability, and efficiency. Software architecture ensures the presence of quality in a software application by implementing structures such as architectural and design principles and patterns and by avoiding antipatterns while applying these constructs. Studies indicate improvements in quality factors using design and architectural refinements. Accordingly, for the application that lacks these quality attributes, it can be redesigned or refactored to incorporate these features of software quality.

15.2 Software architecture patterns

Well-designed architecture helps software applications to scale, better maintain, and respond to increased changing requirements. Software architecture patterns define an application's fundamental characteristics and behavior that make it scalable, modular, and maintainable. Without a standard architecture in mind, developers usually are lost in the code's complexity, especially when the application size gets large. It eventually creates unorganized codes and a lack of coherence among various components of an application [1]. This section discusses the three most used software development architectures: client-server pattern, layered pattern, and model-view-controller pattern. Each of these architectural patterns follows the principle of divide and conquer and separation of concerns.

System Assurances. https://doi.org/10.1016/B978-0-323-90240-3.00015-1

15.2.1 Client-server pattern

Most of the implemented web applications today are based on client-server architecture. This pattern divides a software application into two main areas, where one acts as the client and the other as the server. In the client-server architecture pattern, a client sends the request through a web browser, and the server responds based on the request received from the client, as illustrated in Fig. 15.1. In this pattern, the client needs to know the server's address, whereas the server does not need to know the client's location. In some cases, the client can be used as a server and the server as a client. Some of the standard protocols that the client and server are associated with include the File Transfer Protocol (FTP), Simple Mail Transfer Protocol (SMTP), and Hypertext Transfer Protocol (HTTP) [2,3].

15.2.1.1 Components of client-server architecture

The components of client-server architecture include:

a. **Client:** the requester of the processes either through a web browser interface or chat client, email client, etc.
b. **Server:** the receiver of the requests from the clients. It processes the requests, gathers the required information, generates the results, and sends them back to the client. Once the server completes processing the client request, it tackles other requests such as the FTP server, chat server, and database server.

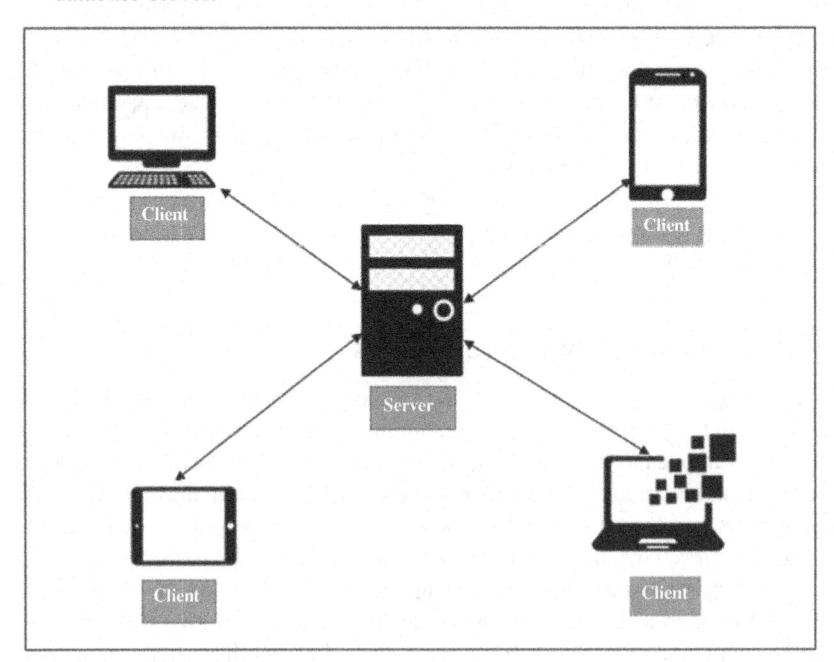

FIG. 15.1 Client-server model.

15.2.1.2 Advantages of client-server architecture

The client-server architecture-based design has several benefits that include:

a. Easy maintenance is the primary advantage of the client-server pattern. The roles are distributed among several standalone systems; hence, maintenance of the clients is independent.
b. The client-server architecture promotes increased scalability. The application scalability is one of the significant advantages considering the latest application development trends as the demand for technological improvement is rapidly increasing.
c. Data is centralized in the client-server architecture, where multiple clients can access it from various locations. The possibility of shared resources and services makes it easier to manage, modify, and reuse software modules [4].

15.2.2 Layered architecture pattern

The layered architecture pattern is among the most common architectural patterns used at the enterprise level, consisting of a hierarchy of layers, as shown in Fig. 15.2. The services communicate between these layers via ports. A layer is modeled as a relation between used and provided services, whereas a layer composition is defined by means of relational composition [5].

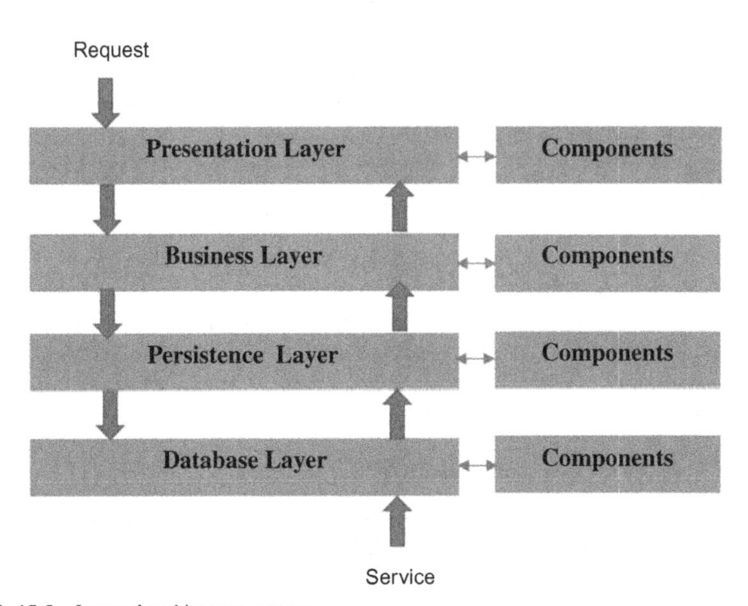

FIG. 15.2 Layered architecture pattern.

15.2.2.1 Components of layered architecture

The layered architecture pattern helps encapsulate and decouple the application by grouping the distinct segments using their functional role. This pattern follows the separation of concerns principle, resulting in improved testability and maintainability of the application. However, involving too many layers may deteriorate performance. Hence, it is advisable to limit the number of layers to approximately four. The four main layers utilized in the layered architecture pattern consist of the presentation layer, the business layer, the persistence layer, and the database layer.

a. **Presentation layer:** The presentation layer is the user interface (UI) layer representing the website or application layout, including web pages, forms, and interactions of APIs.
b. **Business layer:** The business layer contains the application's business logic, including the security, access, and authentication tasks. This layer is responsible for performing specific business rules associated with the user requests coming from the presentation layer.
c. **Persistence layer:** The persistence layer is liable for data presentation. It includes the data access object, object relational mappings, and other modes of presenting persistent data at the application level [6].
d. **Database layer:** The database layer is responsible for storing application data and retrieving it.

15.2.2.2 Advantages of layered architecture

The layered architecture-based design has numerous benefits that include [7,8]:

a. Layers are separated from one another, allowing developers to manage and maintain them independently. It helps in simplifying the application infrastructure as each layer is isolated.
b. Each layer manages a single aspect of the application, making it easier for managing the code. Layers can also be independently tested.
c. The layered architecture approach is particularly good for cloud-based applications that require a flexible and portable architecture.
d. This approach can also be applied to manage legacy systems as the application's architecture can be broken into separate independent modules.
e. The layered approach also provides a swift web solution without much complexity.

15.2.3 Model-view-controller (MVC) pattern

The model-view-controller (MVC) pattern is an excellent example of separation of concerns in an object-oriented paradigm. It divides the application into three logical components, the model, the view, and the controller, as shown in Fig. 15.3. The approach is now widely used in web applications. The MVC pattern ensures the rule of low coupling and high cohesion. The MVC architecture

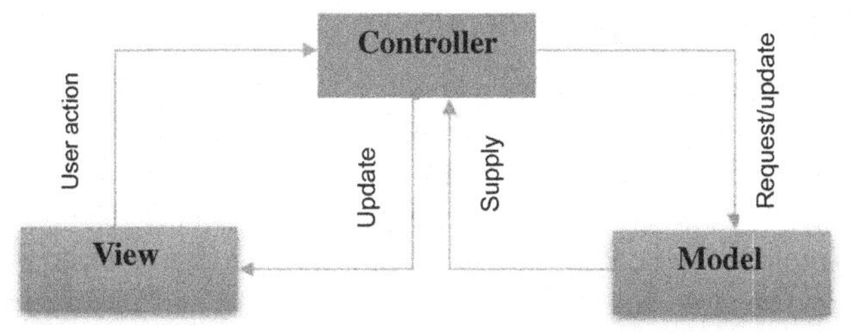

FIG. 15.3 Model view controller pattern.

may be difficult to understand at first but offers several benefits to the architects and designers. The basic idea of MVC is to divide the application into three interconnected components to separate the logic using the ways information is presented to and accepted by the client [9].

15.2.3.1 Components of model-view-controller (MVC) architecture

The model-view-controller (MVC) pattern is a three-layered pattern. The first layer is for user input logic, the second layer is for business logic, and the third is for user interface logic [10,11].

a. **Model:** The model represents the data and the core functionality of an application. It is commonly used to insert, retrieve, or update data into the database associated with the application.
b. **View:** As the name represents, the view deals with how the data is presented on the user interface. Users can interact with the application components through this interface.
c. **Controller:** The controller is responsible for handling the user requests and acts as an intermediator between the model and view.

15.2.3.2 Advantages of model-view-controller (MVC) architecture

The MVC architecture-based design offers several benefits that include [9,12]:

a. The MVC architecture helps to manage the software complexity issues by dividing the processes into three separate components.
b. As it does not use server-based forms, developers will have complete control over the application behavior.
c. The MVC architecture provides the ability to present multiple views for the model. It is an essential feature of contemporary applications as the demand for new ways of accessing the application increases day by day.
d. MVC-based applications are highly maintainable as each component can be managed independently.
e. MVC is instrumental in creating extensible and scalable applications.
f. The reusability and testability of the components can also be increased by applying the MVC pattern.

15.3 Software design principles

Object-oriented design principles provide established guidelines to construct quality solutions. Object-oriented experts and evangelists have identified numerous principles and design heuristics. However, five of these are considered significantly crucial for an object-oriented design [13]. These five design principles include Single Repository Principles, Open-Closed Principle, Liskov Substitution Principle, Interface Segregation Principle, and Dependency Inversion Principle. These principles are collectively termed as SOLID considering the first letter of their names. Following is a brief description of these principles.

15.3.1 Single responsibility principle

The single responsibility principle (SRP) states that each system module should only serve one and only one actor. The most important thing in this principle is the separation of concerns. Each actor should have his/her own implementation stored in different classes. This principle is useful to separate the concerns and perform any changes if necessary.

15.3.2 Open-closed principle

The open-closed principle (OCP) refers to the software entities and artifacts that are open for extension and closed for modification. This principle can be applied using different levels of abstraction (using interfaces) in order to protect different modules from the changes.

15.3.3 Liskov substitution principle

The Liskov substitution principle (LSP) refers to the ability of a subtype to replace the parent type without affecting the intended operation. For example, if a billing application contains a license type, this type can be substituted by a personal license type and a business license type. In other words, the billing application respects the Liskov substitution principle as it does not depend on the subtypes, but only uses the parent type which is license [13].

15.3.4 Interface segregation principle

The interface segregation principle (ISP) refers to removing unnecessary dependencies by different classes. This principle is followed when a class containing different implementations for different components is split into specific interfaces for each component. It is useful as it helps to prevent recompilation and redeployment of all components if a change occurs.

15.3.5 Dependency inversion principle

The dependency inversion principle (DIP) refers to the separation of the high-level modules from low-level ones. This principle can be applied if the components depend only on abstractions (interfaces) and not on the concrete implementations.

15.4 Software design patterns

Design patterns are object-oriented software design practices for solving common design problems [14]. Software architecture often appeared as a set of classes that can become too complicated to be easily understood. These complexities include complicated interactions and coupling between objects within the software system. Design patterns may reduce these complexity problems by offering proven solutions. As design patterns are tested and reusable solutions for reoccurring problems, they act as pre-made blueprints customized to solve these design issues [15].

15.4.1 Essential elements of design patterns

A design pattern consists of four essential elements that include the pattern's name, problem, solution, and consequences of using that pattern, as shown in Fig. 15.4 [16].

A pattern name should describe the pattern's problem, solution, and consequences in a word or two. It is essential as it provides a high level of abstraction and makes it easier to think and communicate design with others. Taking the Mediator design pattern, for instance, it is visible from its name that it creates an object to act as a mediator to manage complex communication between objects; thus, the term Mediator is devised.

A problem decides the usability of a pattern as it provides context for the pattern, similar to a set of conditions for the pattern to be applied. This set of conditions determines the applicability of a pattern to an existing problem [17]. In the Mediator design pattern, one of the conditions to apply the said

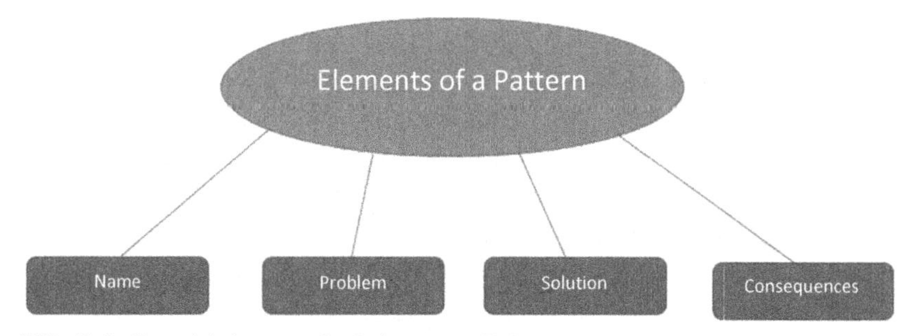

FIG. 15.4 Essential elements of a design pattern [16].

pattern is when communications between objects are defined but in complex ways, resulting in unstructured interdependencies between objects that are hard to comprehend. One of the primary examples of such a situation would be the presence of high coupling in a given system. If a given system is highly coupled, implementing the Mediator design pattern is recommended as it reduces the unstructured interdependencies and complex communications between objects.

A solution is not a concrete explanation of how the pattern should be applied. It explains the elements that make up the design, such as their relationships, responsibilities, and collaborations [16]. Thus, instead of explaining the pattern in a concrete manner, the pattern itself acts as a template in which it can be applied in various situations by providing a description of how the elements should be arranged to solve a given problem.

The consequences of a pattern include the description of the pros and cons when applying this pattern [16]. Studies suggest that not all designs provide the best solution [18]. Thus, one must consider the consequences of applying the pattern as they are critical in evaluating other alternatives present in the body of knowledge. The consequences of a given pattern can impact the system's quality attributes (extensibility, reusability, portability, etc.). Listing the consequences of each pattern and comparing them may assist in deciding the best pattern for a specific problem.

15.4.2 Classification of design patterns

Since software design is a challenging phase in the software development life cycle, many design problems can frequently occur in every project. Therefore, design patterns come in place to overcome these reoccurring design problems. Design patterns are well-tested solutions to many well-known design problems. The design patterns are also categorized into three main categories, which are creational design patterns, structural design patterns, behavioral design patterns. Each design pattern category contains many design patterns that can be used depending on the problem and situation [19].

15.4.2.1 Creational design patterns

The creational design patterns are responsible only for the creation of the objects depending on the situation. These patterns give different ways of creating objects to reduce the complexity of the design and creation mechanisms [20].

15.4.2.2 Structural design patterns

The structural design patterns simplify the design and make relationships between different system's components to have different functionalities [20].

15.4.2.3 Behavioral design patterns

Since communication is essential in software systems, the behavioral design patterns serve to simplify communication and produce more flexibility while communicating [20].

15.4.3 Gang of four design patterns

Gamma et al. [16] provide a catalog of 23 design patterns with their detailed descriptions including classification, intent, motivation, applicability, structure, implementation, and consequences. These design patterns are commonly referred to as gang of four (GoF) design patterns. A concise introduction of these GoF design patterns is presented in Table 15.1, which exhibits the verity and versatility of these patterns for various design problems.

15.4.4 Benefits of using design patterns

In a typical software development project, many changes occur throughout the development process. It is arduous to ensure that the changes made would not introduce new defects into the system as there are way too many uncertainties. However, design patterns are remarkably flexible to the changes as they are tested solutions to the problems that highly mitigate the risk of emerging new defects. Some common benefits of using design patterns are listed below:

a. Design patterns are able to reduce the development time for a software project [21].
b. Design pattern can be used to develop more secure and more reliable software systems.
c. With the tested, proven development paradigm and reusable solution for recurring problems, the system development process is accelerated and the software quality is improved.
d. Design patterns offer a highly stable solution for design problems. As design patterns provide tested and proven solutions for specific problems, the solutions they offer are comparatively very stable and free from bugs [22].
e. Since the design has been proven to work, it saves time during the design phase and reassures the quality of the system design. The probability of defects being met along the development life cycle is minimized as the project has been designed and tested for the long run [16].
f. Through the pattern's high-level abstraction, designers can communicate effectively with other designers to discuss the pattern that is best suited for the project given. This saves more time and costs to be allocated for the design phase.

TABLE 15.1 Gang of four (GoF) design patterns.

Category	Design pattern	Description
Creational	Abstract factory	Create instances for families of classes
	Builder	Separates construction process from representation
	Factory method	Defer instantiation to sub-classes
	Prototype	Creates clone of a class to produce a new object
	Singleton	Restricts object creation to a single instance only
Structural	Adaptor	Convert the interface of a class into another interface client expects
	Bridge	Decouple an abstraction from its implementation
	Composite	Compose objects into tree structures to represent part-whole hierarchies
	Decorator	Provides a way to dynamically modify the behavior of an object without sub-classing
	Façade	Provide a unified interface to a set of interfaces in a subsystem
	Flyweight	Addresses sharing to support large numbers of fine-grained objects efficiently
	Proxy	Provides placeholder for another object to control access to it
Behavioral	Chain of responsibility	An object takes on the job of finding which object can satisfy a client's request
	Command	Encapsulates request as an object
	Interpreter	Defines representation for language elements and interpret it with an interpreter object
	Iterator	Allows traversing an aggregate object sequentially
	Mediator	Define an object that encapsulates how a set of objects interact to promote loose coupling
	Memento	Saves internal state of an object without violating encapsulation
	Observer	Notifies the dependent objects automatically when the object changes state
	State	Distributes state specific logic across classes that represent an object's state
	Strategy	Allows an algorithm to vary independently from the clients that use it
	Template method	Defines skeleton of an algorithm in a method
	Visitor	Defines new operations without changing the class

15.5 Software design antipatterns

Design antipatterns are the bad design practices and the weak solutions used when building a software system [23]. Software quality can be directly affected by design, either positively or negatively. The design patterns positively impact the software quality, while the design antipatterns negatively affect the software quality. Antipatterns are generally caused either by lack of experience or by using the design patterns in the incorrect place [24].

Software quality is an imperative concern for many software development companies as clients want the acquired software system to be agile and perfect. Different quality factors or attributes must be taken into consideration in order to attain high-quality software. Some of the crucial quality attributes include performance, reusability, testability, and reliability. Many of these attributes are negatively impacted by the high complexity of the software. Software performance is described by the ability of the software system to be stable and responsive in a particular environment and workload [25]. Reusability is the ability to efficiently reuse previous software components, design, and implementation in the new development projects [25]. Reliability refers to the ability of a software system to work without any failure in a specified environment within a specific period [25]. Software complexity refers to the high level of interactions among the internal structures of the software system. The complexity is generally related to the software design as the coupling between the system's modules gradually increases due to the interconnected links between the different classes, which negatively affects the software system's quality [26].

15.5.1 Some common antipatterns

The following is a brief description of some of the common design antipatterns.

15.5.1.1 Blob antipattern

The Blob antipattern is a bad practice in the design that refers to the use of one big class to manage and perform many operations. The Blob antipattern reduces the cohesion of the software system and dominates all the processing and decision-making in the software [27]. This antipattern is considered bad practice as it violates the object-oriented design principles, eliminates the reusability, and complicates the testability of the class. Moreover, it consumes many resources for simple operations as the class is entirely loaded into the system's memory [24].

15.5.1.2 Spaghetti Code antipattern

The Spaghetti Code is a design antipattern that refers to the misuse of object-oriented principles. This antipattern consists of using methods with lengthy implementations and no parameters but is only limited to the usage of global variables [28]. In general, the Spaghetti Code eliminates important object-oriented concepts like inheritance and polymorphism [29].

15.5.1.3 Poltergeist antipattern

The Poltergeist, also known as Gypsy Wagons, is an antipattern that can be found on systems containing classes with a short life cycle and minimal operations and responsibilities [30]. These Poltergeist antipatterns are usually used to invoke some other system's classes or methods and waste the system's resources as long as they are used [24].

15.5.1.4 Functional decomposition antipattern

Functional decomposition is an antipattern that consists of writing code without thinking in an object-oriented manner. This antipattern lets classes look like a function with a single purpose. The classes perform one single operation with all the data attributes that are private and used only inside the class. This antipattern also eliminates the important object-oriented concepts like inheritance and polymorphism, which results in reduced reusability [28].

15.5.1.5 Swiss Army Knife antipattern

The Swiss Army Knife antipattern refers to the class that contains many responsibilities. This antipattern is characterized by a large number of methods and interface implementations [28].

15.5.2 When design patterns turn into antipatterns

Developers need to recognize the trade-offs between design factors when applying the patterns to the design. Generally speaking, a class having more than one role or responsibility violates the single responsibility principle as it potentially causes low cohesion and complicates the process of fixing vulnerabilities. However, some of the design classes with one role still negatively impact the system's integrity. For instance, when the composite pattern is applied, it affects the integrity of the software system as it can potentially make the design overly general, which increases the security problems. Therefore, using the composite pattern indicates that the user values transparency more than security [16].

Similarly, the Singleton design pattern also impacts applications' overall integrity as it violates the open-closed principle. Singleton ensures one class has only one object instantiated, indicating that it does not allow inheritance as a private constructor is used. When inheritance is avoided, the open for extension principle will be violated. Changing the existing code that functions perfectly sounds risky, and it might cause the emergence of new threats and bugs which the attacker can exploit to perform any unauthorized access. Besides, allowing inheritance in the Singleton class does not make sense as when inheritance is permitted, there will be more than one instance for the class, making the class not a Singleton class anymore. Moreover, by using a reflection attack, the constructor of the Singleton class can be altered from private to public, indicating that the constructor can be called from outside, which then increases the risk of emerging new threats.

15.6 Conclusion

This chapter discusses the fundamental building blocks of software architecture that include software architecture patterns, software design principles, software design patterns, and antipatterns. Among the primary steps for refinement of software design is its conversion in a more dynamic, adaptable, and flexible format. These building blocks play a pivotal role in improving software quality and avoiding practices that generate a faulty design. Design principles outline the most common scientifically derived approach for constructing a flexible and robust design. The use of design principles provides established guidelines to produce quality solutions. Design patterns rely on proven object-oriented design approaches to construct reliable and reusable solutions for the associated design problems in a system. One of the key purposes of most design patterns is to create a simple solution that is extendible and loosely coupled. Hence, the design produced as a result of applying design patterns eventually reduces the interdependencies between program components and makes them easier to extend and maintain. Software architecture patterns represent a more abstract layer of software design than design patterns and define an application's fundamental characteristics and behavior that makes it scalable, modular, and maintainable. Antipatterns identify the improper application design, which leads to increased development time and can become a bottleneck for the application's performance. They provide a legitimate way to document the problem scenarios and their solutions which assists the designers in watching out for similar situations to avoid bugs and other design constraints that hinder performance. In typical software development, these architectural and design constructs need to be applied in sequence to ensure quality software generation. Therefore, if these three elements of object orientation are not followed sequentially, it increases the chance of inducing flaws and defects in the design.

References

[1] M. Richards, Software Architecture Patterns, first ed., O'Reilly Media, Inc., 2015
[2] H.S. Oluwatosin, Client-server model, J. Comput. Eng. 16 (1) (2014) 67–71.
[3] A. Sharma, M. Kumar, S. Agarwal, A complete survey on software architectural styles and patterns, Procedia Comput. Sci. 70 (2015) 16–28, https://doi.org/10.1016/j.procs.2015.10.019.
[4] A. Senson, A. Burton, T. Boulanger, Software Architecture & Software Design Patterns for Startups, RIC Centre, 2021. https://riccentre.ca/software-architecture-design-patterns/.
[5] D. Marmsoler, A. Malkis, J. Eckhardt, A model of layered architectures, in: Formal Engineering Approaches to Software Components and Architectures (FESCA'15), 2015, pp. 47–61, https://doi.org/10.4204/EPTCS.178.5.

[6] A. Wickramarachchi, Software Architecture Patterns. Layered Architecture | by Anuradha Wickramarachchi | Towards Data Science, 2017. https://towardsdatascience.com/software-architecture-patterns-98043af8028 (Accessed 22 November 2021).
[7] V.S. Sharma, P. Jalote, K.S. Trivedi, Evaluating performance attributes of layered software architecture, in: G.T. Heineman, I. Crnkovic, H.W. Schmidt, J.A. Stafford, C. Szyperski, K. Wallnau (Eds.), Component-Based Software Engineering. CBSE 2005, Lecture Notes in Computer Science, vol. 3489, Springer, Berlin, Heidelberg, 2005, https://doi.org/10.1007/11424529_5.
[8] C. Liyan, Application research of using design pattern to improve layered architecture, in: 2009 IITA International Conference on Control, Automation and Systems Engineering (case 2009), IEEE, 2009, July, pp. 303–306.
[9] M.U. Khan, T.V. Rao, XWADF: architectural pattern for improving performance of web applications, Int. J. Comput. Sci. 11 (2) (2014) 105–113.
[10] T. Dey, A comparative analysis on modeling and implementing with MVC architecture, in: IJCA Proceedings on International Conference on Web Services Computing (ICWSC), vol. 1, 2011, November, pp. 44–49.
[11] A. Leff, J.T. Rayfield, Web-application development using the model/view/controller design pattern, in: Proceedings Fifth IEEE International Enterprise Distributed Object Computing Conference, IEEE, 2001, September, pp. 118–127.
[12] A. Majeed, I. Rauf, MVC architecture: a detailed insight to the modern web applications development, J. Sol. Photoenergy Syst. 1 (1) (2018) 1–7.
[13] R.C. Martin, Clean Architecture: A Craftsman's Guide to Software Structure and Design, Prentice Hall, 2018.
[14] M.O. Onarcan, Y. Fu, A case study on design patterns and software defects in open source software, J. Softw. Eng. Appl. (2018) 249–273, https://doi.org/10.4236/jsea.2018.115016.
[15] A. Shvets, Dive into Design Patterns, 2019. Available at: https://www.goodreads.com/book/show/43125355-dive-into-design-patterns (Accessed 22 November 2021).
[16] E. Gamma, R. Helm, J. Ralph, J. Vlissides, Design Patterns—Elements of Reusable Object-Oriented Software, Addison-Wesley Professional, 1995.
[17] X. Ferre, N. Juristo, A. Moreno, I. Sánchez, A software architectural view of usability patterns, in: 2nd Workshop on Software and Usability Cross-Pollination (at INTERACT'03) Zurich (Switzerland), 2003, September.
[18] P. Wendorff, ASSET GmbH, Assessment of design patterns during software reengineering : lessons learned from a large commercial project, IEEE Xplore (2001) 77–84.
[19] S. Hussain, J. Keung, A.A. Khan, Software design patterns classification and selection using text categorization approach, Appl. Soft Comput. 58 (2017, September) 225–244, https://doi.org/10.1016/j.asoc.2017.04.043.
[20] A. Anand, G. Bansal, Interpretive structural modelling for attributes of software quality, J. Adv. Manag. Res.14 (3) (2017) 256–269, https://doi.org/10.1108/JAMR-11-2016-0097.
[21] R. Subburaj, G. Jekese, C. Hwata, Impact of object oriented design patterns on software development, Int. J. Sci. Eng. Res. 6 (2) (2015) 961–967.
[22] A. Ampatzoglou, G. Frantzeskou, I. Stamelos, A methodology to assess the impact of design patterns on software quality, Inf. Softw. Technol. 54 (4) (2012) 331–346, https://doi.org/10.1016/j.infsof.2011.10.006.
[23] M. Abidi, F. Khomh, Y. Guéhéneuc, Anti-patterns for multi-language systems, in: Proceedings of the 24th European Conference on Pattern Languages of Programs—EuroPLop '19, 2019, pp. 1–14, https://doi.org/10.1145/3361149.3364227.
[24] S. Lujan, F. Pecorelli, F. Palomba, A.D. Lucia, V. Lenarduzzi, A Preliminary Study on the Adequacy of Static Analysis Warnings with Respect to Code Smell Prediction, MaLTeSQuE, September 2020.

[25] A. Anand, G. Bansal, Interpretive structural modelling for attributes of software quality, J. Adv. Manag. Res. 14 (3) (2017, August) 256–269, https://doi.org/10.1108/JAMR-11-2016-0097.
[26] N. Vanitha, R. ThirumalaiSelvi, SICCAT: software inheritance coupling complexity analysis tool, Int. J. Eng. Tech. 4 (2) (2018) 62–67. Available at: http://www.ijetjournal.org.
[27] S. Hussain, et al., Methodology for the quantification of the effect of patterns and anti-patterns association on the software quality, IET Softw. 13 (5) (2019) 414–422, https://doi.org/10.1049/iet-sen.2018.5087.
[28] N. Vavrová, V. Zaytsev, Does Python smell like Java? Tool support for design defect discovery in Python, Art Sci. Eng. Program. 1 (2) (2017, April), https://doi.org/10.22152/programming-journal.org/2017/1/11.
[29] S. Lujan, F. Pecorelli, F. Palomba, A.D. Lucia, V. Lenarduzzi, A Preliminary Study on the Adequacy of Static Analysis Warnings with Respect to Code Smell Prediction A Preliminary Study on the Adequacy of Static Analysis Warnings with Respect to Code Smell Prediction, MaLTeSQuE, 2020, September.
[30] S.R.A. Al-Rubaye, Y.E. Selcuk, An investigation of code cycles and Poltergeist anti-pattern, in: 2017 8th IEEE International Conference on Software Engineering and Service Science (ICSESS), November, vol. 2017, 2017, pp. 139–140, https://doi.org/10.1109/ICSESS.2017.8342882.

Chapter 16

Online condition monitoring and maintenance of photovoltaic system

Neeraj Khera
Department of ECE, Amity University, Noida, Uttar Pradesh, India

16.1 Introduction

Condition-based maintenance is used to detect the system degradation failures, thereby allowing causal stress to be eliminated or controlled prior to any significant deterioration in the component physical state. The need for PV systems as automated backup power sources for the effective operation of critical systems in hospitals, server rooms, hotels, banks, and many other industries requires the use of highly reliable batteries and power converters. The reliability problem for wear out failures in the above components in PV systems can be solved from their performance degradation perspective. The in-circuit identification of degradation failures in these life-limiting components is a challenge that motivates to develop a real-time condition monitoring and maintenance system. The real-time in-circuit condition monitoring based on the parametric degradation of high failure rate components enables the formulation of a condition-based maintenance strategy. In recent times, the real-time online condition of components is monitored from their parametric degradation data using soft computing techniques.

Chang et al. [1] presented the use of electrical detection method and fuzzy algorithm for fault detection of induction motors. Khera and Khan [2] implemented the condition monitoring of in-circuit aluminum electrolytic capacitor using an artificial neural network. The application of advanced inspection techniques such as infrared thermography (IRT) for condition monitoring of equipment in the industrial processes have resulted in their early failure detection in a non-contact and non-invasive manner [3]. IRT is a fast, reliable, and cost-effective technique to identify fault types in different electrical systems such as motors, generators, transformers, PV modules, circuit breakers, etc. Jadin and Taib [4] discussed the applications of IRT for diagnosing the reliability

System Assurances. https://doi.org/10.1016/B978-0-323-90240-3.00016-3

of electrical equipment, including their thermal anomalies and methods of measurement. Huda and Taib [5] presented the predictive diagnosis system for detecting invisible thermal defects in electrical equipment using IRT technology. Shen et al. [6] presented fault diagnosis of insulators based on processing the infrared images taken by an infrared thermal imager mounted on the unmanned aerial vehicle. The real-time diagnosis data of the insulator is also remotely communicated to the operator. P'erez and Daviu [7] and Ródenas et al. [8] demonstrated the potential of infrared thermography as a valuable tool for fault diagnosis in induction motors.

The main aim of this work is to develop a condition monitoring system for identifying the failures in critical components of the stand-alone PV system. Early detection of precursors of failures in these components would allow their condition-based maintenance and provide sufficient time for the controlled shutdown of the PV system, thereby reducing the costs of outage time and repair of PV system in safety critical applications. The power electronic converter and VRLA battery are the most critical and life-limiting components of the stand-alone PV-based system. As per MIL Handbook 217F (revision F) standards, electrolytic capacitors and switching transistors have high failure rates compared to other electronic components and they together constitute more than 90% of failures of power electronic converters. Apart from power converters, the VRLA battery is also one of the most life-limiting components of the PV system and therefore, the condition of aluminum-electrolytic capacitors and MOSFET in power converters and VRLA battery as secondary power supply source in stand-alone PV needs to be monitored at different operating conditions. In a large PV system, rechargeable VRLA battery is connected with each other to form battery string and therefore forms the series reliability system where the reliability of whole PV system depends on the reliability of a single VRLA battery. In the proposed work, the condition monitoring of both VRLA battery and power electronic converter is presented in two sections. In Section 1, an intelligent scheme for predictive fault diagnosis in VRLA battery based on infrared (IR) thermal imaging and the fuzzy algorithm is presented for scheduling its preventive maintenance. IR images of pristine and aged VRLA batteries in uninterrupted power supply applications are acquired using an IR camera at different discharging cycles. Image processing of IR images is performed for the detection of faults. In order to intelligently classify the faults a fuzzy inference system (FIS) is developed. The proposed scheme for automatic diagnosis and classification of faults in VRLA battery is implemented using LabVIEW 2015 software. Based on the occurrence of major faults in VRLA battery, an alert signal is sent to intended users at both onsite and remote locations. In Section 2, the online technique for condition-based maintenance of power electronic converter due to parametric degradation of its high failure rate components is presented. The accelerated aging test (ALT) is performed on aluminum-electrolytic capacitors and power MOSFETs during in-circuit operation in power converter (inverter) to characterize their aging process and it has been concluded from experimental results that their actual in-circuit life is much different than stated by manufacturers in their datasheet.

16.2 Condition monitoring of VRLA battery in PV system

VRLA battery has better thermal and electrical properties in comparison to other lead acid batteries [9] and therefore is a widely used lead acid battery in PV systems. Recently, several studies are done to determine the aging mechanisms in VRLA battery. Pascoe and Anbuky [10] have presented the simulation model using which the VRLA battery of desired ratings is simulated under arbitrary operating conditions (discharge rate, ambient temperature, end voltage, charge rate, and initial state of charge) to determine its charge-discharge behavior. IRT is a widely used technique for condition-based maintenance of electrical systems but the application of IRT to the fault detection and classification in lead acid batteries is not much explored. The parametric degradation in VRLA battery occurs with the lapse of time, therefore its condition monitoring for preventive maintenance is needed in order to improve the reliability of battery application systems. In this section, the automatic detection and classification of faults in VRLA battery for scheduling its preventive maintenance in a single battery stand-alone PV system application is presented using IR thermal imaging and fuzzy algorithm. The small variation in thermal patterns of IR images provides indications for early failures due to impending cells of the battery. For this purpose, several IR images are acquired at a different discharging interval of the already used (aged) and healthy (pristine) six-cell sealed VRLA battery of 12 V, 7.2 Ah, and 20 h rating that is discharging through a constant 60 W Edison lamp load in a single battery stand-alone PV system application. Thermal image processing of thermograms for aged and pristine VRLA battery is done to diagnose the degradation failures of aged battery. The thermal signature parameters signifying the condition of the aged VRLA battery are obtained at different discharge intervals. A fuzzy inference system (FIS) is developed to determine the severity of deterioration on the basis of the thermal signature parameters. Proposed FIS is developed using Fuzzy logic tool of LabVIEW 2015. The output information for the nature of the fault is provided to both the onsite and remote users. The developed automatic fault detection and classification system will, therefore, aids the maintenance engineer to schedule the preventive maintenance task depending on the condition of VRLA battery. The proposed fault classification technique can also be used for any type of battery application involving different lead acid batteries like VRLA battery, flooded lead acid battery or polymer lead acid battery. Therefore, using proposed technique, the reliability of systems having the lead acid battery as a critical component can be enhanced. In Section 16.2.1, the implementation of intelligent condition monitoring of VRLA battery is presented. From the IR images, the automatic diagnosis and classification of faults in VRLA battery has been performed using Visual Assistant and Fuzzy logic tool in LabVIEW 2015 software.

In Section 16.2.2, the results obtained are used to categorize the nature of faults for predictive maintenance of VRLA battery. In Section 16.2.3, web-based condition monitoring of battery is presented.

16.2.1 Implementation of predictive fault diagnosis of VRLA battery in PV system

IRT and image processing for qualitative categorization of faults in electrical equipment into three different categories based on the temperature difference (ΔT_r) between maximum hotspot and reference temperature is widely used recently. For this purpose, the IR image is converted in to the grayscale image from which the regions of hotspot and reference temperature are selected from the white (bright) and black (dark) regions of the image respectively. The fault detection algorithm based on ΔT_r presented in the above studies also minimizes the error due to emissivity of material. Emissivity is defined as the ratio of energy radiated by the object to the energy emitted by the blackbody at a certain temperature. The total radiated IR power by non-blackbody is given by the following relation (16.1) [11]:

$$L = \varepsilon \sigma T^4 \tag{16.1}$$

where

L = total radiance of object;
ε = emissivity of object;
σ = Stefan-Boltzmann constant; and
T = absolute temperature of object.

In this section, the implementation of an intelligent scheme for automatic detection and classification of faults in VRLA battery is presented. At different discharging states of both pristine (healthy) and aged (deteriorated) VRLA batteries, thc IR images are acquired using an infrared camera (FLIR E4). FLIR E4 camera has multi-spectral dynamic imaging enhancement feature that combines thermal image data with the visible light image to provide clear IR images. From the information of standard emissivity value of target (lead and oxidized lead) electrodes of VRLA battery that is 0.43 at 38°C, the IR camera has internal circuitry to accurately determine the hot region temperature of electrodes [12] from the following relation (16.2):

$$P_{rad} = K' \left(\varepsilon_t T_t^4 - \varepsilon_s T_a^4 \right) \tag{16.2}$$

where

P_{rad} = total heat power received from the electrode;
ε_t = emissivity of the electrode;
T_t = absolute temperature of the electrode;
ε_s = emissivity of the IR sensor;

T_a = absolute ambient temperature; and
K' = empirical factor containing Stefan-Boltzmann constant.

The use of IRT for automatic fault diagnosis of lead acid battery offers the advantage of detecting the early failures in a fast, non-contact and non-invasive manner. Therefore, the present work is focused on the determination of the qualitative nature of the fault in VRLA battery used in stand-alone PV system from IRT and Fuzzy logic techniques. For the implementation of the proposed scheme, the pristine and aged six-cell sealed VRLA battery (12 V, 7.2 Ah, and 20 h rating) at their 100% SOC are used one after another in a single battery stand-alone PV system application to supply stored electricity to a constant load (Edison lamp of 60 W). Using the IR camera, IR images are acquired at the start of discharging; after 2 h, and after 4 h of discharging for both pristine and aged VRLA batteries. For every IR image, the IR camera provides the emissivity corrected maximum, minimum, and average hot region temperature values. Using the Vision Assistant tool of LabVIEW 2015 software, thermal image processing on acquired IR images are performed at different discharging intervals to obtain the color intensity matching scores. The values of matching score (M) and maximum hot region temperature (T) for aged battery are used as input thermal signature parameters for proposed FIS to determine the severity of fault in VRLA battery at different discharging intervals. The block diagram of the proposed diagnostic system for implementation of automatic predictive fault diagnosis in a six-cell sealed VRLA battery using IRT and fuzzy algorithm is shown in Fig. 16.1.

The proposed scheme for automatic diagnosis and classification of faults in VRLA battery is simultaneously implemented in LabVIEW 2015 software. The output information defining the condition of the battery is displayed on the front panel of the LabVIEW and is stored in an MS Excel file with the time stamp at the hard disk of the host computer for further reliability analysis of VRLA battery. To facilitate the remote condition monitoring of the battery, the front panel information is continuously provided to the remote user using the web publishing tool of LabVIEW.

16.2.1.1 *Fault detection using IR image*

The acquired raw IR images are stored in the flash memory card using the USB interface of the IR camera. The raw IR images are normalized to have equal pixels size of 100 × 150 pixels and are processed using the Vision Assistant tool in LabVIEW 2015 software. At different discharging intervals, the color intensity matching for aged and pristine (reference) VRLA battery is performed. For this purpose, color intensity of normalized thermograms is compared pixel-by-pixel and the respective matching scores for different discharging intervals of VRLA battery are obtained. Matching score as thermal signature parameters signifies the similarity between color intensity values in thermal patterns of IR images of healthy and aged VRLA batteries. The block diagram VI to obtain

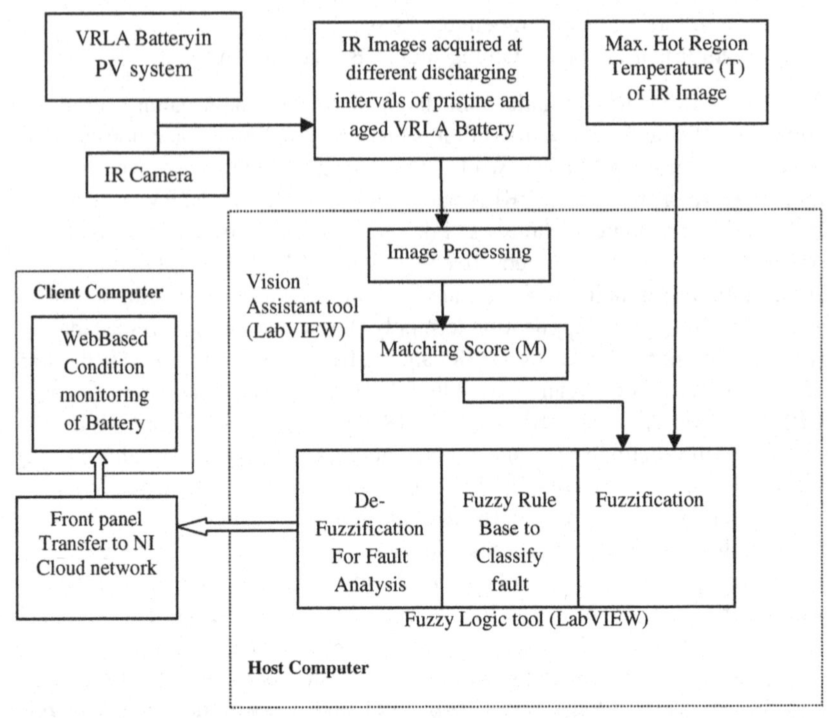

FIG. 16.1 System block diagram.

the matching score as a thermal signature parameter from the color intensity matching of normalized IR thermograms is shown in Fig. 16.2.

Along with the matching score, the emissivity corrected maximum hot region temperature of IR thermogram for aged VRLA battery is taken as another thermal signature parameter for implementing the intelligent and automatic Fuzzy logic-based classification of faults.

16.2.1.2 Fault classification and analysis using Fuzzy logic

Fuzzy logic enables to operate with imprecise information, linguistic variables and rules similar to the way of thinking of human operators. Fuzzy inference system (FIS) is developed using LabVIEW software. FIS has two inputs and one output. The obtained inputs, i.e., matching score (M) and emissivity corrected maximum hot region temperature (T) at different discharging intervals for aged battery are fuzzified into different sets of membership functions. Fuzzy logic-based fault classification process characterizes a set of rules for mapping fuzzified input to fuzzified output. The proposed FIS is implemented using Fuzzy logic tool of LabVIEW 2015 to classify the severity of fault in aged VRLA batteries at different discharging intervals. Considered linguistic variables for matching scores are good (G), average (A), poor (P), and very poor

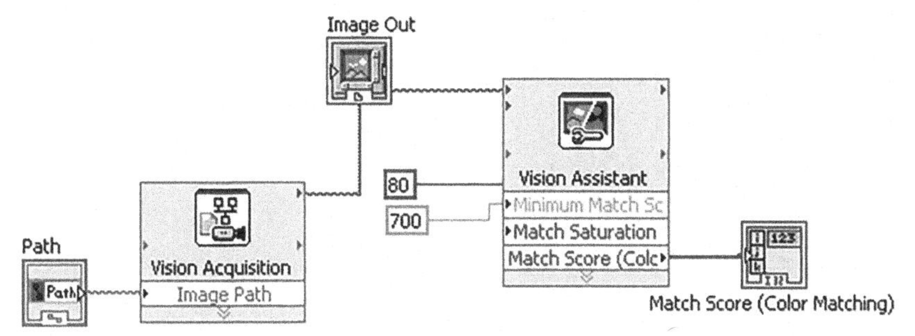

FIG. 16.2 Block diagram VI for fault detection using IR images.

(VP). The relationship of matching scores with membership function is shown in Fig. 16.3.

Similarly, considered linguistic variables for maximum hot region temperature (T) are low (L); medium (M); high (H), and very high (VH). The relationship of input variable T with membership function is shown in Fig. 16.4.

Fuzzified inputs are mapped to fuzzified output using a rule base. Fuzzy output signifies severity of the fault and its considered linguistic variable are critical fault (CR), major fault (MJ), minor fault (MN), and no fault (NF). The relationship of fuzzy output, i.e., fault type with the membership function is shown in Fig. 16.5. As the output value reduces, the fault type in the target battery becomes more severe.

A Fuzzy rule base is defined based on the knowledge of fault types arising from the thermal patterns of the battery. The Fuzzy rule base consists of 12 Fuzzy if-and-then rules that are given in Table 16.1.

Fault analysis from de-fuzzification is done to determine the fault type in aged VRLA battery using the center of area method as shown in Fig. 16.6.

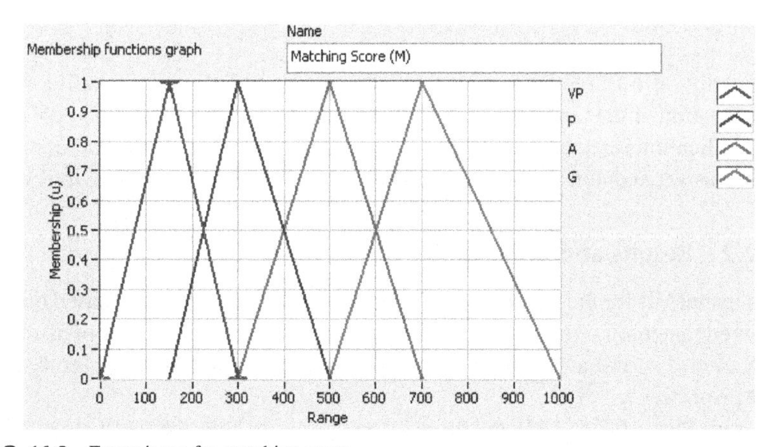

FIG. 16.3 Fuzzy input for matching score.

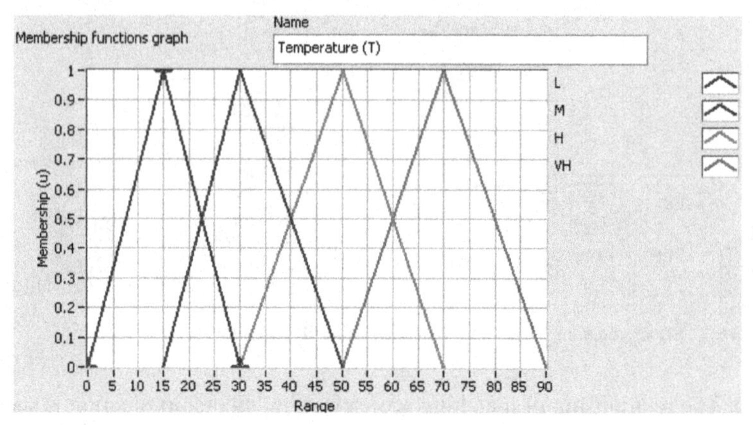

FIG. 16.4 Fuzzy input for maximum hot region temperature.

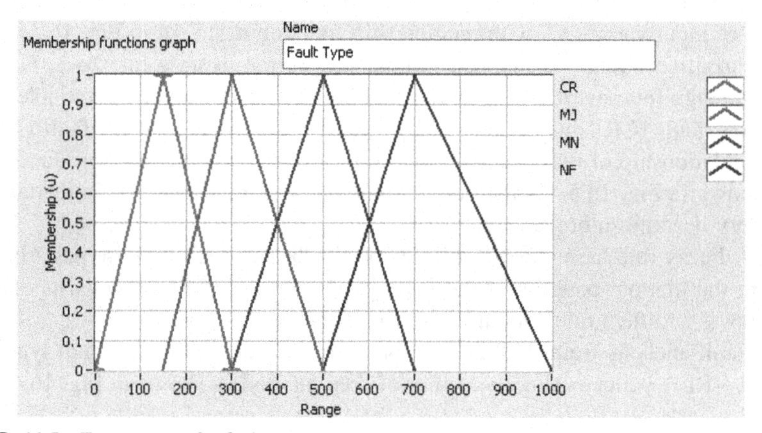

FIG. 16.5 Fuzzy output for fault type.

Finally, at different test conditions, the proposed FIS for qualitative fault classification of the target (aged) VRLA battery used in the stand-alone PV system application is tested to determine the severity of fault in a fast, non-contact, non-invasive, and non-destructive manner at the different discharging intervals.

16.2.2 Results and discussion

Front panel VIs for the condition monitoring of aged VRLA battery based on the obtained matching score as a thermal signature parameter at the start of discharging; after 2 h; and after 4 h of discharging are shown in Figs. 16.7, 16.8, and 16.9 respectively.

From Figs. 16.7 to 16.9, the more is the similarity between thermal patterns of two IR images, the more will be the matching scores (M). From Fig. 16.9, the

TABLE 16.1 Fuzzy rule base for fault diagnosis in battery.

	Input 1: Matching score		Input 2: Temperature		Fault type
IF	M is VP	and IF	T is VH	then	CR
IF	M is VP	and IF	T is H	then	CR
IF	M is VP	and IF	T is M	then	MJ
IF	M is VP	and IF	T is L	then	MJ
IF	M is P	and IF	T is VH	then	MJ
IF	M is P	and IF	T is H	then	MJ
IF	M is P	and IF	T is M	then	MJ
IF	M is P	and IF	T is L	then	MJ
IF	M is A	and IF	T is M	then	MN
IF	M is A	and IF	T is L	then	MN
IF	M is G	and IF	T is H	then	MN
IF	M is G	and IF	T is M	then	NF

matching score of the aged VRLA battery after 2 h of discharging shows that the middle cell has developed more variations in the thermal pattern due to the generation of heat inside the VRLA battery. The joule heating further spreads over the plate of adjacent cells after 4 h of the discharge as shown in Fig. 16.10. Therefore, with the lapse of discharging time, the aged VRLA battery deteriorates faster due to a reduction in its capacity which can be quantitatively obtained from the matching score in a non-invasive manner using IRT. Using the obtained values of matching scores (M) and the maximum hot region temperature (T) at different test conditions; the testing of developed FIS is performed to classify the severity of the fault. The testing of developed FIS after 4 h of discharge of aged VRLA battery is performed using the obtained

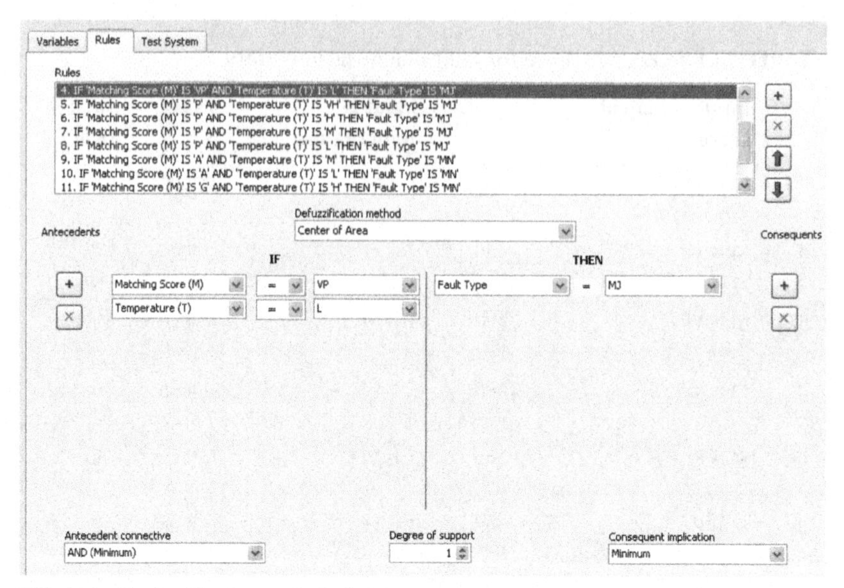

FIG. 16.6 De-fuzzification to obtain fault type.

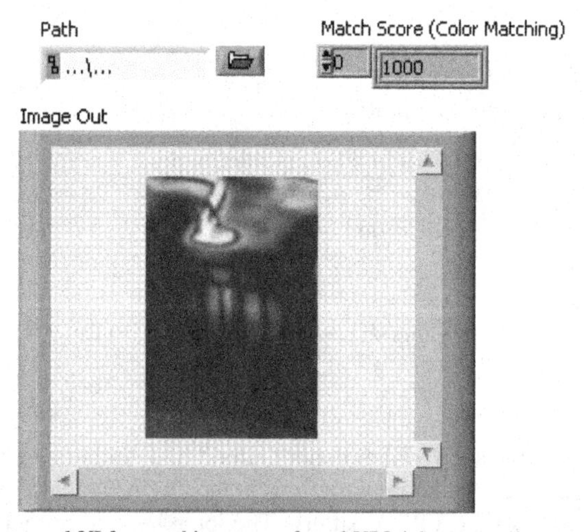

FIG. 16.7 Front panel VI for matching score of aged VRLA battery at the start of discharging.

thermal signature parameters values, i.e., matching score (M) of 407 (shown in Fig. 16.10) and maximum hot region temperature of 49°C and is shown in Fig. 16.10.

From Fig. 16.10, rule numbers 6, 7, and 9 of the Fuzzy rule base are invoked to map the fuzzified inputs to fuzzified output that is finally de-fuzzified. The crisp output value for fault type lies in between major fault with membership

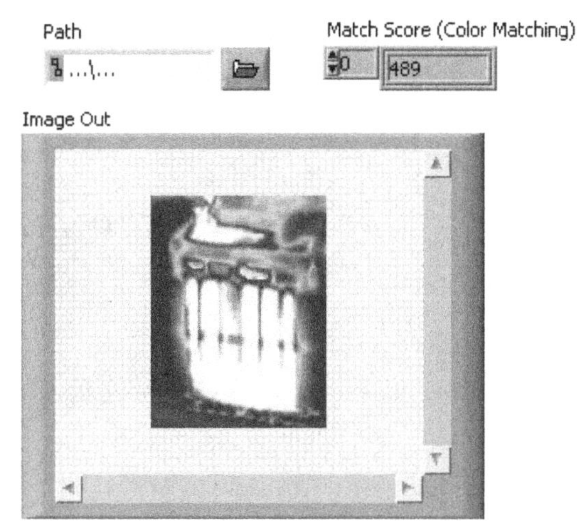

FIG. 16.8 Front panel VI for matching score of aged VRLA battery after 2 h of discharging.

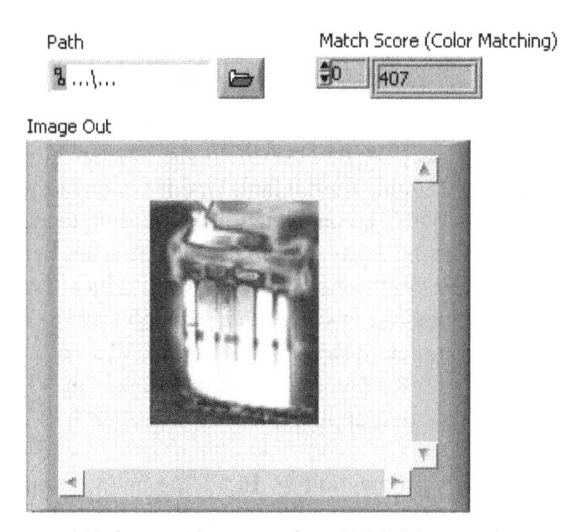

FIG. 16.9 Front panel VI for matching score of aged VRLA battery after 4 h of discharging.

function 0.72 and the minor fault with membership function 0.18. Therefore, after 4 h of discharge of aged VRLA battery, the fault is classified to be more in the region of the major fault and is significant to decide that the battery should be checked for the problem of heat dissipation in its cells. In the similar manner, the Fuzzy logic-based predictive fault diagnosis of target (aged) VRLA battery is implemented at different discharge intervals by obtaining values of matching score and the emissivity corrected maximum hot region temperature as input parameters. The IR image processing for fault detection and Fuzzy logic-based

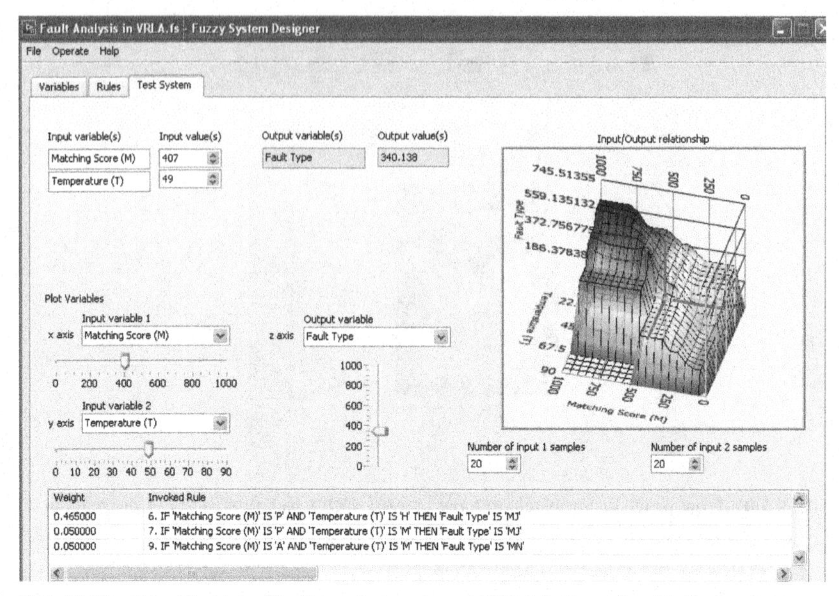

FIG. 16.10 Classification of fault type in target aged VRLA battery after 4 h discharging.

classification and analysis of fault type discussed in the previous section are implemented simultaneously as mentioned in the system block diagram in Fig. 16.2. The crisp output value for the fault type is compared with the threshold value which is set at 350. The onset of a major fault in the target battery is qualitatively diagnosed as the fuzzy output value approaches below the threshold level. On detecting the occurrence of the major or critical fault in the target battery, a warning alert message is sent to the intended users at both onsite and remote locations. The complete system block diagram VI to intelligently implement the proposed scheme for predictive fault diagnosis in VRLA battery is used in stand-alone PV system application using LabVIEW 2015 software is shown in Fig. 16.11.

From block diagram VI shown in Fig. 16.12, the fuzzy output values defining the condition of battery diagnosed from the severity of the fault are stored in MS Excel file with the time stamp at the hard disk of the host computer. Using the stored database, further analysis for the actual operating condition of VRLA battery can be done. Front panel VI of the proposed system is shown in Fig. 16.12.

From Fig. 16.12, whenever the major or critical fault in the target battery is detected by monitoring the crisp output value for fault type to be lower than the threshold value, a warning indication signal is generated in front panel VI and is sent to the intended users at both onsite and remote locations. At the onsite location, the warning signal is serially transmitted to the digital output pin 5 of a low-cost microcontroller board (Arduino Uno) through the virtual instrument

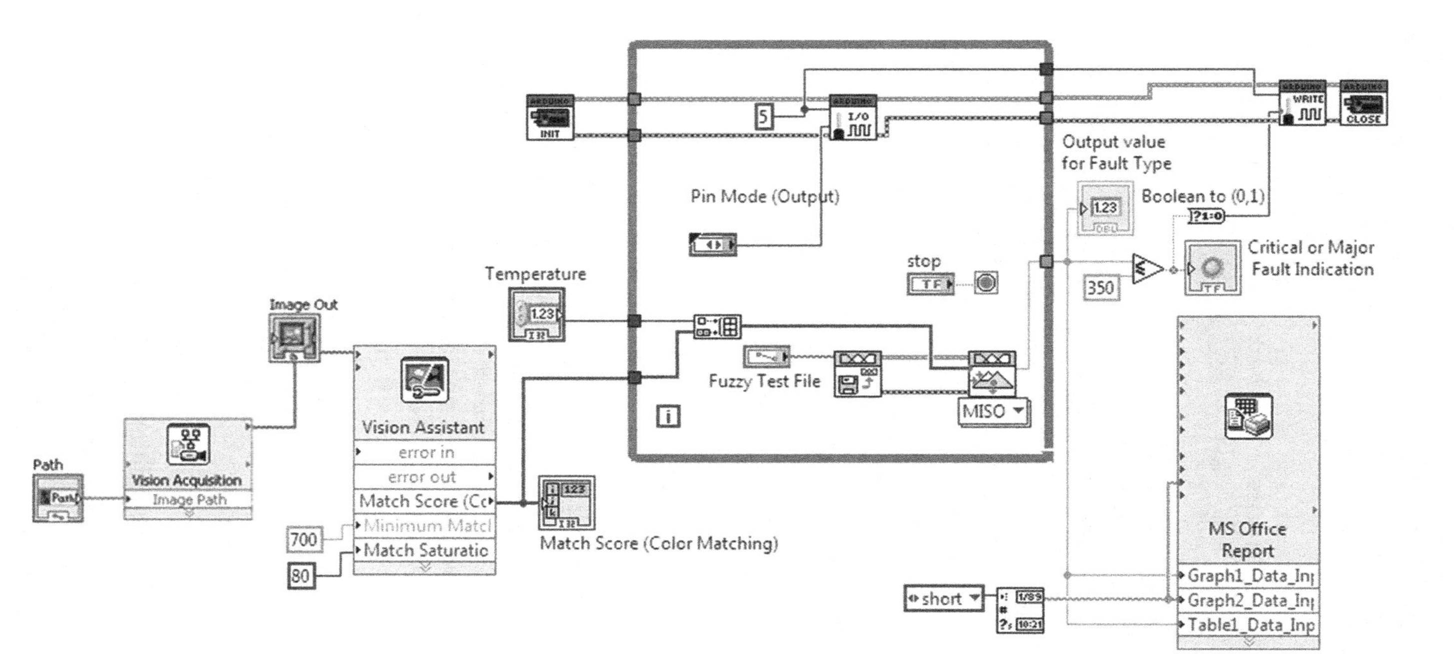

FIG. 16.11 System block diagram VI.

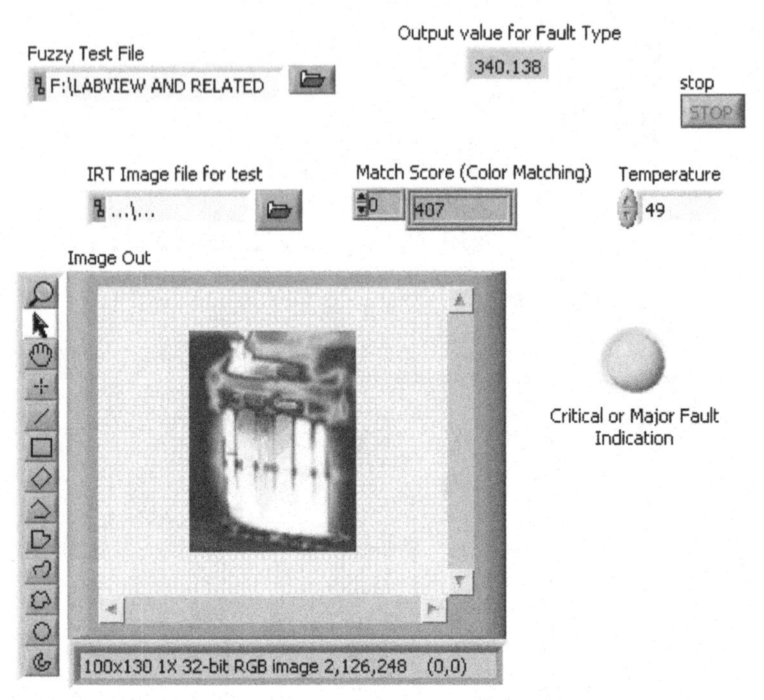

FIG. 16.12 System front panel VI.

software architecture (VISA) driver of LabVIEW software. Microcontroller board is interfaced with a transistor to drive piezoelectric buzzer and is programmed to generate an audio alert for the onsite operator on receiving the warning signal at the output pin. At a remote location, on identifying the critical fault condition, an alert email is sent to the intended user account from LabVIEW software. In a similar manner as above, the condition for other batteries connected in the string for large stand-alone PV system applications can also be monitored to detect their fault severity in a fast and non-invasive manner. The proposed technique is simple and effective in comparison to the monitoring of electrical failure signature parameters that requires the assembly of sensors and instruments for recording the data and communication interfaces for telemonitoring purpose. It will also aid the system designers to compare the reliability of similarly rated VRLA batteries from different manufacturers before selecting them in battery applications.

16.2.3 Web-based condition monitoring of battery

In order to facilitate the remote condition monitoring of the battery, the front panel information is continuously provided to the remote user using the web publishing tool of LabVIEW [13]. Therefore, based on the severity of fault data,

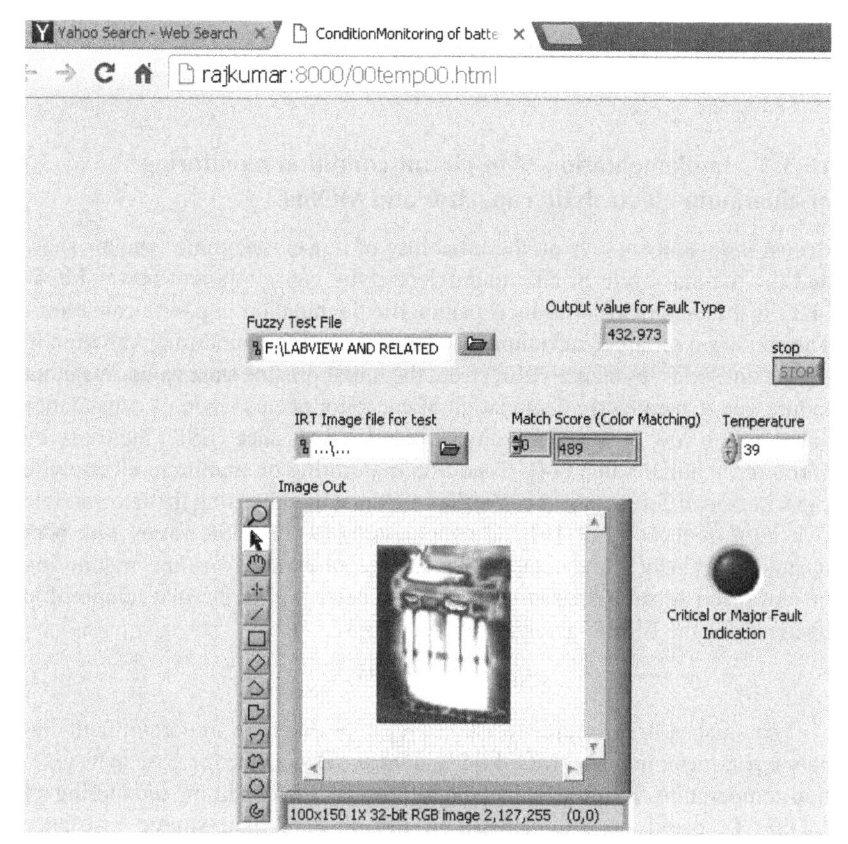

FIG. 16.13 Web-based condition monitoring of battery.

the maintenance engineer can remotely monitor the current condition of the battery within the system and can schedule the preventive maintenance task. The web-based condition monitoring after 2 h of discharge of aged VRLA battery is shown in Fig. 16.13.

16.3 Condition monitoring of aluminum electrolytic capacitor and MOSFET of power converter in PV system

In this section, condition monitoring algorithms are implemented in real-time to simultaneously determine the in-circuit wear out condition of high failure rate components, i.e., aluminum-electrolytic capacitors and power MOSFETs in the power converter (inverter) of the PV system. To obtain the in-circuit condition of target devices and to perform onsite and remote condition-based maintenance, a target electrolytic capacitor from Rubycon BXA is taken at the passive power factor correction section and MOSFETs STP 7NK60Z is taken in

H-bridge inverter section of the ECG circuit. The implementation of the proposed scheme is presented in Section 16.3.1.

16.3.1 Implementation of in-circuit condition monitoring of aluminum-electrolytic capacitor and MOSFET

Recent industrial surveys on the reliability of power electronic systems show that the thermal cycle in aluminum-electrolytic capacitors and power MOSFETs is the major cause for their parametric degradation in power converters. The condition of target capacitors can be estimated by monitoring variation in equivalent series resistance (ESR) from the initial pristine state value. Wear out failure due to parametric degradation of capacitor occurs when its capacitance value falls below 80% or its equivalent series resistance (ESR) increases by 100% of its initial value [14]. Condition monitoring of aluminum-electrolytic capacitors in different power converters are done by subjecting them to variable switching frequency [15–18] and temperature [19–24]. ESR varies with both operating frequency and temperature and requires the real-time monitoring of root mean square (rms) voltage and rms current. The thermal relationship obtained for the ESR is given in relation (16.3),

$$\mathrm{ESR} = 3.346\left(e^{-0.0043T}\right) \tag{16.3}$$

The on-state drain-source resistance $R_{ds(on)}$ value is an important fault signature parameter in power MOSFET that increases with the increase in its junction temperature. Empirical degradation model for condition monitoring of MOSFETs based on the increase in the on-state drain-source resistance (R_{dson}), is given in relation (16.4),

$$R_{dson}\left(T_j\right) = R_{dson}(25°\mathrm{C}) \cdot (1 + \alpha/100)^{T_j - 25°\mathrm{C}} \tag{16.4}$$

In-circuit condition monitoring of target MOSFET and electrolytic capacitors in the power converter of the PV system is performed by obtaining variation in their failure signature parametric values from the initial standard values. A low-cost microcontroller board is programmed for data acquisition and test circuit control. Thermal data for ambient and case temperature is acquired precisely using the linear integrated circuit-based temperature sensors LM35 and digital form of data values are serially communicated to the NI LabVIEW software. Data points are acquired after every 1 h using time to wait function in the loop. Aging algorithm is implemented as per relations (16.3) and (16.4).

16.3.2 Results and discussion

The real-time in-circuit data acquisition and front panel for the proposed scheme are given in Figs. 16.14 and 16.15, respectively.

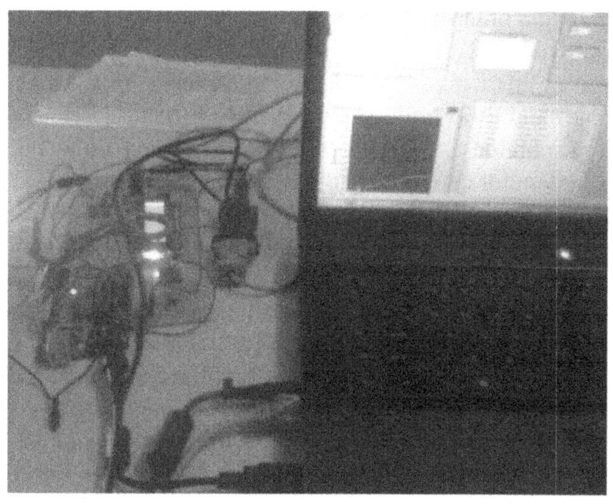

FIG. 16.14 In-circuit condition monitoring of capacitor and MOSFET.

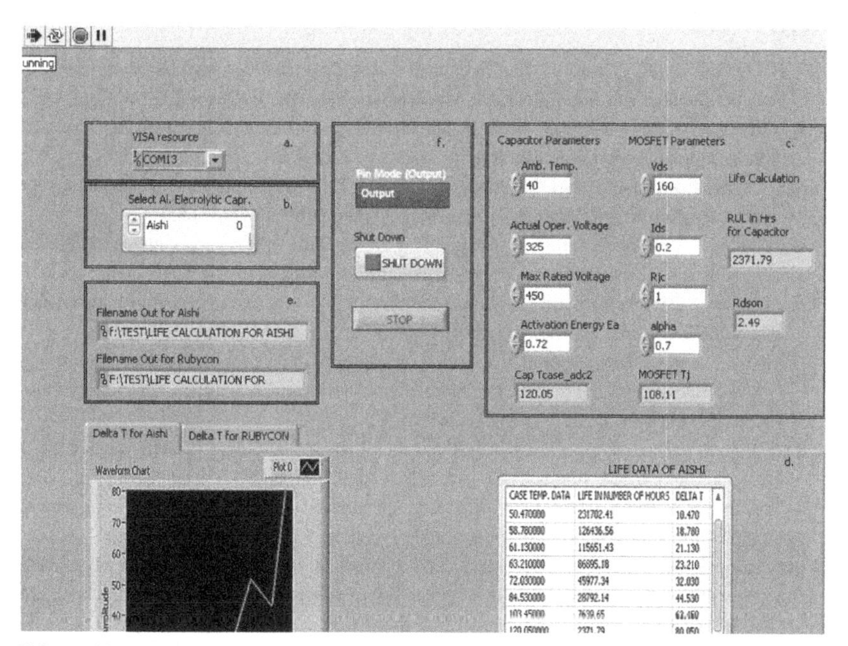

FIG. 16.15 Front panel VI.

16.4 Conclusions

Condition monitoring and predictive fault diagnosis for the high failure rate components of the system has become an essential part of maintenance strategy due to the potential advantages to be gained from reduced maintenance costs and increased availability. Condition-based maintenance is used to detect the system degradation failures, thereby allowing causal stress to be eliminated or controlled prior to any significant deterioration in the component physical state. The main aim of this work is to develop the condition monitoring system for identifying the failures in critical components of the stand-alone PV system. The condition monitoring algorithms are implemented using the intelligent embeddable system for determining the real-time in-circuit degradation in failure signature parameters of the above components at accelerated aging conditions. The proposed real-time in-circuit embeddable condition monitoring and maintenance techniques will eliminate the need for expensive test bed hardware and measurement systems that are used in earlier studies. A simple, non-contact, non-destructive, and non-invasive preventive fault diagnostic system using infrared thermography and fuzzy algorithm for detecting and classification of the severity of fault in VRLA battery used in a stand-alone PV system and the online technique for condition-based maintenance of power electronic converter due to parametric degradation of its high failure rate components are presented in this chapter.

References

[1] H. Chang, S. Lin, C. Kuo, C. Hsieh, Induction motor diagnostic system based on electrical detection method and fuzzy algorithm, Int. J. Fuzzy Syst. 18 (5) (2016) 732–740.

[2] N. Khera, S. Khan, Prognostics of aluminum electrolytic capacitors using artificial neural network approach, J. Microelectron. Reliab. 81 (2018) 328–336.

[3] S. Bagavathiappan, B. Lahiri, T. Saravanan, J. Philip, T. Jayakumar, Infrared thermography for condition monitoring—a review, Infrared Phys. Technol. 60 (2013) 35–55.

[4] M. Jadin, S. Taib, Recent progress in diagnosing the reliability of electrical equipment by using infrared thermography, Infrared Phys. Technol. 55 (4) (2012) 236–245.

[5] A. Huda, S. Taib, Application of infrared thermography for predictive/preventive maintenance of thermal defect in electrical equipment, Appl. Therm. Eng. 61 (2013) 220–227.

[6] Z. Shen, L. Xi, Q. Bing, H. Hou, Research on insulator fault diagnosis and remote monitoring system based on infrared images, in: International Workshop on Wireless Technology Innovations in Smart Grid, 2017, pp. 1194–1199.

[7] D. P'erez, J. Daviu, Application of infrared thermography to failure detection in industrial induction motors: case stories, IEEE Trans. Ind. Appl. 53 (3) (2017) 1901–1908.

[8] M. Ródenas, R. Royo, J. Daviu, J. Folch, Use of the infrared data for heating curve computation in induction motors: application to fault diagnosis, Eng. Fail. Anal. 35 (2013) 178–192.

[9] B. Hariprakash, S. Martha, S. Ambalavanan, S. Gaffoor, A. Shukla, Comparative study of lead-acid batteries for photovoltaic stand-alone lighting systems, J. Appl. Electrochem. 38 (1) (2008) 77–82.

[10] P. Pascoe, A. Anbuky, A VRLA battery simulation model, Energy Convers. Manag. 45 (7–8) (2004) 1015–1041.

[11] F. Adamčík, A. Bréda, T. Lazar, M. Puheim, Diagnostics of complex systems using thermography, in: 12th IEEE International Symposium on Applied Machine Intelligence and Informatics, 2014, pp. 109–113.

[12] K. Keränen, et al., Infrared temperature sensor system for mobile devices, in: 2nd IEEE International Conference on Electronics System Integration Technology, 2008, pp. 809–814.

[13] S. Khan, T. Islam, N. Khera, A. Agarwala, On-line condition monitoring and maintenance of power electronic converters, J. Electron. Test. 30 (6) (2014) 701–709.

[14] R. Kötz, P. Ruch, D. Cericola, Aging and failure mode of electrochemical double layer capacitors during accelerated constant load tests, J. Power Sources 195 (2009) 923–928.

[15] K. Yao, W. Tang, W. Hu, J. Lyu, A current-sensor-less online ESR and C identification method for output capacitor of buck converter, IEEE Trans. Power Electron. 30 (12) (2015) 6993–7005.

[16] A. Amaral, A. Cardoso, On-line fault detection of aluminium electrolytic capacitors, in step-down DC–DC converters, using input current and output voltage ripple, Power Electron. 5 (3) (2012) 315–322.

[17] M. Gustavo, A. Juan, H. Carlos, A. Amaral, A. Cardoso, An online and noninvasive technique for the condition monitoring of capacitors in boost converters, IEEE Trans. Instrum. Meas. 59 (8) (2010) 2134–2143, https://doi.org/10.1109/TIM.2009.2032960.

[18] K. Tsang, W. Chan, Simple method for measuring the equivalent series inductance and resistance of electrolytic capacitors, Power Electron. 3 (4) (2010) 465–471, https://doi.org/10.1049/iet-pel.2009.0146.

[19] A. Dehbi, W. Wondrak, Y. Ousten, Y. Danto, High temperature reliability testing of aluminum and tantalum electrolytic capacitors, Microelectron. Reliab. 42 (6) (2002) 835–840.

[20] F. Perisse, P. Venet, G. Rojat, Reliability determination of aluminium electrolytic capacitors by the mean of various methods. Application to the protection system of the LHC, Microelectron. Reliab. 44 (2004) 1757–1762.

[21] K. Abdennadher, P. Venet, G. Rojat, J. Rétif, C. Rosset, A real-time predictive-maintenance system of aluminum-electrolytic capacitors used in uninterrupted power supplies, IEEE Trans. Ind. Appl. 46 (4) (2010) 1644–1652.

[22] R. Jánó, D. Piticã, Accelerated ageing tests for predicting capacitor lifetimes, in: IEEE Symposium for Design and Technology in Electronic Packaging, 2011, pp. 63–68.

[23] F. Perisse, P. Venet, G. Rojat, J. Rétif, Simple model of electrolytic capacitor taking into account the temperature and aging time, Arch. Elektrotech. 88 (2) (2006) 89–95.

[24] N. Khera, S. Khan, T. Islam, A.K. Agarwala, An intelligent technique for condition based self-maintenance of aluminum electrolytic capacitors, Int. J. Syst. Assur. Eng. Manag. 7 (1) (2016) 25–34.

Chapter 17

Fault diagnosis and fault tolerance

Afaq Ahmad and Sayyid Samir Al Busaidi
Department of Electrical and Computer Engineering, Sultan Qaboos University, Muscat, Oman

17.1 Introduction

Testing is an essential part of system design. A process of testing ascertains a reliable and dependable mission operation of the designed system within a desired set of constraints. To ensure a digital system product of high quality and to maintain it with the service of fault-tolerant system, it is essential to incorporate testing process in every manufactured/assembled device and component of the digital system product. However, the testing process becomes difficult as the complexity of electronic integrated circuits (ICs) grows (see Fig. 17.1) [1–3]. As the scale of integration of digital circuits continues to increase through advances in semiconductor technology, it seems inevitable that future hardware systems will become even more complex, and therefore increasingly liable to design faults. The current hardware design testing and techniques seem to be inadequate to handle this problem. Therefore, testing becomes complex, expensive, and sometimes with no solution. A general rule of thumb is that capital costs run in the range of 50% of the overall IC test cost in the industry, so looking at capital costs is an essential analysis for manufacturing test [4].

The observed semiconductor technology trend impact is that the increasing complexity of digital systems also has been accompanied by a growing awareness of the need for efficient fault diagnosis techniques. This book chapter, which is devoted to "fault diagnosis and fault tolerance," provides a brief overview of the testing of digital circuits. The role of fault tolerance mechanism in the testing process for achieving better testability goals is to enhance availability and to reduce the maintainability cost of digital system products. This chapter is for the targeted audience of practicing engineers, students, and academics. There are five sections excluding the part of Introduction and Bibliography. The second and third sections provide brief knowledge about digital system modeling and fault models, respectively. The topics of fault diagnosis—Test

System Assurances. https://doi.org/10.1016/B978-0-323-90240-3.00017-5

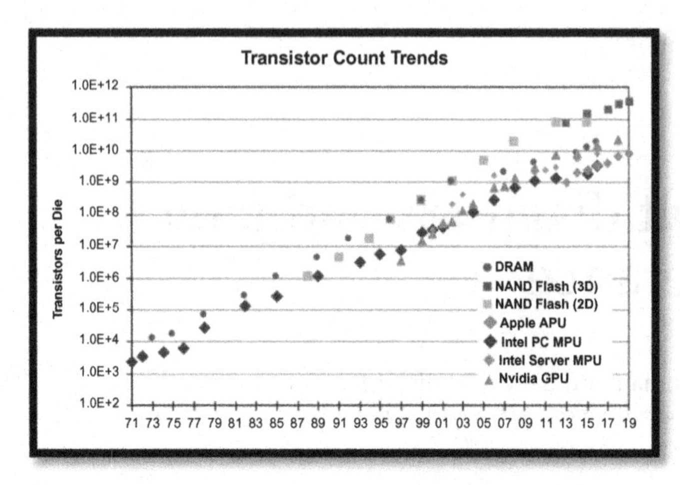

FIG. 17.1 Transistor count trends continue to track with Moore's Law [1].

procedures and fault diagnosis—fault tolerance are presented in Sections 17.4 and 17.5, respectively. The concluding remarks—future directions are discussed in Section 17.6.

"Successful engineering is all about understanding how things break or fail."

—Henry Petroski.

To understand how digital systems break or fail it is essential to understand the models of digital systems in depth. The ensuing section is oriented toward digital system modeling.

17.2 Digital systems modeling

Digital systems design requires rigorous modeling and simulation analysis that eliminates design risks so as to avoid potential harm to consumers. Since digital systems designer is responsible for the integrity of the system design as a whole, it must encompass the interconnectedness of the various aspects of the system. Hence, modeling of a digital system plays an important role in the system's design, synthesis fabrication, analysis, and testing. Mainly the modeling belong to three of the categories namely.

(1) Behavioral,
(2) Functional, and
(3) Structural.

A behavioral model is basically defined by a functional model but it also associates with timing relationships. Modeling of a digital system plays an important role in system design, synthesis fabrication, analysis, and testing. Mainly the functional level modeling can further be classified as follows:

17.2.1 Functional modeling at the logic level

The simplest way to represent a digital system of combinational nature is by its Truth table [5]. An example of a typical Truth table of function $F(x_1, x_2, x_3)$ is shown in Table 17.1.

Assuming binary input values, a digital system realizing a function $F(x_1, x_2, x_3,\ldots,x_n)$ of n variables require a table with 2^n entries. The data structure representing a Truth table is, therefore, an array V of dimension 2^n. The table is arranged with all possible input combinations of n-bit binary representations of decimal values varying from 0 to $2^n - 1$. From the perspective of testing of digital circuit, a typical example of Primitive cubes model representation of Table 17.1 is given in Table 17.2. In Table 17.2, d represents do not care, i.e., the input may be 0 or 1.

Sequential circuit in a digital system is a finite state sequential function. The function can be modeled as a sequential machine which is represented as State tables [6] and Flow tables (State diagram) [7,8]. In digital system modeling,

TABLE 17.1 An example of truth table.

x_1	x_2	x_3	F
0	0	0	1
0	0	1	1
0	1	0	0
0	1	1	1
1	0	0	1
1	0	1	1
1	1	0	0
1	1	1	0

TABLE 17.2 Primitive cube model of Table 17.1.

x_1	x_2	x_3	F
d	1	0	0
1	1	D	0
d	0	D	1
0	1	1	1

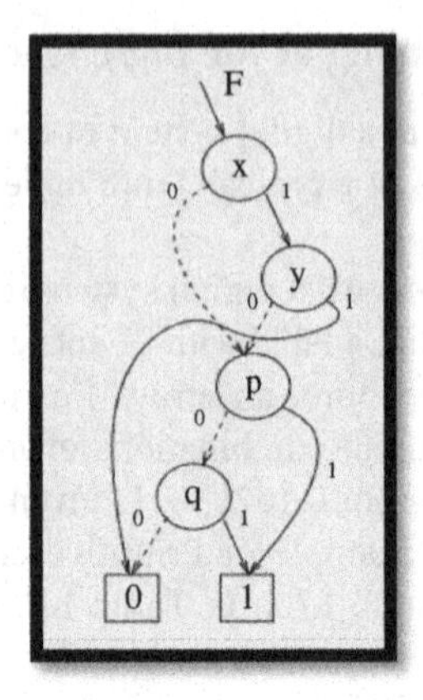

FIG. 17.2 Binary decision diagrams Boolean function $F = (x' \ y')p.q$.

also, exists Binary Decision Diagrams technique [9,10] which is basically a graph model representation of the function of a digital circuit. Depicted in Fig. 17.2 is the binary decision diagrams Boolean function $F = (x' \ y')p.q$.

17.2.2 Programs as functional models

Besides, the common features of the modeling techniques are based on the data structure (truth table, primitive cubes, state table, binary decision diagram) that is interpreted as model-independent program. A different approach to model a function of a digital circuit is to directly program the circuit model in any programming language code.

17.2.2.1 Functional modeling at the register level

Register transfer levels (RTLs) provide models at the registers and the instruction set levels, Data and control words are stored in registers and memories organized as arrays of the registers. To learn more about RTLs readers may refer to Refs. [11–16].

17.2.2.2 Structural models

In this type of modeling, an entity is described as a set of interconnected components. A connectivity language specifies the input-output lines of the system, its component, and their corresponding input-output signals [17].

17.2.2.3 Level of modeling

The level of modeling refers to the hierarchal complexity of the system from top to bottom or bottom to top. The levels may be system, device, interfaces, circuit, and components. Further, primitive components are used in the model such as Gate level or Transistor level or Standard cell.

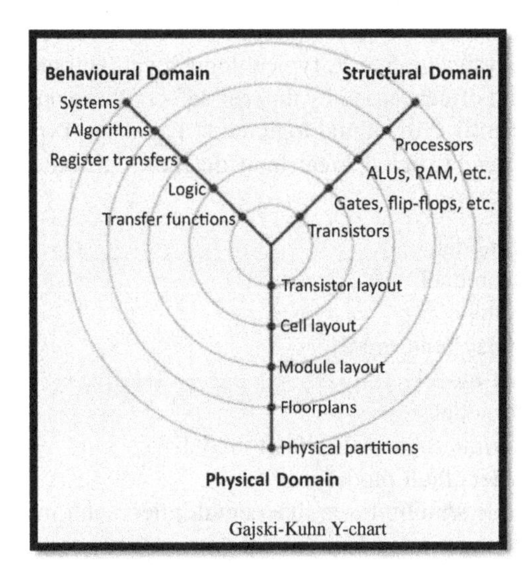

FIG. 17.3 Gajski-Kuhn Y-chart [18].

Just to give a summary of what we discussed in this section about the design flow of digital systems, we present it in the form of a figure which is named hereafter as Fig. 17.3. The chart in the figure is called as Gajski-Kuhn Y-chart [18].

17.3 Fault models

If anything can go wrong, it will …

Murphy's Law.

A digital system fault model helps the designer or user in predicting the consequences of particular faults. The basic assumptions regarding the nature of logical faults in digital systems testing are referred to as a fault model [19–23].

An observed error is an instance of an incorrect operation of the digital system which is being tested. Causes of observed errors may be design errors, fabrication errors, fabrication defects, or physical failures. Design errors can be detected and corrected at an early stage of the design process by simulating the design. Fabrication defects are not directly attributed due to human error but rather they resulted due to an imperfect manufacturing process. Physical failures occur during the lifetime of a system due to component wear out and/or environmental factors. In general, physical faults do not allow direct mathematical treatment of testing and diagnosis. The solution is to deal with logical faults, which are a convenient representation of the effect of physical faults on the operation of the system. Physical faults are classified as

Permanent—It is always present after their occurrence.
Intermittent—It exists only for some duration.
Transient—It occurs temporarily due to some changes in an environmental factor.

In digital logic fault diagnosis, fault models are used to model the failure behavior caused by a physical defect. A typical logic level fault model can simplify simulating the fault effect caused by the real defect. Based on the fault models, defects can be identified by digital logic fault diagnosis. Typical fault models widely used in current digital logic fault diagnosis procedures are given as follows:

Stuck-at fault model
Bridging faults model
Open fault model
Propagation delay fault model
Transition fault model
Memory fault model
Microprocessor/microcontroller fault models
Registers/counters fault model
Encoders/decoders/multiplexer/adders/multipliers fault models
Programmable logic array fault model
Application specific integrated circuits (ISICs) fault model
FPGA fault model
Cell internal fault model

17.3.1 Stuck-at fault model

The stuck-at fault model has been used successfully for describing the permanent faulty behavior on a line in the digital circuit caused by the defect. With a stuck-at fault on a line, the correct value on that line appears to be stuck at a constant logic value, either 0 or 1, referred to as *stuck-at-0* or *stuck-at-1*. The stuck-at-0 represents a short defect between the signal line and the ground line, while the stuck-at-1 could represent a short defect between the signal line and the power line. As an example, Fig. 17.4 demonstrates testing of an AND and OR gates for the *stuck-at-0* or *stuck-at-1* faults respectively.

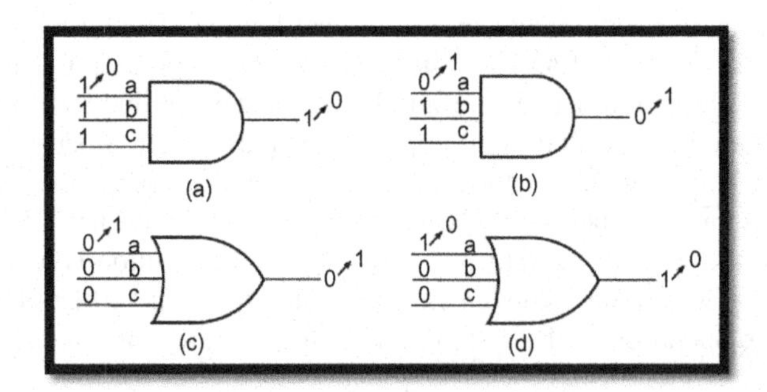

FIG. 17.4 Testing AND and OR gates for stuck-at faults.

17.3.2 Multiple stuck-at faults model

Any no. of circuit wires can have stuck-at faults at any given time. Since there are three total classifications for every wire (good, s-a-1 faulty, s-a-0 faulty), therefore, for a circuit with "n" wires, the total number of multiple stuck-at faults $3^n - 1$.

17.3.3 Bridge fault model

Bridging faults are those faults that involve a short between two signal lines in the digital circuit. The logic behavior of a short defect between signal lines is commonly represented by the bridge fault model. The bridge fault model that models the logic values of the shorted lines as logic AND or OR logic values of these two faulty nodes is referred to as wired-AND/wired-OR bridge fault model. The dominant bridge fault model was proposed for the bridge defects in which one line is assumed to dominate the logic value on the other line. Usually, the bridge fault mode captures the short defect between one signal line with another signal line instead of a power or ground line. The bridging fault can be modeled with an additional AND or OR gate, as illustrated in Fig. 17.5, where AS and BS denote the sources for the two shorted signal nets and AD and BD.

17.3.4 Open fault model

The open fault models the defect by assuming there is an interconnection on a signal line. Usually, the open fault can model defects such as electrical open, break, and disconnected via in a circuit. Open fault can result in state-holding,

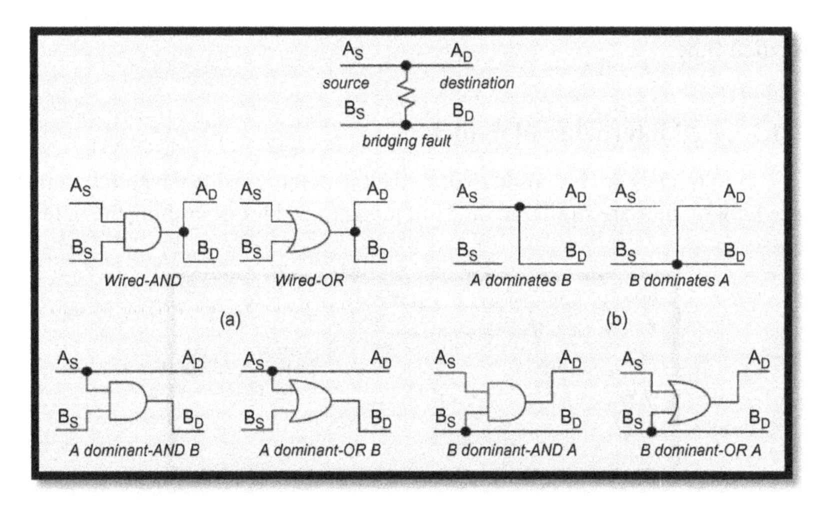

FIG. 17.5 Bridging fault models.

intermittent, and pattern-dependent fault effects which are more complex. Stuck-at-0 open or stuck-at-1 open are often used in logic diagnosis.

17.3.5 Path delay fault model

A delay fault in digital systems causes excessive delay along a path such that the total propagation delay falls outside the specified limit. Delay faults have become more prevalent with decreasing feature sizes. There are two types of different delay fault models, the gate-delay fault and the other model is a path-delay fault. An example of a gate delay model is shown in Fig. 17.6. Delay faults require an ordered pair of test vectors to sensitize a path through the logic circuit and to create a transition along that path in order to measure the path delay. The example of Fig. 17.6, considers the circuit where the fault-free delay associated with each gate is denoted by the integer value label on that gate. The two test vectors, v1 and v2, shown in the figure are used to test the path delay from input x2, through the inverter and lower AND gate, to the output y. Assuming the transition between the two test vectors occurs at time $t = 0$, the resulting transition propagates through the circuit with the fault-free delays shown at each node in the circuit such that it is expected to see the transition at the output y at time $t = 7$. A delay fault along this path would create a transition at some later time, $t > 7$.

17.3.6 Transition fault model

The transition fault model is used to model the delay fault that leads to the transition from the gate input to its output falling outside the specified timing limit. By the transition types, there are two transition fault models: slow-to-rise fault model and slow-to-fail fault model. The slow-to-rise (slow-to-fail) fault assume that the transition from 0 to 1 (1 to 0) cannot reach the output within the specified time.

17.3.7 Cell internal fault model

Some fault models are proposed to describe the defects inside the cell. Usually, those fault models are similar to the fault models used for describing the defects

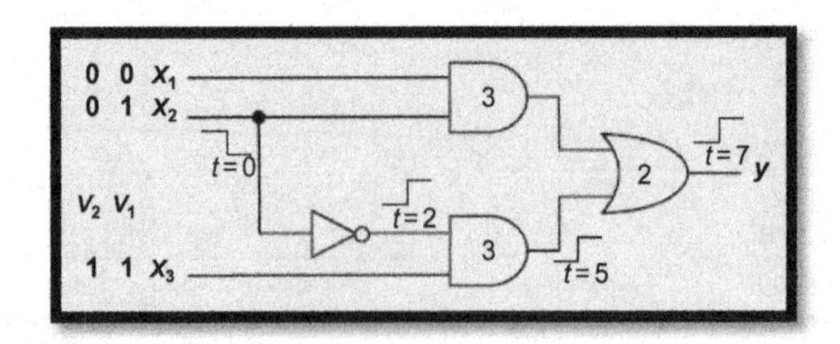

FIG. 17.6 Path delay fault test.

in inter-gates. Instead of modeling the defects between gates, the internal defect models, such as stuck-at fault, stuck-open fault, resistive-open fault, and short/bridge, represent the internal defects existing between transistors.

The most widely used fault model for digital logic fault diagnosis is the stuck-at fault model for its simplicity. Using the stuck-at fault model to simulate logic diagnosis, it is essential to get first a set of possible defective gates with stuck-at faults in its inputs or outputs. Based on the results, complex defects can be identified by applying more sophisticated fault models such as the bridge fault model and net open fault model. For the logic diagnosis using the stuck-at fault model, the diagnosis algorithms can be classified into two categories. The first category is called cause-effect analysis in which a presimulated fault dictionary for all the faults with all the test patterns and then the fault dictionary is looked up to find a set of candidates that can best match the test fails by the failing device observed on the tester. The second category is the effect-cause analysis which derives possible faulty locations by directly examining the failure syndrome of the failing chips.

17.4 Fault diagnosis test procedures

When testing a digital logic circuit, we apply a stimulus to the inputs of the device and check its response to establish that it is performing correctly. The input stimulus is referred to as a test pattern. The response of the circuit under test (CUT) is evaluated by comparing it to an expected response which may be generated by measuring the response of a known good circuit, or by simulation on the computer (see Fig. 17.7). If the CUT passes the test, we cannot say categorically that it is a "good" circuit. The only conclusion that we can draw from the device passing a test, is that the device does not contain any of the faults for which it was tested. It is important to note at this point, a device may contain a huge number of potential faults, some of which may even mask each other under specified operating conditions. The designer can only be sure that the device is 100% good if it has been 100% tested, this is rarely possible in real-life systems.

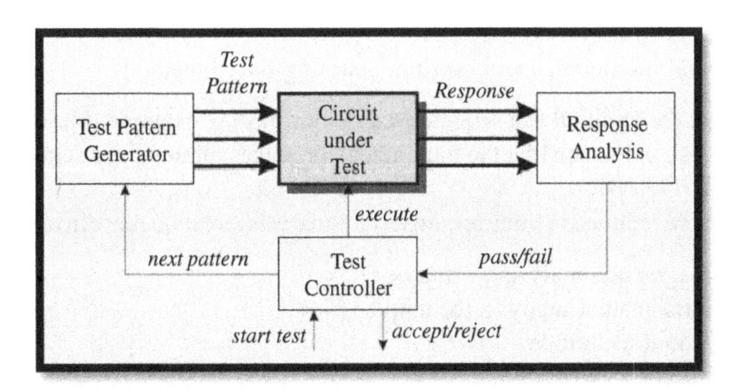

FIG. 17.7 Typical test architecture.

TABLE 17.3 Exhaustive test pattern application time.

N	Number of test patterns	Test time
20	$2^{20} = 1.048576 \times 10^6$	0.1048576 s
40	$2^{40} = 1.09951 \times 10^{12}$	30.54198966 h
60	$2^{60} = 1.15292 \times 10^{18}$	3655.890108 yr

Exhaustive testing can be applied for small combinational logic blocks which contain no redundant logic. However, for large circuits, it is impossible to use exhaustive test patterns (2^N) where N is the number of inputs. Shown in Table 17.3 is the calculated test time for $N = 20, 40, 60$ with 10 MHz test pattern application frequency.

Exhaustive testing methodology is also a very inefficient test strategy; most of the test patterns are actually redundant. Various testing methods have been researched [24–32] for determining significant test patterns, in order to obtain a minimum set of test patterns. Listed below are some test pattern-finding strategies to diagnose the faults of digital logic systems.

Fault table (matrix) method.
Path sensitization method.
Critical path method.
D-Algorithm method.
PODEM method.
Boolean difference method.
Automated test pattern generation (ATPG) method.
Divide and conquer method.
Automatic test equipment (ATE).
Pseudorandom test pattern generating method.
Genetic algorithm method.
Neural network-based method.
Artificial intelligence and machine learning-based method.

How to get the minimal test sequences, a demonstrating example is presented in Fig. 17.8. It can be seen that the fault matrix produces minimal test sequences as {001,010,011,100}.

In test procedures, various response data analysis techniques are in use such as

Serial signature analysis technique
Parallel signature analysis technique
Ones count technique
Transition count technique
Syndrome count technique
Spectral response analysis technique

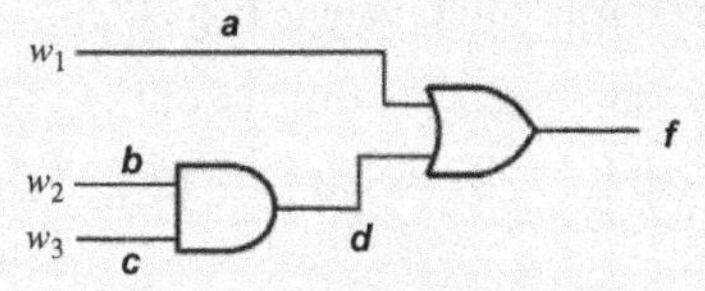

Test	Fault detected									
$w_1w_2w_3$	a/0	a/1	b/0	b/1	c/0	c/1	d/0	d/1	f/0	f/1
000		√						√		√
001		√		√				√		√
010		√				√		√		√
011			√		√		√		√	
100	√								√	
101	√								√	
110	√								√	
111									√	

FIG. 17.8 Minimal test sequences via fault matrix.

17.5 Fault diagnosis and fault tolerance

Fault avoidance and fault tolerance are the main approaches used to increase the reliability of VLSI circuits. Fault avoidance relies on improved materials, manufacturing processes, and circuit design. A manufacturing defect is a finite chip area with electrically malfunctioning circuitry caused by errors in the fabrication process. A chip with no manufacturing defect is called a good chip. The fraction (or percentage) of good chips produced in a manufacturing process is called the yield. Yield is denoted by the symbol Y. Given in Eq. (17.1) is the relationship between Y, number of defects (n), and the probability of defects (P).

$$Y=(1-P)^n \tag{17.1}$$

As given in Eq. (17.2), fault coverage (FC) for a given set of test vectors is defined as the percentage of stuck-at faults detected by test vectors. Fault coverage is also termed test coverage (T). Table 17.4 provides the FC values for each of the test sequences of the fault matrix shown in Fig. 17.8.

In test procedures, 100% (100%) fault coverage may be impossible due to the presence of undetectable faults. Thus, The fault detection efficiency (FDE) can be known by Eq. (17.3). Defect coverage is defined as the percentage (%) of real defects detected by test vectors. Defect level (DL) or reject rate (RR) can be computed via Eq. (17.4).

$$\text{FC}=\frac{\text{Number of detected faults by a test sequence}}{\text{Total number of faults}} \tag{17.2}$$

TABLE 17.4 An example of FC values (Fig. 17.7).

Test (w1w2w3)	000	001	010	011	100	101	110	111
FC (%)	30	40	40	40	20	20	10	10

TABLE 17.5 An example of relative values of Y, FC, and DL.

Y (%)	50	75	90	95	99	90	90	90	90
FC (%)	90	90	90	90	90	90	95	99	99.9
DL (per million)	67	28	10	5	1000	10,000	5000	1000	100

$$\text{FDE} = \frac{\text{Number of detected faults}}{\text{Total number of faults} - \text{Number of undetected faults}} \quad (17.3)$$

$$\text{DL} = \text{RR} = (1 - Y)^{(1-\text{FC})} \quad (17.4)$$

We present Table 17.5 to demonstrate some typical interrelations between Y, FC, and DL.

17.5.1 Fault tolerance

As an example considering a printed circuit board (PCB) with 40 integrated circuits (ICs), each with FC of 90% and with 90% of Y, has a DL of 34.4%. Or it can be said that there exist a probability of 344,000 defective ICs out of per million (PM) ICs. If we dig below the levels of IC or above the levels of PCB, a rule of ten governs the cost of testing. Rule of ten suggests that the cost to detect faulty part increases by an order of magnitude as we move from Material → Wafer → Transistor → Gate → IC → Device → PCB → Module → System → Field operation. Hence, testing needs to be performed at all of these levels.

As digital systems become increasingly large and complex, their reliability and availability qualities play a critical role in supporting next-generation science, engineering, and commercial applications. In digital system design, fault tolerance can be used to increase system reliability. By incorporation of fault tolerance mechanism, the digital system continues to operate without failure despite the presence of faults. Therefore, the adoption of fault diagnosis strategies alone, in general, is insufficient. There is also an upper limit for improvement of a component or system reliability due to design methodology, cost limitations, and available manufacturing techniques. Indeed this is the most

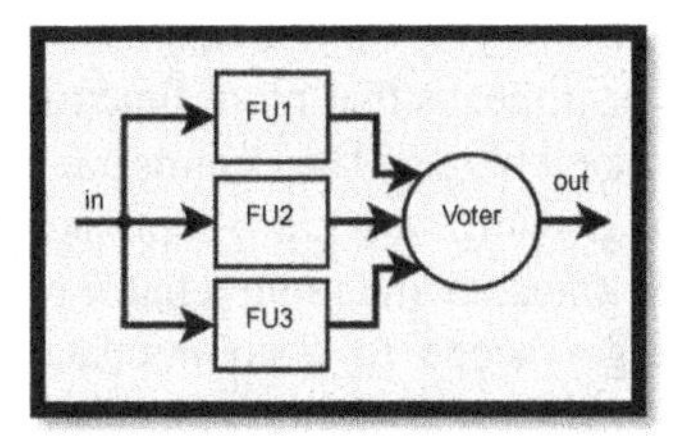

FIG. 17.9 Architecture of TMR.

important reason behind the implementation of designs taking another approach called fault tolerance.

Real-time systems are often used in hazardous or remote applications, such as aircraft and spacecraft, where the systems are highly susceptible to errors due to radiation. In such systems, the required levels of reliability are usually attained through replication of hardware such as architectures based on a duplex system, n-modular redundancy or the application of self-correcting codes. Triple modular redundancy (TMR) uses hardware redundancy to mask any single design failure by voting on the result of three identical copies of the digital systems function unit (FU). TMR is a popular technique used in many fault-tolerant schemes. The architecture of a TMR system can be seen in Fig. 17.9.

17.6 Conclusions

Heavy risks are involved in the application of systems where digital systems are used. Most of the applications are based on online real time processing which requires high reliability, availability, and serviceability. Some of such application sectors are, for instance, railway control, satellites, avionics, telecommunications, control of critical automotive functions, medical electronics, industrial control, etc. In such applications of digital systems, the complexity of the system increases, but higher serviceability is often required. In that case, it might be impossible to shorten the time required for fault repair by the usual test program and manual diagnosis. Also, in such application domains pressure for low-cost products is also one of the targets.

Therefore, there is a corresponding increased demand for cost-effective online test techniques. These needs will increase dramatically in the near future since very deep submicron and nanotechnologies will impact adversely noise margins and make mandatory online testing solutions. Thus, it is needed to consider the test as codesign process. Each functional block in the design should be built with its testing procedure in mind, either for self-testing, or to ease testing of the entire system [33,34]. When the system is put together, the individual test subsystems should work harmoniously toward testing the entire system. Some of the dedicated solutions that are being used in test fields are design for test (DFT) and built-in self-test (BIST).

Delivering more transistors in the same area means the circuitry can be made smaller, saving on cost, or it means that more functionality can be added to a chip without having to make it bigger. The circuits are becoming more complex but compact. The ICs are growing with more depth and density. Hence, allows us to put more circuitry in a smaller and more reliable package. The era of nanotechnology provides bigger scope to accommodate test and fault-tolerant circuitry.

Acknowledgment

The authors would like to express their great appreciation and gratitude to Sultan Qaboos University, Muscat, Oman for providing research facilities, technical supports, and a research environment that enabled us to complete this research task.

References

[1] Transistor Count Trends Continue to Track with Moore's Law, Online: Accessed on 19 March 2021. https://anysilicon.com/transistor-count-trends-continue-to-track-with-moores-law/.

[2] A. Ahmad, Automotive Semiconductor Industry—Trends, Safety and Security Challenges, in: 8th International Conference on Reliability, Infocom Technologies and Optimization (Trends and Future Directions) (ICRITO 2020), IEEE Conference Record Number 48877, Amity University, Noida, India, June 4–5, 2020, IEEE Xplore proceedings, 2020, pp. 1373–1377.

[3] Ahmad A., (Keynote/invited—talk), Challenges for test and fault-tolerance due to convergence of electronics, semiconductor systems and computing, IEEE Conference: 2017 International Conference on Infocom Technologies and Unmanned Systems (Trends and Future Directions) (ICTUS), Amity University, Int'l Academic City, Dubai, UAE, 18–20 Dec., IEEE Xplore proceedings, 2017, pp. 64–68.

[4] K. Flamm, Measuring Moore's law: evidence from Price, cost, and quality indexes, Internal document, University of Texas at Austin, USA (Nov. 2017) 1–44.

[5] L. Wittgenstein, Tractatus Logico-Philosophicus Proposition 5.101, 1922.

[6] D.A. Huffman, The synthesis of sequential switching circuits, J. Franklin Inst. 257 (March and April 1954) 169–190.

[7] T. Booth, Sequential Machines and Automata Theory, John Wiley and Sons, New York, 1967.

[8] E.J. McClusky, Introduction to the Theory of Switching Circuits, McGraw-Hill, 1965.

[9] C.Y. Lee, Representation of switching circuits by binary-decision programs, Bell Syst. Tech. J. 38 (1959) 985–999.

[10] S.B. Akers Jr., Binary decision diagrams, IEEE Trans. Comput. C-27 (6) (June 1978) 509–516.

[11] R. Duley, D.L. Dietmeyer, A digital system design language (DDL), IEEE TC C-17 (1968).

[12] J.R. Duley, D.L. Dietmeyer, Translation of a DDL digital system specification to Boolean equation, IEEE TC C-18 (1968).

[13] R.L. Arndt, D.L. Dietmeyer, DDLSIM—A digital design language simulator, Proc. NEC 26 (1970).

[14] M.R. Barbacci, A comparison of register transfer languages for describing computers and digital systems, IEEE Trans. Comp. C-24 (1975) 137–150.

[15] M. Shahdad, An overview of VHDL and technology, in: 23rd Design: Automation Conference, 1986, pp. 320–326.

[16] M. Shahdad, et al., VHSIC hardware description language, Computer 18 (2) (Feb. 1985).

[17] H.Y. Chang, G.W. Smith Jr., R.B. Walford, LAMP: system description, Bell Syst. Tech. J. 53 (8) (October 1974) 1431–1449.
[18] D. Gajski, R. Kuhn, Guest Editors' introduction, "New VLSI tools", Computer 16 (12) (December, 1983) 11–14.
[19] M.A. Breuer, General survey of design automation of digital computers, Proc. IEEE 54 (1966) 1708–1721.
[20] J.A. Abraham, W.K. Fuchs, Fault and Error Models for VLSI, Proc. IEEE 74 (May 1986) 639–654.
[21] M.A. Breuer, S.J. Chang, S.Y.H. Su, Identification of multiple stuck-type faults in combinational networks, IEEE Trans. Comput. C-25 (Jan. 1976) 44–54.
[22] C.W. Cha, A testing strategy for PLAs, in: Proc. 15th Design Auto. Conf, 1978, pp. 83–89.
[23] R. Bennitts, Progress in design for test: a personal view, IEEE Des. Test Comput. 11 (01) (1994) 53–59.
[24] A. Ahmad, Achievement of higher testability goals through the modification of shift register in LFSR based testing, Int. J. Electron. (UK) 82 (3) (1997) 249–260.
[25] H.Y. Chang, An algorithm for selecting an optimum set of diagnostic tests, IEEE Trans. Electron. Comput. EC-14 (1965) 706–711.
[26] F. Hadlock, On finding a minimal set-of diagnostic tests, IEEE Trans. Electron. Comput. EC-16 (1967) 674–675.
[27] W.H. Kautz, Fault testing and diagnosis in combinational digital circuits, IEEE Trans. Computers C-17 (1968) 352–366.
[28] T.J. Powell, A procedure for selecting diagnostic tests, IEEE Trans. Computers C-18 (1969) 168–175.
[29] Boyce, A. H. Computer generated diagnosing procedures for logic circuits, Joint Conference on Automatic Test Systems, IERE Conference Proc. No. 17, April 1970, pp. 333–346.
[30] D.B. Armstrong, On finding a nearly minimal set of fault detection tests for combinational logic nets, IEEE Trans. Electron. Comput. EC-15 (1966) 66–73.
[31] M.S. Abadir, J. Ferguson, T.E. Kirkland, Logic design verification via test generation, IEEE Trans. CAD 7 (Jan. 1988) 138–148.
[32] M. Fujita, T. Kakuda, Y. Matsunaga, Redesign and automatic error correction of combinational circuits, in: G. Saucier (Ed.), Logic and Architecture Synthesis, Elsevier Science Publishers B.V., North-Holland, 1991, pp. 253–262.
[33] A. Ahmad, D. Al-Abri, Design of an optimal test simulator for built-in self-test environment, J. Eng. Res. 7 (2) (2010) 69–79.
[34] A. Ahmad, A.H. Al-Habsi, Design of a built-in multi-mode ICs tester with higher testability features- a most suitable testing tool for BIST environment, J. IETE Tech. Rev. 15 (3) (1998) 283–288.

Chapter 18

True power loss diminution by Improved Grasshopper Optimization Algorithm

Lenin Kanagasabai
Department of EEE, Prasad V. Potluri Siddhartha Institute of Technology, Vijayawada, Andhra Pradesh, India

18.1 Introduction

The power loss lessening problem is very important in the power system. Many conventional methods [1–5] and swarm-based methods such as moth-flame optimization technique, particle swarm optimization, and Ant Lion Optimizer [6–15] are applied. Grasshopper Optimization Algorithm hybridized with self-adaptive differential algorithm known as Improved Grasshopper Optimization Algorithm (IGSD) for active power loss reduction. The Grasshopper algorithm has been modeled based on the natural actions of Grasshopper [16]. The movement of the Grasshopper has been mathematically formulated to design the algorithm. Then in the proposed approach self-adaptive differential approach has been hybridized with Grasshopper Optimization Algorithm. In the self-adaptive differential algorithm mutation, crossover and selection are the main operators. Then the crossover probability (CR) and scaling factor (SF) control the exploration and exploitation. The performance of the algorithm has been improved through hybridization. Exploration and exploitation have been balanced. Proposed Improved Grasshopper Optimization Algorithm (IGSD) is assessed in IEEE 57and 300 bus test systems. Lessening of power loss is attained.

18.2 Problem formulation

Loss lessening is a key objective and it is defined as

$$F = P_L = \sum_{k \in Nbr} g_k \left(V_i^2 + V_j^2 - 2V_iV_j\cos\theta_{ij} \right) \tag{18.1}$$

System Assurances. https://doi.org/10.1016/B978-0-323-90240-3.00018-7

$$F = P_L + \omega_v \times \text{voltage deviation} \tag{18.2}$$

$$\text{Voltage deviation} = \sum_{i=1}^{Npq} |V_i - 1| \tag{18.3}$$

Parity

$$P_G = P_D + P_L \tag{18.4}$$

Disparity

$$P_{\text{gslack}}^{\min} \le P_{\text{gslack}} \le P_{\text{gslack}}^{\max} \tag{18.5}$$

$$Q_{gi}^{\min} \le Q_{gi} \le Q_{gi}^{\max}, \quad i \in N_g \tag{18.6}$$

$$V_i^{\min} \le V_i \le V_i^{\max}, \quad i \in N_B \tag{18.7}$$

$$T_i^{\min} \le T_i \le T_i^{\max}, \quad i \in N_T \tag{18.8}$$

$$Q_c^{\min} \le Q_c \le Q_C^{\max}, \quad i \in N_C \tag{18.9}$$

18.3 Improved Grasshopper Optimization Algorithm

Grasshopper Optimization Algorithm is designed by replicating the natural events of Grasshopper. Grasshopper's progression is based on social communication, wind, and gravity. Location of Grasshopper is given by

$$X_i = S_i + G_i + A_i \tag{18.10}$$

Social communication during the movement of grasshopper is defined by

$$S_i = \sum_{\substack{j=1 \\ j \neq 1}}^{N} s(d_{ij})\widehat{d_{ij}} \tag{18.11}$$

$$d_{ij} = |X_j - X_i| \tag{18.12}$$

$$\widehat{d_{ij}} = (X_j - X_i)/d_{ij} \tag{18.13}$$

$$S(r) = fe^{-r/t} - e^{-r} \tag{18.14}$$

Gravity force during the movement of grasshopper is mathematically written as

$$G_i = -g\widehat{e_g} \tag{18.15}$$

Wind advection during the movement of grasshopper is defined by

$$A_i = U\widehat{e_w} \tag{18.16}$$

Then the Position of Grasshopper is defined by

$$X_i = \sum_{\substack{j=1 \\ j \neq 1}}^{N} S(|X_j - X_i|)(X_j - X_i)/d_{ij} - g\hat{e_g} + U\hat{e_w} \tag{18.17}$$

Exploration and exploitation is balanced by

$$X_d^i = c\left(\sum_{\substack{j=1 \\ j \neq 1}}^{N} c\frac{ub_d - ib_d}{2} S(|X_j - X_i|)(X_j - X_i)/d_{ij}\right) + \widehat{T_d} \tag{18.18}$$

$$c = c_{\text{maximum}} - I\frac{c_{\text{maximum}} - c_{\text{minimum}}}{L} \tag{18.19}$$

Grasshopper Optimization is hybridized with self-adaptive differential algorithm known as Improved Grasshopper Optimization Algorithm (IGSD) for active power loss reduction. The proposed IGSD algorithm increases the exploration competence and population diversity will be maintained in the last phase of the iterations. In the self-adaptive differential algorithm mutation, crossover and selection are the main operators. Then the crossover probability (CR) and scaling factor (SF) control the exploration and exploitation.

Then the mutation operation is mathematically written as

$$\text{Mt}_i^{t+1} = y_{r1}^{t+1} + \text{scaling factor} \times (y_{r2}^t - y_{r3}^t) \tag{18.20}$$

Crossover operation articulated by

$$\text{Cr}_i^{t+1} = \begin{cases} \text{Mt}_i^{t+1} & \text{if random} \leq \text{CR} \\ y_i^t & \text{if random} > \text{CR} \end{cases} \tag{18.21}$$

With respect to minimization condition, the selection is defined by

$$y_i^{t+1} = \begin{cases} \text{Cr}_i^{t+1} & \text{if } f(\text{Cr}_i^{t+1}) < f(y_i^t) \\ y_i^t & \text{otherwise} \end{cases} \tag{18.22}$$

Then by self-adaptive crossover probability (CR) and scaling factor (SF) are described as

$$\text{SF}_i^{t+1} = \begin{cases} \text{SF}_{\text{Lower bound}} + \text{random}_1 \times \text{SF}_{\text{Upper bound}} & \text{if random}_2 < \tau_1 \\ \text{SF}_i^t & \text{otherwise} \end{cases} \tag{18.24}$$

$$\text{CR}_i^{t+1} = \begin{cases} \text{random}_3, & \text{if random}_4 < \tau_2 \\ \text{CR}_i^t & \text{otherwise} \end{cases} \tag{18.25}$$

a. Start
b. Population arbitrarily initialized in the exploration space
c. Most excellent search agent $\widehat{T_d}$ has been initialized
d. Fitness values of the Grasshoppers are computed
e. Fix the population and highest amount of iterations
f. while (end creteria not met($t < t_{\text{maximum}}$))
g. Boundary conditions are checked
h. Most excellent exploration agent is found means then modernize $\widehat{T_d}$ and fitness value
i. Compute $c = c_{\text{maximum}} - I\frac{c_{\text{maximum}} - c_{\text{minimum}}}{L}$
j. Middling fitness of the entire population is evaluated
k. for every grasshopper($i = 1 : n$)
l. if $(f_i > \overline{f})$ then the position of the grasshopper updated by

$$X_d^i = c\left(\sum_{\substack{j=1 \\ j \neq 1}}^{N} c\frac{ub_d - ib_d}{2} S\left(\left|X_j - X_i\right|\right)\left(X_j - X_i\right)/d_{ij}\right) + \widehat{T_d}$$

m. for every exploration agent compute SF and CR by

$$\text{SF}_i^{t+1} = \begin{cases} \text{SF}_{\text{Lower bound}} + \text{random}_1 \times \text{SF}_{\text{Upper bound}} & \text{if random}_2 < \tau_1 \\ \text{SF}_i^t \text{ otherwise} \end{cases}$$

$$\text{CR}_i^{t+1} = \begin{cases} \text{random}_3, & \text{if random}_4 < \tau_2 \\ \text{CR}_i^t & \text{otherwise} \end{cases}$$

n. Then compute mutation, selection, and crossover by

$$\text{Mt}_i^{t+1} = y_{r1}^{t+1} + \text{scaling factor} \times (y_{r2}^t - y_{r3}^t)$$

$$\text{Cr}_i^{t+1} = \begin{cases} \text{Mt}_i^{t+1} & \text{if random} \leq \text{CR} \\ y_i^t & \text{if random} > \text{CR} \end{cases}$$

$$y_i^{t+1} = \begin{cases} \text{Cr}_i^{t+1} & \text{if } f\left(\text{Cr}_i^{t+1}\right) < f\left(y_i^t\right) \\ y_i^t & \text{otherwise} \end{cases}$$

o. End if
p. End for
q. End while
r. Output the optimal solution
s. End

18.4 Simulation study

Proposed Improved Grasshopper Optimization (IGSD) is verified, in IEEE 57 and 300 bus systems [17]. Tables 18.1 and 18.2 give the limits. Loss assessment is shown in Tables 18.3 and 18.4. Figs. 18.1 and 18.2 show the assessment of loss.

TABLE 18.1 Limits.

Parameter	Min value (PU)	Max value (PU)
V(G)	0.950	1.10
T	0.90	1.10
Volt .A.R	0.00	0.200

TABLE 18.2 Limits of parameters.

Parameter	Q minimum (PU)	Q maximum (PU)
1	−140	200
2	−17	50
3	−10	60
6	−8	25
8	−140	200
9	−3	9
12	−150	155

TABLE 18.3 Loss assessment.

Parameters	Base case [18]	MPSO [19]	PSO [20]	CGA [21]	AGA [22]	IGSD
V(G)-1	1.040	1.093	1.083	0.968	1.027	1.022
V(G)-2	1.010	1.086	1.071	1.049	1.011	1.014
V(G)-3	0.985	1.056	1.055	1.056	1.033	1.002
V(G)-6	0.980	1.038	1.036	0.987	1.001	1.013
V(G)-8	1.005	1.066	1.059	1.022	1.051	1.001
V(G)-9	0.980	1.054	1.048	0.991	1.051	1.010
V(G)-12	1.015	1.054	1.046	1.004	1.057	1.010
T-19	0.970	0.975	0.987	0.920	1.030	0.911

Continued

TABLE 18.3 Loss assessment—cont'd

Parameters	Base case [18]	MPSO [19]	PSO [20]	CGA [21]	AGA [22]	IGSD
T-20	0.978	0.982	0.983	0.920	1.020	0.910
T-31	1.043	0.975	0.981	0.970	1.060	0.913
T-35	1.000	1.025	1.003	NR[a]	NR[a]	1.004
T-36	1.000	1.002	0.985	NR[a]	NR[a]	1.002
T-37	1.043	1.007	1.009	0.900	0.990	1.010
T-41	0.967	0.994	1.007	0.910	1.100	0.911
T-46	0.975	1.013	1.018	1.100	0.980	1.012
T-54	0.955	0.988	0.986	0.940	1.010	0.923
T-58	0.955	0.979	0.992	0.950	1.080	0.940
T-59	0.900	0.983	0.990	1.030	0.940	0.934
T-65	0.930	1.015	0.997	1.090	0.950	1.011
T-66	0.895	0.975	0.984	0.900	1.050	0.920
T-71	0.958	1.020	0.990	0.900	0.950	1.012
T-73	0.958	1.001	0.988	1.000	1.010	1.013
T-76	0.980	0.979	0.980	0.960	0.940	0.942
Tap 80	0.940	1.002	1.017	1.000	1.000	1.020
QC 18	0.1	0.179	0.131	0.084	0.016	0.143
QC 25	0.059	0.176	0.144	0.008	0.015	0.132
QC 53	0.063	0.141	0.162	0.053	0.038	0.114
PG (MW)	1278.6	1274.4	1274.8	1276	1275	1272.60
QG (MVAR)	321.08	272.27	276.58	309.1	304.4	272.32
Reduction in power loss (%)	0	15.4	14.1	9.2	11.6	23.49
Power loss (MW)	27.8	23.51	23.86	25.24	24.56	21.268

[a]NR, *not reported.*

TABLE 18.4 Comparison of real power loss.

Item	Technique EGA [23]	Technique EEA [23]	Technique CSA [24]	IGSD
Loss (MW)	646.2998	650.6027	635.8942	612.1407

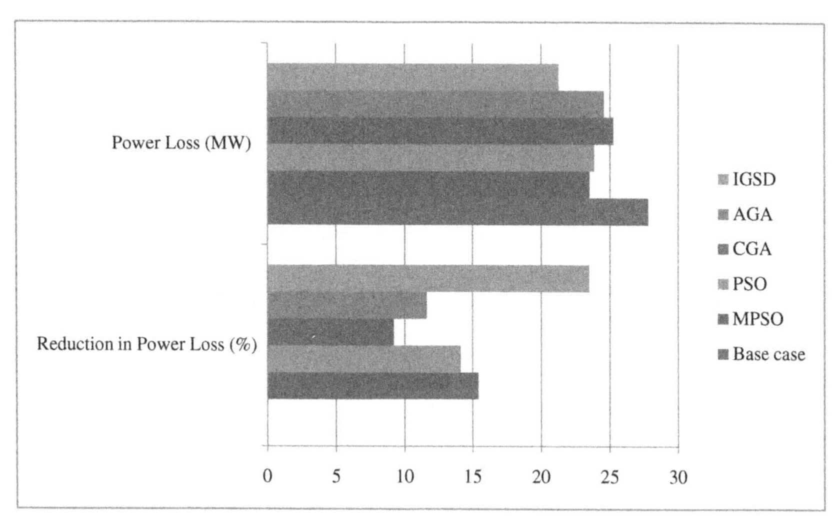

FIG. 18.1 Comparison of parameters.

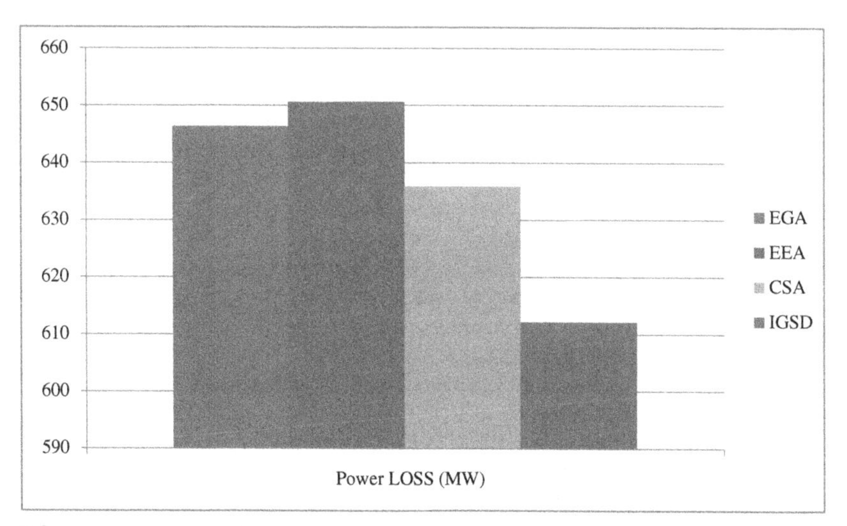

FIG. 18.2 Assessment of loss.

18.5 Conclusions

Improved Grasshopper Optimization Algorithm (IGSD) lucratively solved the loss lessening problem. IGSD increases the searching competence and population diversity has been maintained in the last phase of the iterations. In the self-adaptive differential algorithm mutation, crossover and selection are the main operators. Then the crossover probability (CR) and scaling factor (SF) controlled the exploration and exploitation. The soundness of Improved Grasshopper Optimization Algorithm (IGSD) is confirmed in IEEE 57 and 300 bus test systems. Results show that the IGSD abridged the loss efficiently.

References

[1] K. Lee, Fuel-cost minimisation for both real and reactive-power dispatches, in: Proceedings Generation, Transmission and Distribution Conference, vol. 131 (3), 1984, pp. 85–93.

[2] N. Deeb, An efficient technique for reactive power dispatch using a revised linear programming approach, Electr. Power Syst. Res. 15 (2) (1998) 121–134.

[3] M. Bjelogrlic, Application of Newton's optimal power flow in voltage/reactive power control, IEEE Trans. Power Syst. 5 (4) (1990) 1447–1454.

[4] S. Granville, Optimal reactive dispatch through interior point methods, IEEE Trans. Power Syst. 9 (1) (1994) 136–146.

[5] N. Grudinin, Reactive power optimization using successive quadratic programming method, IEEE Trans. Power Syst. 13 (4) (1998) 1219–1225.

[6] R. Ng Shin Mei, Optimal reactive power dispatch solution by loss minimization using moth-flame optimization technique, Appl. Soft Comput. 59 (2017) 210–222.

[7] G. Chen, Optimal reactive power dispatch by improved GSA-based algorithm with the novel strategies to handle constraints, Appl. Soft Comput. 50 (2017) 58–70.

[8] E. Naderi, Novel fuzzy adaptive configuration of particle swarm optimization to solve large-scale optimal reactive power dispatch, Appl. Soft Comput. 53 (2017) 441–456.

[9] A. Heidari, Gaussian bare-bones water cycle algorithm for optimal reactive power dispatch in electrical power systems, Appl. Soft Comput. 57 (2017) 657–671.

[10] M. Morgan, Benchmark studies on optimal reactive power dispatch (ORPD) based multi-objective evolutionary programming (MOEP) using mutation based on adaptive mutation adapter (AMO) and polynomial mutation operator (PMO), J. Electr. Syst. 12 (1) (2016).

[11] R.N.S. Mei, Ant Lion Optimizer for optimal reactive power dispatch solution, J. Electr. Syst. (2016) 68–74.

[12] P. Anbarasan, Optimal reactive power dispatch problem solved by symbiotic organism search algorithm, in: Innovations in Power and Advanced Computing Technologies, 2017, pp. 1–8.

[13] A. Gagliano, Analysis of the performances of electric energy storage in residential applications, Int. J. Heat Technol. 35 (Special issue 1) (2017) S41–S48.

[14] M. Caldera, Survey-based analysis of the electrical energy demand in Italian households, Math. Model. Eng. Probl. 5 (3) (2018) 217–224.

[15] M. Basu, Quasi-oppositional differential evolution for optimal reactive power dispatch, Electr. Power Energy Syst. 78 (2016) 29–40.

[16] J. Wu, H. Wang, N. Li, P. Yao, Y. Huang, Z. Su, Y. Yu, Distributed trajectory optimization for multiple solar-powered UAVs target tracking in urban environment by adaptive grasshopper optimization algorithm, Aerosp. Sci. Technol. 70 (2017) 497–510, https://doi.org/10.1016/j.ast.2017.08.037.

[17] IEEE, The IEEE-test systems. http://www.ee.washington.edu/trsearch/pstca/.2019/01/21.
[18] M. Vishnu, Sunil, An improved solution for reactive power dispatch problem using diversity-enhanced particle swarm optimization, Energies 139 (1) (2020 December) 2–21.
[19] A.N. Hussain, Modified particle swarm optimization for solution of reactive power dispatch, Res. J. Appl. Sci. Eng. Technol. 15 (8) (2018) 316–327.
[20] C. Dai, W. Chen, Y. Zhu, X. Zhang, Seeker optimization algorithm for optimal reactive power dispatch, IEEE Trans. Power Syst. 24 (3) (2009 May) 1218–1231.
[21] P. Subbaraj, P.N. Rajnarayan, Optimal reactive power dispatch using self-adaptive real coded genetic algorithm, Electr. Power Syst. Res. 79 (2) (2009 December) 374–381.
[22] S. Pandya, R. Roy, Particle swarm optimization based optimal reactive power dispatch, in: Proceeding of the IEEE International Conference on Electrical, Computer and Communication Technologies (ICECCT), March 5–7, IEEE, Coimbatore, India, 2015, p. 2006.
[23] S.S. Reddy, et al., Faster evolutionary algorithm based optimal power flow using incremental variables, Electr. Power Energy Syst. 54 (2014) 198–210.
[24] S. Surender Reddy, Optimal reactive power scheduling using cuckoo search algorithm, Int. J. Electr. Comput. Eng. 7 (5) (2017) 2349–2356.

Chapter 19

Security analytics

Vani Rajasekar[a], J Premalatha[b], and Rajesh Kumar Dhanaraj[c]
[a]*Department of CSE, Kongu Engineering College, Erode, Tamilnadu, India,* [b]*Department of IT, Kongu Engineering College, Erode, Tamilnadu, India,* [c]*School of Computing Science and Engineering, Galgotias University, Greater Noida, India*

19.1 Introduction

The emergence of smart devices and the pervasiveness of the Internet have opened the way for network attacks by intruders, leading to cyberattacks, financial losses, health-care data theft, and cyber wars. With the significant expansion of web access, the rate]of cyberattacks on companies and government departments has risen exponentially, resulting in the disruption of business processes, undermining the credibility and economic support of these corporations. Network security analytics has therefore now become a significant cause of focus and has received significant attention among scientists, especially in the field of anomaly detection that is deemed significant for network security. Preliminary research, however, has shown that current methods for detecting intrusion attempts are not successful enough, especially to monitor them in real time [1]. The latest cyber assault incident at Target Corp shows how income and investor capital can be seriously impacted by these security vulnerabilities. However, despite the growing rate of attacks on government and corporate structures, companies have dropped behind given the limited resources and poor security policies in ramping up their defenses. A comprehensive framework for understanding security risks should be supported by security analytics. The ability to define, understand, and predict the essential features of a process is a universal principle of cyber situational awareness. The concept of defining situational awareness is specified in Fig. 19.1.

Cybercrimes have risen exponentially in recent years, with malicious hackers actively finding new ways to bypass security solutions to obtain unauthorized access to networks and applications [2,3] the most common cyberattacks are defined as follows:

System Assurances. https://doi.org/10.1016/B978-0-323-90240-3.00019-9

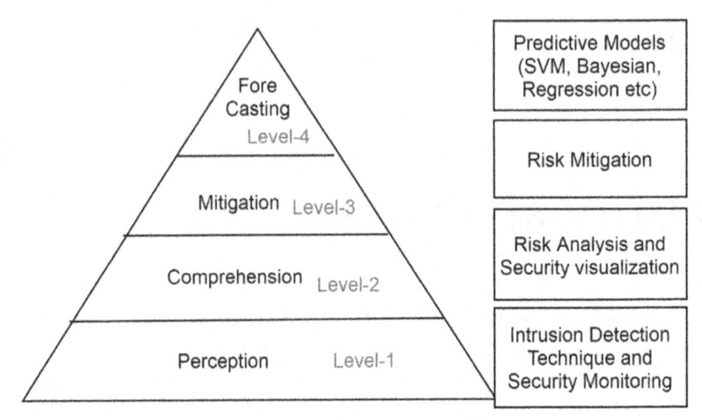

FIG. 19.1 Security analytics models.

19.1.1 Phishing

Through replicating e-communication primarily via email and web spoofing, phishing fraudulently obtains sensitive personal information. Fraudulent emails guide users to compromised web pages that attract sensitive information for email spoofing. Fraudulent websites resemble legitimate websites during web spoofing to deceive users into inputting information. To combat this threat, several antiphishing strategies are in organizational use.

19.1.2 Spamming

Sending unsolicited email messages to multiple destinations is spamming. The amount of spam is expected to be 95% of all private emails by 2019. Munging is authentication of access and sorting of content which are important antispam techniques. Munging makes an email address unaccessible for intruders. Access filtering recognizes spam depending on IP and email ids, whereas web filtering recognizes standardized text patterns for spam detection in emails.

19.1.3 Denial-of-service (DoS) attack

A DoS attack renders a device or some other network asset unavailable to the expected consumers thereof. A significant number of decentralized hosts, e.g., botnet, are launching these attacks. Multiple defensive strategies such as intrusion detection, puzzle answer, packet sniffers, etc. have been designed to reduce DoS attacks.

19.1.4 Malware attacks

Malware is software that is designed to carry out and spread malicious activities, such as viruses, worms, and key loggers. For replication, viruses need human involvement; worms are themselves propagating, whereas Trojans are non-self-replicating. Malware harm involves data or software corruption, spyware download, manipulating personal credentials or storage on the hard disc, etc.

19.1.5 Botnets

Botnets are clusters of infected systems infected by malware controlled by an opponent. To monitor these zombies (bots) and organize them into a channel called the bot system, hackers use bot software fitted with an integrated centralized management system.

19.1.6 Website threats

Attacks to the website apply to attackers abusing, infecting, and implicitly targeting vulnerabilities on legitimate websites. The few methods used to infect legitimate websites are SQL injection, fraudulent ads, and results page redirection.

The substantial damage done by these cyber threats has led to cyber security programs being developed and implemented. Cyber security includes the methods, procedures, and methodologies that thwart illegal or fraudulent cyber threats to defend one or more computers from any form of harm on any network. Cyber security's key objectives are to (1) collect and exchange information safely for correct decision-making, (2) detect and fix vulnerabilities inside applications, (3) avoid unauthorized access, and (4) protect sensitive information. More recently, compared to traditional sources of detecting bad signatures, the priority of cyber security has transitioned to monitoring network and Internet activity for the diagnosis of bad actions.

19.2 Different classes of security analytics

An attack graph is an abstract notation of the various scenarios and routes an attacker can to breach a system's protection with numerous vulnerabilities. It is one of the major methods for risk and security analysis that considers the relationship among various vulnerabilities. Various attack modeling developed by the analyst are built by hand and it was, therefore, a tedious and error-prone operation, especially if the computation time was very high. In particular, conventional cyber security focuses on malware capture by screening traffic coming against absolute numerical that only detect threats of restricted reach that have already been experienced previously. The production of signatures also continues to lag well behind the application of methods for cyberattacks. Thus, an attacker can easily make strategies such as intrusion detection, encryption, and antivirus software inefficient. In the existence of data analytics within, this situation has become more essential. Megabytes and terabytes of information exchanged daily between nodes make it much easier for hackers to access any network, essentially mask their existence and effectively cause serious harm. The different categories or classes of security analytics are as follows:

- Core and structural analytics
- Probability-based analytics
- Time-based analytics
- Lifecycle models of vulnerability

19.2.1 Core and structural analytics

These are measures of aggregation that usually do not use any framework or dependence to measure network security. In order to assess security measures, security metrics use the underlying structure of the attack graph to aggregate the security characteristics of individual devices. The examples of core and structural metrics are total vulnerability measure, Langweg Metric, short path metrics, total path metrics, mean metrics, etc. The total vulnerability is the combination of existing and aggregated vulnerability metrics. Langweng describes the resistance to malware attacks of core metrics. The shortest path metrics of the structural class is used to provide the shortest path to reach the target node for an attacker. The total path metric is used in structural metrics for identifying the number of paths that exist to target an attacker. The mean metric is for computing the mean of all paths available to reach the target for an attacker.

19.2.2 Probability-based analytics

These metrics connect various chances to specific entities in order to measure the channel's aggregated operation to ensure. Attack graph-based probabilistic (AGP) and Bayesian network metrics are some examples that fall into this category. Each module in the network in the AGP reflects a weakness being exploited and a likelihood score is allocated. The allocated score indicates the probability of an attacker leveraging the exploit because all prerequisite criteria are met. The chances for the attachment graph are revised based on new data and previous probabilities associated with the network in BN-based metrics.

19.2.3 Time-based analytics

These metrics calculate how easily a network can be breached or how quickly preemptive steps can be taken by a network to respond to attacks. Mean time to breach (MTTB), mean time to recovery (MTTR), and mean time to first failure are typical metrics in this class (MTFF). MTTB reflects the overall amount of money a red team spends breaking into a scheme divided by the total number of vulnerabilities that were exploited during that span. Until being breached by an attacker, MTTR tests the estimated length of time it takes to restore a device to a stable state. MTFF is the time taken for a device to reach a corrupted or broken state for the first time since the system was initialized. The downside to all these groups of measures is that they take a much more static set of security analysis and therefore do not utilize the specificity of the CVSS metric system to determine the overall complex security situation and help identify important optimization nodes.

19.2.4 Lifecycle models of vulnerability management

The lifecycle of vulnerability management is designed to help organizations to identify vulnerabilities in computer system security; prioritize resources; review, monitor, and remedy the weaknesses, and confirm that they have been removed. A weakness in computer security is a data breach or limitation that enables an attacker to reduce the risk management of a device. Vulnerability includes three elements: a deficiency in the system, the access of an attacker to the weakness, and the ability of the attacker to use a tool or strategy to manipulate the weakness. The steps in vulnerability life cycle management are given in Fig. 19.2.

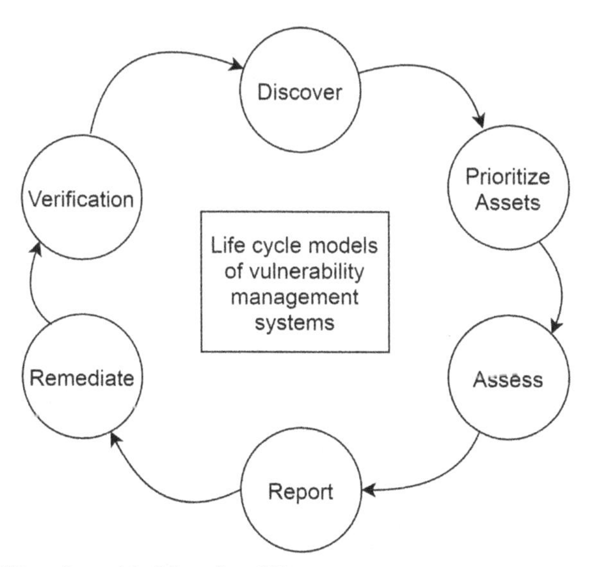

FIG. 19.2 Life cycle model of the vulnerability management system.

19.2.4.1 Discover

To exploit weaknesses, inventory all resources across the organization and create host information, including software and open services. Establish a baseline for the network to identify weaknesses in protection on a daily automatic schedule.

19.2.4.2 Prioritize assets

Categorize resources into classes or product lines and assign definitely benefits a business value based on their containment to the activity of your business.

19.2.4.3 Assess

Determine a basic risk profile such that you can remove risks based on property criticality, danger of weakness, and classification of assets.

19.2.4.4 Report

Assess, according to your security strategy, the amount of security dangers involved with your properties. Record a safety plan, track suspicious behavior, and identify security flaws.

19.2.4.5 Remediate

Prioritize and repair vulnerabilities in compliance with company risk. Develop controls and display progression.

19.2.4.6 Verify

Verify by follow-up reviews whether the risks have been removed.

19.3 Framework of cyber security analytics

This section defines the various concepts involved in developing a framework for cyber security analytics. From attack analysis and vulnerability assessment, an attack graph can be generated and it is possible to visualize the current and future security state metric.

19.3.1 Architecture

19.3.1.1 Common vulnerability scoring system

A fair and open new norm for determining the severity of computer network security flaws is the common vulnerability scoring system (CVSS). CVSS consists of three classes of metrics, foundation, temporal, and environmental, each comprising a metric set. Compared to other weaknesses, it aims to assess the risk involved, so efforts can be prioritized. The findings are based on a variety of tests based on expert evaluation. CVSS consists of three types of metrics such as temporal metric, base metric, and environmental metric. The temporal metrics evaluate the actual state of vulnerability or code availability strategies, the presence of any fixes or work-arounds, or the trust one has in a threat definition [4,5]. The base metric calculates the effect of a positive exploit of the security weakness on the target device on availability. This checks whether to manipulate the vulnerability, an attacker needs to be authorized to the target device or not.

19.3.1.2 Attack graph model

Attack charts are logical diagrams that illustrate how a property, or goal, might be threatened. In a number of applications attack graphs are being used. They were used in the field of computer science to describe vulnerabilities to computer systems and potential attacks to explain these threats. Their use, however, is not limited to analyzing traditional information systems. They are commonly used for the study of vulnerabilities against counterfeit electronic devices in the fields of defense and aerospace. Multilevel graphs comprising of a single root, leaves, and children are attack graphs. From the bottom-up approach, child nodes are a particular set of conditions to allow the direct parent node real; the attack is effective when the root is fulfilled. Only its direct child nodes can satisfy each node (Fig. 19.3).

19.3.1.3 Stochastic graph model

A stochastic model is a method for predicting statistical properties of possible outcomes by accounting for random variance in one or more parameters over time. For a selected duration, the random variance is typically based on variations found in historical data using standardized techniques. The knowledge of the strategy is predicted based on a collection of random variables, and the result is noted. Then, with a new discrete random variable, this is achieved again. This method is actually replicated thousands of times. Distribution of results is available for the rest, which not only indicates the most likely estimate but also what ranges are rational. The stochastic model uses an absorbing Markov chain for vulnerability research and an effect analysis Markov reward model.

19.3.2 Representation of models

This section describes the modeling of the attack graph using Markov chain with two various properties. The attack graph should contain a minimum of one target state or attack state. And also the attack graph should transits from every nonabsorptive state to absorptive state. A Markov chain is a stochastic model that describes a series of potential events in which the likelihood of each perfect contrast only on the state attained in the preceding event. A countless infinite series offers a discrete-time Markov chain in which the chain transfers state at different time stages.

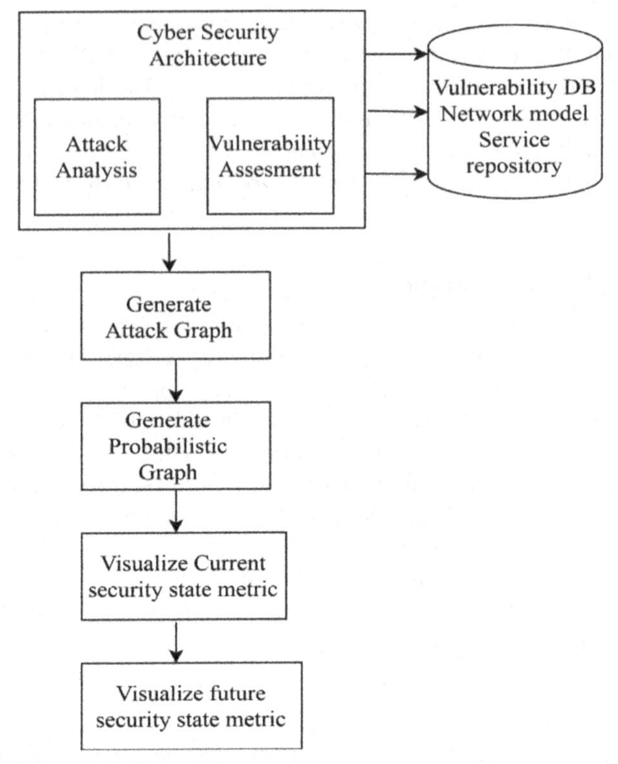

FIG. 19.3 Cyber security framework.

19.3.2.1 Markov model

A state in the Markov model that is not possible to leave is called an absorbing state whereas a state in the Markov chain that is not an absorbing state is called a transient state. State 4 in the following figure represents an absorbing state whereas states 1, 2, and 3 are called nonabsorbing or transient states (Fig. 19.4).

The formula to calculate the Markov chain's transition probabilities was also provided by normalizing the CVSS vulnerability scores over all the transitions beginning from the source state of the attacker. Two key aspects can also be analyzed. The exploitability measures and impact measures are also been considered in this proposed approach. The exploitability measure is calculated as follows:

$$\text{Exploitability measure} = \text{Access vector} * \text{Access Complexity} * \text{Authentication factor}.$$

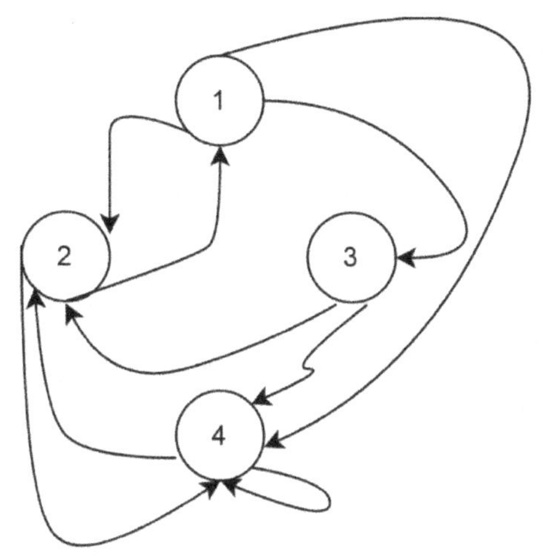

FIG. 19.4 Markov model.

19.3.2.2 Nonhomogeneous model

The transient weight score we observed in the above example was a feature of the period since a widely trusted information security provider revealed the weakness. The CVSS reported temporal values are independent and are not sufficient for use in a nonhomogeneous model. Using a distribution that would take the duration of the vulnerability as input would be more fitting. Therefore, measuring the temporal weight rating of the weaknesses is part of the attack graph model.

$$\text{Model} = (1 - k^*a/t),$$

where t is the duration of vulnerability and a, k is the constant value.

19.3.2.3 Node rank analysis

The key nodes in the attack graph where the intruder is most likely to visit are crucial to recognizing. The requisite systems can be patched for enhanced protection based on this insight. The distinctive feature identified with the fundamental matrix is that, provided that it began in the transient condition, each variable gives the estimated amount of time that the system is in the transient state. When evaluating the security environment of the network, this framework is vital for a security engineer. In the Fundamental matrix, each component knows the predicted number of times that an intruder will visit a state since he began at the state. Through evaluating this matrix, if the intruder is more likely to attack or visit a state that is important to the business process, we can classify points in the network where weaknesses need to be fixed.

19.4 Big data security analytics

Big data is data whose abstraction through conventional data storage frameworks, algorithms, and query frameworks prevents it from being handled, questioned, and analyzed. The big data analytics (BDA) environment is associated with the collection of value from big data, i.e., nontrivial, largely undiscovered, tacit, and potentially valuable insights. These perspectives have a direct effect on the current strategic plan being decided or exploited and motivate what is called the "From information to solution" policy. At first, data to be processed is chosen and preprocessed from real-time sources of big data. Usually, this is called extract transform load and is an intensive operation that can facilitate positive to 60% of BDA's effort, e.g., accounting for inaccurate, incomplete and lacking values, normalizing, discrediting, and minimizing data, ensuring box plots, clustering algorithms, statistical consistency of data, and hypothesis testing. Security analytics offers a "richer" security context compared to traditional approaches by distinguishing what is "ordinary" from what is "irregular," i.e., distinguishing the patterns produced by authorized customers from those produced by malicious or unauthorized users.

19.4.1 Analyzing big data sources and security model

For security analytics, the definition of data is broad and can be divided into passive and active sources. The passive data sources include

1. Computer-based information, e.g., regional IP location, health documents for cyber security, keyboard typing and web analytics patterns, WAP data.
2. Physical data sources such as date and place of physical web access.
3. Human resource data such as security of the organization and security and privacy of officers.
4. Data from external sources such as external IP address, routers, external threats, etc.

The active data sources of big data imply

1. Personal credentials of users such as username, password, smart card, etc.
2. One time password information for remote and online access.
3. Digital certificates for exchanging key information between sender and receiver.
4. Biometrics of users such as fingerprint recognition, facial recognition, palm print recognition, iris recognition, voice recognition and handwriting recognition, etc.
5. Information from social media such as facebook, twitter, internal, and external official networks

19.4.1.1 Big data security areas

Big data arises in the amount of IP-equipped data sources from this enormous escalation. It is really just the word an organization gathers for all the existing evidence in a given field with the intention of identifying hidden trends and

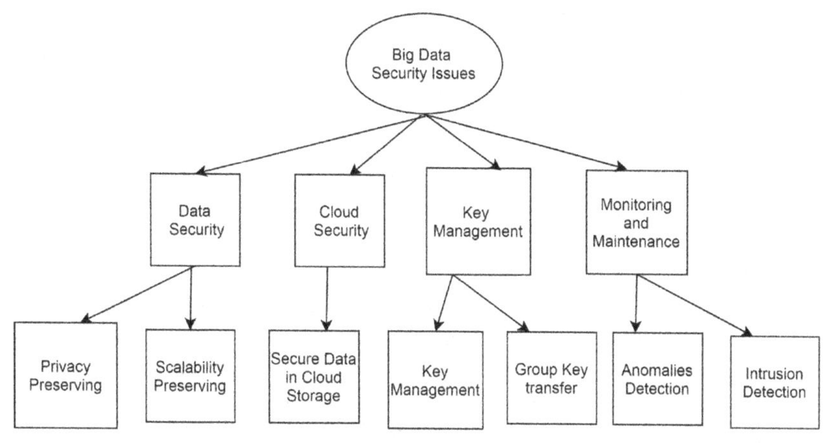

FIG. 19.5 Big data security areas.

patterns within it. Many of the resources associated with big data and smart analytics, sadly, are platform-independent. They are also not planned as a primary feature with protection in mind, leading to even more security problems for big data. In the case of websites using different insecure Content Management Systems, such as Word Press, these risks are much worse, including theft of online encrypted data, malware, XSS attacks or distributed denial-of-service attacks that might crash a server. In particular, attacks on the big data processing of an entity may cause significant financial consequences such as damages, costs of lawsuits, and fines or sanctions. There are three main best practices or obstacles for big data protection that can determine how an enterprise establishes its BI security. A mix of conventional security tools, recently designed toolsets such as plugins malware scanners, and smart processes for tracking security over the life of the platform are needed to protect big data platforms. Big data is supported like never before by companies, using effective technology to create decision-making, recognize opportunities, and increase efficiency. But a whole collection of big data security issues comes with the huge rise in data use and consumption. The major security areas of big data are specified in Fig. 19.5.

19.4.1.2 Big data security analytics model

(A) Big data engines:

To manipulate new routing streams, such engines integrate Hadoop, specialized hardware, and analytics software. As mentioned above, these engines are normally paired with a warehouse that enables network administrators to use warehouse resources for BDA tasks, e.g., ETL, workflows, monitoring services, etc.

(B) Monitoring services:

These systems monitor and evaluate network traffic with activity and risk identification discovered by BDA, or those provided by security professionals based on the experiences.

(C) Data warehousing:

In an organizational repository, all information security information should be stored with system performance metrics stored in fact columns and data queried in data elements such as time, location, customer, etc. This is critical because solutions for data mining (BDA) are now highly paired with a warehouse, such as Oracle, SQL Server, DB2, etc. Various managers need specific analytics to create their own warehouse storage process for security analytics.

(D) Various security controls:

These are used in real-time to provide effective security steps, such as extra user authentication, blocking highly questionable transactions, calling clients while the process is ongoing, etc. Quality management between various places or regions and the efficient execution of requests using BDA frameworks should be facilitated by robust security architecture.

(E) ETL (Extract, Transform, and Load) tool and data source:

These tools may achieving the organization goals ETL at two levels, i.e., at a broader level over a full data source, and a particular level over collected features in possible to qualify one or more BDA strategies on it. Such popular names are Talend, Pentaho, Informatica, etc. Data can be obtained from many locations and there should be no effects on the frequency and sophistication of sources of data on the finished products of BDA.

19.4.2 Security analytics for threat detection

A threat refers to something that has the ability to damage a computer device or network, as the term applies to cyber security. Threats reflect the possibility for attacks to exist; attacks are the process of reducing into a device or network or damaging it. The advanced persistent threat (APT), a more powerful version of threat, appeared several years ago. The threat is pervasive and persists in your infrastructure for a long period, as the name implies, giving hackers a longer window to act. The identification of threats is the mechanism by which you identify threats on your network, systems or software. The concept is to identify threats as attacks until they are exploited. For example, vulnerabilities on an application may or may not have been used in an intrusion. Security teams have also changed their emphasis from so-called compromise indicators such as malware intrusion to strategies, tactics, and procedures. The goal is to capture the bad actor by searching for telltale strategies and finding proof in the process of implementing a hazard.

19.4.2.1 Examples of threat detection

(A) Credentials of users:

Often, malicious hackers are not after you, but after your information. To get into the networks that you have exposure to, they want your login and password. Opening a door with a key is much better than picking a lock or breaking a window. By manipulating the overlay network, some attackers can use a tactic called privilege enhancement to grant themselves reasonable concerns. To get to what they really are seeking, they then use these expanded rights.

(B) Personal information:

Some offenders want personal details, such as a name and address or a driver's license number that they can use to impersonate the legitimate user and use this information to apply for credit cards, to open bank accounts to hack the data and money, etc.

(C) Intellectual property:

Commercial espionage is alive and active. To enhance their own growth, national states are planning to grab proprietary information. By taking full advantage of what their competitors know, competitors are looking to gain a benefit or fill a void in their offerings. Workers are at risk of stealing sensitive personal secrets or even being skipped over for advancement out of provocation. Companies need to secure their product designs, consumer records, marketing strategies, workflows, and more.

(D) Ransomware:

For years, offenders have been extorting businesses and individuals digitally. Ransomware where gateway or server information is encrypted and a ransom requested to decrypt them and distributed DoS attacks where information floods operating systems or networks with fake traffic before the ransom are paid are their two most powerful arms.

19.4.2.2 BDA for mitigating threat detection

Threat detection and accident investigation are the most commonly used security analytics, which are of great concern to both economic and defense organizations. The emphasis is on detecting and studying both known and unknown cyberattack trends, which are expected to have a huge impact on the effectiveness of identifying hidden threats more quickly, monitoring attackers, and predicting potential cyberattack assaults with increasing precision.

(A) Network services:

Detection and prediction of server manipulation-related irregular patterns, For example, irregular or abrupt changes in specification, noncompliance with a predetermined policy, etc.

(B) Network traffic:

Detecting and forecasting, along with dubious sources and destinations, patterns of irregular traffic.

(C) Web transactions:

Detecting and interpreting irregular user access trends, especially in the use of vital resources or operations.

(D) Source of network:

Designed to detect and identify any machine's irregular use habits, e.g., connected to the form of data sent, processed, and retrieved by the source.

19.4.3 Security analytics solution

The solutions for security analytics are proposed as follows:

(A) Streamlining of security analytics solution:

This is the most critical issue facing executives at C-level. It is desirable to complete the above steps within a reasonable level of capital initially, because of the ambiguity that follows BDA. These measures should also be treated independently of the existing workflow, with coordination only taking place where possible. For example, 5–6 months should be provided to a group of 4–5 people together with the software engineer to incorporate the technology solutions as centralized management. Once executives are pleased with the results, steps can be taken to progressively infuse the advanced analytics into the existing workflow.

(B) Developing business strategy for security analytics solution:

The first step is to develop a business plan for the advanced analytics implementation. In order to evaluate its viability, effect, and benefit for their own companies, CIOs and CTOs need to initially develop domain awareness by looking at a few of the effective analytics-based technology solutions.

(C) Security analytics programs:

C-level leaders can only know the true effect of threat detection if they grasp BDA's technological information. Hence, it is especially recommended to attend BDA workshops and training programs. In particular, several open software analytical tools, such as Rapid Miner, allow users to use simple drag-and-drop features to interact with a variety of BDA techniques. Hands-on observations on security data sets using these tools will provide in-depth awareness of the anticipated BDA outcomes and help improve the analytics strategy.

(D) Data Management infrastructure and platform:

This should essentially allow sources from different sources to be incorporated seamlessly, along with ETL design tool, data aggregation, database management, data storage, and execution time mechanisms. The framework should be versatile for both expansion and alteration, such as changing the ETL tools. This should encourage experimenting with a variety of methods, techniques, and algorithms for BDA. The data repository will be equipped with an appropriate transmission device to obtain the data and to store BDA.

(E) Implementation of network and suspicious layer:

By using the series of specialized models that were mined by BDA, this layer controls the network streams at running time. It can also use some kind of security expertise provided by design engineers based on personal experience, or innovation of different forms of external threats from certain government databases. As channel sources are monitored 24/7, this layer must always be "live." It subsequently intimidates the "Suspicion Alert" if any suspicious activity is found [6]. All steps that can be implemented to protect cyber security are enforced by this layer, such as alerting authorities of an ongoing criminal offence, locking access. It should be guaranteed that this component, even if the analytics platform is not used at a particular time, is still "live."

19.5 Security analytics for IoT

The main advantages of the Internet of things (IoT) have increased the understanding of circumstances and optimized consumption of resources. The intruder can use the infected Internet of things (IoT) devices to obtain unauthorized connection and can also use these systems as a jump-off point to target other devices. IoT strong password capability enable unauthorized connection over SSH and thus offer the intruder the opportunities to open their luck by implanting malicious software. It will lead to liberal access to the device through such malicious code transplantation. IoT devices separating various components on protected networks [7]. However, the intruder can also use a malware-implanted computer against external devices to initiate DDOS attacks. The remote monitoring of the IoT system must be safely stored and transmitted for review. Only devices that have the upgrade capabilities built into them can be revised or upgraded for security controls from them).

19.5.1 Interactive model for IoT

Using conventional techniques that are vulnerable to error, existing IoT devices are secured. Many sectors use small sensing instruments that are susceptible to security theft and personal data manipulation. Three components of data are concerned with the interactive IoT model: (1) security and privacy mechanism, (2) fast response, and (3) quality of data. The proposed model is presented in Fig. 19.6.

(A) Perception layer

Most sensing equipment are mounted in locations where it can be easily accessed and tracked. The intruder can gain access to the equipment quickly and can damage these devices physically. Currently, the most significant perception layer security problems include protecting objects from being intercepted by unauthorized users, being protected from fake objects, DoS attack, etc. The authentication method includes the Gateway

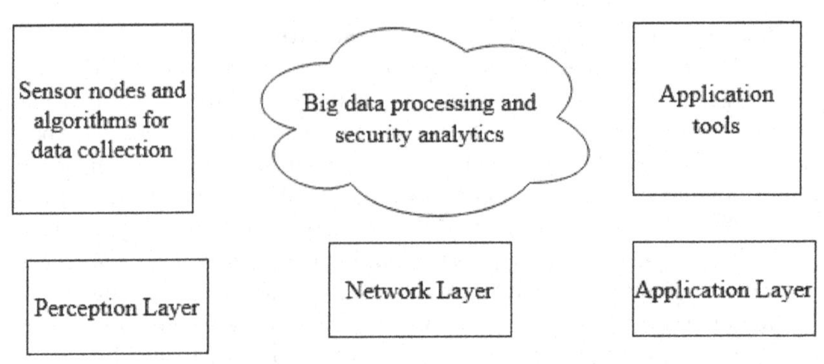

FIG. 19.6 Interactive model for IoT.

application for checking sensors for transmitting data. Fake object recognition would become easy to recognize by inspecting the components of the wearable sensors they send. Safe packet filtering is provided by the firewall to boost data security from different cloud providers.

(B) Network layer

If large numbers of objects, including false objects, submit data at the same time, this can lead to a DoS attack on the network. Security issues for the network layer involve reducing accessibility issues by providing integrity and confidentiality, preserving consumer privacy, preserving DoS, and man in the middle attacks. The IoT system is faced with data transfer problems that can be recovered later.

During its transfer and at the moment of recovery, the data may be attacked. In this model, the network layer passes data to the cloud infrastructure to process information where data is transmitted by creating a protected gateway to follow the suggested verification system by defining an entity and verifying the collected data within the network.

(C) Application Layer.

Visualization is demanded by end users to view research reports and to get their questions answered.

Data is handled from the beginning and there will be no issues with data processing. End users can imagine data that is real, safe, secured, and according to user expectations by using packet transmission techniques and methods. Techniques for data delivery reduce latency, increase processing system's performance as well as provide faster data processing. Authentication and limiting data access, coping with vast volumes of data, including data recovery and authentication mechanisms are the challenges of application layer protection. Sensing systems constantly transmit data, so it is difficult to store large quantities of data. It is therefore difficult to secure this vast quantity of data. Each entity can have better interactions to improve privacy by following this method and can obtain services in a more secure manner.

19.5.2 Ontology-based security modeling

Ontological security is described as a level of order and consistency in one's perceptions. The cyber security approach includes the ontology and security process automatically to find appropriate solutions and service information according to produced notifications. Furthermore, ontology uses logical capabilities to define consistency of the knowledge base, accuracy of feature vectors, and assumptions using rules. This approach extracts implicit facts from present understanding and can be categorized into the line of reasoning, directive reasoning, or deductive or inductive reasoning. As a consequence of logic, certain ontology verification processes happen. The process includes

- Verification of the validity of ontology and the foundation of science.
- Verifying the unintentional class relation.
- Classifying instances of classes automatically.

Securing data describing the document or segment of data in which annotation is done for data protection is the primary class of ontology. There is a hierarchy of similar subcategories in this class, such as Secret Data, which can then be categorized as hidden data or encrypted data. Neither of the classes is disjointed under secure data, so any document or component can be identified as any number of those classes. The next class hierarchy refers to the group of access control [8]. This class defines access control mechanisms that limit access to certain data to an authorized party. Ontologies have a promising way of addressing these problems. Ontology is usually meant to solve challenges or answer database questions. This is done with the assistance of structured data simulations to provide accuracy and entailment. It is suggested that an ontology structure for threat modeling to enhance automated threat modeling. The goal of this system is to enhance automated modeling of threats by solving the two previously described problems: lack of awareness of the domain and imbalanced granularity. With modeling approach built on broad information, the ontology-based paradigm is developed. By applying and using the modeling system, the usefulness of the generated models is assessed.

Data integration is a key aspect of ontology, and data generated and static data can be integrated into a single model by combining huge databases into a single coherent structure while preserving the identity of member functions over time. Ontology may also be used to combine different languages for domain definition, to preserve coherence, continuity and quality control between subject language representations and to optimize model analysis. Semantic methods serve as an instrument for evaluating models of enterprise architecture, where ontology describe the conceptual models and draw appropriate conclusions about the models. The creation phase of the ontology system begins with conceptual modeling and design stages, during which reusable ontology patterns are developed. The last stage in which ontology patterns are introduced is the process of deployment. By solving modeling issues for the domain classes and their attributes, content patterns solve conceptual ontology design issues. In the ontological model, material patterns can be repeated as basic components.

19.5.3 Smart home—Security analytics IoT case study

In recent decades, smart home systems have gained great popularity as they improve convenience and quality of life. Smartphones and microcontrollers control many smart home systems. Using wireless communication methods, a mobile application is used to regulate and monitor home functions. The basic components of the advanced smart home interconnected framework are classic smart home, Internet of Things, big data, and rule-based event processing. Each component adds to the proposed formulation its core attributes and developments. IoT offers internet access and remote mobile device control combined with a range of sensors.

19.5.3.1 Components of smart home

Sensors are the main components used to gather internally and externally home data and used for measurement of home conditions. These sensors are attached to the home directly and also the home-to-home sensors connected. Processors are used for conducting synchronized and local acts. The sensor information is then analyzed by the local server. A set of elements of software wrapped as APIs that allow it to be implemented by external applications because it follows the format of predefined parameters. Such an API can process data from sensors or handle actions that are required [9]. Actuators are used for the implementation and execution of server commands or other monitoring systems. This translates the necessary operation to the syntax of the command; it can activate the device. The role checks if any rule is correct during the processing of the data received from the sensors. In such a scenario, a command may be initiated by the system to the correct device processor. Database for the storage of the processed sensor data obtained. Data interpretation, data presentation, and simulation can also be included.

19.5.3.2 Smart home services

(A) Home conditions measurement:

A conventional smart home is fitted with a series of sensors, such as temperature, humidity, brightness and proximity, for calculating home conditions. Each sensor is designed to capture one calculation or more. One sensor can measure temperature and relative humidity, while other sensors determine the light proportion for a particular area and the distance from it to every object exposed to it.

(B) Home appliances management:

Develop a cloud service that will be hosted on a cloud platform for managing home appliances. The management service enables the user to monitor the performance of smart actuators, such as lights and fans, related to home appliances. Smart actuators are machines that perform acts such as turning things on or off or modifying an operating system, such as switches and controls.

(C) Controlling home appliances:

For publicly accessible doors, home access systems are widely used. A database with the recognition characteristics of authorized persons is used by a common home appliances method. When an individual enters the access control system, the identifying attributes of the person are gathered immediately and compared to the database.

19.6 Security analytics in anomaly detection

Detection of anomalies is a data mining phase that detects data points, events, and findings that deviate from the normal behavior of a dataset. Important occurrences, such as a technological glitch, or future opportunities, such as a shift in customer behavior, maybe signaled by anomalous results. To enhance anomaly detection, artificial intelligence is increasingly being used. It is now simpler than ever for businesses to accurately calculate every single aspect of business operation with all the analytics systems and different management tools available. This involves the operating efficiency of components of software and facilities, as well as key performance indicators that measure the organization's progress. Data patterns reflecting business as normal are inside this dataset. An unforeseen shift is known to be an anomaly within these data patterns or an occurrence that does not adhere to the anticipated data pattern. An exception, in other words, is a divergence from business as normal. Effective identification of anomalies relies on the ability to interpret time series data in real time effectively. A series of values over time is comprised of time series data. That implies that each point is usually a pair of two things, a time stamp for the measurement of the metric and the value at those things compared with that metric. Depending on the business model and use case, it is possible to use time series information anomaly detection for useful metrics such as:

- Web applications
- Mobile applications
- Active users
- Number of transactions

Data anomaly detection in the time sequence must first establish a baseline for acceptable behavior in primary KPIs. Statistical anomaly detection [10,11] systems can monitor seasonal variation within key measurements of cyclical activity trends with the baseline acknowledged. In one data plot, a manual approach may help classify seasonal data.

19.6.1 Techniques in anomaly detection

(A) Statistical method for anomaly detection:

Deleting data points that deviate from standard statistical properties of a sample, including mean, median, mode, and quantizes, is the easiest approach to detecting data anomalies. Let us assume that an anomalous data point interpretation is one that diverges from the mean by a certain confidence interval. It is not exactly easy to traverse mean data over data series, since it is not static. To determine the mean across the data points, you will need a rolling window. This is theoretically referred to as a rolling or growth model and is meant to smooth out short-term variations and demonstrate long-term fluctuations. The knowledge includes noise that may be identical to abnormal behavior since the difference between normal and abnormal behavior is not always specific. As malicious adversaries continually modify themselves, the concept of abnormal or normal may often modify. The threshold dependent on the moving average also does not always apply.

(B) Machine learning method for anomaly detection:

The machine learning-based approach uses the K-nearest algorithm to determine the anomaly. The closet data points are also determined by the numerical or categorical scores. The two most commonly used algorithms in this approach are K-nearest neighbor (KNN) and the relative data density approach [12]. K-NN is an easy, nonparametric slow learning method used in distance measures such as Euclidian, Manhattan, or Hamming distance to perform classification based on the similarity. The relative data density approach also called as local outlier factor is based on distance metric that uses the distance of reachable. Another effective process for detecting anomalies is a support vector machine. Typically, an SVM is synonymous with supervised learning, although there are variants that can be used to describe anomalies as unmonitored issues. In attempts to cluster the usual feature vectors using the training set, the algorithm learns soft boundaries and then tunes itself to recognize the anomalies that fall beyond the learned region using the test examples.

19.6.2 Research challenges in anomaly detection

Research will enable administrators to classify the triggers of the anomalies or the affected devices by creating multiple models including different factors selected according to predetermined meaningful reasons. "Volume" and "velocity" are the two attributes of big data that have the greatest impact on the problem of computational complexity [13,14]. The problem of high dimensionality not only leads to problems in identifying anomalies but also brings additional problems as data increases and arrives at speed as unrestrained data streams. The most common methods to resolve the high-dimensionality issue are to define the most prominent features known as the method of variable selection or to merge parameters into a smaller collection of new distinct concepts as the method of dimensionality reduction. There are many drawbacks to the velocity feature of big data, such as connectedness, arrival rate, and the concept of infinite, but they are negligible when solving problems of high dimensional space or anomaly detection. When data streams have multidimensional characteristics, the issues that arise are caused by the high dimensionality and data stream drifting. An unsupervised, virtual dimensionality learning approach can be used for anomaly detection from nonstationary strong data streams to solve these two challenges at the same time.

An angular subspace anomaly detection technique is designed to locate low-dimensional feature space defects from high-dimensional data sets by choosing failure subspaces and measuring vector angles that measure the local arbitrary number of information instance in its subspace projection. Outlier detection is a tool for identifying patterns in the knowledge that do not correspond to expected behavior. Since, based on certain measurements, an outlier can be described as a data point that is very different from the rest of the data. There are many schemes of outlier identification. Based on its effectiveness and how it can resolve the issue of anomaly. For a specific user or data collection, one may be better than others, depending on the context. A generative approach constructs a model purely based on examples of ordinary training data and then assesses each test case to see how well the model matches. A discriminatory approach, on the other hand, seeks to differentiate between normal and abnormal classes of data. Both types of data are used in discriminative methods to train systems.

References

[1] S. Tayyaba, S.A. Khan, M. Tariq, M.W. Ashraf, Network security and internet of things, in: Industrial Internet of Things and Cyber-Physical Systems: Transforming the Conventional to Digital, IGI Global, 2020, pp. 198–238.

[2] V. Rajasekar, J. Premalatha, K. Sathya, Cancelable Iris template for secure authentication based on random projection and double random phase encoding, Peer-to-Peer Netw. Appl. (2021), https://doi.org/10.1007/s12083-020-01046-6.

[3] K.S. Luzgina, G.I. Popova, I.V. Manakhova, Cyber threats to information security in the digital economy, in: Biologically Inspired Cognitive Architectures Meeting, Springer, Cham, 2020, November, pp. 195–205.

[4] C. Iwendi, Z. Zhang, X. Du, ACO based key management routing mechanism for WSN security and data collection, in: IEEE International Conference on Industrial Technology (ICIT), 2018, pp. 1935–1939.

[5] M.M.N.K.A.R. Javed, M.U. Sarwar, S. Khan, C. Iwendi, Analyzing the effectiveness and contribution of each axis of tri-axial accelerometer sensor for accurate activity recognition, Sensors 20 (8) (2020) 2216.

[6] I.S. Comşa, S. Zhang, M.E. Aydin, P. Kuonen, Y. Lu, R. Trestian, G. Ghinea, Towards 5G: a reinforcement learning-based scheduling solution for data traffic management, IEEE Trans. Netw. Serv. Manag. 15 (4) (2018) 1661–1675.

[7] I.S. Comşa, R. Trestian, G.M. Muntean, G. Ghinea, 5MART: a 5G sMART scheduling framework for optimizing QoS through reinforcement learning, IEEE Trans. Netw. Serv. Manag. 17 (2) (2019) 1110–1124.

[8] V. Rajasekar, J. Premalatha, K. Sathya, Multi-factor signcryption scheme for secure authentication using hyper elliptic curve cryptography and bio-hash function, Bull. Polish Acad. Sci. Tech. Sci. 68 (4) (2020).

[9] M. Mittal, C. Iwendi, S. Khan, A. Rehman Javed, Analysis of security and energy efficiency for shortest route discovery in low-energy adaptive clustering hierarchy protocol using Levenberg-Marquardt neural network and gate, Trans. Emerg. Telecommun. Technol. 13 (2020) e3997.

[10] I.M. Alsmadi, G. Karabatis, A. Aleroud (Eds.), Information Fusion for Cyber-Security Analytics, Springer International Publishing, Switzerland, 2017.

[11] M. Siponen, R. Willison, Information security management standards: problems and solutions, Inf. Manage. 46 (5) (2009) 267–270.

[12] E. Quatrini, F. Costantino, G. Di Gravio, R. Patriarca, Machine learning for anomaly detection and process phase classification to improve safety and maintenance activities, J. Manuf. Syst. 56 (2020) 117–132.

[13] M. Du, Application of information communication network security management and control based on big data technology, Int. J. Commun. Syst. (2020). e4643.

[14] X. Zhou, Y. Hu, W. Liang, J. Ma, Q. Jin, Variational LSTM enhanced anomaly detection for industrial big data, IEEE Trans. Ind. Inform. 17 (5) (2020) 3469–3477.

Chapter 20

Stochastic modeling of the mean time between software failures: A review

Gabriel Pena[a], Verónica Moreno[a], and Néstor Barraza[a,b]
[a]*Department of Sciences and Technology, University of Tres de Febrero, Caseros, Argentina,*
[b]*School of Engineering, University of Buenos Aires, Autonomous City of Buenos Aires Argentina*

20.1 Introduction

Modeling the stochastic software failure detection process turned an important issue in engineering in the 1970s. Many software reliability models have been proposed since then where the Software Reliability Growth Models (SRGM) based on the nonhomogeneous Poisson process constitutes an important group, see, for example, Ref. [1, 2]. A diffusive process was recently proposed in Ref. [3]. A recent survey has been published in Ref. [4]. During the last decades, software has become a critical component in many technological projects. New development and testing methodologies were introduced in the discipline of software engineering with the purpose of obtaining more reliable products with quality assurance. Those methodologies, for example, Agile, which involves modern practices like test-driven development (TDD), Scrum, etc., introduce new challenges to software reliability modeling. Since the main characteristic in Agile procedures is that development is performed simultaneously with testing in an iterative cycle, it is a quite different situation from that the SRGM were developed when waterfall was the prevailing software development methodology. We then draw attention to the revision of the software reliability growth concept in order to consider the importance of a first increasing failure rate stage. Therefore, contagious stochastic processes have been introduced in software reliability, see, for example, Ref. [5]. As opposed to the nonhomogeneous Poisson processes, the failure rate in models based on contagion depends not just on time, but also on the number of failures previously detected, the contagion phenomena could model this way interaction between programmers and testers, or sharing and reuse of code. The recently introduced

System Assurances. https://doi.org/10.1016/B978-0-323-90240-3.00020-5

contagion model is a modification of the Polya stochastic process, a different functional form of the failure rate was proposed in order to obtain either a positive or negative concavity in the mean number of failures curve, in order to be able to model increasing failure rate or reliability growth stages, as it is explained in Refs. [6, 7] and references therein. Mean time between failures (MTBF) and mean time to failure (MTTF) are important and useful metrics in reliability analysis, see, for example, Ref. [8]. In the particular case of software reliability, those parameters are useful to determine the time to release, make adjustments in the development and testing processes, when evaluated after delivery to evaluate the performance of a given release and to set up requirements for future releases. In a reliability growth case, we expect a continuously increasing MTBF, however, taking into account an important increasing failure rate first stage as explained before, we should expect a decreasing MTBF at first, the so-called "infant mortality" stage in hardware reliability. Then, the well-known bathtub curve from hardware reliability is also applicable to software reliability. We consider then that the way this metric is obtained, its properties, how it compares between different models and its goodness of fitting real projects are important issues to be studied. We focus then in this chapter on the analysis of the MTBF metric as obtained from several models, that is, reliability growth models and also the recently proposed model based on contagion. We show the results of comparison of their performance when applied to real projects developed under Agile practices. In order to present a clear and self-explained content, we expose a complete mathematical foundation about the way the metrics are obtained. Since either the contagion-based model as those based on nonhomogeneous Poisson processes are birth processes, we present a mathematical background section on this issue.

20.2 Mathematical background

20.2.1 Birth processes

Birth processes are a particular case of branching processes and continuous time Markov chains, and they can also be considered as one of the various possible generalizations of Poisson processes. As shown, for example, in Ref. [9], a series of proposed axioms are translated into a set of ordinary differential equations whose solution gives the probability mass function (PMF) of the considered process. In the present work, we start with the same three axioms stated in Ref. [9] with a small relaxation: we allow the function λ_r to depend not only on the number of failures r but also on the elapsed time t. This allows us to model a wider class of processes with a more general growth pattern.

By following the same steps, a system of infinite recursive ODEs can be derived from those postulates. For the sake of simplicity, we take this system to be our starting definition. Moreover, we will assume an initial population of 0 individuals at time 0, which is indeed the case in most real scenarios of application.

Definition 20.1 (Birth process). Let $P_r(t)$, $r \in \mathbb{N}_0$, denote the probability of the system being in state r at time t, and let 0 be the initial population, that is,

$$P_0(0) = 1, \tag{20.1}$$

$$P_r(0) = 0 \;\; \forall r > 0. \tag{20.2}$$

Then, for any $t > 0$, $r \geq 0$ the probability mass function of the process, the so called *birth process*, is given by

$$P'_r(t) = -\lambda_r(t)\, P_r(t) + \lambda_{r-1}(t)\, P_{r-1}(t), \tag{20.3}$$

$$P'_0(t) = -\lambda_0(t)\, P_0(t), \tag{20.4}$$

where $\lambda_r(t)$ is a function, called *intensity function* or *failure rate*, that is allowed to depend on both r and t. For $r < 0$ and any t, we have $P_r(t) = 0$.

Existence and uniqueness of the solution follows directly from the theory of differential equations [10] only requiring the intensity $\lambda_r(t)$ to be a continuous function of t. The solution can be expressed explicitly as a system of integral equations (see Ref. [11]), which may be easier to solve in some cases:

$$P_r(t) = \int_0^t \left(e^{-\int_x^t \lambda_r(u)du} \right) \lambda_{r-1}(x) P_{r-1}(x) dx, \tag{20.5}$$

$$P_0(t) = e^{-\int_0^t \lambda_0(u)du}. \tag{20.6}$$

As it is mentioned in Grandell [12], p. 60, in order that $P_r(t)$ are well defined, it is a sufficient condition that

$$\sum_{r=0}^{\infty} \max_{0 \leq s < t} \lambda_r^{-1}(s) = \infty. \tag{20.7}$$

Remark. Eq. (20.6) is usually known as *exponential waiting time*.

20.2.1.1 Process classification

Most known point processes can be obtained as special cases of the birth process, by imposing different restrictions to the intensity function. We highlight four cases:

- Nonhomogeneous birth processes (NHBP): These are birth processes whose failure rate $\lambda_k(t)$ depends on both k and t. The dependence of the intensity on the elapsed time t implies nonhomogeneity. The contagion models studied in the literature [13, 14] fall into this category.

- Homogeneous birth processes: These are described by an intensity $\lambda_r(t) = \lambda_r$ that does not depend on t. They are also called pure birth processes [9].
- Nonhomogeneous Poisson processes (NHPP): A well-known generalization of the regular Poisson process, where the intensity $\lambda_r(t) = \lambda(t)$ does not depend on r but is not constant over t.
- Ordinary Poisson processes: The intensity $\lambda_r(t) = \lambda$ is a constant, that is, depends on neither r nor t.

All the processes described here are Markov point processes that fall under Definition 20.1. Nonhomogeneous Poisson processes also present the stronger property of having independent increments [14].

The classification is depicted in Fig. 20.1.

20.2.1.2 Probability mass function

Although a general expression for the PMF of a birth process is not available, there exist several particular cases (including the ones we require here) where it is possible to obtain one.

1. Nonhomogeneous Poisson processes: If $\lambda_r(t) = \lambda(t)$, then the PMF [14] is given by

$$P_r(t) = \frac{\mu(t)^r}{r!} e^{-\mu(t)}, \quad (20.8)$$

with $\mu(t) = \int_0^t \lambda(x)dx$. Eq. (20.8) also gives the correct PMF expression of the ordinary Poisson process, obtained by making $\lambda(t)$ constant over t.

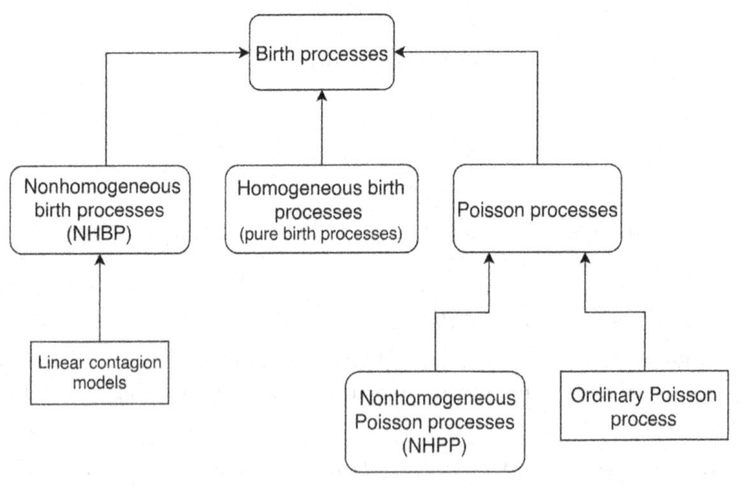

FIG. 20.1 Classification of birth processes.

2. Linear contagion processes: These are NHBPs whose intensities can be expressed as

$$\lambda_r(t) = (mr + b)f(t), \quad m \neq 0, \tag{20.9}$$

where $f(t)$ is any continuous function and $m \neq 0$ and b are real constants. In this case, the PMF can be expressed as

$$P_r(t) = \frac{\Gamma\left(r + \frac{b}{m}\right)}{\Gamma\left(\frac{b}{m}\right)} \frac{e^{-\mu_r(t)}}{r!} \left(e^{\Delta(t)} - 1\right)^r, \tag{20.10}$$

where $\mu_r(t) = \int_0^t \lambda_r(x)dx$ and $\Delta(t) = \mu_{r+1}(t) - \mu_r(t)$ independent of r.

Remark. $\Delta(t)$ not depending on r is equivalent to $\lambda_r(t)$ having the linear form described in Eq. (20.9). The proof can be found in Ref. [14]; we have also shown in Ref. [15] that a particular case of this expression holds for the contagion process described there. The well-known Yule and Polya processes [9, 16, 17] are also special cases of these, and their PMF can be obtained using Eq. (20.10).

20.2.2 The mean value function

The mean value function (MVF) of a birth process, let us call it $M(t)$, is extremely useful for the applications and, in most cases, quite easy to find. The MVF is defined as

$$M(t) = \sum_{r=0}^{\infty} r\, P_r(t). \tag{20.11}$$

The following theorem gives a useful tool to calculate $M(t)$.

Theorem 20.1. *Let M*(t) *be the mean value function (20.11). M(t) is the solution of the differential equation*

$$M'(t) = \sum_{r=0}^{\infty} \lambda_r(t)\, P_r(t), \tag{20.12}$$

provided that $M(t)$ is finite and $\lim_{r \to \infty} r\, \lambda_r(t)\, P_r(t) = 0$.

Proof. Multiplying Eq. (20.3) by r and summing up to R yields

$$\sum_{r=0}^{R} r\, P_r'(t) = -\sum_{r=0}^{R} r\, \lambda_r(t)\, P_r(t) + \sum_{r=0}^{R} r\, \lambda_{r-1}(t)\, P_{r-1}(t) \tag{20.13}$$

$$= -\sum_{r=0}^{R} r\, \lambda_r(t)\, P_r(t) + \sum_{r=0}^{R-1} (r+1)\, \lambda_r(t)\, P_r(t) \tag{20.14}$$

$$= -R\, \lambda_R(t)\, P_R(t) + \sum_{r=0}^{R-1} \lambda_r(t)\, P_r(t). \tag{20.15}$$

Letting $R \to \infty$, the first term of the right member vanishes and we get

$$\sum_{r=0}^{\infty} r\, P'_r(t) = \sum_{r=0}^{\infty} \lambda_r(t)\, P_r(t). \tag{20.16}$$

The theorem will be proved if the derivative $P'_r(t)$ can be extracted from the sum, that is, if

$$\sum_{r=0}^{\infty} r \frac{\mathrm{d}P(t)}{\mathrm{d}t} = \frac{\mathrm{d}}{\mathrm{d}t} \sum_{r=0}^{\infty} r\, P_r(t). \tag{20.17}$$

Although this is not necessarily true, the condition is clearly met if the series in the right member of Eq. (20.17) converges and the series on its left side uniformly converges, which is the case for the models considered here. Hence Theorem 20.1 follows.

It is straightforward to check that linear contagion processes and NHPPs satisfy $\lim_{r\to\infty} r\, \lambda_r(t)\, P_r(t) = 0$. Finiteness of $M(t)$ is a little trickier, but in both cases, it can be shown by using D'Alembert's criterion for the convergence of series. Hence the conditions of Theorem 20.1 hold in all the cases of our interest.

20.2.3 Mean time between failures

In this section, we will analyze birth processes on the time domain.

20.2.3.1 Time domain analysis

We begin by defining random variables corresponding to failure times.

Definition 20.2. (Time to the) kth failure. Let $N(t)$ be a birth process with intensity $\lambda_r(t)$ and probability $P_r(t)$. Let k be any positive integer. Then we define for every k the r.v. $T_k > 0$ to be the exact time in which the kth failure is detected. Moreover, we note as Ω_k the event $T_k < \infty$, that is, $P(\Omega_k)$ is the probability of the kth failure to be detected at any finite time.

The distribution of the T_k variables can be derived from the process $N(t)$. We state Theorem 20.2.

Theorem 20.2 *Let T_k be an r.v. defined as in Definition 20.2. If the kth failure is detected within a finite time, then that time has a (conditional) distribution function given by*

$$F_K(t|\Omega_k) = \frac{1 - \sum_{r=0}^{k-1} P_r(t)}{1 - \sum_{r=0}^{k-1} \lim_{x\to\infty} P_r(x)}. \tag{20.18}$$

The reason to compute the conditional CDF instead of the marginal is because those probabilities are not defined if Ω_k^c occurs, that is, if the failure is never detected at any finite time. This consideration allows a clean definition of the probabilities without any technical difficulties.

Proof. By the definition of conditional probability:

$$F_k(t|\Omega_k) = \frac{P(T_k < t,\, \Omega_k)}{P(\Omega_k)} = \frac{P(T_k < t,\, T_k < \infty)}{P(T_k < \infty)} = \frac{P(T_k < t)}{P(T_k < \infty)}. \tag{20.19}$$

The numerator of Eq. (20.19) can be written as $1 - P(T_k > t)$, and this last probability equals that of having k or less failures at time t. Hence the numerator is proved.

The denominator can be calculated as

$$P(T_k < \infty) \ = \lim_{t\to\infty} P(T_k < t) = \lim_{t\to\infty} (1 - P(T_k \geq t)) \tag{20.20}$$

$$= 1 - \sum_{r=0}^{k-1} \lim_{t\to\infty} P_r(t), \tag{20.21}$$

which completes the proof.

In the particular case of the NHPPs, the denominator of Eq. (20.18) has a simple expression. We show this in the following theorem.

Theorem 20.3. *Let Ω_k be defined as in Definition 20.2. For NHPPs:*

$$P(\Omega_k) = \frac{\gamma(L,k)}{(k-1)!}, \quad L = \lim_{t\to\infty} \mu(t), \tag{20.22}$$

where γ(L, k) is the incomplete Gamma function:

$$\gamma(L,k) = \int_0^L z^{k-1} e^{-z} dz. \tag{20.23}$$

Proof. We are going to demonstrate this by induction. Since we already know that in a NHPP the PMF is given by Eq. (20.8), we can combine it with Eq. (20.21) to obtain an explicit $P(\Omega_k)$. Thus, we have to prove the following equation:

$$1 - e^{-L} \sum_{r=0}^{k-1} \frac{L^r}{r!} = \frac{\gamma(L,k)}{(k-1)!}. \tag{20.24}$$

It is immediate that for $k = 1$, $\gamma(L,1) = \int_0^L e^{-z} dz = 1 - e^{-L}$. For the inductive step, we assume that Eq. (20.22) is valid for k and check that then it holds for $k + 1$

$$\frac{\gamma(L,k+1)}{k!} = \frac{1}{k!} \int_0^L z^k e^{-z} dz. \tag{20.25}$$

Integrating by parts yields

$$\frac{\gamma(L,k+1)}{k!} = \frac{1}{k!}\left(-L^k e^{-L} + k\int_0^L z^{k-1}e^{-z}dz\right) \tag{20.26}$$

$$= \frac{-L^k e^{-L}}{k!} + \frac{\int_0^L z^{k-1}e^{-z}dz}{(k-1)!} \tag{20.27}$$

$$= \frac{-L^k e^{-L}}{k!} + \frac{\gamma(L,k)}{(k-1)!}. \tag{20.28}$$

Finally, by applying the inductive hypothesis we get

$$\frac{\gamma(L,k+1)}{k!} = \frac{-L^k e^{-L}}{k!} + 1 - e^{-L}\sum_{r=0}^{k-1}\frac{L^r}{r!} \tag{20.29}$$

$$= 1 - e^{-L}\sum_{r=0}^{k}\frac{L^r}{r!}, \tag{20.30}$$

as we intended to prove.

Differentiating Eq. (20.18) and replacing Eq. (20.3) yields a more convenient expression for the pdf

$$f_k(t|\Omega_k) = \frac{\mathrm{d}F_k(t|\Omega_k)}{\mathrm{d}t} = \frac{\lambda_{k-1}(t)P_{k-1}(t)}{1 - \sum_{r=0}^{k-1}\lim_{x\to\infty} P_r(x)}. \tag{20.31}$$

In the special case of NHPPs, by Theorem 20.3, we can write the pdf of T_k as follows:

$$f_k(t|\Omega_k) = \frac{(k-1)!\lambda(t)P_{k-1}(t)}{\gamma(L,k)}, \tag{20.32}$$

with γ and L as in Theorem 20.3. By substituting Eq. (20.8), we get the known expression [1]

$$f_k(t|\Omega_k) = \frac{\lambda(t)[\mu(t)]^k}{\gamma(L,k)}e^{-\mu(t)}. \tag{20.33}$$

In Ref. [15], we derive another distribution of interest, that is T_k conditioned to knowing $s = T_{k-1}$ (Eq. 20.34).

$$f_k(t|T_{k-1} = s, \Omega_k) = \frac{\lambda_{k-1}(t)e^{-\int_s^t \lambda_{k-1}(x)dx}}{1 - e^{-\int_s^\infty \lambda_{k-1}(x)dx}}. \tag{20.34}$$

The denominator of Eqs. (20.18), (20.31) takes the value 1 when there is zero probability of the failure k never being detected. Some cases, like that of the ordinary Poisson process and the contagion model described in Section 20.3.1, present this property. The following theorem provides a way to identify those in all the cases concerning this work.

Theorem 20.4. *Consider a linear contagion process with $\lambda_k(t)$ of the form specified by Eq. (20.9). Let T_k be defined as in Definition 20.2. For $P(\Omega_k) = P(T_k < \infty)$ to equal one it is necessary and sufficient that*

$$\mu_r(\infty) = \lim_{t\to\infty} \int_0^t \lambda_r(x)dx = \infty \tag{20.35}$$

for all $r < k$.

Proof. Recalling the definition of $\Delta(t)$:

$$\Delta(t) = \mu_{r+1}(t) - \mu_r(t) \tag{20.36}$$

$$= \int_0^t (mr + m + b)f(x)dx - \int_0^t (mr + b)f(x)dx \tag{20.37}$$

$$= m\int_0^t f(x)dx \tag{20.38}$$

$$= \frac{m}{mr+b}\mu_r(t), \tag{20.39}$$

it follows from Eq. (20.10) that

$$P_r(t) = \frac{\Gamma\left(r+\frac{b}{m}\right)}{\Gamma\left(\frac{b}{m}\right)} \frac{e^{-\mu_r(t)}}{r!}\left(e^{\Delta(t)}-1\right)^r \tag{20.40}$$

$$= \frac{\Gamma\left(r+\frac{b}{m}\right)}{\Gamma\left(\frac{b}{m}\right)} \frac{e^{-\mu_r(t)}}{r!} e^{\Delta(t)r}\left(1-e^{-\Delta(t)}\right)^r \tag{20.41}$$

$$= \frac{\Gamma\left(r+\frac{b}{m}\right)}{\Gamma\left(\frac{b}{m}\right)} \frac{e^{-\mu_r(t)\left(1-\frac{mr}{mr+b}\right)}}{r!}\left(1-e^{-\Delta(t)}\right)^r. \tag{20.42}$$

If $\mu_r(\infty) = \infty$, then the exponentials $e^{-\mu_r(t)\left(1-\frac{mr}{mr+b}\right)}$ and $e^{-\Delta(t)}$ tend to zero. Hence $P_r(t)$ has limit zero for every $r < k$, which implies that $\sum_{r=0}^{k-1} P_r(t)$ has limit zero. Thus the condition is sufficient.

To see that it is necessary, we assume $\mu_r(\infty) < \infty$. Then $e^{-\mu_r(t)}$ and $e^{-\Delta(t)}$ can never have limit zero, which implies $\sum_{r=0}^{k-1} P_r(t)$ has a positive limit, a contradiction. This proves Theorem 20.4.

Remark. The necessary condition holds even if not every $\mu_r(\infty) < \infty$; it suffices that at least one of them accomplishes it in order for the sum to have nonzero limit.

A similar result holds for the nonhomogeneous Poisson processes as shown below:

Theorem 20.5. *Consider a NHPP with intensity $\lambda(t)$. Let T_k be defined as in Definition 20.2. For $P(\Omega_k) = P(T_k < \infty)$ to equal one for every k it is necessary and sufficient that*

$$\mu(\infty) = \lim_{t\to\infty} \int_0^t \lambda(x)dx = \infty. \tag{20.43}$$

The proof follows the same steps as that of Theorem 20.4, taking into account that the process PMF is now given by Eq. (20.8) instead of Eq. (20.10).

20.2.3.2 MTTF and MTBF

Two mean values of interest can be calculated: the *mean time to failure* (MTTF) and the *mean time between failures* (MTBF). Study of the latter is the main objective of this work.

To begin with, we shall state formal definitions for both.

Definition 20.3 (MTTF and MTBF). Let T_k be defined as in Definition 20.2. The mean time to the *k*th failure is the mean value:

$$MTTF_k = E[T_k]. \tag{20.44}$$

Let $X_k = T_k - T_{k-1}$ for any $k > 1$. The mean time between $k-1$ and *k*th failures is the mean value:

$$MTBF_{k-1,k} = E[X_k], \tag{20.45}$$

provided that $MTTF_k$ and $MTTF_{k-1}$ are both finite.

Calculation of the MTTF is straightforward just by taking the expectation of T_k, since in practical cases its pdf is known (Eq. 20.31). A more convenient method can be used to obtain the MTBF of processes with $P(\Omega_k) = 1$ for all *k*.

Theorem 20.6. *Let T_k be defined as in Definition 20.2. If $\lim_{t\to\infty} t\, P_k(t) = 0$ and $P(\Omega_k) = 1$ for all $k > 1$, then we have*

$$MTBF_{k-1,k} = \int_0^\infty t\, P_k(t)dt, \quad k > 1. \tag{20.46}$$

Proof. We begin by writing the definition of $MTTF_{k-1}$ and replacing Eq. (20.3)

$$MTTF_{k-1} = \frac{1}{P(\Omega_{k-1})} \int_0^\infty t\, \lambda_{k-2}(t)\, P_{k-2}(t)dt, \tag{20.47}$$

$$P(\Omega_{k-1})MTTF_{k-1} = \int_0^\infty t\left[P'_{k-1}(t) + \lambda_{k-1}(t)\, P_{k-1}(t)\right]dt, \tag{20.48}$$

$$P(\Omega_{k-1})MTTF_{k-1} = \int_0^\infty t\, P'_k(t)dt + P(\Omega_k)MTTF_k. \tag{20.49}$$

Since by hypothesis $P(\Omega_k)$ and $P(\Omega_{k-1})$ are 1

$$MTBF_{k-1,k} = MTTF_k - MTTF_{k-1} = \int_0^\infty t\, P'_k(t)dt. \tag{20.50}$$

Integrating by parts and using $\lim_{t\to\infty} t\, P_k(t) = 0$ yields Eq. (20.46).

To finalize this section, we will study an alternative definition for the MTBF, that is, in practice, much easier to obtain. Since most software failure reports record the time every failure is detected, this information can be used to enhance the calculation by conditioning. Suppose we know the $k-1$th failure was detected at time s; we wish to calculate the model's prediction for the time up to the next failure, which is $MTBF_{k-1,\,k}$, using our knowledge of $T_{k-1} = s$. This motivates the following definition.

Definition 20.4 (Conditional MTBF). Let T_k be defined as in Definition 20.2. Assume that $s = T_{k-1}$ is known. Then, we define the conditional mean time between the $k-1$ and kth failures as

$$MTBF^*_{k-1,k}(s,k) = E[T_k | T_{k-1} = s] - s, \tag{20.51}$$

where the expectation should be computed by Eq. (20.34).

This definition has three main advantages. First, since we are using additional information, the resulting $MTBF^*$ is a better estimate of the real data. Second, expectation of the conditional density given by Eq. (20.34) is much easier to calculate than that of the marginal density (Eq. 20.31). And last, Definition 20.4 can be used in cases where the MTTF is infinite, which implies that the regular MTBF cannot be calculated by Definition 20.3. Moreover, the denominator of Eq. (20.34), which is the probability of Ω_k conditioned to the knowledge of T_{k-1}, is also easy to calculate.

20.3 Application

In this section, we will use different models based on birth processes to study the failure detection process on two software projects developed in recent years under agile methodologies. The datasets were gently provided by the agile software development company Grupo Esfera S.A. [18]. We begin by presenting the models.

20.3.1 Models considered

Table 20.1 contains a nonexhaustive list of software reliability models based on birth processes studied in the literature. Except for the linear contagion model,

TABLE 20.1 Software reliability models based on birth processes.

Model	$\lambda_r(t)$
Musa-Okumoto	$\frac{ab}{1+bt}$, $a, b > 0$
Goel-Okumoto	abe^{-bt}, a, $b > 0$
Delayed S-shaped	ab^2te^{-bt}, a, $b > 0$
Logistic	$\frac{abe^{-b(t-c)}}{[1+e^{-b(t-c)}]^2}$, $a, b, c > 0$
Linear contagion	$\rho\frac{1+\frac{\gamma}{\rho}r}{1+\rho t}$, $\gamma, \rho > 0$

the rest are traditional models based on NHPPs. The Musa-Okumoto and Goel-Okumoto models are most suitable to model a reliability growth behavior, but evidence presented in Ref. [7] suggests that they cannot model an increasing failure rate. In that work, we have shown that due to this, those models are not good for describing agile projects like the ones we analyze here. The Delayed S-shaped and Logistic are NHPPs with the capacity of modeling datasets with two different stages of growth (as it can be seen in Ref. [7], their mean value curves have an inflection point). Finally, the linear contagion model we consider is obtained as a special case of Eq. (20.9). An interesting feature of this model is that it is able to model both reliability growth and increasing failure rate behaviors.

More information about contexts in which some models may perform better the rest can be obtained by applying Theorems 20.4 and 20.5. The delayed S-shaped and logistic models have nonzero probability of the population arriving at a finite stationary state, while the Musa-Okumoto and linear contagion models do not: every finite kth failure has probability of being detected, which suggests that these processes are better suited to model populations that may grow with no limits. This is indeed the case of failures detected in an Agile project: since new code is constantly added, the failure population continues to grow nonstop. Processes like the delayed S-shaped and the logistic are more suitable to model populations with an upper limit; however, contagion model has proved to perform quite well in the early stages of growth within a finite population, that is, in the segment of the time axis where the upper limit is much larger than the process state (see Ref. [15]).

20.3.2 Experiments

We consider two datasets corresponding to software projects developed under Agile. The application of a model consists mainly on the calibration of the parameters, which can be done by different methods. In this work, the calibration is performed by a nonlinear least-squares fitting, using the Levenberg-Marquardt algorithm. A comparison of calibration methods is presented in Ref. [7], where we found least-squares to perform better than maximum likelihood. Using this estimation procedure, the model is, of course, sensitive to the quality of the data. The presence of outliers, for example, results on a jump in the cumulative values, altering this way the outcomes. The same situation arises when the measuring methodology changes "on the run"; in this cases, a multi-stage analysis might be a better approach, see, for example, Ref. [19]. This kind of "sudden changes" in the behavior of the data is detected by the model, typically producing less accurate fits (which can be quantified, e.g., by a lower R^2 coefficient) and should be carefully treated in order to prevent wrong results.

The MTBF calculation is done by Definition 20.3 for the NHPP-based models. The linear contagion model requires a more careful analysis. By replacing Eq. (20.10) in Eq. (20.44) it can be seen that $E[T_k]$ diverges, which implies the MTTF cannot be defined. This does not imply that the MTBFs are also infinite, but it makes impossible to calculate them via Definition 20.3. Hence we turn to Definition 20.4 and calculate the conditional MTBF, which yields

$$MTBF^*_{k-1,k}(s,k) = \frac{1+\rho s}{\gamma(k-1)}. \tag{20.52}$$

A complete derivation of this result is done in Ref. [15].

20.3.2.1 Project 1

Fig. 20.2 depicts the resulting interpolated MTBF curves of the three models and the real data from the failure report. The x-axis shows the failure number k, and y-axis shows $MTBF_{k,\,k+1}$. MTBF is measured in days. The data show that most of the failures were detected very close in time (note that there is a region with a high density of points in the lower part of the plot). This makes the comparison more difficult, since the three models follow the data very similarly. There are some outlier values between the 40th and 60th failures, which none of the models detect. Slight differences can be appreciated in the shape of the curves. The delayed S-shaped curve is quite smooth, with a perfectly defined bathtub shape. The logistic curve exhibits a sharp peak at the early part, after which it has several increasing and decreasing stages. The conditional MTBF curve of the contagion model is not as smooth as the others: a clear bathtub envelope can be appreciated as well as some high-frequency noise. It is interesting to observe that this curve is the only one to detect the first datum, which is a rather heavy outlier: this is due to the fact that its information is included in the conditioning, one of the main advantages of this method.

20.3.2.2 Project 2

MTBF curves for this project are shown in Fig. 20.3. MTBF is measured in days. The delayed S-shaped and logistic curves show the same behavior as

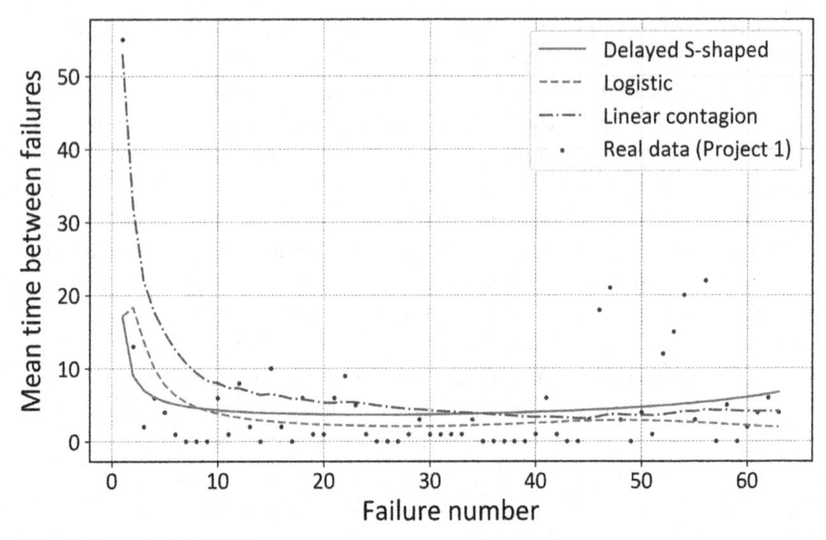

FIG. 20.2 Project 1 MTBF curves.

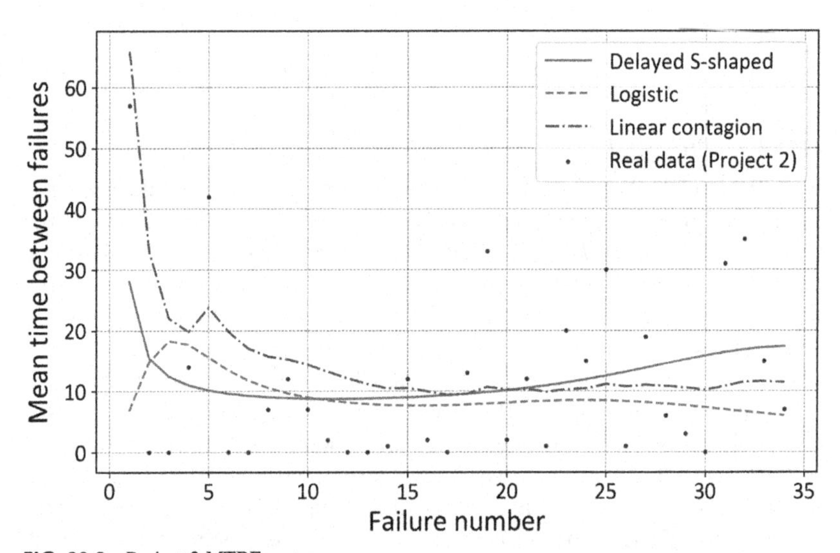

FIG. 20.3 Project 2 MTBF curves.

in Project 1, the first with a smooth bathtub shape and the second with successive increasing and decreasing stages and a peak at the beginning. As can be deduced by the low density of points between the 1 and 10th days, failures are quite scattered in time, which may suggest that this project is reaching a reliability growth stage. This behavior is only captured by the delayed S-shaped model. As in Project 1, the linear contagion model (conditional MTBF) is the only one to detect the early outliers.

20.3.3 Discussion

Results suggest that the behavior of the MTBF in agile software projects does not differ much from that of hardware reliability studies. An intensity failure rate is observed at the beginning, represented by a decreasing stage in the MTBF; this is typical of agile projects, as pointed by previous works on the topic (see Ref. [7]). The models differ, since both delayed S-shaped and logistic curves present an increasing stage in the late part of the dataset, which corresponds to a reliability growth stage in the failure detection process, while the contagion model places the whole dataset in the increasing failure rate phase. This was already observed in Ref. [7], and it is consequence of the absence of an inflection point in the model's mean value curve. Logistic model also presents some minor differences such as the initial peak or the presence of more than two increase-decrease phases. However, the differences are appreciated only in the shape, since numerical values do not differ much at all, remaining most of the time within a short range not far from the real data.

20.4 Conclusions

A summary on the MTTF and MTBF metrics as obtained from several software reliability models described under the framework of nonhomogeneous birth processes was presented. A complete self-contained mathematical background was developed in order to show how those metrics are obtained from the nonhomogeneous birth processes. The well-known software reliability models based on nonhomogeneous Poisson processes and the recently proposed contagion models were analyzed. A comparison of the performance of applying those models to real datasets was also shown and discussed.

Acknowledgments

The authors thank Universidad Nacional de Tres de Febrero for support under grant no. 32/19 80120190100010TF.

References

[1] H. Pham, System Software Reliability, first ed., Springer Publishing Company, Inc., 2010.
[2] M.R. Lyu, Handbook of Software Reliability Engineering, McGraw-Hill, Inc., Hightstown, NJ, 1996.
[3] Deepika, A. Anand, O. Singh, P.K. Kapur, Three-dimensional wiener process based entropy prediction modelling for OSS, Int. J. Syst. Assur. Eng. Manag. 12 (1) (2021) 188–198, https://doi.org/10.1007/s13198-020-01040-4.

[4] T. Shiva, D. Kumar, S. Kumar, Open source software: analysis of available reliability models keeping security in the forefront, Int. J. Inf. Technol. (2019), https://doi.org/10.1007/s41870-019-00293-y.

[5] N.R. Barraza, Software reliability modeled on contagion, in: 2016 IEEE International Symposium on Software Reliability Engineering Workshops (ISSREW), 2016, pp. 49–50, https://doi.org/10.1109/ISSREW.2016.22.

[6] N.R. Barraza, Software reliability modeling for dynamic development environments, in: A. Anand, M. Ram (Eds.), Recent Advancements in Software Reliability Assurance, Advances in Mathematics and Engineering, CRC Press, 2019, pp. 29–37. https://books.google.com.ar/books?id=5p6RDwAAQBAJ. (Chapter 3).

[7] G. Pena, N.R. Barraza, Increasing failure rate software reliability models for agile projects: a comparative study, in: A. Anand, M. Ram (Eds.), Systems Performance Modeling, Applications of Mathematics in Engineering and Information, De Gruyter, 2020. https://www.degruyter.com/document/isbn/9783110607635/html. (Chapter 8).

[8] R. Barlow, F. Proschan, Mathematical Theory of Reliability, Society for Industrial and Applied Mathematics, Philadelphia, PA, 1996. http://epubs.siam.org/doi/abs/10.1137/1.9781611971194.

[9] W. Feller, An Introduction to Probability Theory and Its Applications, vol. 1, third ed., Wiley, 1968.

[10] E. Coddington, N. Levinson, Theory of Ordinary Differential Equations, McGraw-Hill Education, 1955.

[11] K. Sendova, L. Minkova, Introducing the non-homogeneous compound-birth process, Stochastics (2019) 1–19, https://doi.org/10.1080/17442508.2019.1666132.

[12] J. Grandell, Mixed Poisson Processes, vol. 77, CRC Press, 1997.

[13] H. Konno, On the exact solution of a generalized Pólya process, Adv. Math. Phys. (2010), 1–12.

[14] S.A. Klugman, H.H. Panjer, G.E. Willmot, Loss Models: Further Topics, John Wiley & Sons, 2013.

[15] N.R. Barraza, G. Pena, V. Moreno, A non-homogeneous Markov early epidemic growth dynamics model. Application to the SARS-CoV-2 pandemic, Chaos Solitons Fract. 139 (2020) 110297, https://doi.org/10.1016/j.chaos.2020.110297.

[16] G.U. Yule, A mathematical theory of evolution, based on the conclusions of Dr. J. C. Willis, F. R.S., Philos. Trans. R. Soc. Lond. B Contain. Pap. Biol. Character 213 (1925) 21–87. http://www.jstor.org/stable/92117.

[17] O. Lundberg, On Random Processes and Their Application to Sickness and Accident Statistics, Almqvist & Wiksells Boktryckeri-a.-b., 1964. https://books.google.com.ar/books?id=mr4rAAAAYAAJ.

[18] Grupo Esfera S.A., https://www.grupoesfera.com.ar/ (Accessed 22 February 2021).

[19] N.R. Barraza, Software reliability analysis of multistage projects, in: 2019 Amity International Conference on Artificial Intelligence (AICAI), 2019, pp. 67–73, https://doi.org/10.1109/AICAI.2019.8701285.

Chapter 21

Inliers prone distributions: Perspectives and future scopes

K. Muralidharan and Pratima Bavagosai
Department of Statistics, Faculty of Science, The Maharaja Sayajirao University of Baroda, Vadodara, India

21.1 Introduction

There are a plethora of examples of phenomena concerning nature, life, and human activities where the real data do not conform to the standard distributions. In such cases, we either use mixtures of standard distributions of similar types or nonstandard mixtures of degenerate and a standard distribution which maybe again a discrete or continuous one. The literature contains many papers that deal with complete mixtures of distributions of similar types. However, the literature hardly contains papers that provide and deal with special "nonstandard" mixtures that mix discrete (degenerate) and continuous distributions as emphasized in this chapter.

The nonstandard mixture of distributions generally contains inliers, where *inliers* are an observation (or a group of observations) sufficiently small relative to the rest of the observations, which appear to be inconsistent with the remaining dataset. They are either the results of instantaneous failures or early failures, experienced in life testing experiments, survival studies, clinical trials, and many other application areas. The test items that fail at time 0 are called *instantaneous failures*, and the items that fail prematurely are called *early failures*. Kale and Muralidharan [1] have introduced the term *inliers* in connection with the estimation of (π, θ) of early failure model with modified failure time distribution being an exponential distribution with mean θ assuming π known. Following are some of the practical contexts, where inliers can arise as natural occurrences of the specific situations involved, and degeneracy can happen at one or more discrete points and a positive distribution for the remaining lifetimes.

1. The technological components of hardware, intended to function over some time, the failure rate is initially relatively high, and then actually decreases with increasing age. The high failure rate either results in zero lifetime or

System Assurances. https://doi.org/10.1016/B978-0-323-90240-3.00021-7

marginally small lifetimes, otherwise, the lifetime will be of any positive number (usually continuous). Thus, the overall distribution of lifetimes may be represented by using a nonstandard mixture of distribution.

2. In studies of tooth decay, the number of surfaces in a mouth that are filled, missing, or decayed are scored to produce a decay index. Healthy teeth are scored 0 for no evidence of decay. Thus, the distribution is a mixture of a mass point at 0 and a nontrivial continuous distribution of decay score. The problem could be further complicated if the decay score is expressed as a percentage of damage to measured teeth. The distribution should then be a mixture of a discrete random variable (0-for healthy teeth, 1-for all missing teeth) with a nonzero probability of both outcomes and a continuous random variable (amount of decay in the (0, 1) interval).
3. Machines and software are tested for their correctness and perfectness or reliability. Bugs (or errors) in such situations are important to assess the durability and credibility of machines and programs. Zero defects or zero bugs are considered to be good in such situations. If there are bugs, then they can be measured in terms of some discrete measurements. Data on this example contains many zeros (no bugs), few countable bugs, and a large number of bugs constituting a mixture of data.

Practical situations involving degeneracy at one or more points and positive configurations of observations are thus natural phenomena. A univariate probability model or a complete (finite) mixture of distributions may not work in those situations. Hence, probability modeling in such situations may require special attention and treatment. From the above examples, it is seen that the values including zeros and close to zeros are important as well as significant in most cases. For instance, zero bugs in a computer program or electronic machine are all good to judge the efficiency and reliability of the solution, and hence they are significant. Similarly, zero lifetime, zero rainfall (dry day), etc., are all practically not good but again significant as per conditions prevailing. Below, we introduce various inliers-prone models and study their inferences.

The organization of the chapter is as follows: Section 21.2 contains various inliers-prone models. The estimation of model parameters in Section 21.3 and testing-related inferences are given in Section 21.4. The description of the data and its analysis is given in Section 21.5. The issues and future scope are discussed in Section 21.6.

21.2 Inliers prone models

21.2.1 Instantaneous failure models

The model $\mathcal{F} = \{F(x;\theta), x \geq 0, \theta \in \Theta\}$, where $F(x;\theta)$ is a continuous failure time distribution function (FTD) with $F(0) = 0$ is to be suitably modified by mixing a singular distribution at zero to accommodate instantaneous failures. The modified model is represented as

$$G(x;\pi,\theta)=\begin{cases}1-\pi, & x=0\\ 1-\pi+\pi F(x;\theta), & x>0\end{cases} \tag{21.1}$$

with respect to a measure μ which is the sum of Lebesgue measure on $(0,\infty)$ and a singular unit measure at the origin; and $0<\pi<1$. Aitchison [2] was the first to discuss the inference problem of instantaneous failures in life testing. Other authors who have studied this kind of model are Kleyle and Dahiya [3], Jayade and Prasad [4], Vannman [5], Kale [6,7], Muralidharan [8], Muralidharan and Kale [9], Muralidharan and Lathika [10,11], Kale and Muralidharan [12], Adlouni et al. [13], Hazra et al. [14], Knopik et al. [15], and so on.

21.2.2 Early failure model-1

If it is assumed that $\lambda(x)=\lambda=\frac{1}{\theta}$ for all x from an exponential distribution, then the model becomes a constant failure rate model. Under this setup, Miller [16] proposed the early failure model as

$$\lambda(x)=\begin{cases}\lambda_1, & 0\le x<T_0\\ \lambda_2, & T_0\le x\end{cases} \tag{21.2}$$

where $\lambda_1>\lambda_2$. The probability density corresponding to this failure rate is

$$f_X(x;\lambda_1,\lambda_2)=\begin{cases}\lambda_1 e^{-\lambda_1 x}, & 0\le x<T_0\\ \lambda_2 e^{-\lambda_1 T_0-\lambda_2(x-T_0)}, & T_0\le x\end{cases} \tag{21.3}$$

The above model can also be viewed as a model for a shift in the hazard function of the exponential distribution, where the time T_0 becomes the change point.

21.2.3 Early failure model-2

To accommodate early failures, the family $\mathcal{F}$ is modified to $\mathcal{G}_1=\{G(x;\pi,\theta), x\ge 0, \theta\in\Theta, 0<\pi<1\}$, where the cumulative distribution function (CDF) corresponding to $g_1\in\mathcal{G}_1$ is given by

$$G_1(x;\pi,\theta)=(1-\pi)H(x)+\pi F(x;\theta) \tag{21.4}$$

Here $H(x)$ is a CDF with $H(\delta)=1$ for δ sufficiently small, assumed known and specified in advance. Then the modified family $\mathcal{G}_1$ has a probability distribution function (pdf) with respect to measure μ, which is the sum of Lebesgue measure on (δ,∞) and a singular measure at δ as

$$g_1(x;\pi,\theta)=\begin{cases}0, & x<\delta\\ 1-\pi+\pi F(\delta;\theta), & x=\delta\\ \pi f(x;\theta), & x>\delta\end{cases} \tag{21.5}$$

Some of the references which treat early failure analysis are Kale and Muralidharan [1,12], Kale [7], Muralidharan and Lathika [11], Muralidharan and

Arti [17,18], Muralidharan [19], Muralidharan and Bavagosai [20], and the references contained therein. These authors treated early failures as inliers using the sample configurations from parametric models including exponential, Weibull, Pareto, Normal, and Gompertz distribution.

The models in (21.1) and (21.4) can be combined to form the CDF as

$$G(x;\pi,\theta)=\begin{cases}0, & x<d\\(1-\pi)+\pi F(x;\theta), & x\geq d\end{cases} \tag{21.6}$$

Having the corresponding pdf as given in (21.5). The instantaneous failures and early failures can be respectively viewed for $d=0$ and if $d>0$.

21.2.4 Model with inliers at zero and one

In some of the examples discussed above (e.g., 2, 3, and 5), the observations 0 and 1 become a natural occurrence with other positive data points. If these observations are treated as inliers, then, the distribution function of such models can be written as

$$H(x;\pi_1,\pi_2,\theta)=\begin{cases}0, & x<0\\ \pi_1, & 0\leq x<1\\ \pi_1+\pi_2, & x=1\\ \pi_1+\pi_2+(1-\pi_1-\pi_2)\dfrac{F(x;\theta)-F(1;\theta)}{1-F(1;\theta)}, & x\geq 1\end{cases} \tag{21.7}$$

where π_1 and π_2 are the proportion of 0 and 1 observations. This model was first studied by Muralidharan and Bavagosai [21–23] with $F(x;\theta)$ as exponential, Pareto and Weibull distribution.

21.3 Inferences on 0–1 inliers model

Consider the FTD as exponential with density function $f(x;\theta)$ where

$$f(x;\theta)=\frac{1}{\theta}e^{-\frac{x}{\theta}},\ x>0;\ \theta>0 \tag{21.8}$$

then pdf corresponding to the model (21.7) reduces to

$$h(x;\pi_1,\pi_2,\theta)=\begin{cases}\pi_1, & x=0\\ \pi_2, & x=1\\ (1-\pi_1-\pi_2)\dfrac{e^{-\left(\frac{x-1}{\theta}\right)}}{\theta}, & x>1\end{cases} \tag{21.9}$$

The graphical plots of the probability density function for different values of the parameters are shown in Fig. 21.1.

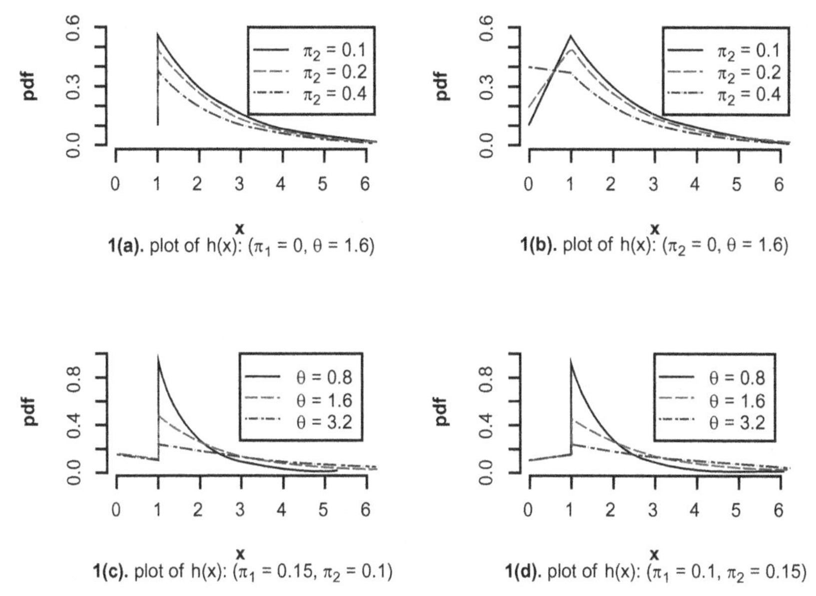

FIG. 21.1 Density function plots.

21.3.1 Parameter estimation

Estimating parameters through various statistical methods is the best way to understand the characteristics of the above models. We use classical methods like maximum likelihood and uniformly most powerful unbiased estimation methods to realize this objective. They are presented below in various subsections.

21.3.1.1 *The maximum likelihood estimation of* $\underline{\theta} = (\pi_1, \pi_2, \theta)$

Let $X_1, X_2, \ldots, X_n$ be a random sample of size n from $h \in \mathcal{H}$ as given in (21.9) and we define

$$I_1(x) = \begin{cases} 1, & x = 0 \\ 0, & \text{otherwise} \end{cases} \text{ and } I_2(x) = \begin{cases} 1, & x = 1 \\ 0, & \text{otherwise} \end{cases}$$

then the likelihood equation can be written as

$$L(\underline{x};\underline{\theta}) = \prod_{i=1}^{n} \pi_1{}^{I_1(x_i)} \pi_2{}^{I_2(x_i)} \left(\frac{(1-\pi_1-\pi_2) e^{-\left(\frac{x_i - 1}{\theta}\right)}}{\theta} \right)^{(1 - I_1(x_i) - I_2(x_i))} \tag{21.10}$$

$$= \pi_1{}^{r_1} \pi_2{}^{r_2} \left(\frac{1-\pi_1-\pi_2}{\theta} \right)^{n - r_1 - r_2} e^{-\frac{1}{\theta} \sum_{x_i > 1} (x_i - 1)} \tag{21.11}$$

where $r_1 = \sum_{i=1}^{n} I_1(x_i)$ and $r_2 = \sum_{i=1}^{n} I_2(x_i)$, denotes the number of zero and one observations respectively. Take logarithm and then Differentiate (21.11) with respect to parameters π_1, π_2, and θ and equate them to zero. WE have MLE of π_1, π_2, and θ as

$$\widehat{\pi}_1 = \frac{r_1}{n}, \widehat{\pi}_2 = \frac{r_2}{n} \text{ and } \widehat{\theta} = \frac{\sum_{x_i>1}(x_i - 1)}{n - \pi_1 - \pi_2} \tag{21.12}$$

Thus, the maximum likelihood estimator (MLE) of the proportion of instantaneous failures is $\widehat{\pi}_1$ and the MLE of proportion premature failures is $\widehat{\pi}_2$ whereas the MLE of mean failure rate in the target population is $\widehat{\theta}$. For asymptotic distribution of parameters see Muralidharan and Bavagosai [21].

21.3.2 Unbiased estimation

Recall, that the model in (21.9) can be expressed as

$$h(x; \underline{\theta}) = a(x)\frac{(h_1(\underline{\theta}))^{C_1(x)}(h_2(\underline{\theta}))^{C_2(x)}(h_3(\underline{\theta}))^{C_3(x)}}{g(\underline{\theta})} \tag{21.13}$$

where, $a(x) = 1$; $h_1(\underline{\theta}) = \frac{\theta\pi_1}{1-\pi_1-\pi_2}$; $h_2(\underline{\theta}) = \frac{\theta\pi_2}{1-\pi_1-\pi_2}$; $h_3(\underline{\theta}) = e^{-\frac{1}{\theta}}$; $g(\underline{\theta}) = \frac{\theta}{1-\pi_1-\pi_2}$; $C_1(x) = I_1(x)$; $C_2(x) = I_2(x)$ and $C_3(x) = (x - 1)(1 - I_1(x) - I_2(x))$. Also $a(x) > 0$ and $g(\underline{\theta}) = \int_{x>1} a(x) \prod_{i=1}^{3} (h_i(\underline{\theta}))^{C_i(x)} dx$. The density in (21.13) so obtained is defined with respect to a measure $\mu(x)$ which is the sum of Lebesgue measure over $(1, \infty)$ a well-known form of a three-parameter exponential family with natural parameters $(\eta_1, \eta_2, \eta_3) = \left(\log\left(\frac{\theta\pi_1}{1-\pi_1-\pi_2}\right), \log\left(\frac{\theta\pi_2}{1-\pi_1-\pi_2}\right), \log\left(e^{-\frac{1}{\theta}}\right)\right)$ generated by underlying indexing parameters (π_1, π_2, θ). Hence $C(X) = (C_1(x), C_2(x), C_3(x)) = (I_1(x),\ I_2(x), (x - 1)(1 - I_1(x) - I_2(x)))$ is jointly complete sufficient for $\underline{\theta} = (\pi_1, \pi_2, \theta)$. For distributional properties see Muralidharan and Bavagosai [21].

Let us now propose some uniformly minimum variance unbiased estimators for the parametric function of the model presented in (21.13).

21.3.2.1 Uniformly minimum variance unbiased estimation of parameters

Let $X_1, X_2, \ldots, X_n$ be a random sample of size n from $h \in \mathcal{H}$ as given in (21.13) which belongs to the three-parameter exponential family of distributions. Therefore, $Z = (Z_1, Z_2, Z_3)$, where $Z_i = \sum_{j=1}^{n} C_j(X_j)$, $i = 1$, 2, and 3 are jointly complete sufficient statistics. The joint pdf of Z is again exponential family and is given by

$$h_Z(z;\underline{\theta}) = \mathrm{P}(Z_1 = r_1, Z_2 = r_2)h(z_3;\theta|Z_1 = r_1, Z_2 = r_2)$$
$$= P(Z_1 = r_1, Z_2 = r_2)h(z_3;\theta|n - r_1 - r_2)$$

Since the distribution of (Z_1, Z_2) is trinomial and is a complete family distribution. Also $(Z_3|Z_1, Z_2)$ is Gamma with parameter $(n - r_1 - r_2, \theta)$ with pdf

$$h(z_3;\theta|\, n - r_1 - r_2) = \frac{z_3^{(n-r_1-r_2-1)}\ e^{-\frac{z_3}{\theta}}}{\Gamma n - r_1 - r_2\, \theta^{n-r_1-r_2}}, z_3 > 0; \theta > 0$$

which depends only on θ and is also a complete family of distribution. Therefore,

$$h_Z(z;\underline{\theta}) = \frac{n!}{r_1!r_2!(n-r_1-r_2)!}\pi_1^{\ r_1}\pi_2^{\ r_2}(1-\pi_1-\pi_2)^{(n-r_1-r_2)}\frac{z_3^{(n-r_1-r_2-1)}\ e^{-\frac{z_3}{\theta}}}{\Gamma n - r_1 - r_2\, \theta^{n-r_1-r_2}},$$
$$0 \le r_1, r_2 \le n; z_3 > 0; 0 \le \pi_1, \pi_2 \le 1; \theta > 0$$
$$= B(z_1, z_2, z_3, n)\frac{\Pi_{i=1}^3 (h_i(\underline{\theta}))^{z_j}}{g(\underline{\theta})^n}$$

where

$$B(z_1, z_2, z_3, n) = \begin{cases} \dfrac{n!}{r_1!\, r_2!(n-r_1-r_2)!}\dfrac{z_3^{(n-r_1-r_2-1)}}{\Gamma n - r_1 - r_2}, & z_3 > 0; r_1 + r_2 - 1 < n \\ 1, & z_3 = 0; r_1 = 0 \text{ or } r_2 = 0 \end{cases} \tag{21.14}$$

$z_i \in T(n) \subseteq \mathbb{R}, \underline{\theta} \in \Omega$. Here $z = (z_1, z_2, z_3, n)$ and $B(z_1, z_2, z_3, n)$ are such that

$$g(\underline{\theta})^n = \int_{z_1 \in T(n)} \int_{z_2 \in T(n)} \int_{z_3 \in T(n)} B(z_1, z_2, z_3, n) \prod_{i=1}^{3} (h_i(\underline{\theta}))^{z_i} dz_1 dz_2\, dz_3$$

Since $E(Z_1) = \sum_{j=1}^n E(C_1(x_j)) = n\pi_1$, $E(Z_2) = \sum_{j=1}^n E(C_2(x_j)) = n\pi_2$, and $E(Z_3) = \sum_{j=1}^n E(C_3(x_j)) = n\theta(1 - \pi_1 - \pi_2)$ which is turn give UMVUE's of π_1, π_2, and θ as

$$\widehat{\pi}_1 = \frac{Z_1}{n} = \frac{r_1}{n},\ \widehat{\pi}_2 = \frac{Z_2}{n} = \frac{r_2}{n}, \text{and}\, \widehat{\theta} = \frac{Z_3}{n(1 - \widehat{\pi}_1 - \widehat{\pi}_2)} \tag{21.15}$$

Note that, the likelihood estimates and minimum variance unbiased estimates coincide everywhere.

21.3.2.2 Uniformly minimum variance unbiased estimation of parametric functions

Following Jani and Singh [24], let $X_1, X_2, \ldots, X_n$ be $n(>1)$ random sample from (21.13), then there exists a UMVUE of $\Phi(\underline{\theta})$ if and only if $\Phi(\underline{\theta})[g(\underline{\theta})]^n$ can be expressed in the form

$$\Phi(\underline{\theta})[g(\underline{\theta})]^n = \int_{z_1 \in T(n)} \int_{z_2 \in T(n)} \int_{z_3 \in T(n)} \alpha(z_1, z_2, z_3, n) \prod_{i=1}^{3} (h_i(\underline{\theta}))^{z_i} dz_1 dz_2 dz_3$$

Thus, the UMVUE of a function $\Phi(\underline{\theta})$ of $\underline{\theta}$ in $h(x; \underline{\theta})$ is given by

$$\psi(Z_1, Z_2, Z_3, n) = \frac{\alpha(Z_1, Z_2, Z_3, n)}{B(Z_1, Z_2, Z_3, n)}, B(Z_1, Z_2, Z_3, n) \neq 0$$

Using the above result, we obtained the UMVUE of different functions of parametric vector $\underline{\theta}$ in the following results.

Result 21.1 **The UMVUE of** $[g(\underline{\theta})]^k = \left(\frac{\theta}{1-\pi_1-\pi_2}\right)^k$**,** $k \neq 0$ as per the model given in (21.13) is

$$\begin{aligned} G_k(z_1, z_2, z_3, n) &= \frac{B(z_1, z_2, z_3, n+k)}{B(z_1, z_2, z_3, n)} \\ &= \frac{[n+1]_k z_3{}^k}{[n-r_1-r_2+1]_k [n-r_1-r_2]_k}, k \leq n-r_1-r_2; r_1+r_2-1 < n \end{aligned}$$

where $(r)_k = \frac{r!}{(r-k)!}$ and $[r]_k = \frac{\Gamma r+k}{\Gamma r}$.

Result 21.2 **The UMVUE of the variance of** $G_k(Z_1, Z_2, Z_3, n)$ **is given by**

$$\widehat{var}[G_k(z_1, z_2, z_3, n)] = G_k^2(z_1, z_2, z_3, n) - G_k^2(z_1, z_2, z_3, n)$$

$$= \left[\frac{[n+1]_k z_3{}^k}{[n-r_1-r_2+1]_k [n-r_1-r_2]_k}\right]^2 - \frac{[n+1]_{2k} z_3{}^{2k}}{[n-r_1-r_2+1]_{2k} [n-r_1-r_2]_{2k}},$$
$$2k \leq n-r_1-r_2; r_1+r_2-1 < n$$

Result 21.3 **The UMVUE of the density (21.9) for fixed** x**, is given by**

$$\phi_x(z_1, z_2, z_3, n) = a(x) \frac{B(z_1 - C_1(x), z_2 - C_2(x), z_3 - C_3(x), n-1)}{B(z_1, z_2, z_3, n)}$$

$$= \frac{(r_1)_{I_1(x)} (r_2)_{I_2(x)} (n-r_1-r_2)_{(1-I_1(x)-I_2(x))} (n-r_1-r_2-1)_{(1-I_1(x)-I_2(x))}}{n[z_3 - (x-1)(1-I_1(x)-I_2(x))]^{(1-I_1(x)-I_2(x))}}$$

$$\left(1 - \frac{(x-1)(1-I_1(x)-I_2(x))}{z_3}\right)^{(n-r_1-r_2-1)}, z_3 > (x-1); r_1+r_2-1 < n$$

Result 21.4 **The UMVUE of the variance of** $\phi_x(Z_1, Z_2, Z_3, n)$ **is given by**

$$\widehat{var}[\phi_x(z_1,z_2,z_3,n)] = \phi_x^2(z_1,z_2,z_3,n) - \phi_x(z_1,z_2,z_3,n)\phi_x(z_1-C_1(x),z_2-C_2(x),z_3-C_3(x),n-1)$$

$$= \phi_x^2(z_1,z_2,z_3,n) - \frac{(r_1)_{2I_1(x)}(r_2)_{2I_2(x)}}{n(n-1)\left(1-\frac{2(x-1)(1-I_1(x)-I_2(x))}{z_3}\right)^{(n-r_1-r_2-1)}}$$

$$\frac{(n-r_1-r_2-1)_{2(1-I_1(x)-I_2(x))}(n-r_1-r_2)_{2(1-I_1(x)-I_2(x))}}{[z_3-2(x-1)(1-I_1(x)-I_2(x))]^2(1-I_1(x)-I_2(x))}, z_3 > 2(x-1); r_1+r_2-1 < n$$

Result 21.5 **For fixed $z = (z_1, z_2, z_3, n)$, the UMVUE of the survival function $S(x) = p(X > x)$, $x \geq 0$ is obtained as**

$$\widehat{S}(x) = \frac{(r1)_{I_1(x)}(r2)_{I_2(x)}(n-r1-r2-1)_{(1-I_1(x)-I_2(x))}(n-r1-r2)_{(1-I_1(x)-I_2(x))}}{n[(n-r1-r2)-(1-I_1(x)-I_2(x))]} Z_3^{(I_1(x)+I_2(x))}$$

$$\left(1-\frac{(x-1)(1-I_1(x)-I_2(x))}{z_3}\right)^{(n-r_1-r_2-1)}, Z_3 > (x-1); r_1+r_2-1 < n$$

Result 21.6 **For fixed $z = (z_1, z_2, z_3, n)$, the UMVUE of $var\left(\widehat{S}(x)\right)$**, is obtained as

$$\widehat{var}\left(\widehat{S}(x)\right) = \left[\widehat{S}(x)\right]^2 - \frac{z_3{}^2(I_1(x)+I_2(x))}{n(n-1)}\left(1-\frac{2(x-1)(1-I_1(x)-I_2(x))}{Z_3}\right)^{(n-r1-r2-1)}$$

$$\left(\frac{(r1)_{2I_1(x)}(r2)_{2I_2(x)}(n-r1-r2-1)_{2(1-I_1(x)-I_2(x))}(n-r1-r2)_{2(1-I_1(x)-I_2(x))}}{[(n-r1-r2)-2(1-I_1(x)-I_2(x))][(n-r1-r2+1)-2(1-I_1(x)-I_2(x))]}\right)$$

$$Z_3 > 2(x-1); r_1+r_2-1 < n$$

21.4 Tests of hypothesis about inliers

Many authors have tried to propose tests for outliers by looking at the data from a specific probability model. Since outliers can be viewed as lower inliers, the technical foundation for proposing tests for inliers can be attempted similarly. See Kale and Muralidharan [25] and the references contained therein. For multiple inliers, one suspects the lower k order statistics $X_{(1)}, X_{(2)}, \ldots, X_{(k)}$ as inliers where k is often assumed known. The inliers are regarded as arising from a different population $\mathcal{Q}$ with a DF $G(x) \in \mathcal{g}$ and pdf $g(x)$ and since smaller observations are suspected it is assumed that $\mathcal{Q}$ is stochastically smaller than the population $\wp$ of $G < F$. Thus, the sample $(X_1, X_2, \ldots, X_n)$ is such that $(n-k)$ of these are distributed as F while k is distributed as G. The identified inliers model of Veale [26] and exchangeable model of Kale [27] are examples of inliers generating models.

Barnett and Lewis [28] point out that none of the models given above take into consideration the fact that usually $(X_{(1)}, X_{(2)}, \ldots, X_{(k)})$ the smallest k observations are suspected as inliers and tested for discordancy. However, for the exchangeable model mentioned above, Kale [27] proves that if $\frac{dG}{dF} = \psi(x)$ is strictly decreasing, then $(X_{(1)}, X_2, \ldots, X_{(k)})$ have the maximum probability of being the inlier observations, i.e., a set of observations is distributed as G. It may be also noted that if $\frac{dF}{dG} = \psi(x)$ is strictly decreasing then the population $\mathcal{Q}$ is stochastically smaller than the population $\wp$. To emphasize the special role that lower k order statistics play in these discordancy tests, Barnett and Lewis [28] have introduced the *labeled slippage model*. The testing for discordancy of inlier observations is now formed as the test of hypotheses problem H_0 against H_1, where H_0 and H_1 can be roughly be formulated as

$$H_0 : X_{(1)}, X_{(2)}, \ldots, X_{(n)} \text{ are from } F \in \mathcal{F} \text{ against}$$
$$H_1 : X_{(1)}, X_{(2)}, \ldots, X_{(k)} \text{ are from } G \in g \text{ and } X_{(k+1)}, X_{(k+2)}, \ldots, X_{(n)} \text{ are from } F \in \mathcal{F} \quad (21.16)$$

In the literature reviewed, two types of tests are proposed for the lower outlier testing problem. The "Block test" studied by Chikkagoudar and Kunchur [29], Kimber and Stevens [30], Lewis and Feller [31] is often used to test for discordancy of k lower outliers in the data in a single hypothesis test. This procedure suffers from masking and swamping effects when too many or too few k inliers are present in the sample. The other type of test is based on the Sequential Procedure, where they can be attempted using *inward* and *outward* procedures. In addition to certain theoretical weaknesses, the inward sequential procedure is not recommended by Kimber [32] and Chikkagoudar and Kunchur [29]. Because of the limitations inherent in the Block test and the inward sequential procedure suffering from swamping and/or masking effects, Rosner [33] suggested using an outward sequential procedure and also calls "inside-out" sequential procedure to the reduced sample. Here, one specifies a maximum number of inliers k. Then the kth smallest inliers are tested first. If this gives a significant result, then k inliers are declared to be discordant. If a nonsignificant result is obtained, then the $(k - 1)$th smallest inliers are tested, and so on. This process is continued until either a significant result obtained or no inliers can be declared discordant. This procedure minimizes the probability and magnitude of both masking and swamping effects. As such, the outward procedure is claimed to be superior over the inward (see [29,32,34]). However, control of type 1 error (the probability of a false alarm) is difficult in the outward procedure. The outward procedure for detecting inliers is described as follows.

Considering a random sample of size n, let n_0 units fail instantaneously and $(n - n_0)$ failure time is obtained. Suppose $X_1, X_2, \ldots, X_{n-n_0}$ are the available lifetimes with corresponding order statistics $X_{(1)}, X_{(2)}, \ldots, X_{(n-n_0)}$, for the null hypothesis H_0, let all the X_i are independent and identically distributed exponential with density $f(x;\theta)$ where is as given by (21.8). The alternative hypothesis is H_1, which is a labeled scale-slippage alternative such that the j smallest observations have arisen from exponential density with pdf $g\left(x;\frac{\theta}{\lambda}\right)$ with $\lambda > 1$. Then the test statistics

$$S_j = \frac{x_{(j+1)}}{\sum_{i=1}^{j+1} x_{(i)}}, j = 1,2,\ldots,(n-n_0-1) \tag{21.17}$$

is an appropriate statistic for testing j lower outliers (inliers). If only the first-order statistics are available, then S_j is the likelihood ratio test statistics. The size α test for up to k inliers is detected as follows:

(i) If $S_k > s_{k,\,(n-n_0)}$, declare the k smallest observations as inliers;
(ii) If $S_i < s_{i,\,(n-n_0)}$, $(i = k, k-1, \ldots, l+1)$ and $S_l > s_{l,\,(n-n_0)}$, declare the l smallest observations as inliers $(l = k-1, k-2, \ldots 1)$;
(iii) If $S_j < s_{j,\,(n-n_0)}$, $(j = 1,2,\ldots,k)$ declare no observations as inliers in the data.

Here, $s_{j,\,(n-n_0)} = s_{j,\,(n-n_0)}(\alpha)$, $j = 1, 2, \ldots, k$ are obtained such that all marginal tests would have equal level β (say), that is

$$P_{H_0}\left(S_k > s_{k,(n-n_0)}\right) = P_{H_0}\left(S_{k-1} > s_{k-1,(n-n_0)}\right) = \cdots = P_{H_0}\left(S_1 > s_{1,(n-n_0)}\right)$$
$$= \beta, \text{say}.$$

and the level β is determined such that $P_{H_0}\left\{\bigcap_1^k \left(S_j < s_{j,(n-n_0)}\right)\right\} = 1-\alpha$, where α is the level of significance. The null joint density of $S_1, S_2, \ldots, S_k$ is

$$f_k(s_1, s_2, \ldots, s_k) = \frac{n!k!}{(n-n_0-k-1)!}\left(1+(n-n_0-k-1)s_k\right)^{-(k-1)}\prod_{j=2}^{k}\left(1-s_j\right)^{j-1},$$

over the region $W_k = \left\{\frac{1}{2} < s_1 < 1;\ \frac{s_j}{1+s_j} < s_{j+1} < 1,\ j = 1,2,\ldots,k-1\right\}$.

The outward sequential procedure test S_j has an advantage over the usual Block test, $T_k = \frac{\sum_{i=1}^{k} x_{(i)}}{\sum_{i=1}^{n-n_0} x_{(i)}}$ of being little influenced by the behavior of the smallest observations. For $k = 1$, the critical value at α level of significance is $s_{1,(n-n_0)} = \frac{(n-n_0)-\alpha}{\alpha(n-n_0-2)+(n-n_0)}$.

21.5 Data analysis

In this section, we consider a real dataset on the NEFT data of the RBI to illustrate the usefulness and effectiveness of the proposed models in identifying the number of inliers and the estimator of the parameters on the distribution. A detailed description of the datasets and other details is given below. To estimate the number of inliers, say 'k' in the data set, maximize the likelihood function (say $L_k(\underline{x}) = \prod_{i=1}^{n} f(x_i)$) for fixed k successively and then determine $\max_{0 \le x \le} \ln L_k(\underline{\theta})^{0 \le x \le}$ and consider $\widehat{k}$ to be that where the likelihood is maximum. Another way of estimating the value of k is by testing the hypothesis about the number of inliers. Readers are advised to refer to various papers of the authors of this article for more details.

The RBI provides data on various aspects of the Indian economy, banking, and finance. The monthly bank-wise data for Electronic Clearing System (ECS), Real Time Gross Settlement (RTGS), and NEFT are available on the RBI website: https://rbi.org.in/Scripts/NEFTView.aspx. For the test, it is preferred to consider the data on an average outward debit of NEFT made in all the banks of India for the month of January-2018. The dataset is very large and, hence, it cannot be reported here. The RBI is an apex body of the Indian banks, evaluates the performance of all banks in terms of NEFT transactions, the volume of transactions, and e-banking related activities. The evaluation is conducted every year to classify banks as performing and nonperforming types. There are 187 banks mentioned in the data set and out of this, one bank ($n_0 = 1$) has an average outward debit of NEFT amount zero. The interest in this data is to know how many banks are nonperforming.

According to the sequential procedure, the observed test statistics are $S_1 = 0.7777132$, $S_2 = 0.477540$, $S_3 = 0.6904526$, $S_4 = 0.4283232$, $S_5 = 0.3020174$. These test statistic values are compared with the critical values are given in Table 21.1.

It can be seen that at a 10% level of significance the test for k up to 5, both S_5 and S_4 is not significant, but S_3 is significant. It indicates that as per the outward sequential procedure is given above, the first three observations are declared as inliers. It can be seen that at a 10% level of significance the test

TABLE 21.1 Critical values for the outward sequential procedure for up to k inliers and associated values of β.

$\alpha = 0.1$	$k = 5$	$k = 4$	$k = 3$	$k = 2$	$k = 1$
$S_{1,186}$	0.977234	0.972731	0.965118	0.950255	0.909491
$S_{2,186}$	0.823096	0.809199	0.788788	0.756235	–
$S_{3,186}$	0.659752	0.645729	0.626066	–	–
$S_{4,186}$	0.538164	0.526590	–	–	–
$S_{5,186}$	0.450919	–	–	–	–
β	0.023420	0.028181	0.036331	0.052617	0.909491

for k up to 4, S_4 is not significant, but S_3 is significant. It indicates that the first three observations are inliers. Whereas, S_3 is significant at a 10% level of significance in the test for k up to 3. It again indicates that the first three observations are inliers. Hence, an average amount of outward debits of National Australia Bank (1.476 NEFT amount), North East Small Finance Bank Limited (5.164071 NEFT amount), and Fino Payments Bank Limited (6.06917 NEFT amount) are declared as inliers (nonperforming banks). In all, the total number of nonperforming banks in the dataset is four including the Export-Import Bank of India, whose average amount of outward debt is zero.

The parameters estimates and the parametric function estimates are presented in Tables 21.2 and 21.3, respectively. The table also shows the standard error of estimators (given in the bracket) and the 95% confidence interval for the estimate of each parameter of the model.

TABLE 21.2 Summary of estimates of parameters of NEFT average amount of outward debits data.

	NEFT data	
	MLE (SE)	*95% CI*
π_{11}	0.00535 (0.00533)	(0.00000, 0.01580)
π_{21}	0.01604 (0.00919)	(0.00000, 0.03405)
θ	789.8033 (58.38394)	(675.37290, 904.23380)

TABLE 21.3 Summary of estimates of parametric functions of NEFT average amount of outward debits data.

Parametric function	UMVUE (Estimate of variance of UMVUE)
UMVUE of $\left[g\left(\theta\right)\right]^{-1} = \frac{1-p_1-p_2}{\theta}, (k = -1)$	0.00123 (8.54e-09)
UMVUE of pdf ϕ_x	ϕ_{100}= 0.00108 (5.14e-09) ϕ_{500}= 0.00066 (3.77e-10) ϕ_{1000}= 0.00035 (5.01e-11)
UMVUE of survival function at time x, S_x,	S_{100}= 0.86374 (0.00015) S_{500}= 0.52108 (0.00062) S_{1000}= 0.27650 (0.00068)

21.6 Inliers-prone distributions: Issues and problems

In literature, many authors have studied their problems taking inliers into account in various contexts: For example, the failure rate function of electronic products is characterized by the bathtub curve, where at the beginning; the failure rate is at a high level and then decreases with time in the early-failure period. After that, the failure rate remains at a stable level for a period, which is called the useful-life period. Then, the failure rate gradually increases again in the wear-out period. Cheng and Sheu [35] attempted to solve the problem of early failure occurrences in conducting parameter estimation for the Weibull distribution in a constant stress partially accelerated life test (CS-PALT) scheme. In the constant stress partially accelerated life test (CS-PALT) experiment, the total test units are first divided into two groups. The items of one of the groups are allocated to a normal condition, and the items of the other group are allocated to a stress condition. Each unit is run at a constant level of stress until the unit fails or is censored. In literature CS-PALT models with various types of distributions have been proposed, such as Weibull, Burr XII, Rayleigh, Exponential, and Inverted Weibull distributions.

In clinical trials, issues such as consent withdrawal, procedure misfit, etc., would rightly attribute to instantaneous failures. Survival data with instantaneous events are not uncommon in epidemiological and clinical studies, and for this reason, herein a general methodology under the proportional hazards (PH) model is developed for the analysis of interval-censored data subject to instantaneous failures. To analyze time to event data arising from clinical trials and longitudinal studies, etc., the event time of interest is not directly observed but is known relative to periodic examination times; i.e., practitioners observe either current status or interval-censored data. In some such studies the observed data also consists of instantaneous failures; i.e., the event times for several study units coincide exactly with the time at which the study begins. In light of these difficulties, Gamage et al. [36] focused on developing a mixture model under the proportional hazards (PH) assumptions, which can be used to analyze interval-censored data subject to instantaneous failures.

When the failure times are exactly observed, as is the case in reliability studies, it is common to incorporate instantaneous failures through a mixture of parametric models, with one being degenerate at time zero; e.g., see Muralidharan [8], Kale and Muralidharan [1], Murthy et al. [37], Muralidharan and Lathika [11], Pham and Lai [38], and Knopik [39]. In the case of interval-censored data, seen commonly in epidemiological studies and clinical trials, accounting for instantaneous failures becomes a more tenuous task, with practically no guidance available in the existing literature. Arguably, in the context of interval-censored data, one could account for instantaneous failures by introducing an arbitrarily small constant for each as an observation time and subsequently treat the instantaneous failures as left-censored observations. For the analysis of interval-censored data subject to instantaneous failures a new mixture model is proposed, which is a generalization of the semiparametric PH model studied by Wang et al. [40].

Financial institutions are at the core of the functioning of the economy, hence, to quantify their exposure to operational risk, banks should take into account the fact that there are huge differences between the behavior of the central part and the tail of the distribution of losses. This is especially true in the case of losses characterized by the so-called low frequency-high severity losses. Adlouni et al. [13] used the instantaneous failure model to estimate the extent of the exposure of a Moroccan bank to operational risk when zero losses are recorded.

Long series of days or decades without rainfall allows one to determine the probabilities of adverse developments in agriculture (droughts). This could be the basis for forecasting crop yields in the future. An instantaneous failure model describing the amount of precipitation and taking into account periods without rain is studied by Bojar et al. [41].

21.7 Future scopes

Since inlier models are nonstandard and incomplete mixtures, closed-form expressions for descriptive statistics are not available in general and hence, the usual inferences are not smooth. Judging a good inliers model itself can pose many problems of completeness and identifiability of the model. In some cases, a specific model for positive observations also poses challenges for estimation in flexible form. This kind of problem is also common in the testing of hypotheses concerning the parameters and detection of many inliers present in the model. In such cases, we resort to the numerical evaluation of the likelihood equations.

As such the literature does not include many studies on inliers-prone models. The least we can find in the literature are some studies on lower outliers (inliers) detection-related studies. They are also not articulated in a proper statistical way. The testing procedures when the number of instantaneous failures is unknown as well as there are more than three or more inliers also pose lots of problems in terms of getting the distribution of the test statistics. For more inliers, the existing procedures may not work well because of masking and swamping effects. This is a challenge to work with.

A very important limitation of inliers study is that, if the number of discrete points (zeros or ones) exceeds more than half of the total number of observations, then the estimation can create problems and in case estimators are available, then those estimators may not be sensible enough to interpret. Although we have not come across such practical situations, we are not ruling out.

Another scope for research is to study the generalization of the inliers-prone model when there are more than two discrete mass and continuous measurements. We strongly believe that there are practical situations where at many discrete points there can be discrete mass points and continuous measures.

There is potential scope for inferences based on the Bayesian approach for constructing a computationally strong model for inliers estimation and detection. Even censoring concepts need a relook into this kind of study.

References

[1] B.K. Kale, K. Muralidharan, Optimal estimating equations in mixture distributions accommodating instantaneous or early failures, J. Ind. Stat. Assoc. 38 (2000) 317–329.

[2] J. Aitchison, On the distribution of a positive random variable having a discrete probability mass at the origin, J. Am. Stat. Assoc. 50 (1955) 901–908.

[3] R.M. Kleyle, R.L. Dahiya, Estimation of parameters of mixed failure time distribution from censored data, Commun. Stat. Theory Meth. 4 (9) (1975) 873–882.

[4] V.P. Jayade, M.S. Prasad, Estimation of parameters of mixed failure time distribution, Commun. Stat. Theory Meth. 19 (12) (1990) 4667–4677.

[5] K. Vannman, On the distribution of the estimated mean from the nonstandard mixtures of distribution, Commun. Stat. Theory Meth. 24 (6) (1995) 1569–1584.

[6] B.K. Kale, Optimal estimating equations for discrete data with higher frequencies at a point, J. Ind. Stat. Assoc. 36 (1998) 125–136.

[7] B.K. Kale, Modified failure time distributions to accommodate instantaneous and early failures, in: J.C. Misra (Ed.), Industrial Mathematics and Statistics, Narosa Publishing House, New Delhi, 2003, pp. 623–648.

[8] K. Muralidharan, Tests for the mixing proportion in the mixture of a degenerate and exponential distribution, J. Ind. Stat. Assoc. 37 (2) (1999) 105–119.

[9] K. Muralidharan, B.K. Kale, A modified gamma distribution with singularity at zero, Commun. Stat. Simul. Comput. 31 (1) (2002) 143–158.

[10] K. Muralidharan, P. Lathika, Statistical modeling of rainfall data using modified Weibull distribution, Mausam, Ind. J. Meteorol. Hydrol. Geophys. 56 (4) (2005) 765–770.

[11] K. Muralidharan, P. Lathika, Analysis of instantaneous and early failures in Weibull distribution, Metrika 64 (3) (2006) 305–316.

[12] B.K. Kale, K. Muralidharan, Maximum likelihood estimation in presence of inliers, J. Ind. Soc. Prob. Stat. 10 (2006) 65–80.

[13] S.E. Adlouni, E. Ezzahid, Y. Mouatassim, Mixed distributions for loss severity modeling with zero in operational risk losses, Int. Appl. Math. Stat. 21 (11) (2011). 14 pages.

[14] A. Hazra, S. Bhattacharya, P. Banik, A Bayesian zero-inflated exponential distribution model for the analysis of weekly rainfall of the eastern plateau region of India, Mausam 69 (1) (2018) 19–28.

[15] L. Knopik, K. Migava, A. Wdzięczny, Profit optimization in operation systems, Polish Maritime Res. 23 (1) (2016) 93–98.

[16] R.G. Miller, Early failures in life testing, J. Am. Stat. Assoc. 48 (1960) 491–502.

[17] K. Muralidharan, M. Arti, Analysis of instantaneous and early failures in Pareto distribution, J. Stat. Theory Appl. 7 (2008) 187–204.

[18] K. Muralidharan, M. Arti, Inlier proness in Normal distribution, Reliab.: Theory Appl. 8 (1) (2013) 86–99.

[19] K. Muralidharan, Inlier prone models: a review, Probst Forum 3 (2010) 38–51.

[20] K. Muralidharan, P. Bavagosai, Some inferential studies on inliers in Gompertz distribution, J. Ind. Soc. Prob. Stat. 17 (2016) 35–55.

[21] K. Muralidharan, P. Bavagosai, Analysis of lifetime model with discrete mass at zero and one, J. Stat. Theory Pract. 11 (4) (2017) 670–692.

[22] K. Muralidharan, P. Bavagosai, A new Weibull model with inliers at zero and one based on type-II censored samples, J. Ind. Soc. Prob. Stat. 19 (2018) 121–151.

[23] K. Muralidharan, P. Bavagosai, A Pareto II model with inliers at zero and one based on type-II censored samples, Reliab.: Theory Appl. 13 (2018) 60–89.

[24] P.N. Jani, A.K. Singh, Minimum variance unbiased estimation in multi-parameter exponential family of distributions, Metro 53 (1995) 93–106.

[25] B.K. Kale, K. Muralidharan, Masking effect of inliers, J. Ind. Stat. Assoc. 45 (1) (2007) 33–49.

[26] J.R. Veale, Improved estimation of expected life when one identified spurious observation may be present, J. Am. Stat. Assoc. 70 (1975) 398–401.

[27] B.K. Kale, Trimmed means and the method of maximum likelihood when spurious observations are present, in: R.P. Gupta (Ed.), Applied Statistics, North-Holland, Amsterdam, 1975, pp. 177–185.

[28] V. Barnett, T. Lewis, Outliers in Statistical Data, third ed., John Wiley & Sons, New York, 1994.

[29] M.S. Chikkagoudar, S.H. Kunchur, Distribution of test statistics for multiple outliers in exponential samples, Commun. Stat. Theory Meth. 12 (1983) 2127–2142.

[30] A.C. Kimber, H.J. Stevens, The null distribution of a test for two upper outliers in an exponential sample, Appl. Stat. 30 (2) (1981) 153–157.

[31] T. Lewis, N.R.J. Feller, A recursive algorithm for null distributions for outliers: I. Gamma sample, Technometrics 21 (1979) 371–376.

[32] A.C. Kimber, Tests for many outliers in an exponential sample, Appl. Stat. 31 (1982) 263–271.

[33] B. Rosner, On the detection of many outliers, Technometrics 17 (2) (1975) 221–227.

[34] U. Balasooriya, V. Gadag, Test for upper outliers in the two-parameter exponential distribution, J. Stat. Comput. Simulat. 50 (3–4) (1994) 249–259.

[35] Y.F. Cheng, S.H. Sheu, Robust estimation for Weibull distribution in partially accelerated life tests with early failures, Quality and Reliability Engineering International, Published online on 25 November in Wiley Online Library, Published online on 25 November in Wiley Online Library, 2015, https://doi.org/10.1002/qre.1928.

[36] P.W.W. Gamage, M. Chaudari, C.S. McMahan, M.R. Kosorok, A proportional hazards model for interval-censored data subject to instantaneous failures, 2018. arXiv: 1804.00096v2.

[37] D.N.P. Murthy, M. Xie, R. Jiang, Weibull Model, Wiley, New York, 2004.

[38] H. Pham, C.D. Lai, On recent generalizations of the Weibull distribution, IEEE Trans. Reliab. 56 (3) (2007) 454–458, https://doi.org/10.1109/TR.2007.903352.

[39] L. Knopik, Model for instantaneous failures, Scient. Probl. Mach. Operat. Mainten. 46 (2) (2011) 37–45.

[40] L. Wang, C.S. McMahan, M.G. Hudgens, Z.P. Qureshi, A flexible, computationally efficient method for fitting the proportional hazards model to interval-censored data, Biometrics 72 (1) (2016) 222–231, https://doi.org/10.1111/biom.12389.

[41] W. Bojar, L. Knopik, J. Żarski, Integrated assessment of business crop productivity and profitability to use in food supply forecasting, in: FACCE MACSUR Mid-term Scientific Conference, Achievements, Activities, Advancement Sassari, April 01-04, 2014.

Chapter 22

Integration of TPM, RCM, and CBM: A practical approach applied in Shipbuilding industry

Rupesh Kumtekar, Swapnil Kamble, and Suraj Rane
Mechanical Engineering Department, Goa College of Engineering, Ponda, Goa, India

22.1 Introduction

Long-term success of any industry depends on how optimal they use their resources. This requires that machines are available at the desired time. Round-the-clock availability of machines is ensured by following the philosophies of effective maintenance strategy. Improper maintenance can affect the quality of products, delivery schedules, employee morale, and product cost, which eventually results in the deterioration of the brand image of the industry and product. Sincere implementation and continuous improvement in maintenance strategies is the need of the hour. This requires maintenance is given high priority as regards other activities in the product life cycle. Depending on the strategies adopted, maintenance can be classified into the breakdown, predictive, and preventive maintenance. Over the years systematic frameworks of maintenance like total productive maintenance (TPM), reliability centered maintenance (RCM), and condition-based maintenance (CBM) were also developed. Each of these maintenance approaches has its own strengths. Total productive maintenance case studies have been reported in various engineering applications such semiconductors [1], manufacturing [2], etc. Reliability centered maintenance has been applied on electric power distribution systems [3–5], gas transmission pipelines [6], maritime operations [7], etc. Emphasis on online health tracking of critical systems makes condition-based maintenance a widely used maintenance strategy and its applications have been reported in laser welding [8], ship pumps [9], etc. Attempts of integrating these stand-alone approaches have also been reported [10,11]. The scope of the work presented in this chapter includes the maintenance of equipment in stationary mode as well as the access of equipment to sensor mounting. The immediate

System Assurances. https://doi.org/10.1016/B978-0-323-90240-3.00022-9

limitation of the work is an increase in the cost of implementation, since it uses health monitoring gadgets and sensors.

The chapter is arranged as follows: Section 22.1 presents an introduction to the chapter. Introduction to maintenance and the proposed combined approach is presented in Section 22.2. Section 22.3 discusses the case study implemented using the proposed approach. The conclusion of the chapter is presented in Section 22.4.

22.2 Maintenance strategies

22.2.1 Total productive maintenance

Total productive maintenance (TPM) is a technique that optimizes the effectiveness of the machinery by reducing the breakdown maintenance and encouraging autonomous maintenance by the operator working on the machine. Total productive maintenance is a teamwork-based preventive and productive approach which involves all levels from the operator to the top executive of the company. Total productive maintenance has proven to be the most successful and effective technique to increase productivity. If the machinery is maintained properly and kept in good working condition, then there is the reduction of breakdown cost, accidents, and manpower. By employing the total productive maintenance concepts, the company can reduce the breakdown maintenance of their equipment. The success of total productive maintenance depends on factors such as employee empowerment, incentive policy, and teamwork. In total productive maintenance, the machine operator performs the routine maintenance including the checking of water, oil, coolant, air level, etc. This happens when required training and motivation are given to the operator. Total productive maintenance is a productive maintenance program that focuses on the following:

(1) Maximizing the overall equipment efficiency,
(2) Establish a planned maintenance system,
(3) Employee participation to initiate corrective activities, and
(4) Empowering employees to perform autonomous maintenance.

The eight pillars of total productive maintenance are focused improvement, autonomous maintenance, quality maintenance, planned maintenance, development management, education and training, health, environment and safety and office maintenance. Overall equipment effectiveness (OEE) is an important measurable metric for measuring the performance of the system. The overall goal of total productive maintenance is to increase overall equipment effectiveness.

22.2.2 Reliability centered maintenance

In recent years, different technologies have put world industrial organization under pressure, to deliver the product well in time and to avoid loss of production due to breakdown. To avoid stoppage of the production, the system or the equipment must be reliable and of high standards so as to achieve the target. In order to achieve this objective, there should be regular maintenance throughout the life cycle of the system. Reliability centered maintenance tries to improve the process and machine reliability, thus reducing life-cycle costs. For an effective reliability centered maintenance technique, we need to look into

(1) function and performance standards,
(2) failures, modes, effects, consequences of function, and
(3) proactive and default task.

22.2.3 Condition-based maintenance

Condition-based maintenance essentially utilizes testing techniques that are nondestructive, uses visual inspection and performance evaluation of the data to evaluate conditions of the machine. Condition monitoring maintenance task periods need to be properly decided and using normal potential-to-functional failure interval project task is obtained. The potential-to-functional interval monitors the frequency with which the predictive task is to be completed. Technological innovation and advancement are frequently implemented to condition-based maintenance structures, which includes better expertise of failure mechanisms, advancements in failure forecasting techniques. The measurement accuracy and criticality of the condition monitoring approach are used since they reduce the reaction time available to eliminate the consequences of the functional failure. The predictive maintenance program will offer factual data of the systems on a real-time basis. In addition to the actual mechanical condition of each machine, the overall performance of the process needs to be analyzed using past data such as mean time-to-failure, mean downtime, etc. To schedule maintenance activities, predictive maintenance utilizes results of each component's performance status as well as overall system status. This information will help to offer maintenance management the real statistics useful for powerful planning and scheduling maintenance strategies. Table 22.1 shows the advantages and disadvantages of the three maintenance approaches.

22.2.4 Proposed methodology

The strength of each of the above maintenance practices can be combined to develop the proposed integrated approach. This provides motivation to develop an integrated maintenance approach which will empower employees to continuously monitor the health of equipment in order to improve its reliability. The proposed methodology involves integrating reliability centered maintenance into total productive maintenance and condition-based maintenance into reliability centered maintenance, at relevant phases. The decision on the appropriate phase for integration can be taken based on the technical and managerial decisions. Based on the detailed literature survey reported in Section 22.1 along with the advantages of each maintenance practice reported in Section 22.2.3, the

TABLE 22.1 Advantages and disadvantages of TPM, RCM, and CBM.

Total productive maintenance	Reliability centered maintenance	Condition-based maintenance
Advantages		
1. Reduction in unplanned maintenance activities. 2. Reduction in equipment downtime. 3. Employees motivated towards undertaking autonomous maintenance. 4. Better synergy in resource handling.	1. Increase in system efficiency. 2. Reduction in maintenance cost. 3. Productivity improvement.	1. Online maintenance process, so no productive time wasted. 2. Reduction in unplanned maintenance activities. 3. Improved reliability of equipment.
Disadvantages		
1. Time-consuming process. 2. Expensive for small organizations.	1. Time-consuming process. 2. Continuous monitoring required.	1. Increased cost of training of employees. 2. Expensive monitoring equipment.

proposed research procedure in implementation of current research work is depicted in Fig. 22.1. The implementation begins with identifying the system for maintenance and its boundaries. Identification can be performed based on cost-value analysis of the system/component or time taken to repair. In the next phase, failure and associated costs are analyzed using techniques such as failure mode effect analysis. Then, the component-level or subsystem level condition monitoring can be implemented. Once the maintenance is completed, a comparison between prior and postmaintenance can be done to verify the effectiveness of maintenance. Once verified, controls can be set on the methods adopted for future use. The benefits of the approach are propagated to all levels of employees so that they are motivated to carry out maintenance effectively.

22.3 Reliability of marine propulsion system: A case study

22.3.1 Phase I: Preparation—System selection and identifying boundaries

The shipbuilding facility under study caters to the maintenance of components of propulsion system such as engine, gearbox, etc. The faster the failed system is

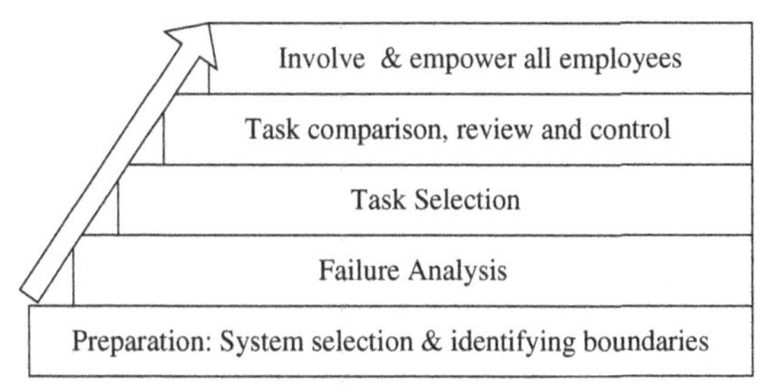

FIG. 22.1 Procedure adopted in the integrated approach.

brought to "as-good-as-new" condition, the higher will be the reliability of the system. The ships coming for overhauling at the maintenance facility can have either 100% failure or degraded performance. For any ship, the propulsion system is the heart of the system which is used to move the ship from one destination to another. In order to obtain propulsion system reliability and to avoid major breakdown of components, reliability issues which were observed in main engine, gear box, and stern gear, etc. of different series of the ship were studied. The purpose of the main engine is to turn the ship's propeller shaft and move the ship through the water. The main engine runs on diesel or heavy fuel depending on the make and model. The main engine consists of a piston, cylinder, piston ring, connecting rod, crankshaft, and valve bearing. The other interrelated subsystem to run the engine are the fuel system, air system, lubrication system, exhaust system, cooling system, control system, etc. Similarly, the gear box consists of gear, shaft, and bearing, and the other interrelated components are flywheel, lubrication system, cooling system, and transmission system which help in the smooth operation of the gearbox. The stern gear system consists of a shaft with the stern tube, propeller, coupling, and cooling system, which helps in performing its task of movement of propeller blades and turning of the ship while moving. Fig. 22.2 shows a schematic representation of a marine propulsion system.

The data of the number of failures of each component is taken, and the total defect that has resulted in the failure of the system is calculated, to get the probability of failure of each component. If 2214 defects were observed due to the failure of different components related to the main engine and the three defects noticed due to fuel contamination in fuel system failure.

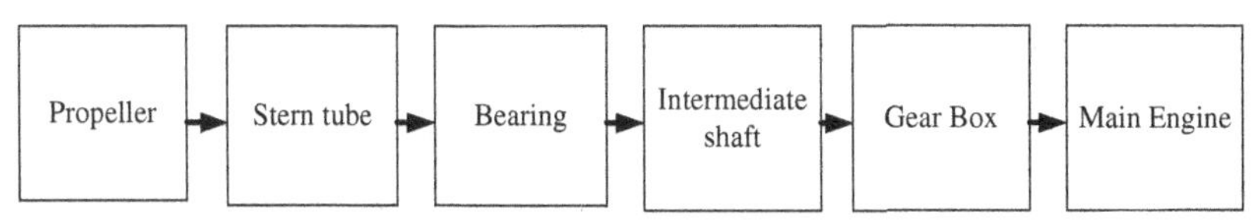

FIG. 22.2 Schematic representation of marine propulsion system.

Therefore, the probability of occurrence of failure due to contamination will be 0.0014. Similarly, other components were also calculated. Table 22.2 provides the worksheet for collating information regarding the number of times the component has failed in performing its function. This information is then used in fault tree analysis (FTA). Fault tree analysis is a graphics tool used for estimating the failure probability of a system. The tree represents failure events and their interdependence relationships. The qualitative and quantitative analysis are involved. Qualitative analysis involves combining failure events using Boolean symbols, and quantitative analysis computes the probability of failure of the system by using expression developed in qualitative analysis. Using fault tree analysis, reliability of main engine was found to be 93.14%, reliability of gear box was 93.97%, and reliability of marine propulsion system was 84.11%. The reliability goals can then be set based on these values. The reliability target was set at 95% for main engine, gear box, and marine propulsion system.

TABLE 22.2 Failure probabilities worksheet of the marine main engine, gear box, and stern gear system.

Category of defects	Description	Nos. of defects	Probability of occurrence
(A) Main engine system			
Category 1	Piston, cylinder, piston ring, connecting rod, crank shaft, valves, bearing failure	6	0.0027
Category 2	Fuel system failure	..	..
2.1	Fuel contamination	3	0.0014
2.2	..	..	..
Category 3	..	..	..
3.1	..	..	..
(B) Transmission system (gear box)			
Category 1	..	..	..
..	..	..	..
(C) Stern gear system			
Category 1	..	..	..
..	..	..	..

22.3.2 Phase II: Analysis

Table 22.3 shows the failure mode effect analysis of main engine where two components considered are fuel system and air system. Initially, the identification number is given to the system or the components and its purpose is stated. The components are analyzed to know what type of system failure and its effect could occur. This approach helps in identifying the possible cause of failure and its effect which can be addressed before failure occurs. The FMEA can be modified based on the age of the equipment or the process change.

The following are the steps taken from the RCM program:

(i) Define specifically and correctly the asset's reliability criteria: This includes all unwanted consequences of failure that can happen in the workplace that must be prevented. Personal and workplace safety, power minimization, loss of production capacity, delay in production, environmental issue, etc. should be taken into account as they are important criteria. These criteria must be approved by the top management and the departments of the organization.

(ii) Implementation and understanding of the reliability centered maintenance consequence failure analysis: The consequence of failure analysis (COFA) logical tree is a complex, logical, and accurate analysis that simplifies analysis to its basic elements. Following the potentially critical guidelines and using the logic tree, the component's economic value is classified, which is then used by reliability centered maintenance program if the corrective maintenance order is high.

(iii) Consequence of failure analysis worksheet: logic tree and the critical guidelines are integrated. The consequence of failure analysis describes all functions of components, functional failure, and failure modes. Table 22.4 shows the worksheet of the consequence of failure analysis.

22.3.3 Phase III: Task selection

A condition-based maintenance strategy is applied at every specified interval. If the engine goes through periodic maintenance, the occurrence of random failure can be reduced. Condition-based maintenance check sheet for A, B, C, and D checks is illustrated in Table 22.5.

Condition-based maintenance frameworks workaround real-time sensor monitoring technologies offering adaptability and cost-savings. Data logger helps to understand the performance of the engine. Data loggers check parameters over time, records, and analyses and perform validations. The results showed improved engine performance which in turn increased the reliability of the engine.

The important metric used in total productive maintenance is overall equipment effectiveness. The higher the overall equipment effectiveness, the better

TABLE 22.3 Failure mode effect analysis.

Identification No.	Item	Function	Failure mode and causes	Operational mode	Local effect	Higher effect	End effect	Failure detection method	Effect
01	Fuel system	Transfer of fuel	Leakage or clogged	Fuel oil supply less	Fuel losses	Fire hazard	Danger of explosion	Direct observation	Yes
02	Air system	To start the engine	Leakage or noise	Air supply less	Air losses	Lack of firing	Stopping of engine	Pressure gauge	Yes

TABLE 22.4 Consequence of failure analysis worksheet.

Sr. No	Component ID and description	Describe all functions of that component	Describe the way each function can fail	Describe the dominant component failure modes for each function failure	Is the occurrence of the failure mode evident?	Describe the system effect for each failure mode	Describe the consequence of failure based on assets reliability	Define the component classification
1	..	..	..	..	..	..	..	..
2	..	..	..	..	..	..	..	..

TABLE 22.5 CBM schedule for marine engine.

A Check	B Check	C Check	D Check
Daily/ weekly inspection	Repeat A check	Repeat A and B	Repeat A, B, and C
Lubrication	Change oil/filter/ bypass filter/ marine gear oil	Fuel system	Fuel system
Fuel system	Fuel system	Change governor oil/clean tank	Clean and calibrate injectors/ replace rocker cover gaskets/ check fuel calibration/replace aneroid bellows
Air system	Check throttle link/ball joint/ filter		Air system
Cooling system	Air system		Clean turbocharger compressor wheel and diffuser/check turbocharger bearing clearance/ tighten manifold nuts of cap screws
	Clean crankcase/air piping		Cooling system
	Cooling system		Change coolant/descale cooling system
	Coolant inhibiter/zinc plugs/belt adjustment		
			Check vibration damper/ Check air compressor/Check safety controls

will be the maintenance strategy employed. The expressions of overall equipment effectiveness and related metrics are given below:

$$\text{OEE} = \text{Availability} \times \text{Performance} \times \text{Rate of quality} \tag{22.1}$$

$$\text{Availability} = \frac{(\text{Total operating time} - \text{Total downtime})\ \text{x}\ 100}{\text{Total operating time}} \tag{22.2}$$

$$\text{Performance} = \frac{(\text{Cycle time x Output of the equipment}) \text{ x } 100}{\text{Operating time}} \tag{22.3}$$

$$\text{Quality rate} = \frac{(\text{Processed quantity} - \text{Defective quantity}) \text{ x } 100}{\text{Processed quantity}} \tag{22.4}$$

22.3.4 Phase IV: Task comparison, review and control

Cost-benefit analysis is performed in order to observe the benefits such as cost savings, reduction in maintenance times, troubleshooting, and fault detection at early-stage diagnosis, accuracy and performance, actual safety, and knowledge of safety.

Table 22.6 illustrates the cost comparison break up for current practices followed at the facility and the proposed approach. If the maintenance of the engine is done as per the recommendation, then it will help to save on the cost by 50%. If the same was resolved or troubleshooted by reactive maintenance, the cost required would be $ 18,500 but with the proposed approach engine is restored for $9300. The frequency of maintenance as well as the costs and time associated with maintenance is reduced. The time saved by using the integrated approach is 75% which helps to complete the operating mission at the set target time. This helps to achieve better performance and efficiency. If the same was resolved or troubleshooted by reactive maintenance the time required would be 357 h, but with an integrated approach, the engine is restored for 90 h. At this stage, the reliability was evaluated for main engine, gearbox, and marine propulsion system and was found to be 96.2%, 95%, and 95.8%, respectively, which was meeting the target set initially.

Table 22.7 illustrates the comparison of overall equipment effectiveness for traditional maintenance practice currently followed and integrated approach. Overall equipment effectiveness has almost doubled with the integrated approach, thus showing the potential of the approach in achieving maintenance objectives.

TABLE 22.6 Cost comparison of current approach and integrated approach.

Failed component	Breakdown/corrective maintenance	Integrated approach
Connecting rod	$18,500	$9300
Liner piston assembly	$9300	
Cylinder block	$4600	

TABLE 22.7 OEE comparison of current methodology and integrated approach.

	Current approach (%)	Integrated approach (%)
Availability	60	86
Performance efficiency	70	82
Quality rate	79	94
OEE	33	66

The following benefits are achieved with the proposed approach:

(1) Savings in parts costs and breakdown reactive maintenance after effects.
(2) Increase in overall operational time, and
(3) Enhanced safety and employee morale.

The benefits of the proposed approach are observed for a longer time so that control on maintenance procedures can be set. The controlled document then acts as a standard operating procedure for future maintenance activities of a similar type.

22.3.5 Phase V: Involve and empower all employees

Most of the shipbuilding industries have preventive and breakdown maintenance for their equipment inside the yard. Based on the study carried out in shipbuilding industry and information gathered, the following steps can be introduced for the successful implementation of maintenance in shipbuilding.

(a) Collection of data regarding the machinery

Initially, the list of equipment or the machinery is to be listed down with details of failure rate and time to repair, so as to create a database regarding the reliability of the equipment. If the data is not available then a system of data collection should be established. This will help in identifying the critical areas in the equipment and plan for the prevention of failure of the equipment.

(b) Training program and overall equipment effectiveness

Training of the employees boosts their morale that will result in higher efficiency during maintenance work. The performance of the equipment is measured through computation of overall equipment effectiveness.

(c) Different methods for reducing the time loss

The manager needs to focus on the improvements that are possible to achieve by means of minimizing the time loss in production for cleaning of the machine, implementation of work, and adjustment. This will help in determining the action plan for reducing the losses due to the failure of the equipment and also to increase the quality.

(d) Obtain opinion of the employees

The views of the employees concerned with the operations and maintenance of the equipment need to be taken while framing any new and upgraded maintenance policy. This will give a feeling of empowerment to them, which will help in achieving the goal of autonomous maintenance.

22.4 Conclusion

The chapter presented a work that aimed at integrating total productive maintenance, reliability centered maintenance, and condition-based maintenance in order to derive benefits of the strength of each of these methodologies. The structured way of implementing this approach will necessarily benefit practitioners implementing this in the industrial world. The integrated approach is applied within the framework of total productive maintenance. The approach begins with identifying the system, subsystems, and system boundaries. The reliability of subsystems is then computed using fault tree analysis to know the current level of system performance. This is followed by failure mode effect analysis to develop a deep understanding of failure mechanisms. Subsequently, reliability centered maintenance program is implemented. This is followed by condition-based maintenance. The integrated approach is evaluated using overall equipment effectiveness. The cost and overall equipment effectiveness are compared for the currently followed methodology and proposed an integrated approach. There were savings in the cost and improvement in overall equipment effectiveness. The reliability of subsystems was also exceeding the target set in the initial phase. The approach can still be refined further by monitoring the health of the system using sensors and an automated maintenance approach. Artificial intelligence and machine learning techniques can also be integrated into maintenance strategies.

References

[1] F.T.S. Chan, H.C.W. Lau, R.W.L. Ip, S. Kong, Implementation of total productive maintenance: a case study, Int. J. Prod. Econ. 95 (2005) 71–94.

[2] M.C. Eti, S.O.T. Ogaji, S.D. Probert, Implementing total productive maintenance in Nigerian manufacturing industries, Appl. Energy 79 (2004) 385–401.

[3] J.H. Heo, M.K. Kim, J.K. Ly, Implementation of reliability-centered maintenance for transmission components using particle swarm optimization, Electr. Power Energy Syst. 55 (2014) 238–245.

[4] D. Piasson, A.A.P. Bíscaro, F.B. Leão, R.S. Mantovani, A new approach for reliability-centered maintenance programs in electric power distribution systems based on a multiobjective genetic algorithm, Electr. Pow. Syst. Res. 137 (2016) 41–50.

[5] B. Yssaad, A. Abene, Rational reliability centered maintenance optimization for power distribution systems, Electr. Power Energy Syst. 73 (2015) 350–360.

[6] K. Zakikhani, F. Nasiri, T. Zayed, Availability-based reliability-centered maintenance planning for gas transmission pipelines, Int. J. Press. Vessel. Pip. 183 (2020) 104105.
[7] A.J. Mokashi, J. Wang, A.K. Verma, A study of reliability-centred maintenance in maritime operations, Mar. Policy 26 (2002) 325–335.
[8] M. Kenda, D. Klobcar, D. Bra˘cun, Condition-based maintenance of the two-beam laser welding in high volume manufacturing of piezoelectric pressure sensor, J. Manuf. Syst. 59 (2021) 117–126.
[9] C. Duan, Z. Li, F. Liu, Condition-based maintenance for ship pumps subject to competing risks under stochastic maintenance quality, Ocean Eng. 218 (2020) 108180.
[10] M. Ben-Daya, You may need RCM to enhance TPM implementation, Mainten. Eng. 6 (2000) 82–85.
[11] M. Braglia, D. Castellano, M. Gallo, A novel operational approach to equipment maintenance: TPM and RCM jointly at work, J. Qual. Maint. Eng. 25 (2019) 612–634.

Chapter 23

Revolutionizing the internet of things with swarm intelligence

Abhishek Kumar[a], Jyotir Moy Chatterjee[b], Manju Payal[c], and Pramod Singh Rathore[d]

[a]Department of CSE, Chitkara University Institute of Engineering and Technology, Chitkara University, Baddi, Himachal Pradesh, India, [b]Department of IT, LBEF, Kathmandu, Nepal, [c]Software Developer Academic Hub, Ajmer, Rajasthan, India, [d]Department of CSE, ACERC, Ajmer, Rajasthan, India

23.1 Introduction

Interesting behaviors are utilized in nature to solve different challenges and they provide an attractive source of ideas for solving actual global problems. The SI method is based on calculations that concentrate on the combined conduct of centralized and self-structured systems to maintain artificial intelligence systems [1]. The following points should be kept in mind for the SI method:

- This method simulates behaviors seen in nature in the activities of some animals and insects, including termites, ants, fish, and birds. The SI method is described through developing local interactions between different individuals, and thus produces intelligent behavior of individuals that relates to different groups [2].
- There are various different SI-based algorithms that have been projected for the management of different problems and have been applied successfully. New problems have been raised requiring increases in efficiency, and this chapter is based on the analysis of different applications and provides brief solutions for problems in a shorter time.

This paper helps the reader to immediately focus on their research by offering them easy access to the associated text. In addition, this paper can be used as a preliminary reading point for examining several SI utilized methods and associated internet-utilized applications.

IoT is the network planned to set the standard according to which each thing around us, including traffic lights or water supply pumps, operates; IoT

System Assurances. https://doi.org/10.1016/B978-0-323-90240-3.00023-0

will be used to convert them into smart objects, with the capability of sensing, processing, and communicating, and thus always being connected for the purpose of efficient management. The field is related to research that may include digital and physical entitles, including both objects and humans, that are interconnected by the internet and so facilitate the grouping of services and applications [3]. Based on the particular realization, such groups of applications and services may provide major challenges to overcome certain problems that need to be based on internet systems, as they are composite and active in nature. However, due to their robustness and elasticity, swarm intelligence may provide a successful design pattern for evaluating different algorithms to deal with these complex problems based on internet systems. In this manner, easy principles can be applied to a person's behaviors and to communication between persons to achieve the best possible goal at the organizational phase. This self-association capability is required to adjust various systems according to the different environmental circumstance in order to efficiently speed up the processing and, thus, provide a flexible way to achieve the sustainability of the system [4].

Thus, internet-related networks provide the capacity for many different applications that contribute considerably to improve our everyday life across many different sectors, such as collecting the medical information of patients utilizing internet-based medical devices; remote monitoring of soil standards such as temperature, humidity, etc. [5] in actual time to increase harvest production; ecological monitoring related to the environment; monitoring of power systems; waste management; water monitoring; disaster alleviation; road and traffic management; and so on. In addition to these applications, internet-based networks should encompass a broad range of applications; however, there are certain boundaries that can restrict these resources, which must be taken into consideration at both the software and hardware stages [6]. Furthermore, there are many limitations, including the features of an individual node, like restricted memory and processing, which thus impair the behavior of the network; these may include different topologies that can alter and constrain the latency provided at the application level. There are many constraints that are directly dependent on the application and most of them are considered as for long term.

Due to the increasing scientific advancement applicable to various applications, there is rapid development in the research field associated to the internet. The heterogeneity of devices and systems is based on the internet, which produces a number of limitations, especially in the research area on the hardware, software, and network phases. Significant research is being done in the device phase, thus providing developers with different solutions to many applications; yet in the majority of cases, there is a lack of consistency amongst manufacturers. Therefore, for the growing number of applications at the level of the organization, the majority of the effort is focused on various resources management techniques, including cluster-based routing, data management, etc. [7]. Although the diversity of internet-based devices enables information to be

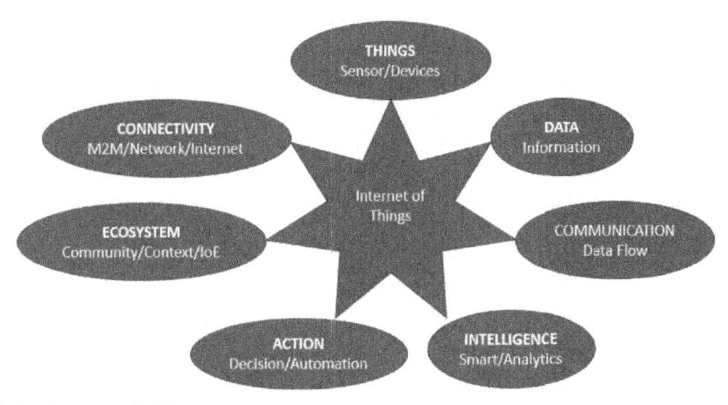

FIG. 23.1 Internet of things.

integrated, connected, and verified to match the people's requirements, virtualization of the internet-based network at the device and system levels also plays a major part in organizing the restricted assets of the IoT (Fig. 23.1) [8].

23.2 Characteristics of IoT

The IoT possesses seven characteristics features, as follows:

(1) **Connectivity**. It does not require more ahead clarification. The sensors, devices are required to be linked: for an item, for one another, actuators, a procedure and for "the internet" or additional types of network.
(2) **Things**. Everything which can be tagged or linked, as like as it is created for linked. From sensors and home equipment to livestock tagged. The devices can have sensors or sensing equipment may be connected to gadgets and objects.
(3) **Data**. It is the glue of the IoT. it is the primary phase for action and intelligence.
(4) **Communication**. There are some types of devices will be connected because it can help for communicate with the information and this information could be examined.
(5) **Intelligence**. The phase of intelligence as in the sensing strength in the IoT devices and the intelligence which is collected by the data analytics (as well as artificial intelligence).
(6) **Action**. It is a result of intelligence. This result may be manual action, action depended upon discussion regarding event (e.g., in weather alteration choices) and automation, often the most crucial part.
(7) **Ecosystem**. The space of the IoT from a paradigm of other communities, technologies, objective and the images in which the IoT enable. The internet of everything dimension, the stage dimension and the required for solid cooperation.

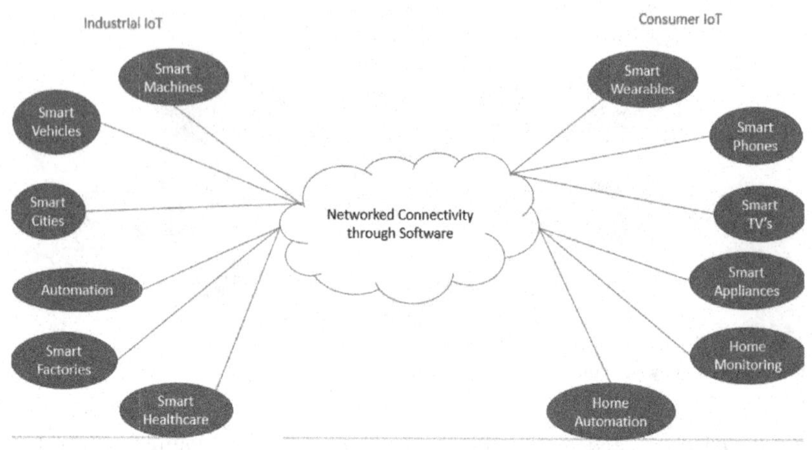

FIG. 23.2 Network connectivity in terms of industrial IoT and consumer IoT.

The IoT encompasses a number of related systems, including the IoTDS (internet of things detection system), IoTS (internet of things and services), IoMT (internet of medical things), IoHT (internet of hospital things), and loTSP (internet of things, services, and people) (Fig. 23.2).

23.3 The consumer IoT

The consumer internet of things (CIoT) is where you can search applications to track individual "components" (components tracking), ranging from your pet to our skateboard, or connect to associated "smart equipment" like a connected refrigerator, light bulb, or washing machine, etc. Wearable devices can be used in this way, and can also be applied for healthcare and manufacturing, to name just two. Whole types of consumer electronics, like smart wrist wear, are of this class, as well as whole types of smart home equipment, such as thermostats or linked parking gate openers. Apps for this type of device have already improved and are becoming smarter. They are also becoming much more independent than those for other types of devices. This is particularly the case with smart wearables.

Consumer IoT has a simple definition: IoT as used for user applications and user-oriented services. Generally, in consumer IoT, the requirements of data volumes and data communication are limited and lower than in other IoT categories. This is the reason that there are many types of technologies—some technologies are especially created for user applications. It is from the smart home connectivity standards to a particular operating system. Ref. [9] explains the elements that accompany the development of IoT in the furnishings and kitchen producing industry. Ref. [10] attempted to give a reasonable, progressively critical understanding of the IoT in big data (BD) structure, related to its issues and challenges, and zeroed in on providing potential game plans utilizing a machine learning (ML) framework. Ref. [11] utilized IoT for building up a framework dependent on IoT which will enable us to check the status of Gas Knob and could spare us from the gas leakage. Ref. [12] presented the latest advancements on IoT and IoC of disseminated articles from 2009 to 2017.

23.4 The industrial IoT

The term "industrial IoT" (IIoT) covers cases of specific industry use in many areas. Some people regard industrial IoT as applying specifically to "heavy" industries, such as manufacturing or utilities; however, the term is also useful in cases such as smart metering and smart cities. We also see it in the form "business IoT," which understands that there are some overlaps with consumer IoT. For example: suppose your home has a smart thermostat and smart energy consumption meter; they, are on the one hand, user applications, as they are for individual use, but on the other hand, from the point of view of the energy supplier, that application it to help sales and improve energy consumption, which is a business use (Fig. 23.3).

23.5 IoT definitions by various companies

Here, we will try to explain the IoT more systematically. The definitions will go from the network, data capture and connected things section in the real result. We will present modern words like the Internet of Services, etc.

- According to Gartner IoT is the network of physical objects which is known as the IoT. It is consisting the embedded technology. These technologies are used to sense or interact and communicate by its internal phase or the external nature. According to the IDC that explain the IoT like a network of specific identifiable final points (or things). Which communicates without interact with human through IP connectivity-whether it is local or world class are on it (https://www.gartner.com/en/information-technology/glossary/internet-of-things).
- In terms of Bosch there are many types of sources available in internet like file sharing, social media and e-commerce for connecting things and devices. After these sources, the next generation of the internet which is IoT. IoT is used for connecting things and devices. These types of devices range to sensors and security cameras to production and vehicles machines. Connecting the devices gets data which opens recent review, industries models and income streams. In turn, the review obtained by this data lead to modern services which may complement the traditional items industry (https://docs.bosch-iot-suite.com/things/things-intro/).
- According to the IBM (Watson), IoT which refers to the increasing range of devices connected to the internet, which captures or originate large amounts of information always. For users, these types of devices comprise sports wearable, mobile phones, air conditioning systems and home heating and

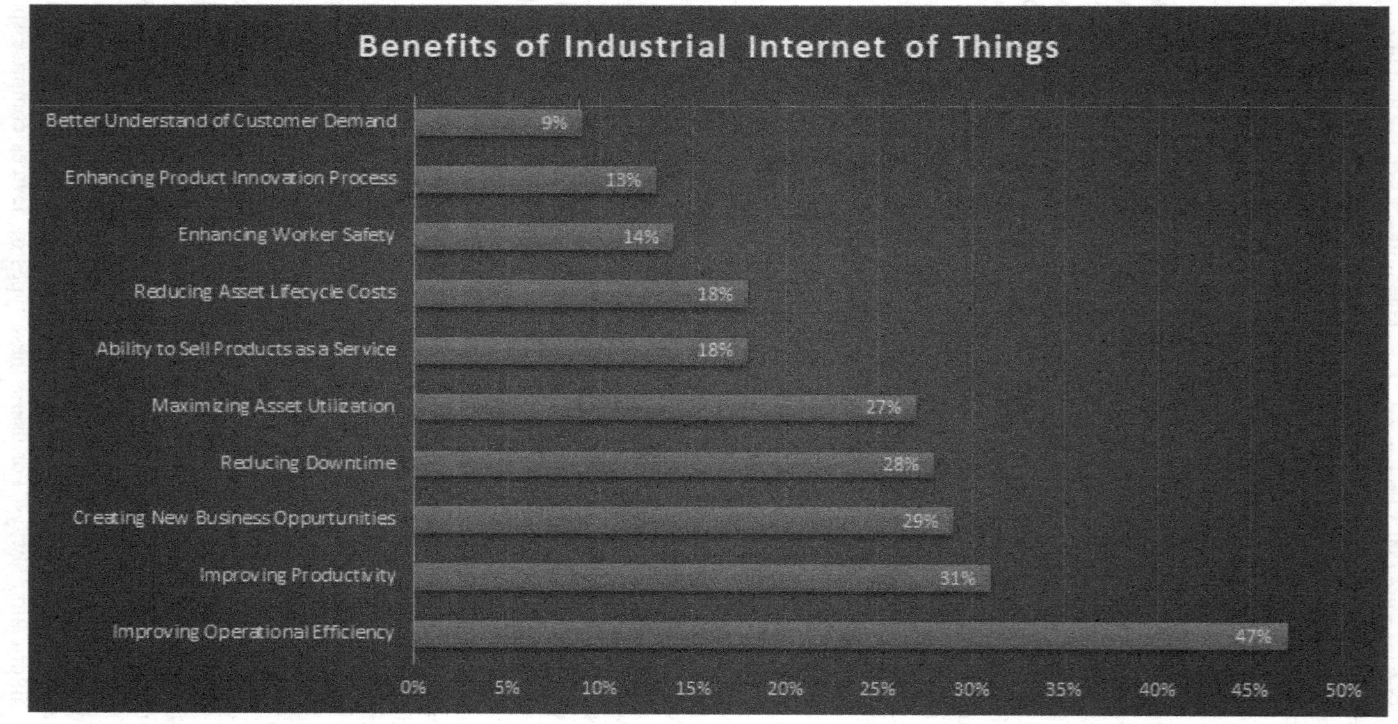

FIG. 23.3 Benefits of industrial IoT.

much. In a business setting, these type of devices and sensors may be established in rebuild appliance, vehicle components, and supply chains (https://www.ibm.com/cloud/internet-of-things).

- In terms of Google, IoT is a proposed evolution of the internet that has network connectivity in everyday objects. It is giving permit them to send and receive data (https://www.iotworldtoday.com/2021/10/18/google-cloud-iot-strategies/).
- According to the Cisco, IoT is an intelligent connectivity of smart devices, which is hope to take big advantages of the capacity, industry development and attribute of life. that is to say, when the objects can understand one another and interact, then these modify how and where and who confirm about our physical globe (https://www.cisco.com/c/en/us/solutions/internet-of-things/overview.html).
- The system is a collection of items which is available in the physical globe. Sensors are fixed to its devices and are linked to the internet through wired or wireless internet connectivity. These types of sensors can utilize different kinds of LAN like: Wi-Fi, NFC, RFID, Zigbee, and Bluetooth. Sensors can keep the wide area connectivity like GPRS, GSM, LTE and 3G etc. according to Lopez Research (http://www.inxero.com/resource/20788-report-lopez-research-introduction-to-iot/social).
- According to the Guardian, in these basic, IoT is easy—These is regarding connecting devices on the internet, so that they can communicate with us, usage, and one another (https://www.theguardian.com/technology/2015/may/06/what-is-the-internet-of-things-google#:~:text=At%20its%20core%2C%20IoT%20is,%2C%20applications%2C%20and%20each%20other).
- On the very essential phase, "IoT" imply devices it can understand about the actual globe fact: such as lighting, temperature, when the absence or presence of individuals or objects, so on—and report that actual globe data, or function at this. At the internet begin produced and consumed through individuals (audio, text, video) instead of most data. Using the machines maximum information will be produced and consumed, communication between us (hopefully) reform the merits of our lives according to Digital Trends.
- Financial Times said that presently, the internet is a computers network with an identity label that has a unique number which is known as Internet Protocol address. The view of the IoT is at connect small devices to each single item to create this identifiable through its self-unique Internet Protocol address. These types of devices can later communicate with one another autonomously (https://www.ft.com/stream/e50ca0ae-2716-428a-8690-be169e5d9e7f).
- According to IERC, IoT is an active global network structure with the self-sustaining abilities. It is dependent at the classic and compatible communication protocols. In this protocol includes the physical and virtual "things" are identified, physical properties and virtual individuality and utilized smart

interfaces and are organized in the data n/w (http://www.internet-of-things-research.eu/about_iot.htm#:~:text=The%20IERC%20definition%20states%20that,use%20intelligent%20interfaces%2C%20and%20are).

- ITU IoT GSI defined IoT as a global infrastructure in the recommendation ITU-T Y.2060 (06/2012). It is defined for information society. It is depended on existing and developed intraoperative information, interconnecting (physical and virtual) Enables to upgrade things and communication technology. IoT creates totally use of things to supply services for all types of application using exploitation of identity, data capture, executing and communication abilities. While it makes sure about the security and privacy needs are met. From a vastly perspective, IoT are considered as a vision with technological and social implications (https://www.itu.int/en/ITU-T/gsi/iot/Pages/default.aspx).
- Forbes Simply put, it is basically the idea of connecting any type of devices to the internet (and/or with one another) with an on and off switch. It is consisting coffee makers, cell phones, headphones, washing machines, wearable tools, lamps, and nearly everything you can consider of. It also enforces to tools of machines, For instance: An airplane jet engine or oil rig drill. like I described, suppose there is an available on and off switch in it, it is likely that it may be a piece of the IoT (https://www.forbes.com/sites/forbestechcouncil/2021/07/01/integrating-the-industrial-internet-of-things-the-benefits-and-challenges/?sh=25db3d22306f).
- According to Margaret Rouse, The IoT is a combination of computing gadgets, objects, automated digital gadgets, animals or public which supply with unique identifiers. It is capable to transmit data on the n/w without need of human-computer interaction or human-to-human (https://internetofthingsagenda.techtarget.com/definition/Internet-of-Things-IoT).
- According to PC Magazine Encyclopedia, connecting the physical global to a mobile devices or computer using the internet. It is consisting the door locks, home equipment's, thermostats, doorbells, security cameras, heating and air conditioning (https://www.pcmag.com/encyclopedia/term/internet-of-things).
- According to SAP, it is a massive network of connected devices through internet. It is consisting the tablets and smart phones, and nearly everything with sensors at this—machines, cars in product devices, oil drills, jet engine, wearable tools, and much. These types of "things" accumulate and transfer data (https://www.sap.com/insights/what-is-iot-internet-of-things.html).
- As per SAS, the IoT is the collection of per day things—From commercial machines to wearable devices, through the built-in sensors, it is collecting the data and take action on this data at a network (https://www.sas.com/en_us/software/analytics-iot.html#:~:text=SAS%20Analytics%20for%20IoT%20enables,Viya%20that%20streamlines%20ETL%20tasks).
- As per Tech World, IoT is an consist that environment which can communicate the physical items over the web. In which steady devices, peoples,

white goods and animals may send and receive data using the web. In which every IoT devices or items consist the specific identifier, and may interact without any individual interference.

- According to Wikipedia, the IoT is refer to the internetworking of physical devices, vehicles, buildings and another objects. It is also including the connected devices and smart devices. This physical device is embedded with software, electronics, actuators, sensor, and network connectivity, which capable those items for integrated and exchange of data (https://en.wikipedia.org/wiki/Internet_of_things).
- As per IoT-A with the RFID technology, the concept of mutual continuity of devices, objects and things in general arises in the global form, and this idea has been extended to the present vision, which envisages a number of ideas of odd objects communicate with the physical environment.
- As per McKinsey and Company, in what is known the IoT, actuators and sensors which embedded in physical items. It is embedded with the physical devices from road way to pacemakers. This physical device connected via wired and wireless n/w. Mostly through the similar IP which connects the internet. These networks churn up large amounts of data flowing into the computer for analysis. The IoT is a computing idea. It is illustrating a about the coming time; whither physical items will be linked at the internet every day and will be capable to identified ourselves to another devices. This word is nearby identified with the method of communication as the RFID, though this also can consist another QR codes, wireless technologies, and sensor technologies etc (https://www.mckinsey.com/featured-insights/internet-of-things/our-insights).
- As per Webopedia, the IoT referenced to each developing network of physical item. Which advantages an Internet Protocol address to internet connectivity? The communication which occurs amid its items and another internet capable devices and systems. It is elaborate internet connectivity. It is elaborate internet connectivity beyond traditional devices like tablets and smartphone, laptop and desktop computers at different area of devices and per day items which use embedded technology to cooperate and communicate with the outer environment, totally through the internet (https://www.webopedia.com/definitions/internet-of-things/).

23.6 The industrial internet

The term "industrial internet" was first used by General Electric (GE), a major player in industry and founder of the Industrial Internet Consortium. Industrial internet is defined by GE as: "Global industrial system linked with advanced computing power, low power sensors, analytics and layers of connectivity which allowed through the internet." Therefore, industrial IoT aka IIoT completely plays an important role in this.

23.7 The internet of everything (IoE)

The "internet of everything" (IoE) is a phrase coined by Cisco as a term to describe the network of connected devices but also to incorporate individuals and processes. There have been small changes in the last few years in the definition of IoE, with Cisco using the current definition: "(IoE) phrase coined by Cisco describe the network of connected devices but also to incorporate individuals and processes." Cisco ceased using the term IoT after 2015–16.

23.8 Cyber physical systems (CPS) and industry 4.0

Another significant term is "industry 4.0." Although this has much in common with business IoT and with GE's industrial internet, the IoT also makes a big contribution to it. A construction block of industry 4.0 is cyber physical systems (CPS). This is a similar to industrial IoT, as has been said by many people; however, they are not equivalent, even if what is meant is fast industrial IoT. Industry 4.0, in the same way as industrial internet, occurs through the commercial marketplace but, when developed by GE, German industry used the term industry 4.0. Two definitions of industry 4.0 are: "In information-depth change of manufacturing in a connected environment of people, information, services, processed, production tools & systems with leverage, generation and utilization of actionable information like a means to see the smart industry & recent manufacturing ecosystems." "One is the Intensive change of information in a connected environment with the use of actionable information with the production, production and use of people, data, processes, systems, services and production tools. Realize the smart industry and recent manufacturing ecosystem (https://www.forbes.com/sites/bernardmarr/2018/09/02/what-is-industry-4-0-heres-a-super-easy-explanation-for-anyone/?sh=75c0cb279788 and https://ptolemy.berkeley.edu/projects/cps/)." Much concerning industry 4.0, It is defining (inclusive the series of industry 4.0 technicality of another) and the definition of CPS.

23.9 The internet of services (IoS)

The internet of services represents a trend within the IoT: What matters ultimately is services. In addition, organizations are increasingly using the IoT to create services and to move to an "as-a-service" structure. That is where the IoS comes in. The term IoS is used primarily in reference to the industrial internet and industry 4.0. In addition, this necessarily affects other regions too, where the end results of an IoT project or customer application are a service. The term IoS is occasionally used by many institutions in various references, including the European Union and the leading German industry 4.0 software firm SAP. It is also used by research industries; the Institute of Electrical Electronics Engineers (IEEE) has released a paper mentioning this. Even so, none of these bodies actually defines the word; however, we are sure that you get the meaning.

23.10 The internet of robotic things (IoRT)

The term "internet of robotic things" (IoRT) was established by ABI Research, and is not much more than a discussion word, indicating what the IoT currently is and how it may develop. All of a sudden, we are looking at the probabilities of the IoT, not in a rather inactive way, but where data from physical items is capable of informing services, insights, and so on. Although data is important and we have described this as an IoT, there is tremendous potential for more complete services—there is much more. It is clear that the IoT isn't a one-way street. Although until now we have not used IoT tools in analytical tasks, with the addition of consolidation probabilities with other systems, internet-capable physical devices can be employed—at least partially autonomously—to do a lot of work. IoRT concerns these phases in terms of robots (such as robot utilization by Amazon in their logistics). As suggested by ABI Research, IoRT indicates the inclusion of a robotic phase in the wider IoT.

23.11 More internet of X terms

There are many dissimilarities in the common topics that we have explained here. The IoTDS (Internet of things, data, and services) is a specific term within industry 4.0. As previously mentioned, other areas discussed in certain fields include the internet of healthcare things (IoHT), the internet of medical things (IoMT), and the internet of things and services (IoTS); furthermore, the industrial business ABB coined the term "internet of things, services, and people" (IoTSP) [13].

23.12 Swarm intelligence

SI is an AI (artificial intelligence) regimen. It is a related to the intelligent multi-agent systems design and takes inspiration from the mass conduct of social insects/animals. Even though modeling of lone individuals within these colonies is unsophisticated, they are capable of achieving complicated tasks in collaboration. Originally, the term SI was used in the 1980s as a malicious term by Beni and Wang [14] to indicate a type of cellular robot system; later, it was utilized for covering a broad area of studies from optimization to social months [15].

23.13 Definitions of SI

The basic social insects of SI are implemented in biological learning of self-united behavior. This is explained like any try for make algorithms, or distributed devices and solve the issues motivated through the social living mass behavior, like, ants, fish, and birds. SI is "an artificial-intelligence approach to problem solving using algorithms based on the self-organized collective behaviour of social insects," according to the online Collins Dictionary (https://www.collinsdictionary.com/dictionary/english/swarm-intelligence). One can divide SI research according to several criteria [16]:

(1) Analyzed system nature analysis: we are learning biological systems and artificial swarm intelligent domain distinguishing from natural SI where individual artwork is agreed.
(2) Objective: we difference amid (I) scientific SI which focuses at the perception of natural swarm system and (II) a scientific stream capable of focusing on the create and execution of artificial flock systems by engineering SI, which solving issues by exploiting findings from.

Instead of a social insect colony, this is a kind of centralized system built of sovereign parts, which were divided in the ecology since easy possible stimulus-reaction character. The principle is that in order to control communication among individuals, actions are executed on the basis of local data without the requirement to learn the universal pattern. The organization's structure allows these simple behaviors to have effect at the colony level by taking place among individuals. Consequently, easy communication among individuals may reconcile a wide range of issues and answer external problems in a flexible and strong manner. In 1959, Pierre-Paul Grassé first proposed the concept of stigmergy to explain the organization of social insects [17]. He saw that insects leave information in the environment (e.g., pheromones) when carrying out a task, which may guide the behavior of others [4] (Fig. 23.4).

23.14 SI and systems intersections

SI is a concept that explains how complicated social behavior can occur through easy, non-integrated individual actions. For instance, bee/ant/wasp/termite colonies show such self-united and mass intelligence [18]. While any member of the colony does not have the strength or skills to make judgments concerning normal integrated actions, ultimately, the colony behaviors are systematically and efficiently following essential rules. This natural and easy algorithmic character has motivated many individuals to attempt to develop equivalent mechanisms in robot communication or artificial intelligence systems, which developing area of research is known as CSI (computational swarm intelligence) [19]. As we can already achieve SI activity between persons [20,21] with the development of the information society, we can discuss how to make SI applicable in the IoT [22–24]. So, we can see that bio motivation is a prosperous fashion in AI research and SI can supply a credible structure for the modulation and production of social behavior, even in computational and electronically linked phases.

Considering the technological probabilities of enhancing the available smart technologies and the promotion of the technological probabilities, we suggest the incoming scenario in which the SI is a section of person's behavior, not only the management of the sub-conscious but actions.

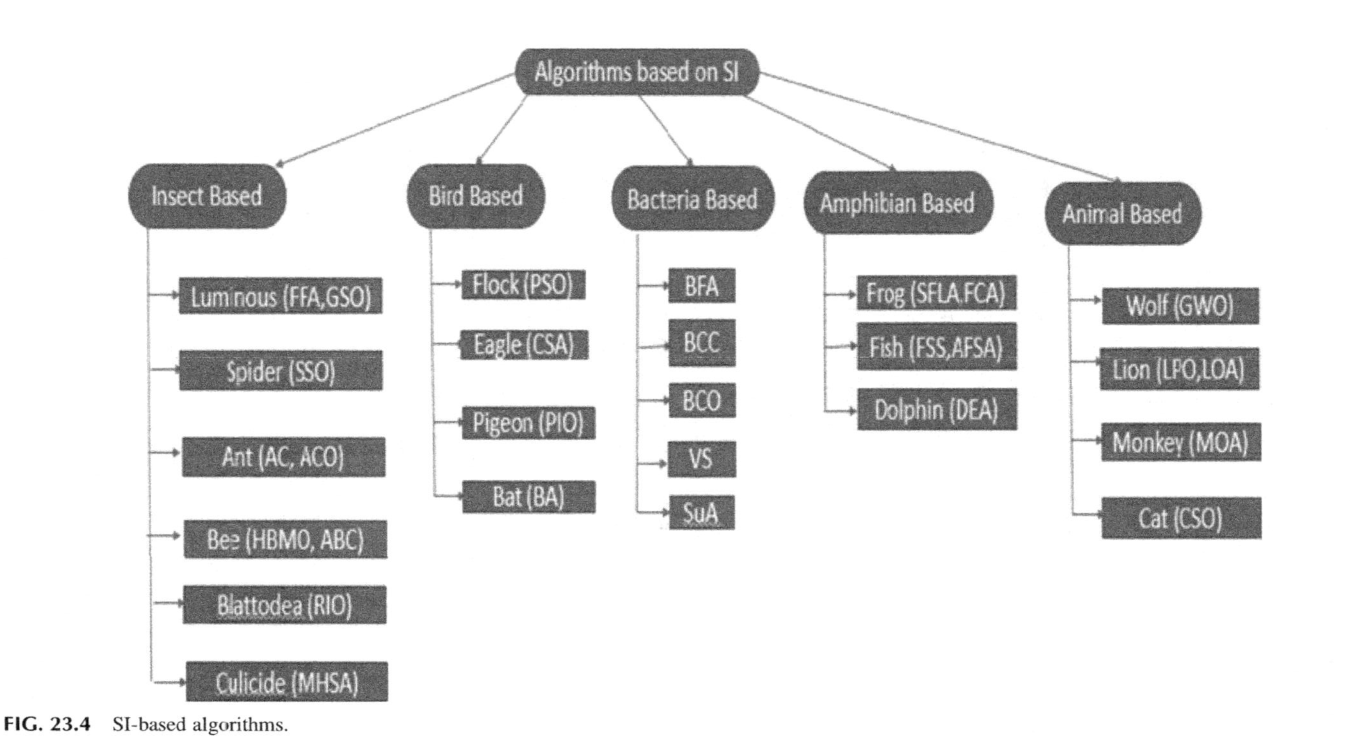

FIG. 23.4 SI-based algorithms.

23.15 Swarm Intelligence and smart gadgets

We look into the present biotic perceptional global like a consequence of different executions throughout a length evolutional changeable growth, with not last phase or fixed accomplishment. Biotic development has been implemented to various billion season at planet globe, after an un supervised actions (except you acknowledge the vast forces like a predicate of those alteration, even if without every objective or determinative guiding), but rather any alternative approaches/action to obtain evolutionary alteration? The attainment in the building up of peptide nucleic acid (PAN) pinpoints the response—"yes" [25]. Probably the major case of perceptional science and their results once fixed into difficult or on the top-down, Artificial Intelligence has been recognized as a smart method with the functional attributes of minds or pivotal nervous systems [26] and (cultural) tokenism [27]. As with those related to contiguous attributes [28], intelligent decisions, some forms of management, mass behavior [18], or the less cognition [29], are not consist as better or adequately energetic models for Artificial Intelligence used to go. Although, Here the reality is that intelligent conduct can also established in much easy phases in life like bacteria [30] or mould [31]. The forward probable query is whether another function is not developed in different biotic cells as an associative and social (physical) pattern, as separate entities, we want to develop? Gratitude for the technical progress, we are able to estimate that a bio motivate structure of information interaction might be enforced into the clever person environments: By devices of wearable, increased bodies, smart cities, vast intelligence, virtual or increased fact, and another new paths to add procedure progress with electronic probabilities. Although, what if we have a tendency to add a brand-new channel of communication? Presently, alone wireless internet with whole the "advantages" of the radio or internet/TV. What if we have a tendency to might include a brand-new world mode of communication during which whole complicated organisms on the earth could distribute multimodal data method? Can it trigger recent phase of develop or might be completely adversity? The integrated info processed through organism F recognize the sense of the life, increases it to seamlessly link organism that we have tendency to might establish that their phase of sense ought to be accumulated through the mixing of wireless devices, as they might be capable of process a lot of integrated info than in a very non-connected state, this is said in Ref. [32]. Even presently, at a particular purpose we have a tendency to are whole cyborgs [33], partially-augmented human's exploitation previous methods to cope with final generation devices.

23.16 Implants & prosthetics

The person body isn't ended, however a momentary framework that had been developed in many years. In a connection with a coupling among the atmosphere and the body [34], Persons have been able to changes their behaviors [35]. Gratitude to the different technical progress: Symbolic models, physical components, language, or social approaches, between different another. Now, gratitude to synthetic bio, genetic engineering, wearable computing, and an extended index of technical progress, person organisms may totally change their minds and their bodies. Explain the internal coupling among embodied and extended pattern [36,37] cognitive system procedure that role to the information procedure on its many phases. From this approach, the cognitive process is based at a multidimensional net of facts, places, etc. [38]. We can get the fundamentally involved cyper-Neal Harbison, an individual who utilizes the sounds to catch the color data [39]. As of late, the organization CyborgNest "North sense" [40] has discharged, which is an outside apparatus that gives its clients a sense about the condition of the North coordination. As per CyborgNest, North Sense is an exo-sense that is intended to development admirably, implying that it sits external the body however is for all time associated. It enables an individual to comprehend the electromagnetic field of the planet. The North sense is generally connected to the upper chest and oscillates gradually when the attractive north is experienced. This experience (to be a component to encounter the North bearing) is to realize that just a couple of winged creatures and creepy crawlies have a place with the animal varieties [41], and is currently exposed to uncommon human spending. We can without much of a stretch envision new sorts of gadgets that upgrade the human experience, increment bodies or faculties, increment the scope of genuine faculties, or even make an intriguing synth condition. This will totally modify our experience and potential outcomes of communicating with different animals. For instance, fake compassion will be conceivable to change to build the utilization of human neuromodulators. Accordingly, we can confirm how we feel on the planet and how we connect among people, modern people or machines, is completely open to modern guidelines and conduct situations.

Another of the main improvements of a mammal body is a body—PC interface. The benefits of non-intrusive or even obtrusive body-to-PC interface were presented a few times and even in prominent writing and movies such "Ghost in the shell" and "Matrix." An abnormal state perspective on the issue could resemble the one communicated in the accompanying situation: we need to give the interface to 86×10^9 neurons and 10^{13} synapses, taking into account [42]. We can see a few issues, particularly seeing the nanometers size of Deoxyribonucleic Acid (DNA) in eukaryotic cells that could build the radio recurrence radiation of this sort of transmitter exceedingly incapable, and accordingly the yield flag could be amazingly powerless, meddling with the current radio recurrence clamor of a mammalian cerebrum. The in-organ arrangement has the benefits of maintaining a strategic distance from the nanometer-scopes of the sizes, yet it is as yet an open inquiry how we could make a compelling transmitting aerial wire dependent at their sources of a mammalian cell, particularly in a manner that is alright for a creature. Few past specialists have termed this probability "The Internet of Bio-Nano Things" [43] or "Nano-Networks" [44]. These recent cyborgs can associate by the neuromuscular interfaces [45] or expansion and installed in a many biotic systems [46], which can be upgraded in a different path [47].

There are deep challenges that these altering can generate our feeling, sensing, and experience for the global can face us into front of a fundamental progresses landscape—as a caste and on a personal phase. It is the Important Fact: To view that on it phase, the personal and mass ways stream in various and now not constantly usual supplementary paths. By a biotic classification perspective [48].

23.17 The swarm behavior of the augmented and quantified human—The next stage of the IoE?

For the past few years, there has been a presentative phenomenon for handling the working of artificial intelligence systems that are likely to be able to deal with information and various decision-making processes that are consistent and accurate, and providing an effective method for an outstanding model known as "SI" model [49]. This model provides an effective way for managing the shared decisions (even for human beings [50]) and joint workflows, which can delegate to robotic applications in a few particular areas [18]. Whenever we think about this bioinspired technique, we generally relate it to the network to provide new ideas for dealing with the combined knowledge of being an individual entity as well as being a part of a wider, shared community. We have improved and advanced devices that can work together with or without our conscious decision. For such instances, we can also permit certain agreements that form teamwork under particular conditions. There are certain structures available that can provide new ways in the network that are useful for managing social interaction. For instance, the protest movement in Hong Kong against Chinese suppression, in which thousands of people avoided the governmental suppression of networks using a phone-to-phone connection application (Firechat) that provide end to end connectivity [51].

There are number of critical features that include:

(1) New technological mechanisms that can improve, promote, and drastically alter the sensing and feeling of the world.
(2) Identifying and increasing the ways in which we can access or create new methods of experiencing the world. For instance, the services which could check the pressure on the sensor through a radar under 360 degrees boundary width.
(3) Lastly, by joining the location could experience the closeness of non-illustration goals and as a result it will increase the efficiency for feedback.

Thus, not only the normally accepted human senses could be considered in order to develop a person's observation capabilities, but thoroughly new senses could be made available to individuals through directly activating specific zones of that person's brain. This development could change not only an individual's perception, but also include devices and sensors that could link us with another's mind to experience different types of sensory input in a more composite manner.

New and innovative smart ideas are effective enough that would relate with the outfits that generally combines with the various rules and attributes according to the closeness of other people's by making a smart and creative organization including the metro services, or a bus. Recent patterns of human swarm behavior become comprehensible when we introduce another language on the larger scale and thus the "language" of communication might involve a form of language which does not contain words. We can also imagine that how much better the "combined intelligence" of the human swarm of that type can be related with communication. We already understand instances of how a large number of individuals could be synchronized, if they share their overall goals, and form an intelligence that work together as individuals (e.g., Ref. [52]). States, businesses, and associations of various kinds can also show how the mode of communication for various multiple individuals can bring about new ideas for logical study under a single entity. However, all of these examples are especially connected with the processes of understanding, which are very limited as far they are concerned. How far are we able to make the crowd more developed with the recent developments in technology of communication and distributed sensors? How will isolated followers react to entities that are related with the artificial intelligence system as part of it, including the non-intelligent communication devices and distributed sensors and actuators [53]? We can now use mathematical calculations for presenting communication skills to the public and evaluating revolutions in government that may be greatly sped up with new ideas [54]. The innovative ideas can be implemented using the internet with the interconnected devices and equipment's, robotics, and rich source of computational power that are mainly available throughout the internet and exceed the edge [55] up to the certain boundary conditions that will alter the way to identify the people's, network, imagine, and feel the difference that not only contributes to the smart infrastructure and but also contribute to the new society. This will have certain implications for the welfare of the public services and their needs, implementing orders under the law, and many other aspects of the social arena. It may also alter aspects of the economy that affect how it works and also include the status of the economy. These new and innovative ideas [56] can contribute to a society in which people are connected end to end entirely with each other, producing new types of superconnected communities that we could call social connectives [57].

23.18 ANT colony-based IoT systems

The ant colony (AC) approach is used in [58] to find a particular path for a routing processing mechanism that involves IoT. The AC approach is used to minimize the number of useful network nodes. Ref. [59] dealt with the difficulty of collecting the data from numerous sensors, operating the processing, and producing effective information. To consume this information efficiently in managing the flow of traffic in big towns, specific agents use the method of H-ABC (hierarchical ant-based control) to utilize the essential information transformations. A recent concept to enhance the efficiency of the framework that is used in the IoT network is suggested in Ref. [60]. The purpose is to collect and gather all the sensors with same reference information in sensor semantic overlay networks (SSONs). Most authors rely on the clustering mechanism, which consists of a number of sensors depending on their kinds to make a semantic overlay network. They cluster the sensors regarding their reference information using an ant-based algorithm known as AntClust. Eventually, they utilize modification to minimize the price of the sensor finding process. Ref. [61] proposed a modified AC algorithm to compute the expectation value in order to determine whether an object is expected or not. Most authors make the alteration to store the value to expected increase and use the phenomena for reducing the expected values as per its needs. Ref. [62] also uses the phenomena of AC to solve the routing problem of city traffic in an IoT system. It is generally represented through the number of agents to solve the problem and thus, every "artificial ant" uses the single computational "vehicle agent," and this city traffic network can be reduced through the diagram-theoretic concept. Ref. [63] uses the ant mechanism for tourists, which is used for guiding the process and to search for things that can be used for various mechanisms of ants. This mechanism enables unique characteristics for tourist navigation through using an artificial mechanism that involves indirect coordination between agents and actions to provide the tourists with many advantages such as:

(1) Unexpected discovery, while providing information for the rest of the agents.
(2) Routing decisions using the information which is supplied through the server.

23.19 ANT colony optimization (ACO)-based IoT systems

A technique to find a series of paths by an automatic mode (in the area of Santander) is suggested in Ref. [64]. This technique uses a mixture of both soft computing and geographic information system techniques.

An ant colony algorithm (ACO) algorithm is used along with node attributes to find out the comparison between different communities to efficiently detect the nodes in Ref. [65]. Ref. [66] suggested a routing algorithm to select a better way of transmitting information within the IoT network by utilizing several networks. ACO algorithms should be implemented in every network that has its own ACO to deregulate the process of routing. Most authors establish an algorithm to manage the utilization of ACO and to define a solution for overlapped zones. The real time analyzer proposed in Ref. [67] for classifying disturbance and the intended recovery model accord with this. The authors have made an improved ACO to obtain recent schedules accurately. They use the classical ACO method to make some adjustments according to the substance produced by the path and also intersections, along with the mutations that are accepted in improved ant colony optimization (IACO) for avoiding early meetings. Ref. [68] established a system based on the actual time waste monitoring system. The ACO was used to calculate a better way to solve the VRP (vehicle routing problem). The route provided by the road network was explained through a graph, which indicates the arcs as roads and vertices as junctions. An algorithm was recommended in Ref. [69] for improving the ACO method through which the clustering based on the indexing are used. The improved ACO is utilized to move slowly and retain the cluster information in the network using the internet. According to the AC, the routing information, the speed, and the position of the cluster resource indexing of the IoT are simplified so that the objective of ACO can be used to establish overall findings and confine findings to resources. A routing algorithm is used in Ref. [70] along with the IoT. The planned algorithm first partitions the IoT as an environment into several subzones based on the network type. Later, this choses the particular ACO algorithm that is appropriate to every network. The planned algorithm mainly considers the routing problem into the overlapped zones that can occur in the different IoT systems. A dual mediator is used for developing the optimized algorithm from dissimilar ACO algorithms.

Another smart waste management system, which is suggested in Ref. [71], is basically dependent on the IoT, including the technologies, and relates to waste collection management in Kayseri, a city in Turkey. Fundamentally, the authors planned to make a garbage container that was equipped with various sensors to measure the phases of temperature of the container, and carbon dioxide ratio within the container. The collected information is transmitted to the waste management software; then the ACO is utilized to determine the most well-organized waste collection path to include each garbage container, which is delivered to the garbage truck drivers using cellular smart tablet devices. Ref. [72] used another concept in which the ACO is utilized to optimize the route for energy utilization—that is, the amount of energy utilized through transmitting the node during transmission and energy consumed by listening node during listen process, which were based on the expected transmission of calculated metrics and the goal function.

23.20 Particle swarm optimization (PSO)-based IoT systems

A model based on IPSO (improved particle swarm optimization) is recommended in Ref. [73]. This model [73] used the IPSO algorithm for quantity used by the broadcasting effect to maintain the accuracy to several body parts including the body fat, body weight, blood oxygen, rate signals of heart in machine medical care include the IoT as the network system computing. The established system can be utilized in hospitals to gather individual information for observation within the physical healthcare information system. A non-linear weight-reducing strategy applying the IPSO method is recommended in Ref. [74]. This method includes the final conclusion for performing the information combination in respect to a multisensory network. This learning helps in developing new monitoring systems using a number of dissimilar sensors to build a network system with the use of the internet through integrating several sensors together to test the signal processing and data combination methods.

Another discussion was introduced to find out that how to apply data mining processes for information removal and computational intelligence for computing the FIoT (future internet of things) as specified in Ref. [75]. The authors also suggest a method using intelligent ant colony optimization in which every sensor is connected with a straightforward data agent where the sensing of every mediator is stimulated through ACO and PSO. In this process, the mediator acts as a swarm through distributing information and knowledge among different particles. Nowadays, authors also use thc example of a smart home to explain the exchange of data and create conclusions based on the detailed concepts of DEFSO (Differential Evolution based Feature Subset Selection Optimization http://citeseerx.ist.psu.edu/viewdoc/download?doi=10.1.1.852.9603&rep=rep1 &type=pdf). Ref. [76] suggested an improved efficiency and intelligent fault-tolerance algorithm (IEIFTA). This method is used to offer a quick improvement mechanism to deal with failure due to power reduction or physical damage by utilizing another path. It mainly chooses the pathway with the best fitness from the most favorable sensor nodes.

A method is suggested in Ref. [77] for maintaining a smart platform for building energy management systems (BEMS). These systems are mainly based on the evolved hierarchical agent organization that permits the lower level mediators for abstracting the data from the direct environment in the form of single value data blocks for the higher level mediators. The PSO optimizer was usually invented to improve the ability for developing the buildings elasticity for making the smart framework. Ref. [78] proposed a novel technique based on the difficulty of finding end of use products for the recovery process. The criterion is usually based on the method of PSO in which the manufacturer represents the raw material inventory level for finding the best solutions according to the encoding method to estimate the total income. Ref. [79] recommended a method for calculating the resources allocation scheme depending on HQPSO (hybrid quantum-behaved particle swarm optimization). The authors thought of using the mobile phone edge computing (MPEC) method for calculating the offloading, made for decision problems. This model solves the problem mathematically using the content of the whole wireless communication system with MEC calculation offloading technology. Then later, HQPSO is used to make the best calculation based on the offloading decision. One method, which is proposed in Ref. [80], for a distribution scheme that determines the optimal place and operating channel for the network sensor devices to manage the intervention and resource distribution, uses the particle optimization method to search for the best position and operating channel between the appropriate network nodes. Ref. [81] recommend a technique for a MOESO (multi-objective element swarm optimization) system for cloud brokering, to search out the most suitable connections among customers and service providers to optimize the energy utilization of service providers, the income of the cloud agent, and the response time for the requirements of the customers. A method is proposed in Ref. [82] for making the separate flows of control systems a network-enabled service, which can be used without the need for a dedicated tool server. In such a process, the level and flow are independently controlled through three essential processes including PSO proportional, PSO essential-copied mechanism, and PSO proportional primary controller correspondingly.

23.21 Artificial bee colony (ABC)-based IoT systems

One method, which is recommended in Ref. [83], uses the hybridized bat algorithm with the ABC algorithm to resolve the problem of radio frequency identification network arrangement. This arrangement basically uses the search process of the bat, which is an enhanced feature including the observer method (BA-OM).

A CMABC (cross-modified artificial bee colony) technique is suggested in Ref. [84] for achieve the best possible solution for services in a suitable time and with high precision. Basically, the CMABC method is used for optimization including the IoT as a network service instantiation. The authors construct a service representation and use the CMABC method to achieve its instantiation in a specified time with high efficiency. A method is suggested by Ref. [85] for improving the access efficiency for various service requests for managing the wireless sensor network (WSN) and creating complete use of the variable resources for accessing the data at the core including the IoT as the network. The authors planned to use the best algorithm to set the immigration depending on the grouping scheme between ABC and chaos search theory. Ref. [86] suggested a method for tackling the service optimization problem (SOP). The purpose is to set the service domain-oriented ABC algorithms based on the optimization method of ABC and to manipulate the service domain features for managing the optimal solutions related to particular services. The management related to vertical handover in heterogeneous WSN's is considered in Ref. [87]. Here, the network selection is controlled and managed through several constraints including end-to-end stoppage, bit error, jitter, packet loss, etc., and is performed via the ABC algorithm to select the best possible network with the least handover stoppage and time.

One method, which is proposed in Ref. [88], is based on the Hadoop artificial bee colony (H-ABC) algorithm to select the characteristics for electronic-health related to a big data internet (IoT) network. The conditions are thus created for implementing the usage for enhanced the MapReduce network and H-ABC algorithm. Basically Hadoop-ABC is used to choose the characteristics and processes of big data sets and the MapReduce network for processing other information with the Hadoop network for maintaining the efficiency and actual-time processing features. The technique recommended in Ref. [89] is for a hybrid-ABC algorithm with the capability to schedule the basic transformations to find the best possible number of disjoint separations for improve the life span of a wireless smart devices network to report the final target application. One novel method, which is proposed in Ref. [90], combines the GSA (gravitational search algorithm) and ABC algorithm to achieve the grouping of clusters based on the head selection. It may include selections such as the energy, distance, delay, weight, and temperature of the network devices during the process of the cluster head selection.

23.22 Bacterial foraging optimization (BFO)-based IoT systems

A method is recommended in Ref. [91] for efficient routing by means of secured energy consumption by combining the OSEAP (optimal secured energy aware protocol) and IBFO (improved bacterial foraging optimization) algorithms. First of all, an effective topology is developed on the basis of network topology. In this, the fuzzy c-means (FCM) clustering algorithm is used, in which all the sensor nodes in the network topology are grouped together to facilitate the protected communication of different messages from the source nodule to the destination nodule assembly, including the key allocation method engaged in OSEAP, and the main selection is carried out with the help of the EBFO. A method is proposed in Ref. [92] for determining the energy consumption for different patterns used in optimizing the home energy management system that utilizes the load changing policy of the demand side management system. The main purpose is to manage the load demand efficiency in a more accurate way to reduce the electrical energy cost and peak to evaluate the average ratio while maintaining customer comfort by organization between home appliances. Consequently, the authors also suggest a mixture of optimization techniques, which combine the GA as well as the BFOA algorithms.

23.23 BAT optimization (BO)-based IoT systems

A technique is recommended in Ref. [93] for modeling the speed and power of a BLDC (brushless DC) motor in which the different parameters are evaluated for the brushless DC motor. Other methods are deep neural network (DNN) and BAT optimization algorithm methods, which are implemented for the evaluation of different parameters of the BLDC motor by IoT. Ref. [94] detailed a new equation for the velocity updating problem, which is combined with the centric policy to optimize the nodule selection by means of the cluster-head to LEACH (low energy adaptive clustering hierarchy) procedure to save energy costs. The authors also proposed the new alternative of BAT algorithm associated with centric strategy. This novel BA alternative is not only capable of increasing optimization but it also increases the presentation and convergence speed for enlarging the universal find space.

23.24 More SI-based IoT systems

Ref. [95] proposed a recommendation mechanism by inventing new innovative adaptive methods for maintaining efficient parking. This mechanism helps drivers to search out parking spaces faster and minimize traffic overcrowding through using a cellular automaton method and cognitive radio network model. The authors also implement the AFSO (artificial fish swarm algorithm) to construct parts of the parking recommendation method. This led to faster development of new parking schemes, and thus, reduction of traffic congestion is achieved under some circumstances. An HBMO (Honey-bee Mating Optimization)-based routing algorithm is recommend in Ref. [96] to be used in respect to cognitive broadcasting sensor networks. The recommended algorithm reduces the chances of package failure due to which the conservation of high link quality is maintained between different sensor nodes in smart grid locations. The method recommended in Ref. [97], which is used to build-up the drainage network, is prepared through sensors and consists of sequences of electronically changing gates that are controlled through a decentralized actual-time system depending on a conversation-based algorithm which helps decrease sewer flooding and combined sewer overflows. A gossip-based algorithm is used to calculate the average level, as measured through alle the agents of a generated network. At the convergence point, the predictable average value is broken for every entrance-agent to change its entrance so as to hold water levels better for that average. The algorithm confirms the fault-tolerance attributes and the system remains working even if an unexpected occurrence significantly changes a few structural attributes, as in the case of difficulty, damages, blockages, and so on. Ref. [98] proposes PIO (pigeon inspired optimization) for DSM (demand side management) in a smart grid. This algorithm is recommended by the authors to achieve agreement for electricity bills, electricity consumption, and peak-load-reduction. The purpose is to reduce the costs and peak to average ratio, and improve customer comfort, through applying DPS (dynamic pricing signals). Ref. [99] proposed a method for evaluating the performance of HEMS (home energy management system) through the combination of GWO (grey wolf optimization) and BFA (bacterial foraging algorithm) techniques, which were designed based on the nature of the grey wolf and bacteria, respectively. For this reason, home appliances are categorized into two different classes on the basis of their power utilization prototype. Grey wolf optimization gives optimized results as well as quick convergence. Ref. [100] recommended an improved cuckoo search (ICS) algorithm to simplify the problem for the visible light communication (VLC) power reporting problem in smart homes. The proposed ICS algorithm uses the anarchy hypothesis to optimize the arrangement for the preliminary solutions depending upon the quality of the various disorganized distributions so that they can be avoided by the non-uniform distribution used by the solutions. In addition, ICS divides the multidimensional solutions into several areas by using the concept of various measurements.

23.25 Towards SI-based IoT systems

In today's modeling world, internet-based systems that are considered as part of the cloud can help a lot, and can provide advantages including the main characteristics of the cloud systems, which are greater scalability, robustness, and elasticity [101]. These key characteristics can conquer several of the key architectural needs of IoT-based systems, which include robustness, elasticity, scalability, and interoperability [102,103]:

- Robustness is a primary characteristic of SI-based IoT systems. The system remains ready even with interruption from the atmosphere or the failure of its individual components. A number of factors are added to make it more robust:
  - (1) Management is decentralized and the damage caused to a specific portion of the system is not likely to stop its process.
  - (2) Individual components are straightforward and thus result in a smaller amount of failure.
  - (3) The sensing process is circulated and hence the system is strong enough to handle local disturbances.
- Elasticity is an important feature, which includes individual components that adjust themselves according to the altering environmental conditions through managing their actions in a way to control work from several directions. Thus, SI-based algorithms can help the devices to reach agreement based on the problem of changeability that could occur in the different conditions due to nearby devices.
- Scalability is possible because SI-based algorithms can maintain tasks that are related to a particular association, including teamwork and self-arrangement, which allows for the cloud function below a large variety of group sizes and thus accurately supports a large number of individuals without significantly impacting performances.
- Interoperability feature includes the network capabilities that are used to manipulate the cluster-based intelligence to choose the best possible pathway that can handle the large variety of heterogeneity and asymmetry assignments related to the advanced efficiency and technologies for communicating devices.

Apart from this, the SI-based algorithms are more appropriate to the processes including the phases such as optimization, development, planning, and design, and also for administration problems. Moreover, these types of problems can be found anywhere in the form of investments, construction, allocation, and so on. In particular, we have selected the problems that are more frequently associated with the applications that are based on the SI-based algorithms, created with the best practices for internet-related based systems. More applications based on algorithms from classification one may comprise:

(1) WSNs, which include direction-finding, data transmission, energy efficient sensor, association, clustering, nodule searching, assignment of sending nodules, shared mobile sensing including the devices, efficient clustering based on the head selection, and charger deployment in wireless rechargeable sensor networks.
(2) Radio frequency identification (RFID), including networking planning, optimization for RFID reader operations, and label design.
(3) Dynamic routing, including railway connection rescheduling, routing for automobiles, and train routing selection.
(4) Forecasting, including the weather conditions, used for the evaluation of water assets, travelling time, estimate of water temperature in prawn cultures, electrical load forecasting, hydropower dams, and time series calculation.
(5) Image processing, including the feature choices of high-dimensional categorization, extraction of photovoltaic module parameters, water distribution and optimal power flow multi-level thresholding, operations involving MRI (magnetic resonance imaging) brain segmentation, comparing image enhancement techniques, bilinear spectral unmixing of hyper spectral images, registering different remote sensing features, and so on.

So, the above conclusion may include the following points:

- First, the performance of SI-based algorithms is powerfully dependent on algorithmic differences and modification parameters; thus, the investigator, based on their treated problems, can consider SI-based algorithms in order to obtain excellent performances.
- Second, swarm robotics is a narrative method that includes the management of large numbers of robots that has emerged for the multi-automation systems. Internet-related networks are faster for designing rather than the SI-based swarm robotics, with its emphasis on the physical personification of individuals and practical communications involving the number of individuals and their environmental surroundings. Thus, devices including the internet can be considered as robots and stimulation including the various coordination mechanisms in swarm robotics, which can be adopted in different IoT-based systems [4,101].

23.26 SI and its applications

SI is considered to be the combined behavior of self-structured and decentralized systems, which may be a conventional or robotics system. Below, we classify the two most widely used SI algorithms.

23.26.1 ACO (Ant colony optimization)

Ants are considered to be the most perfect instance of natural SI. They survive in colonies and place pheromone (chemical substance) markers on the way from their nest to their food source. Individual ants do not have sufficient brainpower to find the shortest way for their food. However, by functioning in colonies, they can perform many complex tasks with no difficulty. The ant colony optimization algorithm is based on the ants' path optimization for finding the shortest path from the food source. The ACO algorithm is mainly used in optimization problems, scheduling problems, the traveling salesman problem, vehicle routing protocol, [104] and so on. The application of SI is a class used in the ant algorithm which identifies the copy behavior of the ant. Similarly, the key iteration in the ant system is updated by the ACO system, which identifies the substance released into the environment.

23.26.2 BCO (Bee colony optimization)

Bees are one of the best examples of normal and active swarms. They show swarm intelligence through distributing their work between many bees. Different tasks are performed by bees, including foraging, stockpiling, honey separation, collecting pollen, recovery, and communication, and they adapt themselves to the changing environment. Bees are impressive in classifying their colonies [105]. Below are the functionalities that are used by the bee colony algorithm:

(1) The Bees are arranged and divided into two classes: employed bees and unemployed bees.
(2) The purpose of the unemployed bees is to explore for the employed bees when the food source has been perceived by the employed bees.
(3) The Employed bees moves from one flower to another flower arbitrarily, and search for the food source.
(4) When the food source is found, the unemployed bees search out the shortest way to the food source and, when they search, they inform other bees through a wiggle dance.
(5) Finally, workers accumulate the nectar from the food source and return to their comb to store the nectar.

23.27 SI application for IoT processes

The process of SI has a number of capabilities that can improve the IoT processes. Many narrative approaches are included in the SI algorithm, which consists of number of particular IoT fields that are described in this chapter. We have selected the three fields where the SI algorithm could offer the best solution:

(1) The difficulty faced by connected cars in vehicle routing with AC optimization.
(2) The mechanism of data routing divided over a sensor network with ACO algorithm.
(3) Methods of cloud computing for data optimization.

23.27.1 Use of SI for connected cars

The vehicle routing problem is considered to be the logistic problem in the SI algorithm. Basically, the problem deals with the running cost reduction for "n" number of vehicles to supply the "m" customers. So, the vehicle routing is the major service for connected cars provided through internet platforms. This type of car has the capability for revealing sensor data to a safe and protected cloud platform over the internet. There are services related to the current platform that include the shortest route to the final fuel station, analytical study of car health, and others. This type of services become more necessary in the background for the emerging independent vehicles, which would produce gigabytes of data per second to generate locally dynamic maps, and for logistics delivery, movement of persons, vehicle health analytics, usage-dependent insurance, etc. Without a human driver to take decisions, independent vehicles will be completely dependent on outside computing platforms to process the data to produce significant information. Through the ant colony optimization (ACO) algorithm, IoT services can be organized in such a way as to offer path construction, track update, and route development methods [104].

23.27.2 Use of SI for data routing

The SI algorithm covers a large variety of applications including the communication as well as the telecommunication networks. In the IoT network, the concept for routing data is used with the centralized cloud covering the large-scale distributed sensor network. It is estimated that around 20–30 Billion sensors will be connected to each other via the internet in 2020 [106] and a quick direction-finding strategy is used for storage of the sensor data that will be used for the communication backbone. The use of the SI algorithm would surely be defined as systematic data routing and storage, which in turn avoid the loss of data. We have used the AntNet algorithm [107] to obtain the process of utilization. This algorithm is a robust routing algorithm that became prominent after the use of other routing algorithms. In this algorithm, routing is mainly done by forward and backward interactions of a mediator that discovers the network. The whole system is relatively complex due to the number of network nodes, including the internet-related network. Each nodule in the network identifies the ants in a forward direction at the proper interval, and thus, the ant will find the route arbitrarily rooted on the routing table. These ants generate the stack that is inherited by the backward ant after the forward ant reaches its destination path. The backward ant reverses the direction of the stack and gets back by the means of the same route. Thus, the node tables get updated based on the trip time for both the forward and backward directions.

23.27.3 Use of SI in cloud computing for data optimization

Data optimization consists of several values including processing, storage, and visualization of information. Around 1.2 zettabytes of internet data is used to generate the devices, which would certainly require storage space in the cloud. With such huge amounts of data, there is loss of performance; thus, the processing time is increased in data processing algorithms, which may utilize the ACO for data optimization, thus minimizing the whole data processing time and effort [108]. The ACO algorithm is recommended for optimization up to large dimensions for rapid data mining of various datasets, and it can provide a number of features, which include high dimensional data, the presentation of the algorithm with big datasets, ability to handle dynamic data that offer nearly actual time data processing abilities, and lastly, the multi-purpose optimization in ACO, which can handle the information coming from several sources. This could offer a capable and strong solution to data optimization in IoT [109].

23.28 Conclusion

Optimization algorithms dependent on SI can have some unmistakable points of interest over conventional strategies. This examination has centered around the method for accomplishing investigation and findings, and the essential parts of developmental administration, such as crossover, mutation, and selection of the fittest. The investigation likewise suggests that there is opportunity to improve. A few calculations, for example, PSO, may need blending and hybridization, and thus, hybridization might be valuable to upgrade its execution. It merits mentioning that the above examination depends on the framework used for consistent advancement issues, and it is not unusual that these outcomes are still legitimate for combinatorial streamlining issues. In any case, care ought to be taken for combinatorial issues where neighborhood may have a distinctive significance, and, in this way, the subspace idea may likewise be unique. Further analysis and future studies may help to provide more elaborate insight. In this chapter we have tried to show how SI is linked with IoT.

References

[1] R.S. Parpinelli, H.S. Lopes, New inspirations in swarm intelligence: a survey, Int. J. Bio-Inspired Comput. 3 (1) (2011) 1–16.

[2] S. Garnier, J. Gautrais, G. Theraulaz, The biological principles of swarm intelligence, Swarm Intell. 1 (1) (2007) 3–31.

[3] G. Fortino, P. Trunfio (Eds.), Internet of Things Based on Smart Objects: Technology, Middleware and Applications, Springer Science & Business Media, 2014.

[4] O. Zedadra, A. Guerrieri, N. Jouandeau, G. Spezzano, H. Seridi, G. Fortino, Swarm intelligence-based algorithms within IoT-based systems: a review, J. Parallel Distrib. Comput. 122 (2018) 173–187.

[5] P.P. Ray, A survey on internet of things architectures, J. King Saud Univ.-Comput. Inform. Sci. 30 (3) (2018) 291–319.

[6] S.K. Das, H.M. Ammari, Routing and data dissemination, in: A Networking Perspective, 2009, p. 67.

[7] F. Liu, C.W. Tan, E.T. Lim, B. Choi, Traversing knowledge networks: an algorithmic historiography of extant literature on the internet of things (IoT), J. Manag. Anal. 4 (1) (2017) 3–34.

[8] S. Zahoor, R.N. Mir, Resource management in pervasive internet of things: a survey, J. King Saud Univ.-Comput. Inform. Sci. 33 (8) (2021) 921–935. ISSN: 1319–1578.

[9] J.M. Chatterjee, R. Kumar, M. Khari, D.T. Hung, D.N. Le, Internet of things based system for smart kitchen, Int. J. Eng. Manuf. 8 (4) (2018) 29.

[10] J. Chatterjee, IoT with big data framework using machine learning approach, Int. J. Mach. Learn. Netw. Collab. Eng. 2 (02) (2018) 75–85.

[11] S. Garg, J. Moy Chatterjee, R. Kumar Agrawal, Design of a simple gas knob: an application of IoT, in: 2018 International Conference on Research in Intelligent and Computing in Engineering (RICE) (pp. 1–3). IEEE, 2018, August.

[12] S. Jha, R. Kumar, J.M. Chatterjee, M. Khari, Collaborative handshaking approaches between internet of computing and internet of things towards a smart world: a review from 2009–2017, Telecommun. Syst. (2018) 1–18.

[13] https://www.i-scoop.eu/internet-of-things/.

[14] G. Beni, J. Wang, Swarm intelligence in cellular robotic systems, in: Robots and Biological Systems: Towards a New Bionics? Springer, Berlin, Heidelberg, 1993, pp. 703–712.

[15] C. Blum, D. Merkle, Swarm intelligence, in: C. Blum, D. Merkle (Eds.), Swarm Intelligence in Optimization, 2008, pp. 43–85.

[16] M. Dorigo, M. Birattari, Swarm intelligence, Scholarpedia 2 (9) (2007) 1462.

[17] P.P. Grassé, La reconstruction du nid et les coordinations interindividuelles chezBellicositermes natalensis etCubitermes sp. la théorie de la stigmergie: Essai d'interprétation du comportement des termites constructeurs, Insect. Soc. 6 (1) (1959) 41–80.

[18] E. Bonabeau, M. Dorigo, G. Theraulaz, Swarm Intelligence: From Natural to Artificial Systems (No. 1), Oxford University Press, 1999.

[19] A.P. Engelbrecht, Fundamentals of Computational Swarm Intelligence, John Wiley & Sons, Ltd., Hoboken, NJ, 2006.

[20] S. Krause, R. James, J.J. Faria, G.D. Ruxton, J. Krause, Swarm intelligence in humans: diversity can trump ability, Anim. Behav. 81 (5) (2011) 941–948.

[21] J. Izquierdo, I. Montalvo, R. Pérez, V.S. Fuertes, Forecasting pedestrian evacuation times by using swarm intelligence, Phys. A Stat. Mech. Appl. 388 (7) (2009) 1213–1220.

[22] L. Atzori, A. Iera, G. Morabito, M. Nitti, The social internet of things (SIoT)–when social networks meet the internet of things: concept, architecture and network characterization, Comput. Netw. 56 (16) (2012) 3594–3608.

[23] O. Bello, S. Zeadally, Intelligent device-to-device communication in the internet of things, IEEE Syst. J. 10 (3) (2014) 1172–1182.

[24] P. Chamoso, F. De la Prieta, F. De Paz, J.M. Corchado, Swarm agent-based architecture suitable for internet of things and smart cities, in: Distributed Computing and Artificial Intelligence, 12th International Conference (pp. 21–29), Springer, Cham, 2015.

[25] P.E. Nielsen, Peptide nucleic acid (PNA): a model structure for the primordial genetic material? Orig. Life Evol. Biosph. 23 (5-6) (1993) 323–327.

[26] D.C. Dennett, Artificial intelligence as philosophy and as psychology, in: Brainstorms: Philosophical Essays on Mind and Psychology, 1978, pp. 109–126.
[27] H.A. Simon, Artificial intelligence: an empirical science, Artif. Intell. 77 (1995) 95–127.
[28] K. Mainzer, From embodied mind to embodied robotics: humanities and system theoretical aspects, J. Physiol.-Paris 103 (2009) 296–304.
[29] P.C. Garzón, F. Keijzer, Cognition in plants, in: Plant-Environment Interactions, Springer, Berlin, Germany, 2009, pp. 247–266.
[30] S. Martel, W. André, M. Mohammadi, Z. Lu, O. Felfoul, Towards swarms of communication-enabled and intelligent sensotaxis-based bacterial microrobots capable of collective tasks in an aqueous medium, in: Proceedings of the IEEE International Conference on Robotics and Automation, Kobe, Japan, 12–17 May, 2009, pp. 2617–2622.
[31] A. Adamatzky, Physarum Machines: Computers from Slime Mould, vol. 74, World Scientific, Singapore, 2010.
[32] G. Tononi, Consciousness as Integrated Information: A Provisional Manifesto, Mar. Biol. Lab. 215 (2008) 216–242.
[33] A. Clark, Natural-born cyborgs? in: Cognitive Technology: Instruments of Mind, Springer, Heidelberg, Germany, 2001, pp. 17–24.
[34] L. Cosmides, J. Tooby, From evolution to behavior: evolutionary psychology as the missing link, in: The Latest on the Best Essays on Evolution and Optimality, The MIT Press, Cambridge, MA, 1987.
[35] L.L. Cavalli-Sforza, A. Piazza, P. Menozzi, J. Mountain, Reconstruction of human evolution: bringing together genetic, archaeological, and linguistic data, Proc. Natl. Acad. Sci. U. S. A. 85 (1988) 6002–6006.
[36] L. Shapiro, Embodied Cognition, Routledge, Abingdon, 2010.
[37] A. Clark, D. Chalmers, The extended mind, Analysis 58 (1998) 7–19.
[38] L.W. Barsalou, Grounded cognition, Annu. Rev. Psychol. 59 (2008) 617–645.
[39] N. Harbisson, I Listen to Color. TEDX Talk, 2012, Available online: http://www.ted.com/talks/neil_harbisson_i_listen_to_color. (Accessed 22 August 2017).
[40] S.D. Guler, M. Gannon, K. Sicchio, Super humans and cyborgs, in: Crafting Wearables, Springer, Berlin, Germany, 2016, pp. 145–159.
[41] W. Wiltschko, R. Wiltschko, Magnetic orientation in birds, J. Exp. Biol. 199 (1996) 29–38.
[42] W.W. Cohen, A computer Scientist's Guide to Cell Biology: A Travelogue from a Stranger in a Strange Land, Springer, New York, NY, 2007.
[43] I. Akyildiz, M. Pierobon, S. Balasubramaniam, Y. Koucheryavy, The internet of bio-nano things, IEEE Commun. Mag. 53 (2015) 32–40.
[44] B. Atakan, O.B. Akan, S. Balasubramaniam, Body area nanonetworks with molecular communications in nanomedicine, IEEE Commun. Mag. 50 (2012) 28–34.
[45] A. Bozkurt, R. Gilmour, D. Stern, A. Lal, MEMS based bioelectronic neuromuscular interfaces for insect cyborg flight control, in: Proceedings of the IEEE 21st International Conference on Micro Electro Mechanical Systems, Tucson, AZ, USA, 13–17 January, 2008, pp. 160–163.
[46] W. Tsang, A. Stone, Z. Aldworth, D. Otten, A. Akinwande, T. Daniel, J. Hildebrand, R. Levine, J. Voldman, Remote control of a cyborg moth using carbon nanotube-enhanced flexible neuro prosthetic probe, in: Proceedings of the 2010 IEEE 23rd International Conference on Micro Electro Mechanical Systems (MEMS), Wanchai, Hong Kong, 24–28 January, 2010, pp. 39–42.
[47] A.D. Jadhav, I. Aimo, D. Cohen, P. Ledochowitsch, M.M. Maharbiz, Cyborg eyes: microfabricated neural interfaces implanted during the development of insect sensory organs produce stable neuro recordings in the adult, in: Proceedings of the 2012 IEEE 25th International

Conference on Micro Electro Mechanical Systems (MEMS), Paris, France, 29 January–2 February, 2012, pp. 937–940.

[48] G. Li, D. Zhang, Brain-computer interface controlled cyborg: establishing a functional information transfer pathway from human brain to cockroach brain, PLoS One 11 (2016), e0150667.

[49] C. Blum, X. Li, Swarm intelligence in optimization, in: Swarm Intelligence, Springer, Berlin, Germany, 2008, pp. 43–85.

[50] J. Krause, G.D. Ruxton, S. Krause, Swarm intelligence in animals and humans, Trends Ecol. Evol. 25 (2010) 28–34.

[51] N. Cohen, Hong Kong protests propel fire chat phone-to-phone app, New York Times 5 (October 2014) 26–28.

[52] E.A. Mennis, The wisdom of crowds: why the many are smarter than the few and how collective wisdom shapes business, economies, societies, and nations, Bus. Econ. 41 (2006) 63–65.

[53] S. Distefano, G. Merlino, A. Puliafito, Sensing and actuation as a service: a new development for clouds, in: Proceedings of the 2012 11th IEEE International Symposium on Network Computing and Applications (NCA), Cambridge, MA, USA, 23–25 August, 2012, pp. 272–275.

[54] H.H. Khondker, Role of the new media in the arab spring, Globalizations 8 (2011) 675–679.

[55] D. Bruneo, S. Distefano, F. Longo, G. Merlino, A. Puliafito, V. D'Amico, M. Sapienza, G. Torrisi, Stack4Things as a fog computing platform for smart city applications, in: Proceedings of the 2016 IEEE Conference on Computer Communications Workshops (INFOCOM WKSHPS), San Francisco, CA, USA, 10–15 April, 2016, pp. 848–853.

[56] G.F. Davis, What might replace the modern corporation: uberization and the web page enterprise, Seattle Univ. Law Rev. 39 (2015) 501.

[57] J. Vallverdú, M. Talanov, A. Khasianov, Swarm intelligence via the internet of things and the phenomenological turn, Philosophies 2 (3) (2017) 19.

[58] Y. Lu, W. Hu, Study on the application of ant colony algorithm in the route of internet of things, Int. J. Smart Home 7 (3) (2013) 365–374.

[59] P. Chamoso, F.D. Prieta, F.D. Paz, J.M. Corchado, 12th International Conference Distributed Computing and Artificial Intelligence, Vol. 373 of Advances in Intelligent Systems and Computing, Springer International Publishing, Cham, 2015.

[60] M. Ebrahimi, E. Shafieibavani, R.K. Wong, C.-H. Chi, A new metaheuristic approach for efficient search in the internet of things, in: 2015 IEEE International Conference on Services Computing, IEEE, 2015, pp. 264–270.

[61] V. Suryani, S. Sulistyo, W. Widyawan, Trust-based privacy for internet of things, Int. J. Electr. Comput. Eng. 6 (5) (2016) 2396.

[62] I. Sabbani, M. Youssfi, O. Bouattane, A multi-agent based on ant colony model for urban traffic management, in: International Conference on Multimedia Computing and Systems (ICMCS), IEEE, 2016, pp. 793–798.

[63] P. Lopez-Matencio, J. Vales-Alonso, E. Costa-Montenegro, ANT: agent Stigmergy-based IoT-network for enhanced tourist mobility, Mob. Inf. Syst. 2017 (2017) 1–15.

[64] O. Cosido, C. Loucera, A. Iglesias, Automatic calculation of bicycle routes by combining meta-heuristics and GIS techniques within the framework of smart cities, in: 2013 International Conference on New Concepts in Smart Cities: Fostering Public and Private Alliances (Smart MILE), IEEE, 2013, pp. 1–6.

[65] M.A. Kowshalya, M. Valarmathi, Detection of sybil's across communities over social internet of things, J. Appl. Eng. Sci. 14 (1) (2016) 75–83.

[66] O. Said, Analysis, design and simulation of internet of things routing algorithm based on ant colony optimization, Int. J. Commun. Syst. 30 (8) (2017).

[67] Y. Jiang, Q. Ding, X. Wang, A recovery model for production scheduling: combination of disruption management and internet of things, Sci. Program. 2016 (2016).
[68] V. Bellini, T.D. Noia, M. Mongiello, F. Nocera, A. Parchitelli, E.D. Sciascio, Reflective Internet of Things Middleware-Enabled a Predictive Real-Time Waste Monitoring System, Springer International Publishing, 2018.
[69] Y. Hong, L. Chen, L. Mo, Optimization of cluster resource indexing of internet of things based on improved ant colony algorithm, Clust. Comput. (2018) 1–9.
[70] G. Mahalaxmi, K.E. Rajakumari, Multi-agent technology to improve the internet of things routing algorithm using ant colony optimization, Indian J. Sci. Technol. 10 (31) (2017) 1–8.
[71] Z. Oralhan, B. Oralhan, Y. Yigit, Smart city application: internet of things (IoT) technologies based smart waste collection using data mining approach and ant colony optimization, Int. Arab J. Inform. Technol. 14 (4) (2017) 423–427.
[72] P. Thapar, U. Batra, Implementation of ant colony optimization in routing protocol for internet of things, in: Innovations in Computational Intelligence, Springer, 2018, pp. 151–164.
[73] W.-T. Sung, Y.-C. Chiang, Improved particle swarm optimization algorithm for android medical care IoT using modified parameters, J. Med. Syst. 36 (6) (2012) 3755–3763.
[74] W. Sung, C. Hsu, IoT system environmental monitoring using IPSO weight factor estimation, Sens. Rev. 33 (3) (2013) 246–256.
[75] C.-W. Tsai, C.-F. Lai, A.V. Vasilakos, Future internet of things: open issues and challenges, Wirel. Netw 20 (8) (2014) 2201–2217, https://doi.org/10.1007/s11276-014-0731-0.
[76] S. Luo, L. Cheng, B. Ren, Practical swarm optimization based fault-tolerance algorithm for the internet of things, KSII Trans. Internet Inf. Syst. 8 (4) (2014) 1178–1191.
[77] L. Hurtado, P. Nguyen, W. Kling, Smart grid and smart building interoperation using agent-based particle swarm optimization, Sustain. Energy Grids Netw. 2 (2015) 32–40.
[78] C. Fang, X. Liu, P.M. Pardalos, J. Pei, Optimization for a three-stage production system in the internet of things: procurement, production and product recovery, and acquisition, Int. J. Adv. Manuf. Technol. 83 (5-8) (2016) 689–710.
[79] S. Dai, M. Liwang, Y. Liu, Z. Gao, Hybrid Quantum-Behaved Particle Swarm Optimization for Mobile-Edge Computation Offloading in Internet of Things, Vol. 1 of Communications in Computer and Information Science, Springer, Singapore, 2018.
[80] W. Jinyi, Y. Qin, A. Shrestha, S.-J. Yoo, Optimization of cognitive radio secondary base station positioning and operating channel selection for IoT sensor networks, in: International Conference on Information and Communication Technology Convergence: ICT Convergence Technologies Leading the Fourth Industrial Revolution, ICTC 2017 2017- Decem, 2017, pp. 397–399.
[81] T. Kumrai, K. Ota, M. Dong, J. Kishigami, D.K. Sung, Multiobjective optimization in cloud brokering systems for connected internet of things, IEEE Internet Things J. 4 (2) (2017) 404–413.
[82] A.L. Sangeetha, N. Bharathi, A.B. Ganesh, T.K. Radhakrishnan, Particles warm optimization tuned cascade control system in an internet of things (IoT) environment, Meas.: J. Int. Meas. Confed. 117 (2018) 80–89.
[83] M. Tuba, N. Bacanin, Hybridized bat algorithm for multi-objective radiofrequency identification (RFID) network planning, in: 2015 IEEE Congress on Evolutionary Computation (CEC), IEEE, 2015, pp. 499–506.
[84] L. Huo, Z. Wang, Service composition instantiation based on cross-modified artificial bee colony algorithm, China Commun. 13 (10) (2016) 233–244.
[85] M. Yi, Q. Chen, N.N. Xiong, The adaptive recommendation mechanism for distributed parking service in smart city, Sensors 11 (16) (2016) 395–413.

[86] X. Xu, Z. Liu, Z. Wang, Q.Z. Sheng, J. Yu, X. Wang, S-ABC: a paradigm of service domain-oriented artificial bee colony algorithms for service selection and composition, Future Gener. Comput. Syst. 68 (2017) 304–319.

[87] M. Khan, S. Din, M. Gohar, A. Ahmad, S. Cuomo, F. Piccialli, G. Jeon, Enabling multimedia aware vertical handover management in internet of things based heterogeneous wireless networks, Multimed. Tools Appl. 76 (24) (2017) 25919–25941.

[88] A. Ahmad, M. Khan, A. Paul, S. Din, M.M. Rathore, G. Jeon, G.S. Choi, toward modeling and optimization of features selection in big data based social internet of things, Futur. Gener. Comput. Syst. 82 (2018) 715–726.

[89] Z. Muhammad, N. Saxena, I.M. Qureshi, C. Wook, Hybrid artificial bee colony algorithm for an energy efficient internet of things based on wireless sensor network, IETE Tech. Rev. 34 (S1) (2017) 39–51.

[90] M. Praveen Kumar Reddy, M.R. Babu, Energy efficient cluster head selection for internet of things, New Rev. Inf. Netw. 22 (1) (2017) 54–70.

[91] P. Reddy, R. Babu, An evolutionary secure energy efficient routing protocol in internet of things, Int. J. Intell. Eng. Syst. 10 (3) (2017) 337–346.

[92] A. Khalid, N. Javaid, M. Guizani, M. Alhussein, K. Aurangzeb, M. Ilahi, Towards dynamic coordination among home appliances using multi-objective energy optimization for demand side management in smart buildings, IEEE Access 6 (2018) 19509–19529.

[93] K. Balamurugan, R. Mahalakshmi, Mathematical analysis of BLDC motor parameter estimation using internet of things, TAGA J. Graphic Technol. 14 (2018) 2429–2436.

[94] Z. Cui, Y. Cao, X. Cai, J. Cai, J. Chen, Optimal LEACH protocol with modified bat algorithm for big data sensing systems in internet of things, J. Parallel Distrib. Comput. 132 (2017) 217–229.

[95] G.-J. Horng, The adaptive recommendation mechanism for distributed parking service in smart city, Wirel. Pers. Commun. 80 (1) (2015) 395–413.

[96] E. Fadel, M. Faheem, V. Gungor, L. Nassef, N. Akkari, M. Malik, S. Almasri, I. Akyildiz, Spectrum-aware bio-inspired routing in cognitive radiosensor networks for smart grid applications, Comput. Commun. 101 (2017) 106–120.

[97] G. Garofalo, A. Giordano, P. Piro, G. Spezzano, A. Vinci, A distributed real-time approach for mitigating CSO and flooding in urban drainage systems, J. Netw. Comput. Appl. 78 (2017) 30–42.

[98] Z. Amjad, S. Batool, H. Arshad, K. Parvez, M. Farooqi, N. Javaid, Pigeon inspired optimization and enhanced differential evolution in smart grid using critical peak pricing, in: International Conference on Intelligent Networking and Collaborative Systems, Springer, 2017, pp. 505–514.

[99] C.A.U. Hassan, M.S. Khan, A. Ghafar, S. Aimal, S. Asif, N. Javaid, Energy optimization in smart grid using grey wolf optimization algorithm and bacterial foraging algorithm, in: International Conference on Intelligent Networking and Collaborative Systems, Springer, 2017, pp. 166–177.

[100] G. Sun, Y. Liu, M. Yang, A. Wang, S. Liang, Y. Zhang, Coverage optimization of VLC in smart homes based on improved cuckoo search algorithm, Comput. Netw. 116 (2017) 63–78.014.

[101] B. Christian, M. Daniel, Swarm Intelligence. Introduction and Applications, Natural Computing Series, Springer, 2008.

[102] I. Yaqoob, E. Ahmed, I.A.T. Hashem, A.I.A. Ahmed, A. Gani, M. Imran, M. Guizani, Internet of things architecture: recent advances, taxonomy, requirements, and open challenges, IEEE Wirel. Commun. 24 (3) (2017) 10–16.

[103] O. Bello, S. Zeadally, Intelligent device-to-device communication in the internet of things, IEEE Syst. J. 10 (3) (2016) 1172–1182.

[104] J.E. Bell, P.R. McMullen, Ant colony optimization techniques for the vehicle routing problem, Adv. Eng. Inform. 18 (1) (2004) 41–48.

[105] L. Dou, M. Li, Y. Li, Q.Y. Zhao, J. Li, Z. Wang, A novel artificial bee colony optimization algorithm for global path planning of multi robot systems, in: 2014 IEEE International Conference on Robotics and Biomimetics (ROBIO 2014), December 2014, pp. 1186–1191.

[106] S.K. Datta, C. Bonnet, N. Nikaein, An IoT gateway centric architecture to provide novel M2M services, in: 2014 IEEE World Forum on Internet of Things (WF-IoT), March 2014, pp. 514–519.

[107] I. Kassabalidis, M.A. El-Sharkawi, R.J. Marks, P. Arabshahi, A.A. Gray, Swarm intelligence for routing in communication networks, in: Global Telecommunications Conference, 2001. GLOBECOM'01. IEEE, vol. 6, pp. 3613–3617, 2001.

[108] S. Cheng, Y. Shi, Q. Qin, R. Bai, Swarm Intelligence in Big Data Analytics, Springer Berlin Heidelberg, Berlin, Heidelberg, 2013, pp. 417–426.

[109] T. Chakraborty, S.K. Datta, Application of swarm intelligence in internet of things, in: 2017 IEEE International Symposium on Consumer Electronics (ISCE) (pp. 67–68). IEEE, 2017, November.

Chapter 24

Security and challenges in IoT-enabled systems

S. Kala[a] and S. Nalesh[b]
[a]*Department of Electronics and Communication Engineering, Indian Institute of Information Technology Kottayam, Kottayam, India,* [b]*Department of Electronics, Cochin University of Science and Technology, Kochi, India*

24.1 Introduction

Internet of things (IoT) is one of the biggest trends in both the industry and the consumer good market in the current years. Applications of IoT spans for a wide range, including home automation, smart cities, transportation, health care, retail, public safety, and many more. Wearable devices for medical and fitness applications are finding more popularity these days. IoT devices connect various components, such as sensors, actuators, etc., which transfer information through internet. Typically, IoT enables the physical world to collect data, process it, and then transmit it through the internet. The number of devices connected to IoT framework is exponentially increasing. Even though IoTs enable our daily life experiences more convenient, it raises several safety issues also. Large amount of sensitive data could to be processed and transmitted in all these applications.

Privacy and security are of great concern for any IT services, but are often not given priority while designing and implementing an IoT system [1–3]. Especially when the framework is used in sensitive applications such as health care and defense, security should be the primary concern. The isolated devices that are connected to IoT may lack protection of its own. IoT framework consists of numerous such devices connected to each other, which brings in several challenges of security and privacy, when compared to managing a single device. Moreover, unlike the security issues in a conventional computer network, the challenges that arise due to the resource constraints, power budgets, and cost are additional in the case of IoT framework.

Connectivity is one of the sources for vulnerabilities to be created in the IoT system, as IoT consists of numerous devices connected to each other and hence authentication is essential before a new device is being connected to the system.

System Assurances. https://doi.org/10.1016/B978-0-323-90240-3.00024-2

Deployment of security measures from end to end in an IoT is a challenging task and hence selection of the appropriate hardware for IoT device is very important in this scenario. IoT has been bound with biometrics in Ref. [4] in order to increase the security level.

Security levels in IoT can be divided mainly into two categories, that is, software level and hardware level. Security measures can be taken either at the software level, where the solutions are relatively cheaper but has the capability to upgrade the services as and when the threats evolve. For example, hacking, illegal access, etc., falls under software level. A malware can enter the system without the user's knowledge and may not affect the functioning of the system, but can collect confidential data such as credit card passwords and similar sensitive data. These attacks can be prevented to some extent by updating the software by using firewalls, etc. Hardware attacks differ from software attacks in many ways. For hardware attacks, the attacker should have a good knowledge about the hardware, whereas in software attacks, software tools are available in internet, for injecting the attacks. Important layers in IoT trust are shown in Fig. 24.1.

As more and more transistors are integrated on to a chip, very large-scale integration (VLSI) fabrication steps have become completely distributed. Integrated circuit (IC) designers rely on other vendors for developing the chips. Also, usage of intellectual property (IP) cores from various vendors and accessing computer-aided design tools from different companies are creating an insecure environment. A malicious threat can be injected at any point of the IC design or even after fabrication. One of the examples of this kind of threat is hardware trojans and is discussed in detail in Section 24.3.

Once the hardware (chip) has been manufactured, the next step is supply chain process. Attacks can be injected even during the process of shipping. It could be repacked so that consumers are not aware of it. In some cases, malwares can be present in hardware devices such as memory cards, USBs, and so on. Circuits such as RFID, application-specific integrated circuit (ASIC), etc. are also attacked to get the information present in it. In some cases, the

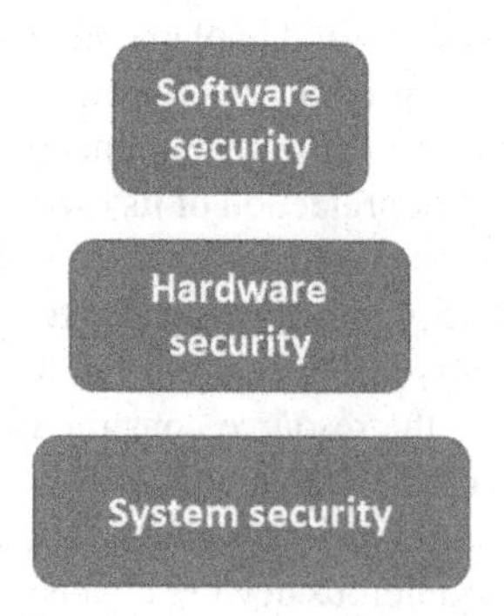

FIG. 24.1 Various layers in IoT trust.

secrecy will be in the form of algorithms in these hardware units, and hence attackers will try to reverse engineer these algorithms.

One of the defending mechanisms against threats in IoT is to add a protection at the very lower level, that is, implementation at hardware or silicon level. Here, the hardware layer is isolated from the software layer and as software layers keep on updating, the keys that are used for software execution on IoT need to be protected on the hardware level. Moreover, data available in hardware implementations which are physically isolated from other interfaces remain stable and cannot be changed, which makes it more trustworthy.

In a hardware-based security system, specialized hardware for cryptography functions can be used to secure the IC. These hardware can be optimized for providing better performance with less area overhead. Hardware-based solutions for security can be effectively used for many applications, where the device or IC is physically accessible to threat.

24.2 Commercialized secure hardware primitive designs

IoT devices have limited processing capability, storage, and power as already mentioned. Most of the cyber-security solutions are implemented on software and hence security of IoT devices needs to be explored widely [5]. Before discussing about the security techniques or methods adopted in IoT, an understanding of various kinds of threats or attacks is essential. Hardware attacks such as trojans (will be discussed in detail in Section 24.3), side-channel attacks (SCA) (will be discussed in detail in Section 24.4), physical probing, etc. are possible, whereas there is another classification called software attacks, where the device is not physically attacked. Here the attacker tries to deny the services, or make distortions in the output or tries to find out the secret key or password existing in a device.

Fig. 24.2 shows various attack vectors in IoT and some of the security techniques deployed using hardware. Malwares such as virus, trojan, etc. work as operating system (OS) level attackers [5]. Integrity violation denotes to access

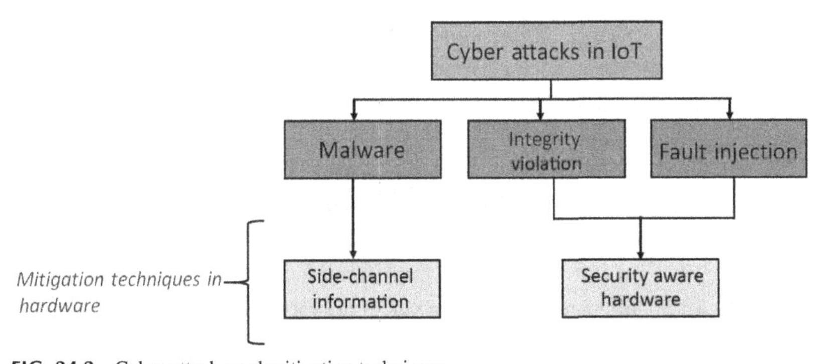

FIG. 24.2 Cyber-attacks and mitigation techniques.

the private or secret keys, which are embedded in the IoT device. In fault injection, the faults are injected during runtime, which requires good understanding of the hardware as well as the software of that particular device. Software-based solutions may not be effective since the coverage of defending mechanism is less there, whereas in hardware-based methods, information at the microarchitecture level is gathered for defending the threats or attacks. Event-based monitoring has gained popularity in malware detection in IoT devices [6].

Secure hardware refers to an integrated circuit (IC) which can protect the secret information and data present in your hardware if attackers enter the device. But even though secure hardware protects the device, there can be gaps for exploitation, for example, from any application perspective. Secured hardware can be deployed either as a stand-alone IC, which acts as a coprocessor for security functionality along with the main processor or it can be integrated into the same IC where the functionality is realized. In the latter case, only single IC will be there. This is shown in Fig. 24.3.

System-on chip vendors include various hardware primitive designs, for securing their devices. Some of those designs are discussed here. For example, (a) in ARM Trustzone a hardware-based isolation of both program data and code are done as discussed in Refs. [7, 8]. (b) Trusted platform module (TPM) is another technique that utilizes cryptographic keys for securing applications on an IC.

Hardware attacks can be either active attacks or passive attacks [9]. In an active attack, the operating conditions are not according to the specifications, where as in passive attacks, the device will work according to its operating specifications. Both these kinds of attacks can be either invasive or noninvasive attacks. Hardware threats on IoT can be classified into various categories, where the main ones which are paid more attention are hardware trojans, side-channel attacks, and reverse engineering. These threats are discussed in the following sections.

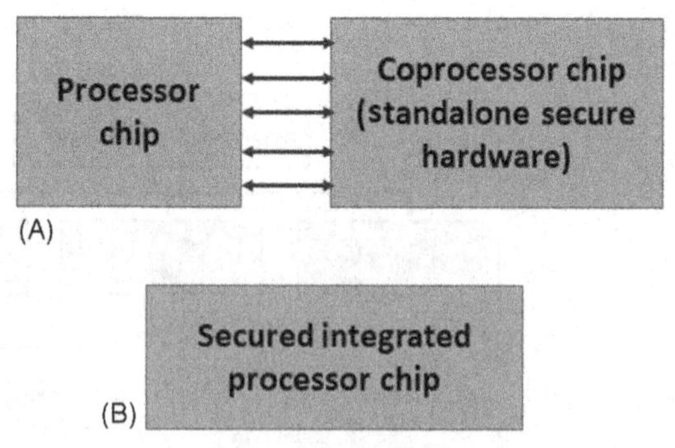

FIG. 24.3 Secured hardware. (A) Security module as an independent IC and (B) security module is integrated.

24.3 Hardware trojan

Integrated circuit designing and manufacturing involve different third-party vendors and raises the concern of security of the ICs. The chip is prone to be attacked at any stage in the supply chain. Hardware trojans are malicious modifications on the actual IC design, where the chip does not perform the functionality which it is supposed to perform. There are several research papers that discuss about hardware trojan models and their detection techniques as given in Refs. [10–14].

Researchers classify trojans under various categories such as combinational and sequential and also based on the triggering functions used, such as analog and digital [10]. Trojan detection techniques can be classified into various categories such as destructive and nondestructive type. Nondestructive techniques are again classified as invasive and noninvasive approaches. One of the classifications of hardware trojan designs include trojan trigger and trojan payload as shown in Fig. 24.4. Various triggering and payload mechanism have been researched over years [11]. The trigger will look into any events happening in the circuit and the payload will be activated once an event has been detected.

Design for trust is an approach used to detect the trojans in hardware [11]. Here, functional testing and side-channel analysis are mainly performed to detect the trojan threats. Hardware trojans could be inserted into the device at any phase of its development. This is shown in Fig. 24.5. Here, different phases of IC design and manufacturing are shown, such as usage of computer-aided design (CAD) tools and intellectual property (IP), developing the layout of design and manufacturing. A trojan can be inserted at any of these phases as shown in Fig. 24.5.

24.4 Side-channel attack (SCA)

SCA is an attack where the physical properties of the hardware device are observed in order to get the functionality of that particular IoT device [15–17]. In this type of attack, side-channel signals which are the physical

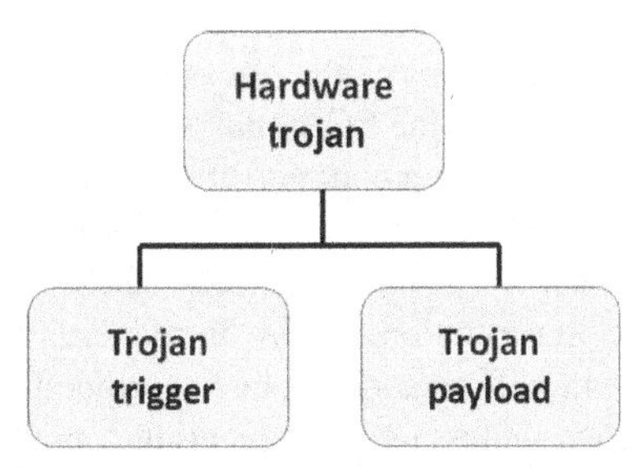

FIG. 24.4 Hardware trojans.

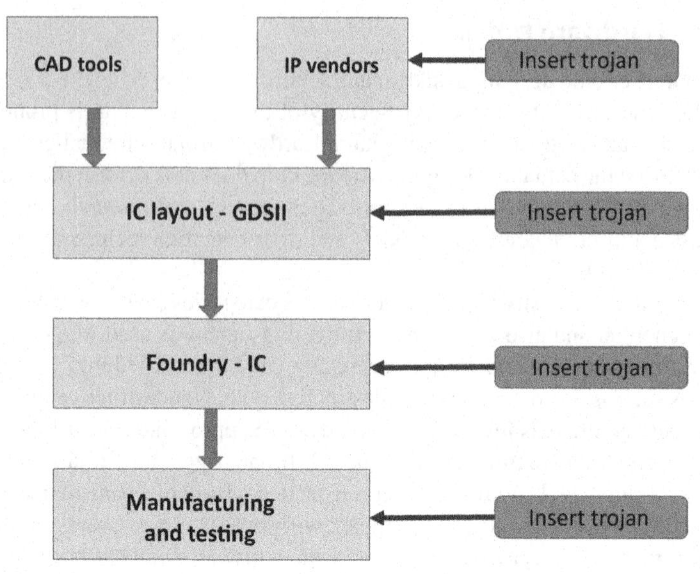

FIG. 24.5 Insertion of hardware trojans.

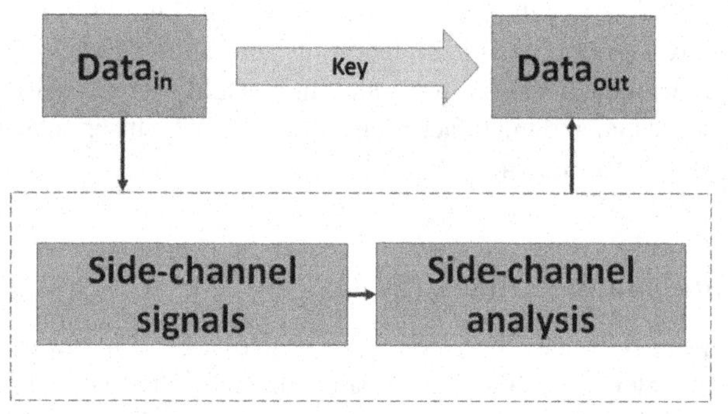

FIG. 24.6 Side-channel attack analysis.

properties, such as power, memory, etc., are analyzed. This is a noninvasive type of attack. Fig. 24.6 shows the SCA analysis.

Differential power analysis is a common attack where the attackers can easily measure the power and electromagnetic information of the device, which it is hold. SCA is more prone to low-cost IoT devices. These attacks can be defended either by physical implementations or by introducing algorithmic changes. Information leakage via side channels can be avoided by physically hiding them, using several logic gates, by implementing a power management unit. This is for power side-channel attacks. But still EM attacks are to be defended, using an EM monitor.

24.5 Reverse engineering

In this type of hardware attacks, the attackers will try to find out what the circuit does. Once the functionality is found out by understanding the algorithms behind it, the algorithm itself will be modified. This type of attack was seen in subway cards.

Other categories in hardware attacks include black-box testing, physical probing, and fault generation. Black-box testing is a noninvasive testing, where the attacker sends an input (or random inputs) to the circuit and receives an output. Based on the behavior of the output, the algorithm used in the circuit can be found out. Physical probing is an invasive method of attack. Here, the attacker inserts a probe to get the information either through the memory or through the bus or other modules or blocks in the circuit. Fault generation is a noninvasive attack, where the technologies can be failed. In this method, memory contents can be changed, or the clock signal can be affected and many more.

Random number generators are another solution to prevent attacks. But they are also power consuming and area expensive. Secret keys for authentication and encryption need to be stored in the chip. Generally, nonvolatile memory storage or physically unclonable functions (PUFs) are used for this purpose [18].

One of the requirements for a secure IoT is owner verification or identification and authentication. Authentication is generally performed using cryptographic keys. Implementation of these cryptographic methods poses constraints in area and power. Hence, another commonly used technique has been proposed, called PUF, which uses lightweight authentication for preventing malicious threats. PUF circuits are widely used to prevent malicious threats. Running these cryptography algorithms in software is not desirable due to its large latency, and memory requirement. Researchers have proposed accelerators for lightweight AES algorithm for reducing power and better utilization of resources.

24.6 Key challenges

IoT devices will have very low storage capacity and computational capacity. They are resource constrained and are also power constrained for some applications. Security and reliability of IoT poses several challenges in this aspect. Autonomous operation of the devices can be one of the requirements where there is less accessibility. Some of the key challenges that may occur while introducing security and privacy to any IoT system are as follows:

- Adding a security module to the system will increase the area overhead, which is undesirable for the tiny “Things” in IoT.
- Cryptographic algorithms with large computational complexity are not suitable for IoT framework.

- Updating the software in the IoT device is essential to keep it secure for long period.
- Battery power is limited in most of the IoT systems. There are cases where the entire IoT set up has to be kept where human access is not possible for long period. Here, a better battery solution is required for proper functioning.

Several machine learning (ML) approaches are also there to detect malwares in IoT, during runtime. But ML algorithms will require training of large dataset, its testing, etc. [19], and is effective if the attack is known priorly. Designing ML algorithm for malware detection is a challenging task. Security aware hardware design is another solution, where one of the techniques is to make a module for injecting timing faults to the device, during runtime.

24.7 Conclusion

Internet of things is a promising emerging technology for a wide variety of application domains including health care, transportation, waste management, smart city, defense mechanism, etc. Since each of the IoT device has varied quality, security aspect needs prior attention. Also if the application domain is sensitive, such as defense system or health-care system, then privacy is of utmost importance. Security and privacy of hardware, especially IoT devices, have become a popular area of research in the recent years. Significant progress has been made in this research topic and several open problems are also there. In this chapter, we discussed several kinds of attacks which an IoT is prone to. Some of the hardware and software threats were analyzed in detail. We have focused more on toward the hardware threats such as hardware trojans, side-channel attacks, and reverse engineering. Main challenges in IoT arises due to the limited computational power and battery power, which restricts highly efficient cryptographic algorithms to be deployed on the device. This has also been discussed in this chapter. Classification of threats, how it is affecting the device, and their defending mechanisms have been explained here.

References

[1] B. Pearson, On misconception of hardware and cost in IoT security and privacy, in: ICC 2019—2019 IEEE International Conference on Communications (ICC), Shanghai, China, vol. 2019, pp. 1–7.

[2] S. Koley, P. Ghosal, Addressing hardware security challenges in internet of things: recent trends and possible solutions, in: IEEE 12th International Conference on Ubiquitous Intelligence and Computing and 2015 IEEE 12th International Conference on Autonomic and Trusted Computing and 2015 IEEE 15th International Conference on Scalable Computing and Communications and Its Associated Workshops (UIC-ATC-ScalCom), Beijing, China, 2015, pp. 517–520, https://doi.org/10.1109/UIC-ATC-ScalCom-CBDCom-IoP.2015.105.

[3] S.R. Rajendran, Security challenges in hardware used for smart environments, in: Internet of Things and Secure Smart Environments, 2020, https://doi.org/10.1201/9780367276706-9.

[4] Z. Guo, N. Karimian, M.M. Tehranipoor, D. Forte, Hardware security meets biometrics for the age of IoT, in: 2016 IEEE International Symposium on Circuits and Systems (ISCAS), Montreal, QC, Canada, 2016, pp. 1318–1321, https://doi.org/10.1109/ISCAS.2016.7527491.

[5] F. Rahman, M. Farmani, M. Tehranipoor, Y. Jin, Hardware-assisted cybersecurity for IoT devices, in: 2017 18th International Workshop on Microprocessor and SOC Test and Verification (MTV), Austin, TX, USA, 2017, pp. 51–56, https://doi.org/10.1109/MTV.2017.16.

[6] A. Tang, S. Sethumadhavan, S.J. Stolfo, Unsupervised anomaly-based malware detection using hardware features, in: International Workshop on Recent Advances in Intrusion Detection, Springer, 2014, pp. 109–129.

[7] Y. Jin, Towards hardware-assisted security for IoT systems, in: 2019 IEEE Computer Society Annual Symposium on VLSI (ISVLSI), Miami, FL, USA, 2019, pp. 632–637, https://doi.org/10.1109/ISVLSI.2019.00118.

[8] C. Lesjak, D. Hein, J. Winter, Hardware-security technologies for industrial IoT: TrustZone and security controller, in: IECON 2015—41st Annual Conference of the IEEE Industrial Electronics Society, Yokohama, Japan, 2015, pp. 002589–002595, https://doi.org/10.1109/IECON.2015.7392493.

[9] F. Tudosa, E. Picariello, L.D.V. Balestrieri, F. Lamonaca, Hardware security in IoT era: the role of measurements and instrumentation, in: 2019 II Workshop on Metrology for Industry 4.0 and IoT (MetroInd4.0&IoT), Naples, Italy, 2019, pp. 285–290, https://doi.org/10.1109/METROI4.2019.8792895.

[10] R.S. Chakraborty, S. Narasimhan, S. Bhunia, Hardware trojan: threats and emerging solutions, in: 2009 IEEE International High Level Design Validation and Test Workshop, San Francisco, CA, USA, 2009, pp. 166–171, https://doi.org/10.1109/HLDVT.2009.5340158.

[11] K. Xiao, D. Forte, Y. Jin, R. Karri, S. Bhunia, M. Tehranipoor, Hardware trojans: lessons learned after one decade of research, ACM Trans. Des. Autom. Electron. Syst. 22 (1) (2016), https://doi.org/10.1145/2906147.

[12] J. Dofe, J. Frey, Q. Yu, Hardware security assurance in emerging IoT applications, in: 2016 IEEE International Symposium on Circuits and Systems (ISCAS), Montreal, QC, Canada, 2016, pp. 2050–2053, https://doi.org/10.1109/ISCAS.2016.7538981.

[13] S. Sidhu, B.J. Mohd, T. Hayajneh, Hardware security in IoT devices with emphasis on hardware trojans, J. Sens. Actuator Netw. 8 (2019) 42, https://doi.org/10.3390/jsan8030042.

[14] S.R. Rajendran, N. Devi, Malicious hardware detection and design for trust: an analysis, Electrotech. Rev. 84 (1–2) (2017) 7–16.

[15] F. Koeune, F.-X. Standaert, A tutorial on physical security and side-channel attacks, in: Foundations of Security Analysis and Design III, vol. 3655, 2005.

[16] Q. Zhang, A. Wang, Y. Niu, N. Shang, R. Xu, G. Zhang, L. Zhu, Side-channel attacks and countermeasures for identity-based cryptographic algorithm SM9, Secur. Commun. Netw. (2018), https://doi.org/10.1155/2018/9701756.

[17] S. Liu, Y. Wei, J. Chi, F.H. Shezan, Y. Tian, Side channel attacks in computation offloading systems with GPU virtualization, in: 2019 IEEE Security and Privacy Workshops (SPW), San Francisco, CA, USA, 2019, 2019, pp. 156–161, https://doi.org/10.1109/SPW.2019.00037. vol.

[18] K. Yang, D. Blaauw, D. Sylvester, Hardware designs for security in ultra-low-power IoT systems: an overview and survey, IEEE Micro 37 (6) (2017) 72–89, https://doi.org/10.1109/MM.2017.4241357.

[19] S. Kala, S. Nalesh, Efficient CNN accelerator on FPGA, IETE J. Res. 66 (6) (2020) 733–740, https://doi.org/10.1080/03772063.2020.182179.

Chapter 25

Provably correct aspect-oriented modeling with UPPAAL timed automata

Jüri Vain, Leonidas Tsiopoulos, and Gert Kanter
Department of Software Science, Tallinn University of Technology, Tallinn, Estonia

25.1 Introduction

Modeling is inescapable prerequisite of model-based techniques. Modeling languages provide conceptual framework for extracting features of concern, assuring correctness by model construction techniques, and creating the opportunities for verification and test automation [1].

The key challenge with model-based development is the complexity of real-world systems that reflects also in models and that makes both the analysis of models and their comprehension by developers difficult. The models of real-world systems such as cyber-physical infrastructures or financial systems, global weather monitoring networks, etc., can easily exceed the capability limits of contemporary software and data analysis methods, not to say about human comprehension. There are two general approaches to deal with systems complexities, at first, increasing the level of model abstractions and, second, applying various modularization techniques. Just to name few of modularization paradigms: program slicing, object orientation, design viewpoints, actors, etc. all of them serve the goal of splitting the system descriptions into smaller and tractable parts so that other components can be partially or completely ignored when modeling and analyzing the components of interest [2].

The focus in this chapter is on aspect-orientation as one possible modularization mechanism for reducing the model construction and analysis complexity. As stated in Ref. [3], aspects are usually defined as units of system decomposition that can be either functional or nonfunctional. An aspect in the requirements is a concern that crosscuts requirements artifacts. Early identification and managing of aspects helps to improve modularity already in the requirements as well as when designing the implementation architecture. Knowing the requirement-level aspects helps the architect to design a better

System Assurances. https://doi.org/10.1016/B978-0-323-90240-3.00025-4

system, whereas, knowing the architecture-level aspects helps producing a more robust implementation [3].

The ideas of aspect-oriented modularization emerged in programming languages at first [4, 5] and have spread over all the software development process giving birth to so-called Aspect-Oriented Software Development (AOSD) paradigm. AOSD as argued in Ref. [6] aims at addressing crosscutting concerns by providing means for their systematic *identification*, *separation*, *representation*, and *composition*. In particular, AOSD focuses on the modularization and composition of crosscutting concerns.

Aspect-oriented modeling (AOM) [7, 8] grew out from the needs of AOSD. To address the problems of complexity and traceability of aspect-oriented (AO) models formally this chapter revisits the key principles of AOM in the context of UPPAAL timed automata (UTA) formalism and develops a method for creating AO models that are correct by construction. The correctness of AO model construction is assured by a set of steps that are built in the method. Namely, these are TCTL temporal logic model-checking steps that are executed in lock-step with AO model construction steps applied.

The approach is exemplified using the home rehabilitation system (HRS) case study. The quantitative characterization of the advantages of the proposed AOM method is presented in the form of analytical estimates of the modeling effort and the time complexity of model-checking needed for verification of weaving correctness.

The rest of the chapter consists of Section 25.2 on related work, Section 25.3 with the preliminaries on AOM and UTA for AOM, Section 25.4 where we present our weaving approach together with the weaving correctness properties, Section 25.5 where we exemplify our approach, Section 25.6 where we elaborate on the usability and effort induced by our approach, and Section 25.7 where we conclude this chapter and discuss future work.

25.2 Related work

In an early work, Jacobson and Ng [9] developed various views of a system by describing system aspects as use cases. The authors of [9] provide the basis of AOM but details about the model engineering are not given. Several UML profiles have been proposed for aspects modeling, for example, Refs. [10, 11]. Studies in Refs. [12–14] provide surveys and assessments of AOM techniques that mainly focus on UML models and heuristic AOM approaches. We target correct-by-construction AOM with advanced tool support while introducing the AOM constructs. Although UTA is less expressive than UML, UTA models having formal semantics are better suited for timed model-checking and test generation.

Due to the several AO UML extensions, extensive research effort has been put into integrating such UML-based approaches with various formalisms in order to be able to check correctness of the models and/or generate tests from the models. For instance, the work by D. Xu was one of the first concerning generation of tests from aspectual use case diagrams [15]. In their approach, authors proposed translating the use case diagrams into AO Petri Nets first, before generating the use case sequences based on transition, state, and use case coverage. This work is related to ours but it is restricted to use Petri Net models and does not consider timing constraints. In Ref. [16], another UML profile was proposed for modeling behavior with aspect state machines and generating automatically test cases. Our approach differs from this work by separating aspects to provide AO metrics that can be used as coverage criteria that are aspect specific. However, testing is out of the scope of this chapter and is addressed in Ref. [17] in more detail. Generally, numerous interesting and valuable research results on AO model-based system development and verification have been published during the last 20 years. Few of such results can be observed in Refs. [18–24]. The main differences between our approach and these works are (i) handling of real-time specifications and (ii) being able to model, verify, and deploy model-based testing within the same framework with strong tool support through the UPPAAL tool suite.

Our first approach for integrating AOM concepts in UTA was presented in Ref. [25]. This work proposed to develop the aspect models by refining locations and edges in the base model. Our extensions to this work are presented in detail in Section 25.4.

Truscan et al. [26] also used UTA and aspects modeling was handled with a set of different weaving operators. It is well known in AOSD that weaving aspects to a base model may cause interference [12, 27]. In order to handle possible aspect interferences, the results of [26] were extended and published in Ref. [28]. This extension was based on the noninterference criteria via assume/guarantee assertions proposed in Ref. [29]. Differently from Iqbal et al. [28], in this article the noninterference of weaving is granted by weavers that are based on superposition refinement and related to them correctness properties. Moreover, instead of propositional and linear temporal logic (LTL) used in Ref. [29], we consider specifications expressed in timed computation tree logic (TCTL) [30] targeting real-time systems.

Hannousse et al. [31] also proposed the use of UTA to verify aspect noninterference but in a different setting and with objectives different from ours. Elementary and composite components with their interfaces as well as aspect components are modeled in an architectural description language based on the fractal component model and then a transformation is performed to create a network of automata that can be verified with the UPPAAL model checker against given system properties in TCTL. If verification fails due to aspect interferences, then different weaving operators modeled as UPPAAL templates composed in parallel with the base and aspect model processes, can be deployed to solve the interferences. We follow a slightly different approach, aiming at compositional verification and testing by applying aspect weaving correctness verification locally which assures that weaving is conservative with respect to the model global properties including aspects noninterference.

This chapter is a substantial extension of previous work [25]. We extend the notion of join point by typing them and introduce aspect weavers for each of the type. We also define correctness criteria of augmented models for showing aspect weaving local correctness at join points that entails noninterference between aspects. We exemplify our approach on the HRS system and present its analytical characterization.

25.3 Preliminaries

25.3.1 Aspect-oriented modeling

In AOM, the core functionality of the system under development is captured in the *base model*. The features such as reliability, security, safety, performance, etc. supporting the core functionality are added to the base model and refined in the form of *advices*. Composing an advice with base model is called *aspect weaving*. The advices can be woven with the base model at several places which are called *join points*. The rules which define under what conditions and where exactly the weaving should take place are defined by *pointcut specifications*.

We elaborate the AOM approach in the semantic framework of UTA. This formalism has been proven to be one of the most relevant formalisms for modeling and verification of real-time systems. UTA has also mature tool support for hierarchical and stochastic modeling and correctness verification by model checking.

This motivates us extending the UTA modeling technique with AO constructs preserving at the same its syntax and semantics. Thanks to this conservative approach, the existing UPPAAL tool set can be fully exploited for AO model simulation and verification purposes. In particular, we capitalize on verification of weaving correctness properties and noninterference checks of advices that makes applying our AO model building and verification approach compositional. In this chapter, the noninterference of aspects is granted by weaving operation that is based on superposition refinement and related to that refinement rules.

25.3.2 UPPAAL timed automata

UTA [32] are defined as a closed network of extended timed automata that are called *processes*. The processes are combined into a single system by synchronous parallel composition. The nodes of the automata graph are called *locations* and directed arcs between locations are called *edges*. The *state* of an automaton consists of its current location and valuation of all variables, including clocks. Synchronous communication between processes is done by synchronization links called *channels*. A channel *ch* relates a pair of transitions in parallel processes where synchronized edges are labeled with symbols for input and output actions (denoted `ch?` and `ch!`, respectively).

Formally, an UTA is given as the tuple (L, E, V, CL, $Init$, Inv, T_L), where

- L is a finite set of locations.
- E is the set of edges defined by $E \subseteq L \times G(CL, V) \times Sync \times Act \times L$, where $G(CL, V)$ is the set of constraints in guards, $Sync$ is a set of synchronization actions over channels and Act is a set of assignments with integer and Boolean expressions and clock resets.
- V denotes the set of variables of Boolean and integer type and arrays of those.
- CL denotes the set of real-valued clocks ($CL \cap V = \varnothing$).
- $Init \subseteq Act$ is a set of initializing assignments to variables and clocks.
- Inv: $L \rightarrow I(CL, V)$ maps locations to the set of invariants over clocks CL and variables V.
- $T_L: L \rightarrow \{ordinary, urgent, committed\}$ defines types of locations.

In this chapter, we present an AOM approach in the semantic framework of UTA. Specifically, AOM terms in the context of UTA can be defined as follows:

- A *base model* is a set of UTA processes that model the core functionality of the system.
- An *aspect model* is an UTA process or a composition of parallel processes that implement a crosscutting concern.
- An *advice model* introduces features and behaviors specific to a given aspect.
- *Weaving* is the process of composing a base model with the advice model.
- *Join points* are model fragments in the base model to which an aspect can be woven.
- A *pointcut* is the set of join points and conditions under which an advice can be woven. A *pointcut expression* is a logic condition which uniquely defines the model fragments (join points) where the weaving is applied.
- A *woven model* is an UTA in which the base model is woven with the intended aspects.

25.4 Provably correct weaving of aspects

In Ref. [25], the weaving of advices to base model is implemented via superposition refinement of UTA structural elements such as locations and edges and called location and edge refinement, respectively. The locations and edges that represent join points are called *atomic* join points to distinguish them from larger base model fragments that can similarly be used as join points. To keep the resulting augmented model still modular, the proposed weaving construct is not simple substitution of the join point element with advice model. Instead, parallel composition of the base model and advice is used where synchronization between them is granted by model construct named *weaver*. However, this parallel composition-based weaving construct does not introduce any extra behaviors in the augmented model that is typically due to interleaving of executions of parallel processes. The equivalence of operational semantic between the models of direct element substitution and weaver-based composition is assured by weaving correctness conditions.

In this chapter, we introduce *nonatomic* join points that capture both data and timing constraints to be satisfied at join points to assure weaving correctness.

25.4.1 Join points

Join points (JP) in AO UTA are structural elements and their connected compositions in the base model to which the advices are woven. In the following, we call *weavers* the model constructs used for weaving. Since weaving presumes adding auxiliary model elements in both, base model and advice, we distinguish two sides of a weaver, base model side and advice model side. In this section, we first revisit the types of JPs defined in Ref. [25] and introduce new types of JP-s that provide more flexibility in defining advices and enable straightforward verification of weaving correctness. By types and composition of UTA structural elements of JPs, the following groups are distinguished:

- *Atomic* JPs are ground-level UTA structural elements such as *locations* and *edges* (weavers of these elements are defined in Ref. [25]).
- *Nonatomic* JPs that capture data constraints as well as timing constraints combined and to be satisfied by advices at join points to assure weaving correctness. Nonatomic JPs include sequential composition of edges and locations.

Any location and edge in AO base model can be an atomic JP without context restrictions. For defining the weavers, at first, for atomic JPs we consider the general case where locations may have 0 to n incoming and 0 to m outgoing edges (Fig. 25.1A). In addition, we do not pose restrictions on the attributes of locations and edges being selected as JPs. For a location being JP (L-JP for short), the invariant Inv specifies conditions which should be satisfied by the advice when woven with this JP. For an edge being atomic JP (denoted E-JP) of AO UTA (Fig. 25.1B), no restrictions to attributes such as I/O action label on channel CH, guard condition Grd and set of updates of clocks and state variables denoted as Asg are posed. It has to be stressed that there is no formal restriction on which of the locations or edges is picked as a JP, so the modeler is free to select JPs purely based on the semantics of the base model and advice to be woven with it.

However L-JP and E-JP taken separately pose some limitations on what can be expressed in advices: L-JP does not allow to explicitly capture in the advice state transformation information and E-JP does not allow capturing duration constraints on modeled actions. Therefore, nonatomic JPs are introduced as sequential compositions of E-JP and L-JP called EL-JP and its symmetric counterpart, composition of L-JP and E-JP called LE-JP (Fig. 25.1). Both LE-JP and

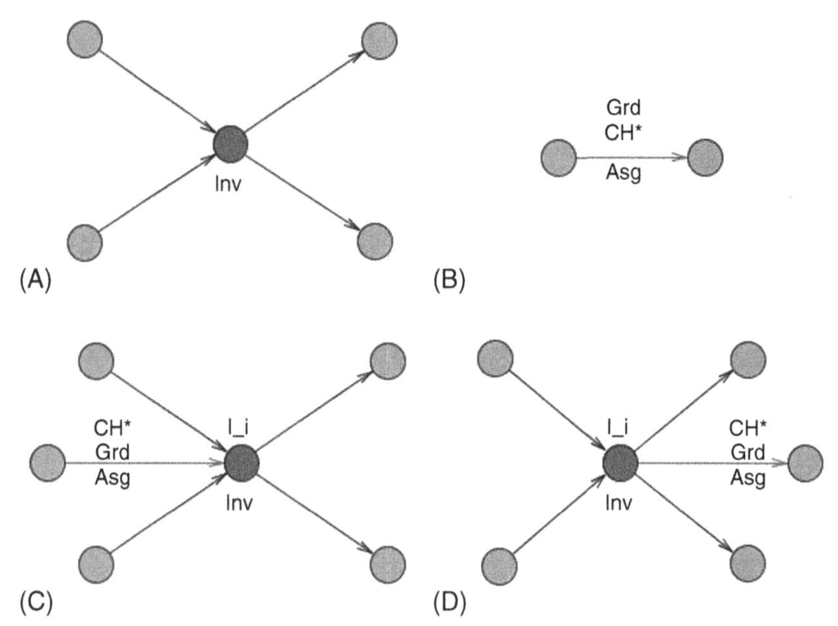

FIG. 25.1 Atomic and nonatomic join points: (A) location as atomic join point of type L-JP; (B) edge as atomic join point of type E-JP; (C) edge-location nonatomic join point (type EL-JP); (D) location-edge nonatomic join point (type LE-JP).

EL-JP have subtypes depending on which action type (input or output action), the channel label CH* denotes on the edge (either input action CH? or output action CH!). Since these subtypes need different weaving constructions, we will denote these join point LE-JP subtypes as LE?-JP and LE!-JP and join point EL-JP subtypes as EL?-JP and EL!-JP, respectively.

25.4.2 Advice weaving

For JP types defined in Section 25.4.1, the weaving is done by means of model construct called *weaver*. A weaver has two sides, base model side (extension of the base model) and advice model side (extension of the advice model), to compose the advice with base model together. In our approach, we require that the weaver preserves the correctness of both the base model and advice after weaving. Therefore, the JP should be harnessed with a weaver so that the semantics of the rest of base and advice models is not violated due to weaving. As shown in Ref. [33], the weaving correctness verification without preserving local correctness at JPs and cross checks of advice noninterference at different JPs turns to be very complex and is often infeasible in larger models. Our choice of JP types and design of weaving constructs is strongly founded on avoiding these concerns.

In the following, we introduce the weavers by JP types. In Fig. 25.2A, L-JP weaver is shown, base model side above and advice model below (weaver

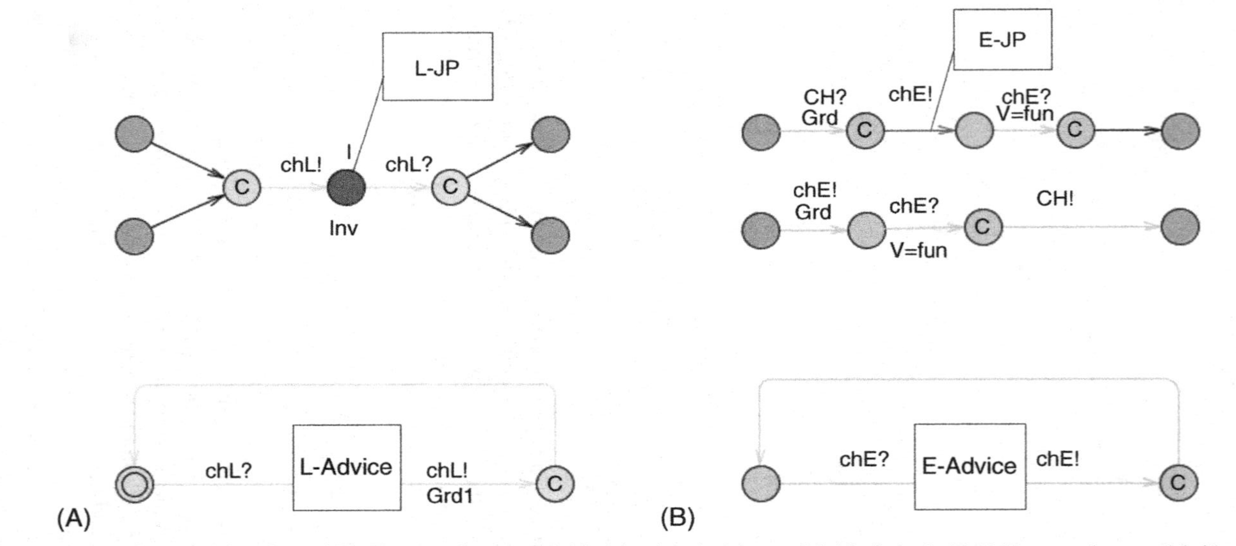

FIG. 25.2 Weavers of atomic join points: (A) L-JP weaver base model side (*above*) and advice model side (*below*); (B) E-JP weaver base model side with input action CH? (*above*), with output action CH! (*middle*) and advice model side (*below*).

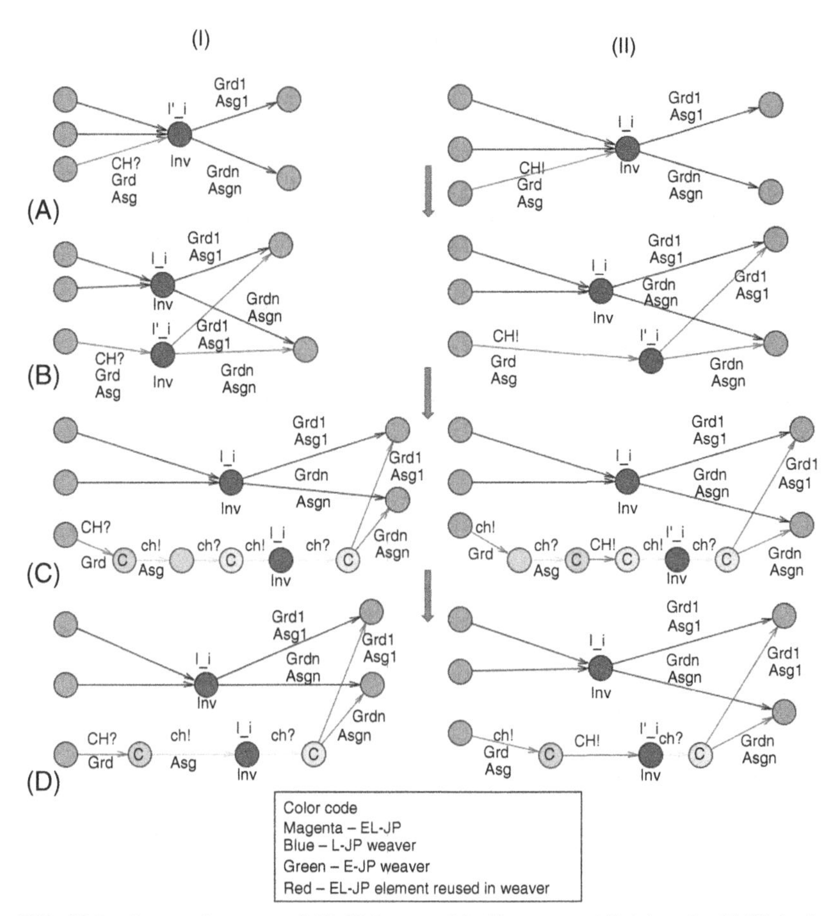

FIG. 25.3 Construction steps of EL-JP base model side weaver: (A) join point highlighted; (B) extracting EL-JP from the base model; (C) composing E-JP and L-JP weavers; (D) optimizing the EL-JP weaver. Column (I): EL?-JP weaver option; Column (II): EL!-JP weaver option.

constructs are outlined with magenta color in figures). Two options of E-JP weaver (E?-JP and E!-JP) on base model side are shown in Fig. 25.2B above and middle, respectively. Note that the advice side of E-JP weaver remains the same for both E-JP options in the base model side.

When weaving advices with nonatomic JPs, consisting of sequential composition of an edge and location or location and edge (Fig. 25.3), the weaver construction steps need to guarantee that synchronization between both weaver sides and via the channel on the edge that is involved in JP are not mixed. If there are updates on the edge labeled with channel's CH coaction, the direction of channel is important to preserve the right order of updates on these edges, that is, according to UTA semantics updates of edge with CH! label are executed before the updates of edge with CH? label.

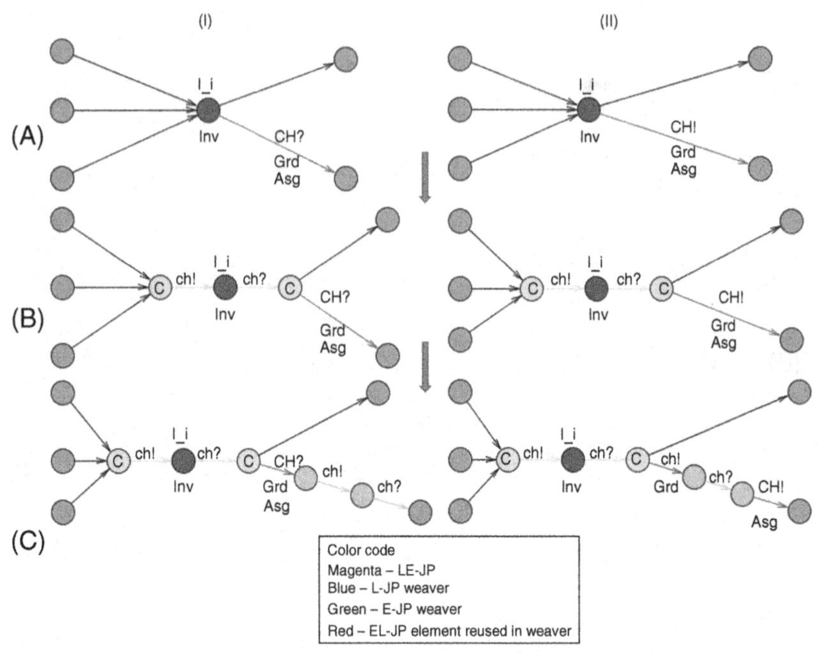

FIG. 25.4 Construction steps of LE-JP weaver base model side: (A) LE-JP highlighted; (B) L-JP weaver introduced; and (C) E-JP weaver introduced. (I) LE?-JP option and (II) LE!-JP option.

The model transformation steps of constructing EL-JP weaver base model side for EL?-JP are depicted in Fig. 25.3A to D column (I) and for EL!-JP base model side in column (II). Without loss of generality, we can leave untouched the attributes of other edges not involved in JP since due to locality of the weaving construct their semantics is not influenced by the weaver. In the first step (Fig. 25.3A and B), the location l of JP is duplicated and the duplicate is introduced as an auxiliary locations l'. This is conservative transformation keeping operational semantics of the base model unchanged because edge e is redirected to l' and duplicates of edges outgoing from l are created so that their source location is l' and their enabling conditions do not change. Provided the base model has been proven correct, though introducing structural changes, its correctness is preserved after this step.

In the next step (Fig. 25.3B and C), edge e is split in two e' and e'' connecting them with an auxiliary committed location to trigger the advice model simultaneously when executing edge e. Edge e' copies all the attributes of e, while e'' is labeled with an auxiliary channel ch! that has coaction in the weaver advice side. Similarly to the L-JP weaver (Fig. 25.2A), a weaver for l' is introduced.

Construction of EL-JP and LE-JP weavers can be considered as a composition of L-JP and E-JP weavers (Figs. 25.3C and 25.4C). Thus, the advice model is composed of L-JP and E-JP advices in separate and their respective weavers remain the same as shown in Fig. 25.2. However, due to the semantics of

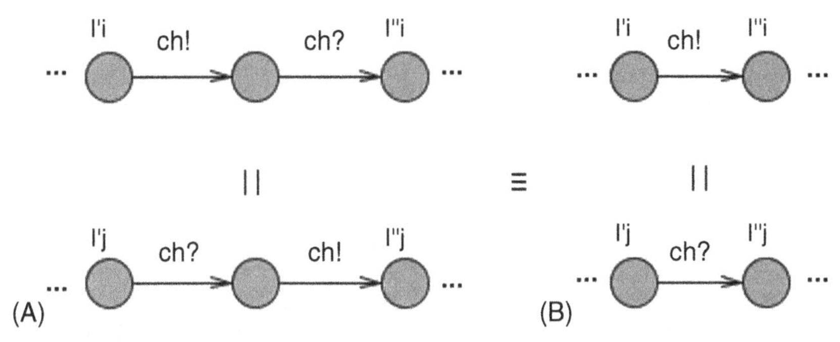

FIG. 25.5 Reduction of nonatomic JP weavers: (A) fragment of atomic JP weavers composition and (B) simplification of atomic weavers composition.

sequential composition of UTA committed locations and connecting their transitions, the weaver simplifications are possible. Here, we apply the semantic equivalence of construct that appears due to sequential composition of weavers shown in Fig. 25.5A and B. Parallel composition of synchronized model fragment with committed locations (Fig. 25.5A) in sequential composition of weavers has overhead that is eliminated in simplification (Fig. 25.5B).

In Fig. 25.6, we demonstrate the reduced form of LE-JP weaver that has been derived from composition of L-JP and E-JP weavers by using equivalence depicted in Fig. 25.5. Simplified forms of weavers introduce less structural overhead and improve the readability of models. Similarly, reduced weaver can be constructed for EL-JP as shown in Fig. 25.7. The application of reduced weavers will be demonstrated in detail on HRS case study in Section 25.5.

25.4.3 Weaving correctness

To assure the correctness of AO models, we follow the correctness-by-construction modeling principles. For an augmented model to be correct, it may not be enough to prove the correctness of base model and advice model taken separately. Even if the weaver preserves the correctness of both, the base model and advice model, they can be semantically noncompatible which emerges when proving behavioral correctness at the JP after weaving. Moreover, weaving one advice may violate the correctness of weaving another advice model. This phenomenon is called *aspect interference*. In the rest of this section, we introduce and explain the AO UTA model correctness conditions and the methods of verifying them in three steps:

1. We show that weavers themselves are correct and do not change the base and advice model semantics when applied at join points.
2. We show what conditions the advice models should satisfy to assure the weaving correctness at join points.
3. We show how the noninterference of advice models woven at different join points is verified.

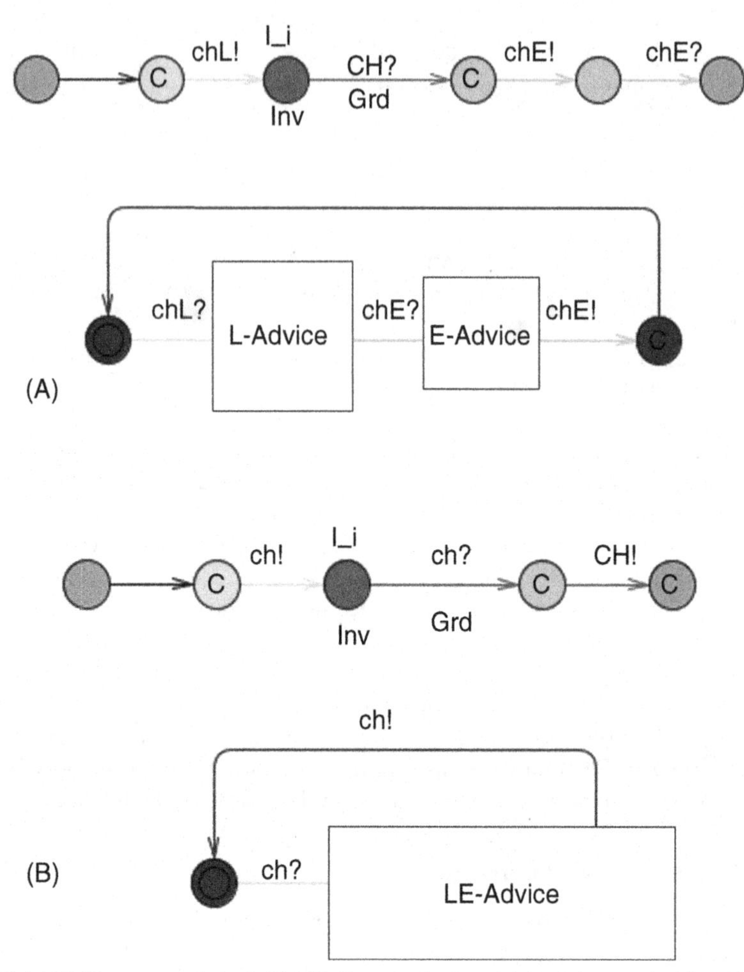

FIG. 25.6 LE-JP weaver optimized. (A) LE?-JP weaver base model side (*above*) and advice side (*below*); (B) LE!-JP weaver base model side (*above*) and advice side (*below*).

25.4.3.1 Correctness of weavers

To assure the correctness of weavers, we construct the weavers by types of JPs they are applied to. To demonstrate that weavers do not influence the woven model's correctness we apply the weavers to "empty" advice and base models showing that weaving does not have side effects on the augmented model, that is, weaver does not violate the local properties at JPs. By "empty" model, we mean that both the base model and advice model woven together are represented by place holding elements only (denoted with gray color in Figs. 25.8–25.10) to which the control is returned after executing the weaver at JP. For verifying LE-JP and EL-JP weavers, the place holders of advice

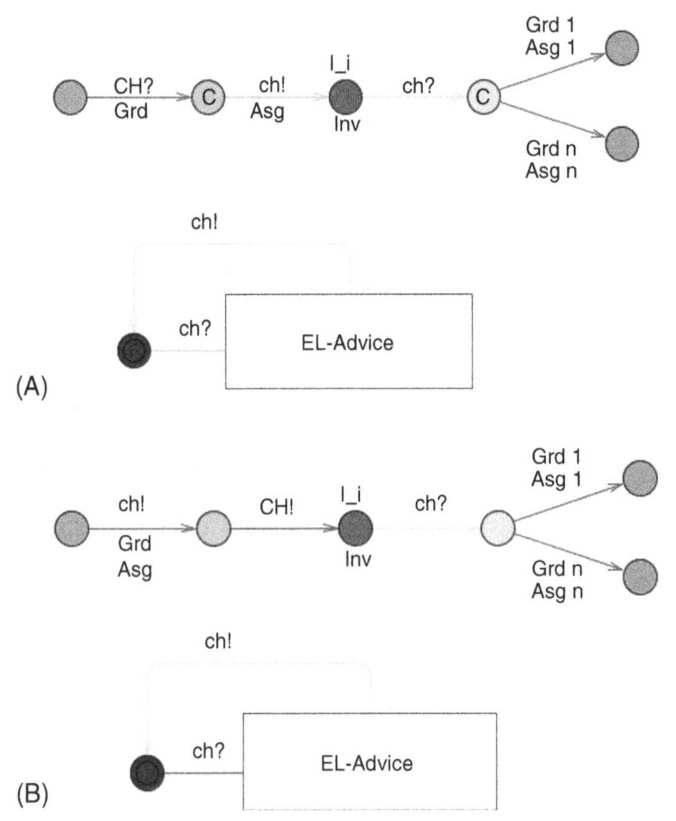

FIG. 25.7 EL-JP weaver optimized. (A) EL?-JP weaver base model side (*above*) and advice side (*below*); (B) EL!-JP weaver base model side (*above*) and advice side (*below*).

models are composed in the same order as their corresponding JPs in base model side. The augmented "empty" models for verifying the correctness of weavers in Figs. 25.3 and 25.4 are depicted in Figs. 25.8–25.10.

Note that for verifying the weaver models, the models must be closed, meaning that each channel should be specified at both ends, that is, input and output actions of the channel must be defined at different processes. Therefore, an auxiliary process has been depicted in Fig. 25.8, which is implicitly assumed to be used for weavers verification also in models of Figs. 25.9 and 25.10.

Given the definitions of weavers, let M_0 denote an empty base model and M_0^A an empty model of advice A, both having initial location l_{pre}^B and l_{pre}^A, respectively. The weavers correctness is verified by applying TCTL model-checking queries (25.1) and (25.2) to augmented models shown in Figs. 25.8–25.10 where locations L_Advice and E_Advice denote place holders for E-JP and L-JP advice models, respectively.

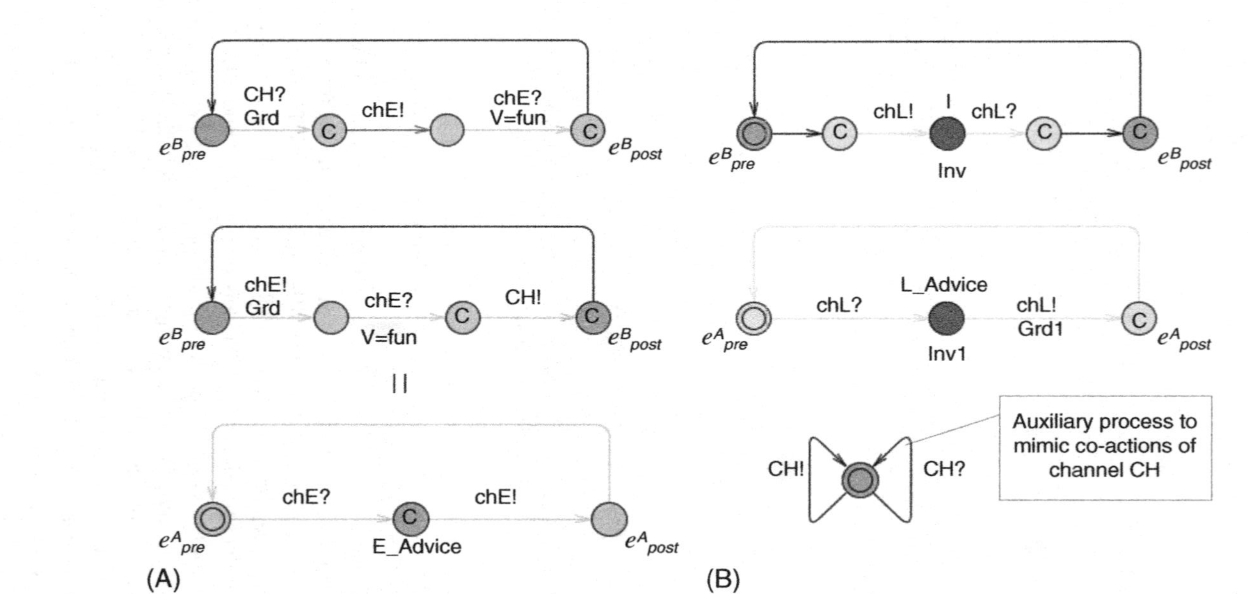

FIG. 25.8 E-JP and L-JP weaver verification configurations ("place holder" locations are colored *gray*). (A) E-JP weaver verification configurations and (B) L-JP weaver verification configurations.

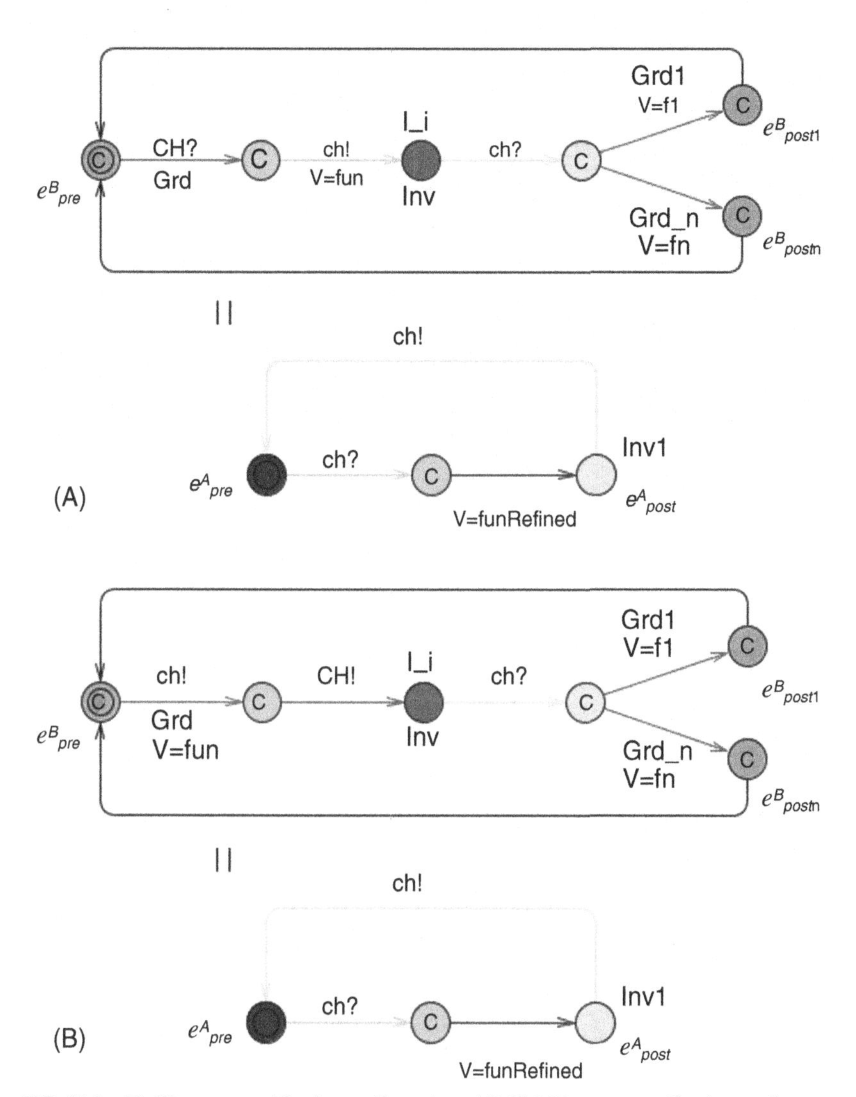

FIG. 25.9 EL-JP weaver verification configurations. (A) EL?-JP weaver verification configuration and (B) EL!-JP weaver verification configurations.

$$\text{A}\square \text{ not deadlock} \tag{25.1}$$

$$\begin{aligned} \text{A}\square \ (&\text{basemodel.lpost imply advicemodel.lpost}) \ \&\& \\ (&\text{advicemodel.lpost imply basemodel.lpost}) \end{aligned} \tag{25.2}$$

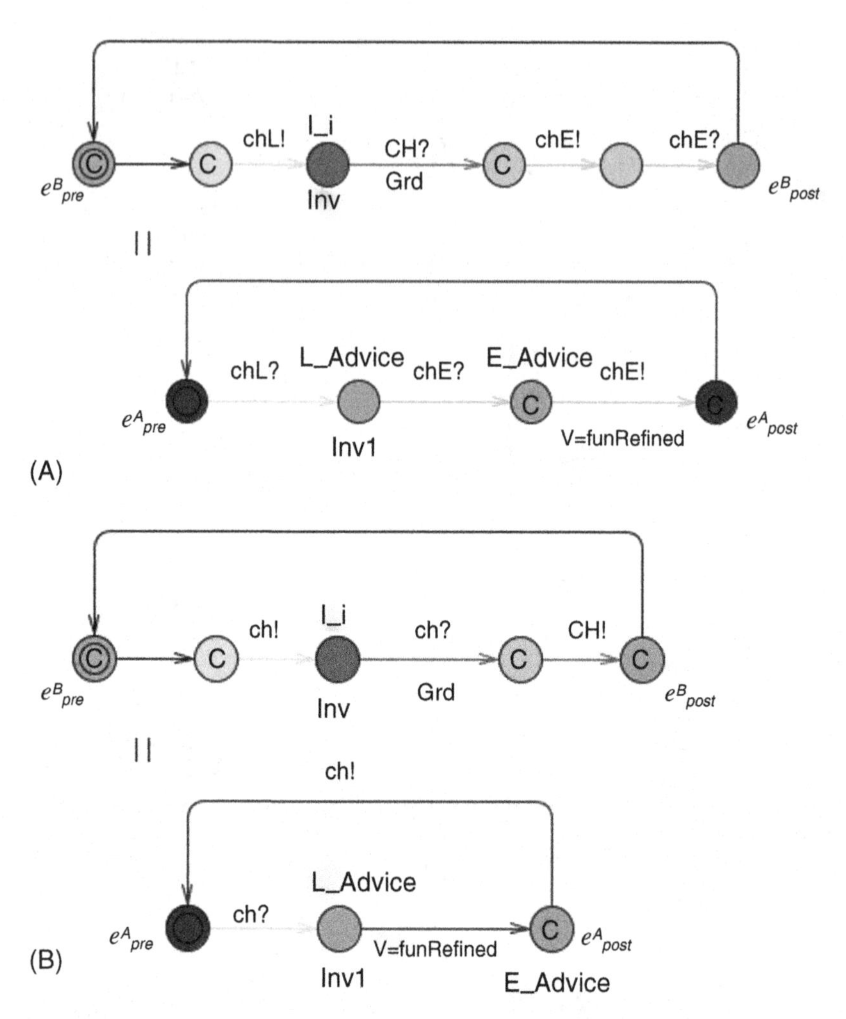

FIG. 25.10 LE-JP weaver verification configurations. (A) LE?-JP weaver verification configuration and (B) LE!-JP weaver verification configurations.

Query (25.1) specifies the property that the weaver does not cause deadlock. deadlock is standard predicate of UPPAAL query language. Query (25.2) specifies that whenever started from initial state (l^B_{pre}, l^A_{pre}) the poststate (l^B_{post}, l^A_{post}) of the weaver in both sides (base model and advice side) is reachable simultaneously. To avoid side effects from the base model and advice to weaver, it is assumed that the invariant Inv of JP location is equivalent to the invariant Inv1 of advice location L_Advice and also the functions *fun* and funRefined in updates are same. It has been verified with UPPAAL model checker that properties (25.1) and (25.2) are valid for all types of weavers defined earlier.

25.4.3.2 Weaving correctness at join points

The next step is to verify conditions the advice models must satisfy to assure the weaving correctness at JPs. Since these correctness conditions are local to JPs, no restrictions are posed to the rest of the base model except those given for JPs. In Ref. [25], we presented weaving correctness properties of atomic JPs as follows. The advices of all JP types must satisfy property P1.

- P1 (*side-effect free updates*): All updates in the advice model must be side-effect free, that is, there are no variables of base model M updated in advice model M^A. It is straightforward to check by static analysis that variables of M do not occur in the left-hand side of any update expression in M^A.

L-JP must satisfy properties P2 and P3:

- P2 (*nonblocking*): This property states that if there is not deadlock at JP in the base model then weaving an advice should not cause JP blocking also in the augmented model. This property can be expressed as an implication between two satisfiability relations

$$(M_{JP} \models \text{not deadlock}) \Rightarrow (M_{JP} \oplus M^A \models \text{not deadlock})$$

where M_{JP} denotes a JP in base model and $M_{JP} \oplus M^A$ denotes an augmented model where advice model M^A is woven to join point M_{JP}, $\oplus$ denotes the weaving operation. Verifying P2 means that the base model should be model checked at first for proving that there is no deadlock in the JP and if this is true then proving that the augmented model does not cause a deadlock in this JP either.

- P3 (*nondivergence*): This property states that if the clock invariant of L-JP specifies an upper bound d of staying in this location then the advice model woven to this JP must satisfy this constraint as well, that is, any trace of M^A should not violate this upper bound d.

$$\forall \text{cl} \in \text{CL(Inv)}\ \exists\ d < \infty : M_{JP} \oplus M^A,\ l^B_{pre} \models l^B_{pre} \rightsquigarrow_d l^B_{post}$$

where CL(Inv) denotes a set of clocks in L-JP invariant Inv, "$\rightsquigarrow_d$" denotes time bounded "leads to" operator of TCTL, l^B_{pre}, l^B_{post} denote prelocation and postlocation of the L-JP weaver.

E-JP advice must satisfy properties P4, P5, and P6:

- P4 (*feasibility of advice model execution paths*): This property states that the weakest precondition wp of any execution path from E-JP weaver advice model side prelocation l^A_{pre} to postlocation l^A_{post} must be consistent with the E-JP guard Grd. Let $\langle l^A_{pre}, l^A_{post}\rangle$ denote a set of all paths from the initial location l^A_{pre} to final location l^A_{post} and $\langle l^A_{pre}, l^A_{post}\rangle_k \in \langle l^A_{pre}, l^A_{post}\rangle$ be kth path in that set, then

$$\forall k \in [1, |< l^A_{pre}, l^A_{post} >|] : wp(< l^A_{pre}, l^A_{post} >_k, l^A_{post}) \Rightarrow \text{Grd}(e_i)$$

- P5 (0-*duration unwinding*): Since the execution of any path in the advice model M^A must be atomic and instantaneous not to break the E-JP semantics (executing the edges in UTA is instantaneous), all locations of M^A must be of type *committed* and/or *urgent*. Like property P1, property P5 can be easily verified by static inspection of location types in M^A.
- P6 (*feasibility*): Whenever the guard condition of E-JP Grd is true, there must exist a feasible path that reaches weaver post location l^A_{post} in E-JP advice model:

$$M_{JP} \oplus M^A,\ l^B_{pre} \models \text{Grd} \leadsto_{d=0} l^A_{post}$$

LE-JP advice must satisfy property P7:

As shown in Fig. 25.6A, the LE?-JP weaver is the only case where advice models of L- and E-advices are composed via an edge labeled with CH channel; therefore, it must be explicitly sequentialized in the weaver as well. Thus, the correctness properties P1–P6 of E-JP and L-JP advices apply also to these partitions taken separately. However, due to the clock constraints of Inv and Grd, the properties P3 and P6 combine to property P7 that strengthens the clock constraints of LE-JP.

- P7 (two-side bounded time window): This property states that the advice cannot have traces shorter in time than d_1 and longer in time than d_2:

$$\forall \text{cl} \in \text{CL}(\text{Inv})\ \exists\ d_1 \le d_2 < \infty : M_{JP} \oplus M^A,\ l^B_{pre} \models$$
$$(l^B_{pre} \leadsto_{\ge d_1} l^B_{post}) \wedge (l^B_{pre} \leadsto_{\le d_2} l^B_{post})$$

EL-JP advice:

Like LE-JP advice must satisfy properties P1–P6 depending on which part of the advice model (L- or E-JP) is considered, but the crucial difference is that property P7 is irrelevant since it cannot be satisfied in case of join point EL-JP. The reason is that when adding lower bound condition to EL invariant Inv it may cause deadlock when clock values at entry to EL-JP violate the time bound specified in Inv. This, in turn, forces extending the correctness conditions to the rest of base model surrounding EL-JPs and eliminates the locality advantage of our verification conditions. Therefore, this option is left out of the scope in current study.

25.4.3.3 Noninterference of advice models

As stated in Ref. [33], proving noninterference of advices is a critical issue regarding verification complexity in large AO models. Thanks to the locality of verification conditions in our approach which concern attributes of only JPs and not of their broader context in the model, our noninterference verification technique is compositional allowing breaking the global verification task into subtasks where aspects noninterference is proven pair wise. Namely, for each JP_i depending on its type the preservation of properties P1–P7 is verified in augmented model configurations where only advice M_i^A of this JP and one advice M_j^A of another JP are woven, that is, the set of augmented model configurations $M \oplus M_i^A \oplus M_j^A$, $i \neq j$ needs to be verified against the correctness properties of JP_i. No doubt, extending the model with another advice increases the complexity of model checking but the growth is linear in the size (the number of edges and locations) of advice models, thus still feasible in practical correctness analysis. In total, having k JPs in the base model and n advices, where generally $k \geq n$ because for some advice, there can be several JPs, and noninterference verification consists of $k(n-1)$ model checking task sets each consisting of properties P1–P7 depending on the type of JP.

25.5 Case study: Home rehabilitation system

The HRS [34] is chosen as the case study to present our AOM approach. HRS is a health system that monitors a patient's health condition online. HRS includes a sensor system for fall down detection as well as physical activity and exercise monitors. HRS functions in various modes but for this chapter we focus on "home exercising" use case.

25.5.1 Model description

We demonstrate the AOM of HRS on the requirements level of abstraction with actors patient, physiotherapist, sensor system, and exercise monitoring. The modeling starts with the base model concerning the requirements level actors and their dependencies. Then, from aspects like *Patient's safety*, *Exercising quality*, and *Exercising performance*, we focus on the weaving of aspect *Exercising quality*. The weaving correctness is verified using model-checking of the properties specified in the previous section.

25.5.1.1 Base model

The interaction of the physiotherapist, patient, and HRS actors is represented with the base model: the physiotherapist, patient, and HRS. The UTA templates *Doctor*, *Patient_physical_condition*, *Patient_exercising*, and *HRS* model them, respectively, as shown in Fig. 25.11. The *Doctor* initiates the training session with the output action *Exercise*. The *Patient* starts exercising for a time interval with lower bound *Ex_Lb* and upper bound *Ex_Ub* if the Patient's condition is *Normal*.

Abnormal activity (e.g., falling down) is modeled with edge *Normal* → *Bad* in template *Patient_physical_condition* and communicated to *HRS* with the

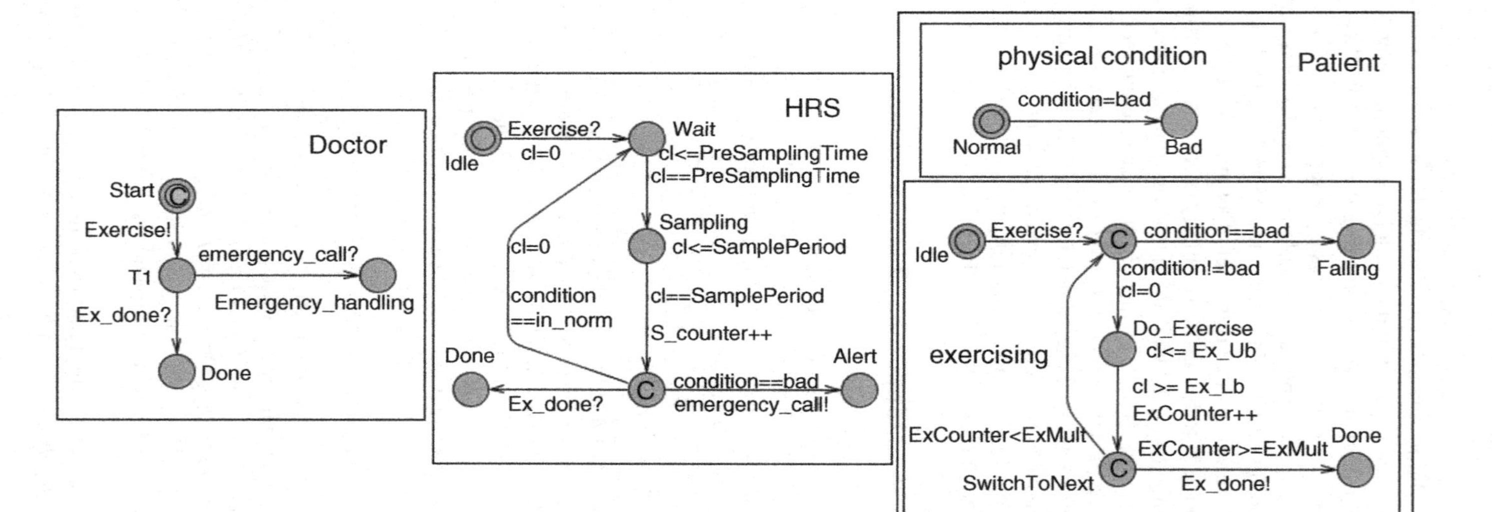

FIG. 25.11 The base model of HRS home exercising and monitoring case.

update of global variable *condition* (*condition* = *bad*). The patient's data sampling by *HRS* is occurring with the synchronization channel *Exercise* when the exercising starts. Data from the patient is sampled periodically with period *SamplePeriod* until the exercise is done (signal *Ex_done*) or the health condition of the patient turns to *Bad.*

In the following, we will present in detail only the weaving of *exercising quality* aspect since the weaving of the other aspects is done in a similar manner.

25.5.1.2 Aspect: Exercising quality

Aspect *exercising quality* models how *HRS* must act if changes of the patient's biometric characteristics occur during exercising and how the deviations from nominal values are communicated back to the patient. This aspect consists of three subaspects, namely *physical condition*, *multiple exercise*, and *exercising quality monitor*.

For *physical condition* advice, three new (sub)states *Normal2*, *Better*, and *Worse* are introduced which refine state *Normal* of the base model template *Patient physical condition*. The aspect advice is woven to the base model with the nonatomic L-JP weaver as shown in Fig. 25.12. In order to quantitatively model the condition of the Patient, *M* numeric values in the ranges *val_N*, *val_B*, *val_W* are generated periodically once in *Tick* period. The value regions *val_N*, *val_B*, *val_W* of body characteristics correspond to states *Normal2*, *Better*, and *Worse*, respectively. These values are generated in the model dynamically by self-loops attached to these states in template *Patient_physical_condition_advice*. The generation of numeric values in state *Bad2* is not shown in this case study.

LE-JP weaver is applied for location-edge weaving of the *multiple exercise* advice to the base model for exercising as shown in Fig. 25.13. The patient can do three different exercises still preserving the original duration of each exercise and the repetition counter. Template *Doctor* is shown in Fig. 25.13 for model visual completion regarding the triggering and ending of the exercise session.

HRS_Sampler base template is woven with advice *quality_monitor* by applying the EL-JP weaver to model the functionality of *HRS* when the sampling data of the patient do not conform to the nominal values, as shown in Fig. 25.14. The nominal values are given by a physiotherapist as an interval [*L_bound*, *U_bound*], defined for *M* different body characteristics being sampled. The online training guidance is done by showing qualitative values *LB_warning*, *normal* and *UB_warning* to the patient through the *HRS* user interface screen (variable *Screen* in Fig. 25.14).

25.5.2 Verification of the aspects' weaving correctness

In this section, we prove by model checking that the weaving of the advice models on the base model is correct.

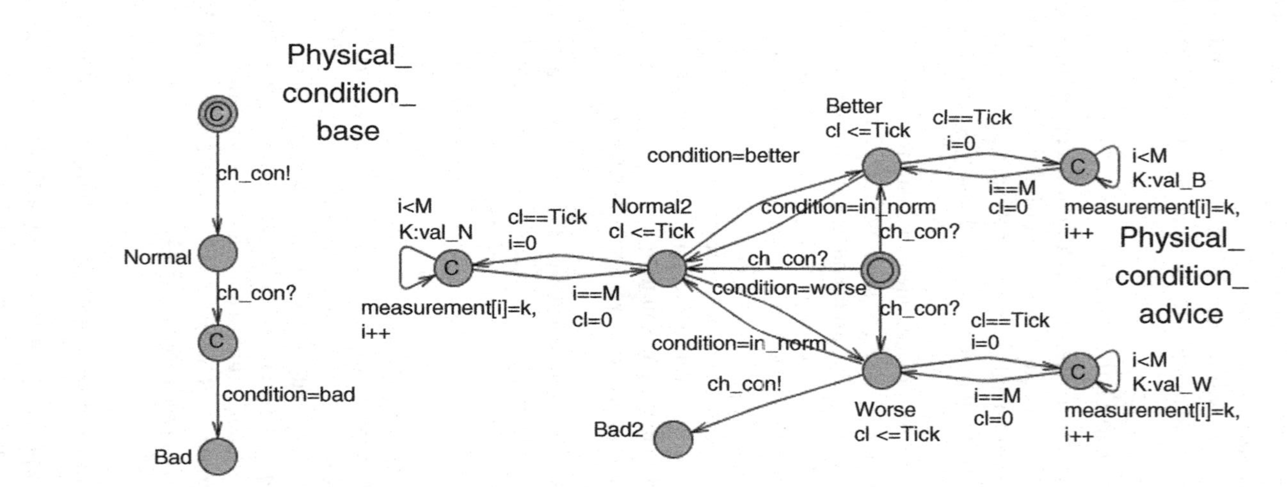

FIG. 25.12 Physical condition weaving for exercising quality aspect model.

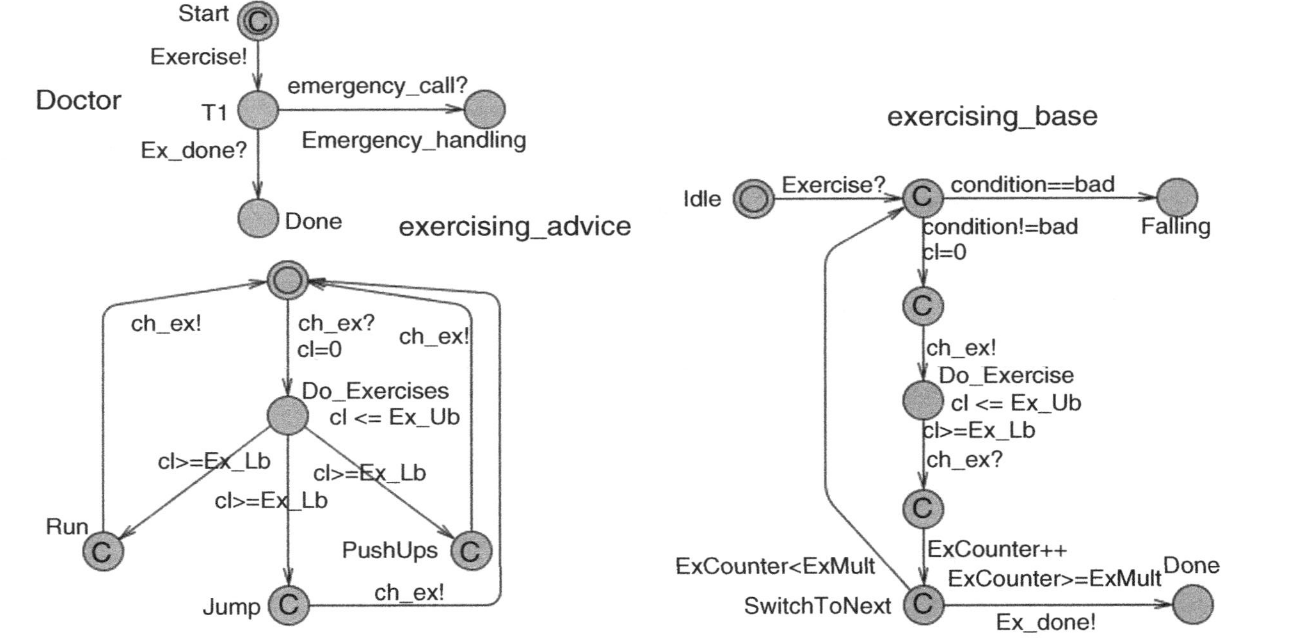

FIG. 25.13 Multiple exercise weaving for exercising quality aspect model.

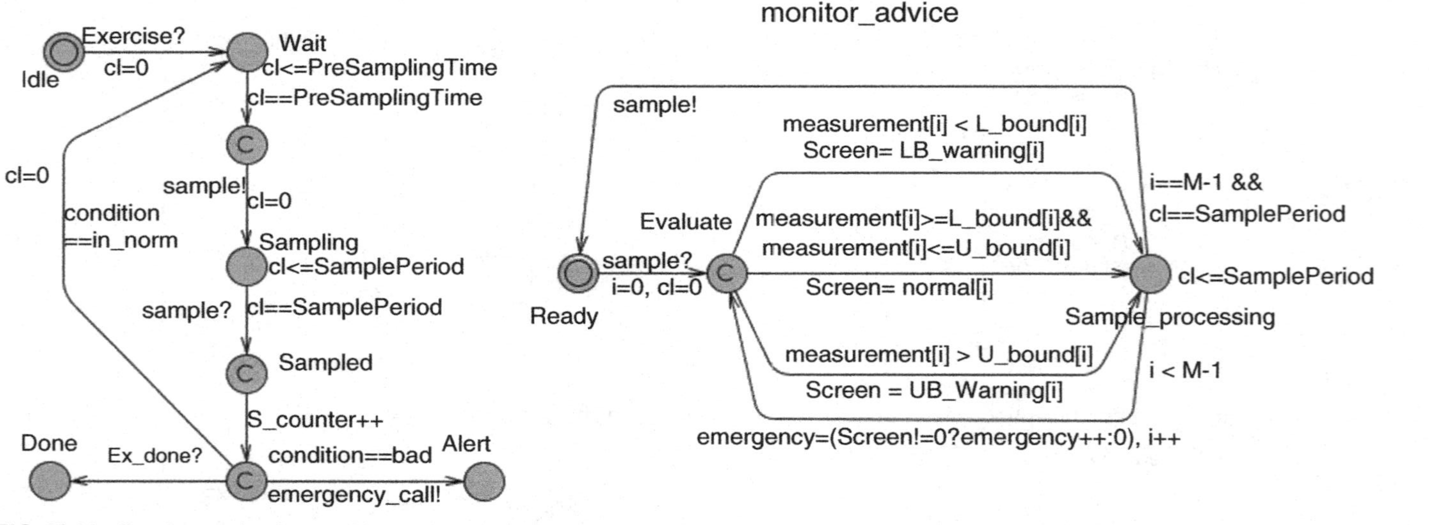

FIG. 25.14 Sampler weaving for exercising quality aspect model.

Regarding the L-JP weaving of aspect for the patient's *physical condition*, it has been proved that the base model is not deadlocking at the join point and also that the advice is not deadlocking at the join point after weaving (Property P2). Property P3 is not applicable since the base model does not have a clock invariant at the join point.

For the LE-JP weaving of the *multiple exercise* aspect, the main property that we proved is property P7 stating that the advice cannot have traces shorter than the lower time bound (*Ex_Lb*) of the traces of the base at the join point and longer than the upper time bound (*Ex_Ub*) of the traces at the join point. For this we checked two "leads to" queries:

$$Patient\_Exercising\_base.Do\_Exercise \rightsquigarrow_{\geq Ex\_Lb} Patient\_Exercising\_base.SwitchToNext$$

$$Patient\_Exercising\_base.Do\_Exercise \rightsquigarrow_{\leq Ex\_Ub} Patient\_Exercising\_base.SwitchToNext$$

For the EL-JP weaving of the *quality_monitor* advice, the main property that we proved is property P4 regarding the feasibility of advice model execution paths. For our woven model, property P4 is transformed to a "leads to" query stating that if the advice model execution reaches location *Evaluate* this leads to the committed location *Sampled* in the base model:

$$HRS\_Sampler\_quality\_monitor\_advice.Evaluate \rightsquigarrow HRS\_Sampler\_base.Sampled$$

25.6 Usability of AO modeling and verification

We use model checking for the verification of AO UTA model correctness. Properties P1–P7 presented in Section 25.4.3.2 define the weaving correctness and depend on the type of join point. As a result of model-checking UPPAAL returns timed witness traces that help tracing back the causes of property violation if either the advice alters the properties of the base model or there is an interference between advices. Thus, the weaving correctness verification effort can be characterized by time and space complexity of verifying properties P1–P7.

It has been shown in Ref. [35] that the worst-case time complexity $\mathcal{O}$ of model-checking TCTL formula ϕ over timed automaton M, with the clock constraints of ϕ and of M in ψ is

$$\mathcal{O}(|\phi| \times (n! \times 2^n \times \Pi_{x \in \psi} c_x \times |L|^2)) \quad (25.3)$$

where n is the number of clock regions, ψ is the set of clock constraints, c_x is the maximum constant the clock x is compared with, L: $L \rightarrow 2^{AP}$ is a labeling function for symbolic states of M. L denotes the product of data constraints over all locations and edges defined in the UTA model and AP is the set of atomic propositions used in guard conditions and invariants.

The space complexity lower bound of TCTL model checking applied on timed automata is PSPACE-hard [30]. However, when proving properties P1–P7 on AO UTA models, the size of symbolic state space is determined dominantly by the size of advice models. Due to compositionality of weaving operations, the number of locations $|L(M^{ao})|$ and edges $|E(M^{ao})|$, as well as the number of variables $|V(M^{ao})|$ of an AO model M^{ao} is not greater than that of a behaviorally equivalent non-AO model M.

From this fact and from formula (25.3), we can yield that any reduction in the number of model elements being irrelevant for verification of current aspect weaving, reduces drastically the number of symbolic states to be explored in model-checking properties P1–P7. Recall that due to the superposition refinement-based weaving method our approach also does not introduce additional interleavings when executing the transitions of base and advice models.

Relying on the given reasoning the following claim can be yielded [36].

Claim (weaving correctness verification effort)

For any reachability property ϕ_i of noninterfering aspects A_j, $i=1,\ldots,n$ that is decidable on $M^B \oplus M^{A_j}$ the model-checking effort **E** (in terms of time or space) is equal or less than the effort of model-checking the property ϕ_i on the non-AO model M, where the semantics of $M^B \oplus M^{A_j}$ is a subset of semantics of M, that is,

$$
\begin{aligned}
&M^B \oplus M^{A_j} \models \phi_i \wedge [[M^B \oplus M^{A_j}]] \subseteq [[M]] \\
&\Rightarrow M \models \phi_i \wedge \mathbf{E}(M^B \oplus M^{A_j} \models \phi_i) \leq \mathbf{E}(M \models \phi_i)
\end{aligned}
\tag{25.4}
$$

Here notation $[[M]]$ denotes the operational semantics (a set of behaviors) of model M. The validity of formula (25.4) stems from the fact that AO models that are composed by adding aspects incrementally or selectively represent subsets of the behavior of the non-AO model of the same system. The performance gain due to the compositionality via AOM is demonstrated with the numerical example in Fig. 25.15 [36].

The plot in Fig. 25.15 is generated using the approximating function $f(m,|\phi|) = m(|\frac{\phi}{m}| \times ((\frac{n}{m})! \times 2^{n/m} \times C \times |\frac{L}{m}|^2))$ which is derived from Eq. (25.3) by applying assumptions that the set of locations and the set of literals of verification formula is partitioned between m aspect models evenly and $\Pi_{x \in \psi} c_x$ is some constant C.

It can be seen from Fig. 25.15 that verification effort measured in terms of the model-checking complexity of formula ϕ is, according to formula (25.4), inverse super-exponential in the number of aspects m and exponential in the number of literals in ϕ. This analysis confirms once again the benefit of the old "divide and concur" principle in computer science meaning specifically that when partitioning the model-checking task to m smaller ones verifiable on each AO model separately instead of one monolithic nonaspect-oriented model, it provides in total exponentially smaller verification effort.

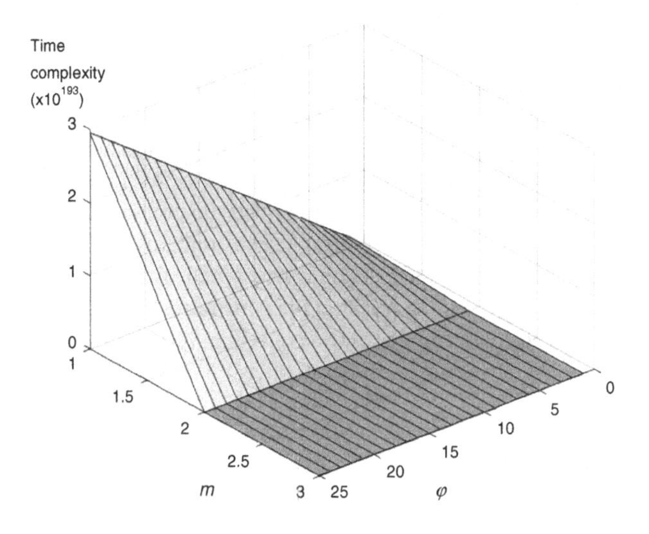

FIG. 25.15 The dependency of verification effort on the number of logic connectives in the weaving correctness formula ϕ and the number of aspects m [36].

25.7 Conclusions and discussion

In this chapter, we presented an aspect-oriented model engineering methodology for UPPAAL timed automata targeting efficient verification of models at each step of model construction using predefined join points and advice weavers. This methodology is based on an aspect-oriented requirements engineering paradigm, which results in three advantages: verification of model construction steps, simple rules for composition, and better comprehension of models. Aspects' weaving was implemented using a set of model superposition refinement operators and related to them verification conditions that guarantee weaving correctness and aspect noninterference. The weaving along its correctness verification is done automatically by running TCTL model-checking queries on woven models. If the weaving correctness conditions are violated due to incompatibility of advices and the base model the resulting witness traces thereafter are used as diagnostics for finding reasons why weaving fails. A set of correctness criteria was elaborated giving meaningful weaving options based on join point types defined as structural units of the base models. The usability of our AO model construction method was demonstrated on the HRS case study. Additionally, the quantitative characterization of the advantages of the proposed AOM method was presented in the form of analytical estimates of the modeling effort and model-checking time complexity.

In general, the cross-cutting concerns occur in multiple join points over the system architecture. Typical examples are safety and security-related aspects that means the same advice can be woven in many places with the base functionality. Since the weaving mechanism in all these cases does not depend on the advices, we have not included multiple join point weaving examples in the case study presented here. The second limitation of our weaving approach is not allowing recursive weaving, that is, weaving an advice with itself. In principle, bounded recursion could be considered provided the pointcut expression has a fixpoint on the depth of nested weavings, but this remains as part of our future work. Another question is whether there is true practical need for that. Also, as part of the future work, we plan relaxing the constraints on how the join points can be chosen in the base model and developing verification conditions for proving correctness of such weaving constructs.

Acknowledgment

This work has been supported by EXCITE (2014-2020.4.01.15-0018) grant.

References

[1] M. Utting, A. Pretschner, B. Legeard, A taxonomy of model-based testing approaches, Softw. Test. Verif. Reliab. 22 (5) (2012) 297–312.

[2] A. Bhave, B.H. Krogh, D. Garlan, B. Schmerl, View consistency in architectures for cyber-physical systems, in: Proceedings of the 2011 IEEE/ACM Second International Conference on Cyber-Physical Systems, Chicago, IL, 2011, pp. 151–160.

[3] G. Georg, I. Ray, K. Anastasakis, B. Bordbar, M. Toahchoodee, S.H. Houmb, An aspect-oriented methodology for designing secure applications, Inf. Softw. Technol. 51 (5) (2009) 846–864, https://doi.org/10.1016/j.infsof.2008.05.004.

[4] R. Filman, T. Elrad, S. Clarke, M. Akşit, Aspect-oriented software development, first ed., Addison-Wesley Professional, 2004.

[5] G. Kiczales, J. Lamping, A. Mendhekar, C. Maeda, C. Lopes, J.-M. Loingtier, J. Irwin, Aspect-oriented programming, in: M. Akşit, S. Matsuoka (Eds.), Proceedings of ECOOP 1997—Object-Oriented Programming, LNCS, vol. 1241, Springer, Berlin, Heidelberg, 1997, pp. 200–237.

[6] S.J. Sutton, Aspect-oriented software development and software process, in: M. Li, B. Boehm, L. Osterweil (Eds.), LNCS, Unifying the Software Process Spectrum—SPW 2005, vol. 3840, Springer, Berlin, Heidelberg, 2006, pp. 177–191.

[7] S. Clarke, E. Baniassad, Aspect-Oriented Analysis and Design, the Theme Approach, Addison-Wesley, 2005.

[8] R.B. France, I. Ray, G. Georg, S. Ghosh, An aspect-oriented approach to early design modelling, IEE Proc. Softw. 151 (4) (2004) 173–185.

[9] I. Jacobson, P.-W. Ng, Aspect-Oriented Software Development With Use Cases, Addison-Wesley Object Technology Series, Addison-Wesley Professional, 2004.

[10] S. Ali, T. Yue, L.C. Briand, Does aspect-oriented modeling help improve the readability of UML state machines? Softw. Syst. Model. 13 (3) (2014) 1189–1221, https://doi.org/10.1007/s10270-012-0293-5.

[11] S. Op de Beeck, E. Truyen, N. Boucke, F. Sanen, M. Bynens, W. Joosen, A study of aspect-oriented design approaches, Department of Computer Science, K.U.Leuven, Leuven, Belgium, 2006. Technical Report CW 435.

[12] R. Pawlak, L. Duchien, L. Seinturier, CompAr: ensuring safe around advice composition, in: M. Steffen, G. Zavattaro (Eds.), Lecture Notes in Computer Science, Proceedings of 7th IFIP International Conference on Formal Methods for Open Object-Based Distributed Systems (FMOODS 2005), vol. 3535, Springer-Verlag, Athens, Greece, 2005, pp. 163–178.

[13] A. Mehmood, D. Jawawi, A quantitative assessment of aspect design notations with respect to reusability and maintainability of models, in: Proceedings of 8th Malaysian Software Engineering Conference (MySEC), IEEE, 2014, pp. 136–141.

[14] F. Pinciroli, J.L. Barros Justo, R. Forradellas, Systematic mapping study: on the coverage of aspect-oriented methodologies for the early phases of the software development life cycle, J. King Saud Univ. Comput. Inf. Sci. (2020), https://doi.org/10.1016/j.jksuci.2020.10.029.

[15] D. Xu, X. He, Generation of test requirements from aspectual use cases, in: Proceedings of the 3rd Workshop on Testing Aspect-Oriented Programs—WTAOP '07, ACM, New York, NY, 2007, pp. 17–22, https://doi.org/10.1145/1229384.1229388.

[16] S. Ali, L.C. Briand, H. Hemmati, Modeling robustness behavior using aspect-oriented modeling to support robustness testing of industrial systems, Softw. Syst. Model. 11 (4) (2012) 633–670.

[17] J. Vain, L. Tsiopoulos, G. Kanter, Aspect-oriented model-based testing with UPPAAL timed automata, in: C. Attiogbe, S. Ben Yahia (Eds.), Proceedings of the 10th International Conference on Model and Data Engineering (MEDI), Springer-Verlag, 2021.

[18] D.X. Xu, O. El-Ariss, W.F. Xu, L.Z. Wang, Aspect-oriented modeling and verification with finite state machines, J. Comput. Sci. Technol. 24 (5) (2009) 949–961, https://doi.org/10.1007/s11390-009-9269-5.

[19] Y. Tahara, A. Ohsuga, S. Honiden, Formal verification of dynamic evolution processes of UML models using aspects, in: 2017 IEEE/ACM 12th International Symposium on Software Engineering for Adaptive and Self-Managing Systems (SEAMS), 2017, pp. 152–162, https://doi.org/10.1109/SEAMS.2017.4.

[20] D. Xu, I. Alsmadi, W. Xu, Model checking aspect-oriented design specification, in: 31st Annual International Computer Software and Applications Conference (COMPSAC 2007), vol. 1, 2007, pp. 491–500, https://doi.org/10.1109/COMPSAC.2007.152.

[21] X. Sun, H. Yu, H. Liang, N. Yang, Modeling and analyzing web application with aspect-oriented hierarchical coloured petri nets, China Commun. 13 (5) (2016) 89–102, https://doi.org/10.1109/CC.2016.7489977.

[22] C. Vidal Silva, R. Saens, C. Del Rio, R. Villarroel, OOAspectZ and aspect-oriented UML class diagrams for aspect-oriented software modelling (AOSM), Ing. Investig. 33 (2013) 66–71.

[23] M. Alférez, N. Amálio, S. Ciraci, F. Fleurey, J. Kienzle, J. Klein, M. Kramer, S. Mosser, G. Mussbacher, E. Roubtsova, G. Zhang, Aspect-oriented model development at different levels of abstraction, in: R.B. France, J.M. Kuester, B. Bordbar, R.F. Paige (Eds.), Modelling Foundations and Applications, Springer, Berlin, Heidelberg, 2011, pp. 361–376.

[24] E.E. Roubtsova, M. Aksit, Extension of petri nets by aspects to apply the model driven architecture approach, in: Preliminary 1st International Workshop on Aspect-Based and Model-Based Separation of Concerns in Software Systems (ABMB); Conference date: 07-11-2005 Through 07-11-2005, 2005.

[25] K. Sarna, J. Vain, Exploiting aspects in model-based testing, in: Proceedings of the Eleventh Workshop on Foundations of Aspect-Oriented Languages—FOAL '12, ACM, New York, NY, 2012, pp. 45–48, https://doi.org/10.1145/2162010.2162023.

[26] D. Truscan, J. Vain, M. Koskinen, Combining aspect-orientation and UPPAAL timed automata, in: A. Holzinger, J. Cardoso, J. Cordeiro, M. van Sinderen, S. Mellor (Eds.), ICSOFT-PT: Proceedings of the 9th International Conference on Software Paradigm Trends, SciTePress, Vienna, Austria, 2014, pp. 159–164.

[27] F. Sanen, R. Chitchyan, L. Bergmans, J. Fabry, M. Sudholt, K. Mehner, Aspects, dependencies and interactions: report on the workshop ADI at ECOOP 2007, in: Proceedings of the 2007 Conference on Object-oriented Technology—ECOOP'07, Springer-Verlag, Berlin, Heidelberg, 2008, pp. 75–90.

[28] J. Iqbal, L. Tsiopoulos, D. Truscan, J. Vain, I. Porres, The Crisis Management System—A Case Study in Aspect-Oriented Modeling Using UPPAAL, Turku Centre for Computer Science, Turku, Finland, 2016. TUCS Technical Reports 1169.

[29] E. Katz, S. Katz, Incremental analysis of interference among aspects, in: Proceedings of the 7th Workshop on Foundations of Aspect-Oriented Languages, FOAL '08, ACM, New York, NY, 2008, pp. 29–38, https://doi.org/10.1145/1394496.1394500.

[30] R. Alur, C. Courcoubetis, D. Dill, Model-checking for real-time systems, in: Proceedings of Fifth Annual IEEE Symposium on Logic in Computer Science, LICS'90, IEEE, 1990, pp. 414–425.

[31] A. Hannousse, R. Douence, G. Ardourel, Static analysis of aspect interaction and composition in component models, in: Proceedings of the 10th ACM International Conference on Generative Programming and Component Engineering, Portland, Oregon, USA, GPCE '11, ACM, New York, NY, 2011, pp. 43–52, https://doi.org/10.1145/2047862.2047871.

[32] G. Behrmann, A. David, K.G. Larsen, A tutorial on UPPAAL, in: M. Bernardo, F. Corradini (Eds.), LNCS, SFM-RT 2004, vol. 3185, Springer-Verlag, 2004, pp. 200–237.

[33] E. Katz, S. Katz, Incremental analysis of interference among aspects, in: Proceedings of the 7th Workshop on Foundations of Aspect-Oriented Languages, FOAL '08, Association for Computing Machinery, New York, NY, 2008, pp. 29–38, https://doi.org/10.1145/1394496.1394500.

[34] A. Kuusik, E. Reilent, K. Sarna, M. Parve, Home telecare and rehabilitation system with aspect oriented functional integration, Biomed. Eng. Biomed. Tech. 57 (2012) 1004–1007, https://doi.org/10.1515/bmt-2012-4194.

[35] J.-P. Katoen, Concepts, Algorithms, and Tools for Model Checking, IMMD, 1999.

[36] K. Sarna, Aspect-Oriented Model-Based Testing (Ph.D. thesis), Tallinn University of Technology, 2018.

Chapter 26

Relevance of data mining techniques in real life

Palwinder Kaur Mangat and Dr. Kamaljit Singh Saini
University Institute of Computing, Chandigarh University, Chandigarh, India

26.1 Introduction

DM is a well-established discipline used with artificial intelligence, machine learning, statistics, and knowledge engineering. Data mining has been built using different technologies. Continuous advancement in hardware and technology had made it possible to store and mine a large scale of structured, unstructured, and semistructured data which is stored in heterogeneous sources.

Data mining is comprised of three scientific disciplines: statistics (uses concepts as standard deviation, regression analysis, factor analysis, cluster analysis, discrimination of data, etc., to study data relationships), artificial intelligence (built upon heuristics and requires vast computing power; this human-like intelligent behavior is done by machines or software), and machine learning (is an evolution of AI). Algorithms are used that learn from data and make predictions [1].

DM is a procedure of information extraction from huge databases, warehouses, and other repositories. This information is used for making the process of decision-making easier for large establishments. Its basic purpose is to find information from the large sets and transfer it to usable structures. Then this information is processed. Local, global, and national officialdoms work with this mined information. Data mining uses different approaches, algorithms, and techniques to find out past trends of data and can be used to predict future trends [2,3].

Knowledge discovery and DM is a process that involves multiple steps—data selection, data collection, cleaning of data, preprocessing of data which includes data cleaning and data integration, data transportation, and DM is used for computational, statistical, or visual methods [4].

KDD is a synonym for data mining, which aims at the discovery of important information from the huge collection of data. KDD systems are interactive tools [5]. KDD process contains different steps [3]—the first step in knowledge

System Assurances. https://doi.org/10.1016/B978-0-323-90240-3.00026-6

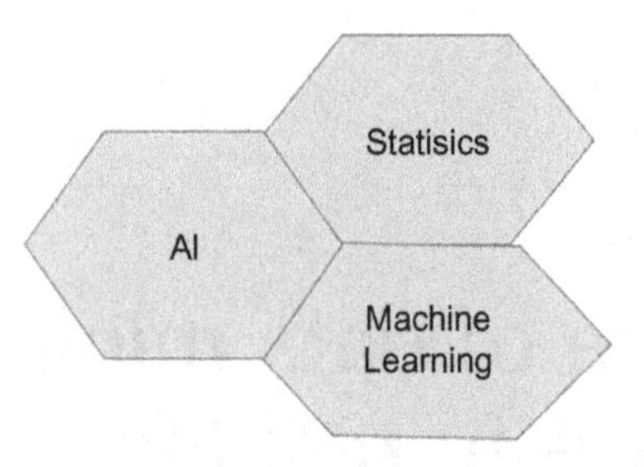

FIG. 26.1 Data mining.

discovery databases process is to understand those fields in which the DM is useful. Then the next step is to evaluate and understand the objectives of the data mining process. The next step after understanding the applications is the KDD process involves selecting, checking, integrating, and preparing the data sets. The third step in KDD processing is data transformation. In this step, data cleaning is done by applying various algorithms. In this step, noisy and erroneous data is removed. After that, missing and unknown values are handled, which is known as data processing.

The fourth step involves pattern discovery from the data and looks for modified types of patterns in post-processing of patterns. Appropriate algorithms are applied in this step (Fig. 26.1). The last step is to interpret and present results for decision-makers. Finally, these results are put into use in various applications such as health-care data, scientific data, skicat systems, and financial applications [5].

26.2 Methodology

To write systematic review, methodology used is based on the guidelines proposed by Barbra Kitchenham [6], which contain three phases. The first phase is planning, the second is conducting, and the last phase is reporting. Every phase contains various steps. All these steps are needed to create a systematic review. This systematic review paper follows various steps—(1) the first step is to identify all the research questions, (2) create a research strategy, (3) list the digital resources that have been used for research, (4) define the criteria on which study has to be done, (5) the next step is to develop the rules that are necessary for assessment of the quality of the selected papers, (6) then define the policies that are needed for extracting the data, and (7) the last step is to extract the data and synthesize this data. Fig. 26.2 demonstrates the research methodology. A detailed description of the steps followed to review is given as follow.

Step 1—Identify research questions

The main thing that has been performed in this paper is to designate data mining, DM techniques, and applications for that literature related to data mining have been studied. This paper has been included that satisfy the following questions:

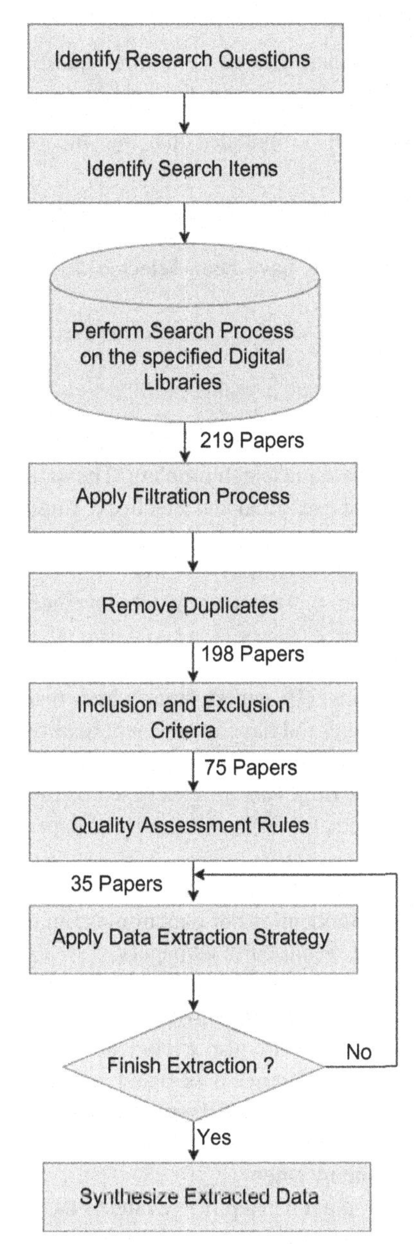

FIG. 26.2 Research methodology [6].

(1) Does the paper define data mining and knowledge discovery?
(2) Does the paper include types of data mining?
(3) Does the paper relate to categories of data mining techniques?
(4) Does the paper include data mining techniques?
(5) Does the paper relate to data mining applications?

Step 2—Research strategy

There are numerous papers related to research questions, so the following guidelines have been adopted to narrow the search:

(1) Only those papers will be included that that include search terms from research questions.
(2) Most common techniques have been selected that are continuously being used in various applications.
(3) Data mining applications have been selected according to their priority in use.
(4) For doing search various search operators was used such as AND, OR, and Parenthesis, etc.

Step 3—Digital resources

This is an important step in research methodology to collect data from different articles according to the research problem. The digital libraries chosen in this step are the Institute of Electrical and Electronic Engineers (IEEE) Conferences and transactions, IRJET, Springer, Elsevier, IJSR in Computer Science, Association for Computing Machinery (ACM).

Table 26.1 recapitulates those papers which have been selected from each source.

Step 4—Study selection criteria

Initial research contains 219 papers from which many of them were not related to research questions and have insufficient quality. So additional filtering was applied as shown in Fig. 26.2. During this filtration process, duplicate documents were removed. Irrelevant articles based on the inclusion and exclusion criteria were discarded. To verify the quality of papers, quality assessment was performed.

Criteria for inclusion included those papers that contain data mining techniques, that describe the applications of data mining. and that are printed from 1996 to 2019 and include recent editions papers.

Exclusion included those articles which include neither data mining techniques nor data mining applications. Those papers which are not categorized as journal or conference papers are also excluded.

After this filtration, 75 conference papers and journal articles were selected for further study. These papers were further reduced to 35 by applying quality assessment that is shown in Fig. 26.2.

Step 5—Quality assessment rules

After study selection, the next step is to evaluate the quality of the selected papers, five quality assessment rules were developed to determine whether these selected papers are relevant to the research or not. Each quality assessment was set as—fully answered, above average, average, below average, and not answered and they were weighted 1, 0.75, 0.50, 0.25, and 0, respectively. The papers have been selected according to their quality and the quality indicators are described as follow:

TABLE 26.1 Number of papers selected per digital resource.

Publisher	Paper selected
IEEE International Conferences	6
IEEE Transactions	9
Elsevier	3
Image (Rochester, N.Y.)	1
International Journal of Modern Engineering Research (IJMER)	2
ACM Int. Conf. Proceeding Ser	1
Data Min. Knowl. Discov. Handb	1
International Journal of Computer Graphics & Animation (IJCGA)	1
IEEE Ind. Manag. Data Sys.	1
Oriental Journal Of Computer Science & Technology	1
International Research Journal of Engineering and Technology (IRJET)	1
International Journal of Scientific Research in Computer Science, Engineering and Information Technology	1
Springer Basel AG	1
IEEE ACCESS	5
International Journal of Engineering Research & Technology (IJERT)	1
Total	35

(1) Are the articles fulfilling the objective of the research?
(2) Are the data mining approaches explained in detail?
(3) Are the data mining applications explained in detail?
(4) Are the data mining types included?

Step 6—Filtration process output

After applying the abovementioned steps, finally, 35 articles were selected. Table 26.2 illustrates the articles that have been selected. In the next columns, the research questions with their answers are given.

Step 7—Data extraction strategies

The objective of this step is to provide answers in semistructured way to the research questions that have been defined in the first step. The data extraction form used in the research methodology is given in Table 26.3.

Step 8—Synthesize of the extracted data

Various methods were discussed in Ref. [6] to do a synthesis of the extracted data. Narrative synthesis is used for results presentation (Figs. 26.3 and 26.4).

TABLE 26.2 Applicability of the selected papers to search questions.

Paper ID Acc. to Ref.	Paper name	Research questions					Year
		1	*2*	*3*	*4*	*5*	
[5]	Data mining: machine learning, statistics, and databases.	1	0	0	0	1	1996
[7]	Fuzzy K-modes algorithms for clustering categorical data.	0	0	0	1	0	1999
[8]	Special section communications.	0	0	0	0	1	2007
[3]	A decision—theoretic approach to data mining.	1	0	0	1	1	2009
[4]	Spatial data mining and geographic knowledge discovery.	1	1	0	0	0	2009
[9]	Intelligent transportation and control systems using data mining and machine learning.	1	0	0	1	0	2011
[10]	Remote health monitoring of heart.	0	0	0	0	1	2011
[1]	Data mining: future trends and application.	1	1	1	1	0	2012
[11]	Learning analytics and educational data mining towards communication and collaboration.	0	1	0	0	0	2012
[12]	Data mining techniques and applications-a decade review from 2000 to 2011.	1	0	1	0	1	2012
[13]	Visual analytics for multimodal social network analysis: a design study with social scixentists	0	0	0	1	0	2013
[14]	Learning analytics and educational data mining towards communication and collaboration.	0	1	0	0	0	2014

TABLE 26.2 Applicability of the selected papers to search questions—cont'd

Paper ID Acc. to Ref.	Paper name	Research questions 1	2	3	4	5	Year
[15]	Data mining and knowledge discovery.	1	0	1	0	0	2014
[16]	Overview of algorithm ED mining for higher education-an application	0	1	0	1	0	2014
[17]	Data mining: the massive data set.	1	1	0	1	0	2015
[18]	A review of data mining techniques	0	1	1	0	0	2015
[19]	A comparative study of classification techniques in data mining	0	0	0	1	0	2015
[20]	Bridging the vocabulary gap between health seekers and healthcare.	0	0	0	0	1	2015
[21]	Multimedia mining research	0	1	0	0	0	2015
[22]	A study of data mining prediction techniques in healthcare sector.	0	0	0	1	0	2016
[23]	Outlier detection: applications and techniques in data mining.	0	0	0	1	1	2016
[24]	Research on technology, algorithm, and application of web mining.	0	0	0	0	1	2017
[25]	An expensive review on data mining methods and clustering models for intelligent transportation system.	0	0	0	1	1	2017
[2]	Applications, current and future trends in diverse sectors.	1	0	0	0	1	2018

Continued

TABLE 26.2 Applicability of the selected papers to search questions—cont'd

Paper ID Acc. to Ref.	Paper name	Research questions					Year
		1	*2*	*3*	*4*	*5*	
[26]	Optimal process mining for large and complex event logs.	1	0	0	1	0	2018
[27]	KnowEdu: a system to construct knowledge graph for education.	0	1	0	0	1	2018
[28]	A new method to compute ratio of secure summations and its application in privacy preserving distributed data mining.	0	0	0	0	1	2019
[29]	Predictive tools in data mining and K-means clustering: universal inequalities.	0	0	1	1	0	2019
[30]	A new method to compute ratio of secure summations and its applications in privacy preserving distributed data mining.	0	1	0	0	1	2019
[31]	Application of data mining in customer relationship management	1	0	0	1	0	2019
[32]	Assessment of career adaptability: combining text mining and item response theory method.	1	0	0	1	0	2019
[33]	A hybrid infrastructure of enterprise architecture and business intelligence analytics for knowledge management in education.	0	0	0	1	1	2019
[34]	Towards comprehensive support for privacy preservation cross organization.	0	0	0	1	0	2019

TABLE 26.3 Data extraction form.

Data item	Value
Paper ID	ID no. of paper
Paper title	Title of selected paper
Author of paper	Name of author or authors of paper
Year of publication	The year in which the paper has been published
Journal/conference	Name of journal or conference in which paper has been published
R Q 1	R Q 1 defined in the Step 1
R Q 2	R Q 2 defined in the Step 1
R Q 3	R Q 3 defined in the Step 1
R Q 4	R Q 4 defined in the Step 1
R Q 5	R Q 5 defined in the Step 1
Any additional notes	Any specific notes regarding the selected paper

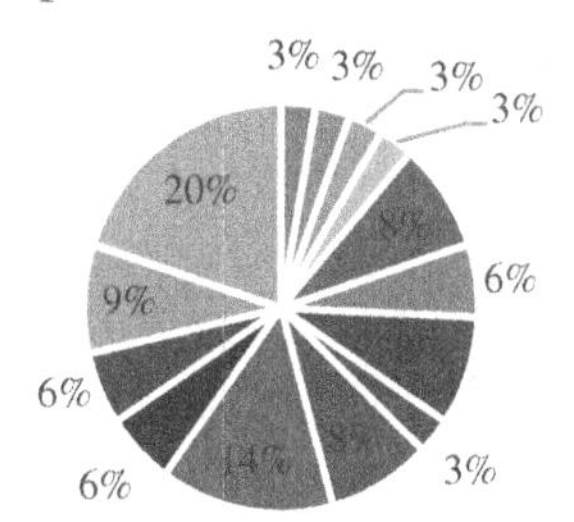

FIG. 26.3 No. of papers selected per year.

26.3 Need of data mining

Due to the new technologies, devices, communication means such as social media, social applications, and use of Internet, the amount of data generated is increasing day by day and unstructured data are available in Zettabytes and Zottabytes. Considering the rate at which mankind has generated 5 billion Gigabytes of data over the past two decades, which is the equivalent of the

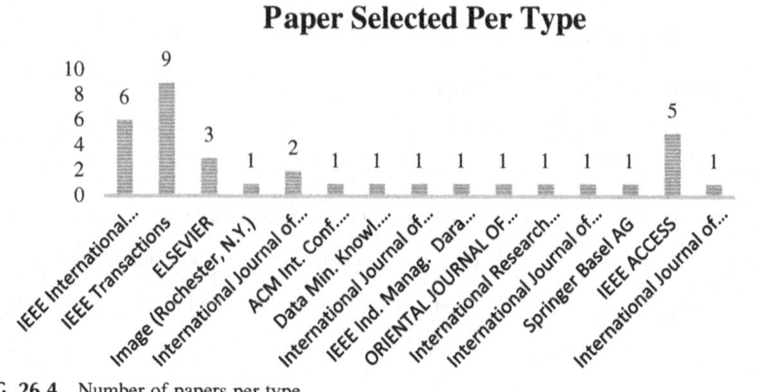

FIG. 26.4 Number of papers per type.

amount of data produced by us from the beginning of time until 2003, the same amount of data would be created every 2 days in 2011 and every 10 min in 2013. According to the Cloud Study, maximum amount of data have been created on the Internet. Nowadays, the amount of data produced is truly mind-boggling. Every second quintillion bytes of data are being created even by those people who are not educated or who do not know how to operate a computer with the use of social sites. So, it is the very tough task to retrieve useful knowledge from this huge data due to which data mining is required.

Data mining allows us to separate unwanted data and unnecessary data from the meaningful data. It makes good use of relevant data and helps in decision-making.

26.4 Types of data mining

26.4.1 Ubiquitous data mining

Nowadays a lot of electronic devices such as iPad, laptops, wearable watches, cell phones, and palmtops are being used due to which a large amount of data is being collected. Ubiquitous DM is all about analyzing data to pull out valuable information by applying special privacy-preserving methods. So it is a tough task to analyze data and is a work that is full of challenges [17]. Communication, security, computation, and many other actors add additional costs to Ubiquitous data mining. So it is the basic objective to mine data while minimizing the cost of ubiquitous presence [1].

26.4.2 Multimedia data mining

Multimedia data mining can be defined as a process that finds patterns in various types of data, including images, audio, video, and animation. Text mining and hypertext/hypermedia data mining are closely related to multimedia data

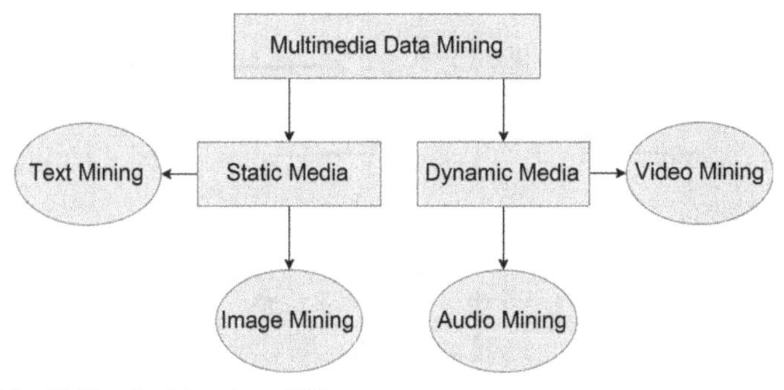

FIG. 26.5 Multimedia data mining [21].

mining because most of the information that is described in these fields are also applicable to multimedia data mining [1,17]. The data mining techniques that are applied to multimedia data are classification, clustering, association rule mining, sequence pattern mining, visualization, support vector machines, instance-based learning algorithms, and decision tree classification algorithms. Multimedia contain a huge amount of data not only with text but also audio, video, and images that arise the need for mining this data (Fig. 26.5).

26.4.3 Distributed/collective data mining

Distributed DM can be defined as mining of the information which is located in different physical locations. DDM is very useful for the multiple nodes that are connected over high-speed networks. This area of data mining is taking great attention because it is a big challenge to mine that data which is located in heterogeneous sites [17,28]. For preserving privacy in distributed data mining, the most important tool is the ratio of secure submissions (RSS) [30]. Various algorithms used with DDM have distributed classifier learning, the combiner scheme, heterogeneous distributed classifier, distributed association rule mining, and distributed clustering (Fig. 26.6).

26.4.4 Constraint-based data mining

Constraint-based DM involves guidance, control, and human involvement in the methods of data. It incorporates different types of constraints on data that guide the process of mining. Constraints are applied to reduce the number of patterns that are offered to the user [1]. To enhance its power, it is combined with multidimensional mining [17]. There are several types of constraints as follow:

Knowledge-type constraints—This knowledge-type constraint lays down the "type of the knowledge" which is to be extracted and it is stated at the starting of the query. Such types of constraints can include classification, clustering, and association.

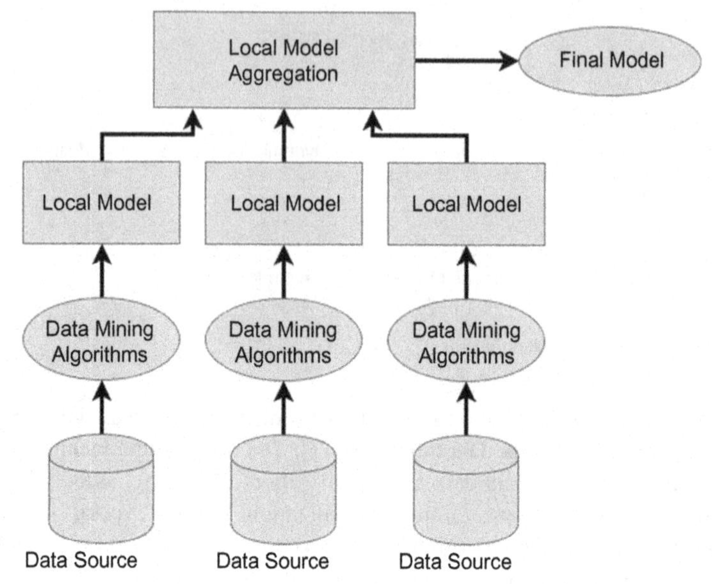

FIG. 26.6 Distributed data mining framework [35].

Data constraints—These types of constraints are used in the data mining query. Constraints can be stated in the form of queries that are similar to the queries specified in SQL.

Dimension constraints—Most of the large scale of knowledge or information that is needed to be mined is stored in the database and the multidimensional warehouses. Due to the large amount of data, it is required to specify constraints.

Interestingness constraints—These constraints are beneficial to determine those variables or measures that are interesting means value for the query and should be contained within the query [17].

26.4.5 Hypertext data mining

Hypertext data mining includes a collection of data, text, hyperlinks, and other types of hypermedia information which is associated with web mining and also with multimedia mining. WWW (World Wide Web) includes material found in online collections and digital libraries, etc. Hypertext and hypermedia include various techniques such as classification (supervised learning) application used in the area of web directories for grouping similar soundings into appropriate categories, clustering (unsupervised learning) uses training data. In unsupervised learning, documents are created based on a similarity then these documents are organized. Semistructured learning uses documents that are labeled and unlabeled and learning from both methods is needed. Social network analysis uses a web that examines networks formed through collaborative associations [1,17].

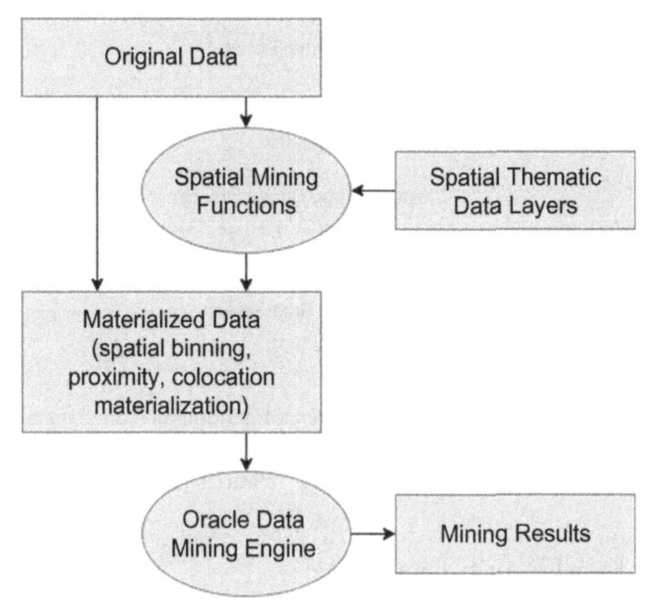

FIG. 26.7 Spatial data mining [36].

26.4.6 Spatial and geographic data mining

Spatial DM can be defined as a process of determining beneficial nonretrieval patterns that are taken from large spatial databases and concise total characteristics to use in certain application domains and provide new understanding. The spatial data includes astronomical data, satellite data, and data related to spacecraft [1,17]. Some of the data mining techniques are used to discover spatial patterns are classification, clustering, association, and outlier detection [4]. Various techniques are used to analyze spatial data (Fig. 26.7).

26.4.7 Phenomenal data mining

Phenomenal DM focuses on finding the associations between data and phenomena and not only the relationships within the data. Basically, it concerns establishing the relationships among the phenomenal underlying data. For example, take the scenario of a grocery store where various customers are coming to purchase groceries. A data mining program might be able to recognize which goods have been purchased by the same customer. These clients are being characterized by different attributes such as age, caste, sex, income distribution, etc. Here the customers got receipts after paying the bill. These receipts are known as data

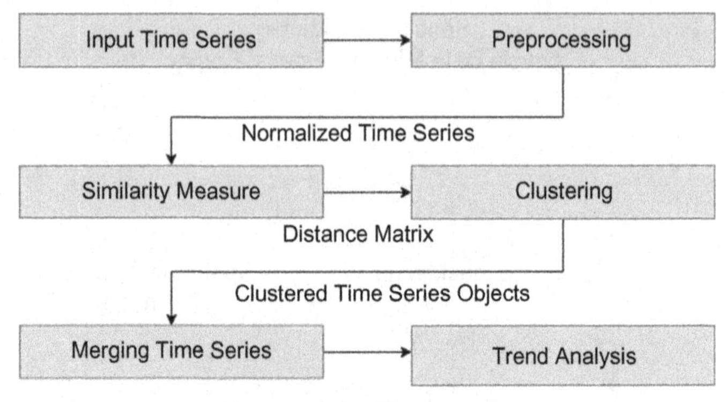

FIG. 26.8 The logic of phenomenal data mining [37].

and customers are known as the phenomena. Phenomena from data are inferred by supplying facts about their relations (Fig. 26.8). The database contains additional fields on existing relations and new relations [1].

26.4.8 Social security data mining

Social security DM is also known as social welfare DM discovers interesting patterns and concessions in social welfare and social security data. Social security DM deals with data mining goal perspective and data mining task perspective. In the first perspective, it handles different business objectives such as debt prevention and the second deals with traditional data mining methods such as classification (Fig. 26.9). Social security data mining has five basic goals:

- Overpayments-centric analysis
- Customer-centric analysis
- Policy-centric analysis
- Process-centric analysis
- Fraud-centric analysis [11].

26.4.9 Educational data mining

Educational DM has an important place in educational software and student modeling which focuses on automated discovery. It places emphasis on examining distinct components and associations between them. Techniques and methods used in it are classifications, clustering, association rule mining, discovery and models, visualization, and Bayesian modeling [14]. Algorithms used in educational DM are logistic regression, decision tree, linear regression, gradient boosted machines, Gauss-Newton algorithm Bayesian, apriori, and *k*-means [16] (Fig. 26.10).

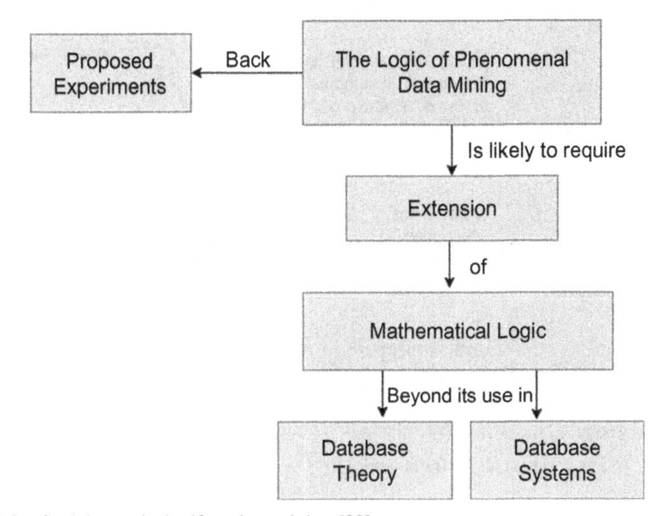

FIG. 26.9 Social security/welfare data mining [38].

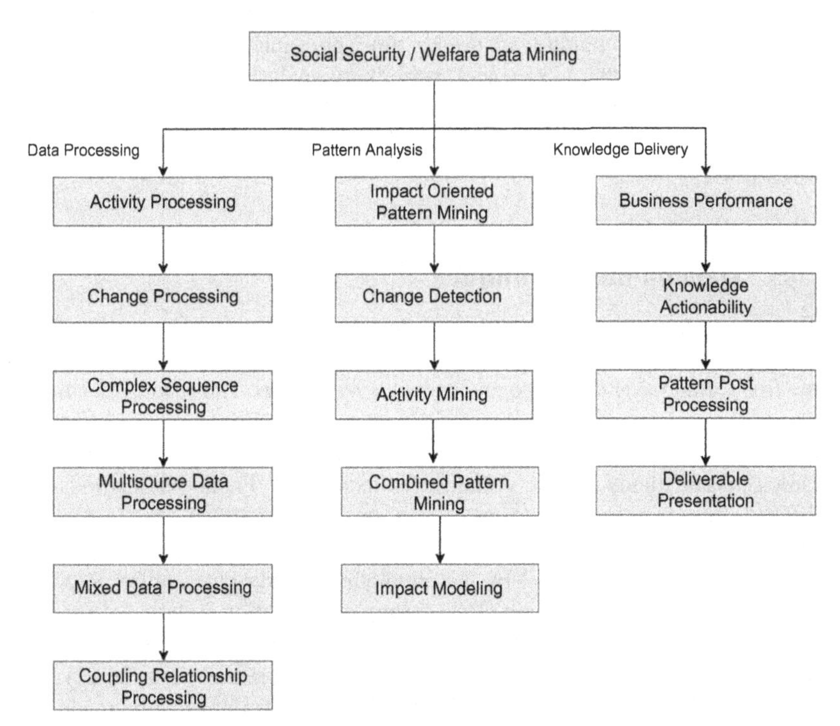

FIG. 26.10 The cycle of applying data mining in educational systems [16].

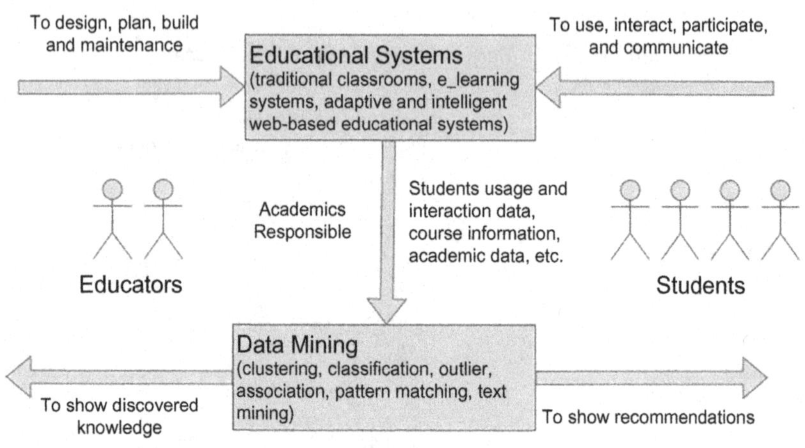

FIG. 26.11 Time series analysis framework [39].

26.4.10 Time series/sequence data mining

It includes mining of data that can be referenced by time. Time series data mining aims to identify movements and components which exist within the data. Different types of techniques are used in time series including similarity search (identification of pattern sequence), periodicity analysis (analysis data which repeat or recur), and sequential pattern mining (it is to identify sequences that occur frequently) [1]. Examples of time series DM include currency exchange rates, production sales, stock prices, and weather data, etc. (Fig. 26.11).

26.5 Data mining techniques

There are mainly two types of data mining techniques—(1) verification-oriented techniques and (2) discovery oriented techniques. The main focus of the first technique is to test the preconceived hypothesis. The second technique focuses on unconventionally discovering new patterns. This technique is further divided into two methods such as descriptive methods and predictive methods. Descriptive methods include visualization techniques. Predictive methods are divided into two types such as regression and classification. Regression involves the relationships between attribute values. It also provides the automatic production of a model by just predicting the attribute values. Records can be assigned to prearranged classes using classification methods of predictive techniques, for example, predictive techniques can be applied in the medical field to classify the patient depending on their diseases (Fig. 26.12). By analyzing the previous treatment of the patients, predictions can be made for similar kind of patients [3].

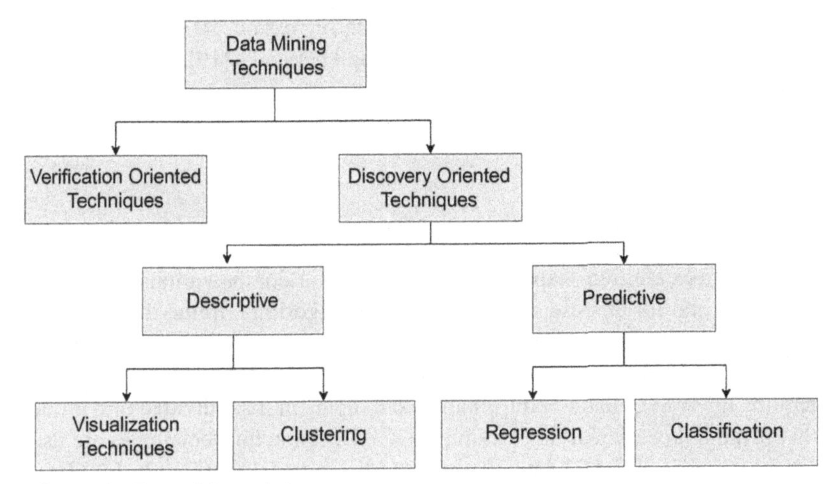

FIG. 26.12 Data mining techniques.

26.5.1 Clustering

Data mining techniques work on different types of data that can be physical or digital and it is stored in various repositories. Clustering analysis is a data mining technique that refers to that data that resemble each other. The clustering technique helps to find the similarities and differences between the data. For transportation data analysis, two types of clustering models are used—supervised models and unsupervised models [25]. One important part of data mining is to make clusters of similar data. Different methods are used for this purpose such as partitioning methods, hierarchical clustering, fuzzy clustering, etc. The fastest method to find clusters in data is K-means clustering (data analysis tool) [29]. Various algorithms are used for clustering categorical data such as the PAM algorithm, fuzzy-statistical algorithm, and dissimilarity measures. For clustering large sets of data, k-means algorithm is best [7].

26.5.2 Classification

Classification technique is capable of processing a huge amount of data [19]. This technique of DM is used to classify unalike types of data in different classes. Important information is retrieved from data using the classification technique of data mining. Classification is a technique in which similar data sets are classified in one set based on their characteristics. In classification, a classifier or model is made to predict the class label attributes. The basic task of classification is to forecast the target class for every kind of data. Classification models are based on the observed data that have been taken from many data sets. When the exact classes are known in advance for the creation of classification procedure from a set of data, it is termed pattern recognition. The mainly used algorithms in classification are—logistic regression, ID3 algorithm, random forest, C4.5 algorithm, and artificial neural networks [19]. A classification model can be used for assigning individuals to credit status on the basis of financial information or for identifying loan applicants as high, medium, and low credit risks [19].

26.5.3 Decision trees

Decision trees are tree-shaped structures that represent the various set of decisions [18] and for this use divide and conquer algorithm. In the tree structure, leaves represent classes and branches represent features that lead to those classes. Each node of the tree contains different attributes. A decision tree does not require any type of domain information and inputs in it are divided into groups that make it an easy and fast technique [22]. Decision tree techniques are used for the classification of a data set and provide a set of rules (Fig. 26.13). These rules are applied to the data set to predict those records that will provide desired output [18]. Decision tree methods include various techniques such as regression trees (CART), classification, and chi-squared automatic interaction detection [1].

26.5.3.1 Association rules

Association is a relation between two or more variables or data sets based on their characteristics. Association relations are mined from various records and they assist in decision-making. A common example of an association rule is shopping basket analysis [31].

26.5.3.2 Prediction

Prediction is a technique to predict what is going to be happen in the future depending on previously gained knowledge and on the basis of this experience

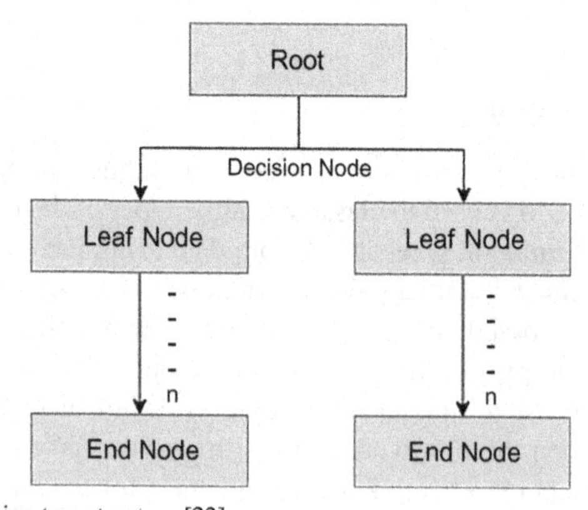

FIG. 26.13 Decision tree structure [22].

and knowledge some unknown results are predicted. Predictor variables are used to focus on a single aspect of data for predicting future data [12]. Some data mining techniques that are based on prediction are—neural network, Bayesian classifier, decision tree, and support vector machine [22]. The most common example of prediction can be seen in traffic prediction. Different approaches are used for finding traffic parameters. Mainly these approaches are categorized into four categories: (1) approaches that estimate and predict actual traffic flow, (2) methods that predict short-term traffic flow, (3) methods that estimate and predict travel time in real-time, and (4) approaches that guess and predict the actual traffic density [9].

26.5.3.3 *Relationship mining*

Relational mining or relational DM is commonly used for relational databases for finding relationships among different variables. Specific patterns are searched in different patterns by using the relationship mining algorithm like a priori and adaboost, etc. [12].

26.5.3.4 *Outlier detection*

The term outlier has been originated from statistics and it can be explained as every information whose appearance is different from the remaining set of data. Outliers are of three types—point outlier, contextual outlier, and collective outlier [23]. Outlier detection compares different values in data sets and finds out the smallest or largest values in them to catch the deviation among different values.

26.5.3.5 *Dynamic prediction-based approach*

The dynamic prediction-based approach which is used for modeling the process of molecules. It is a mathematical model. Two major approaches of dynamic prediction are—first one is the joint modeling of longitudinal and survival data and the second approach is landmark analysis. This approach is used in the medical and stock market.

26.5.3.6 *Text mining*

Text mining is the data mining technique or process which discovers earlier unfamiliar and valuable information from a huge quantity of unstructured text data. This knowledge is then analyzed and processed for operators, so they can receive valid knowledge. Text mining contains various types of text data such as documents, plain text files, messages, HTML files, and e-mails. So, a huge amount of data is mined using text mining. Text mining can be categorized into different categories including—text clustering, text categorization, and document summarization. The common approach used for text mining in text classification is of two types—supervised text classification and unsupervised text classification [32].

26.5.3.7 *Social network analysis*

Social network analysis (SNA) can be defined as a process of analyzing the set of social actors connected by relations. Modern information technologies are making SNA more crucial to connect more easily and in new ways. Nowadays the people in the whole world are connected through social media applications, such as LinkedIn, Facebook, and Twitter. Millions of users are using a huge amount of data that demands software support for statistics, computation, and decision-making [13].

26.5.3.8 *Process mining*

Process mining analyzes any business process and extracts the knowledge that is related to any event log. The process model is used to represent the event log [26]. Various process mining techniques have been proposed by the researchers to take event logs and produce process models without taking any prior information [34].

26.5.3.9 *Data distillation for judgment*

In data distillation for judgment data is represented in an intelligent manner using visualization and summarization techniques. A large amount of data can be efficiently explored and interpreted using it [12].

26.6 Categories of data mining techniques

Data taken from different sources may be related to designs, products, materials, machines, inventories, sales marketing, performance data may include patterns in data sets, associations among them, and dependencies of data sets on other datasets [3] which is totally different and asynchronous. So various data mining techniques are applied depending on the specific type of problems. Different data mining techniques use various algorithms based on which models are made depending on the requirement. There are basically nine categories of data mining techniques [12,18] that have been discussed as follow:

26.6.1 Information systems

Information systems have become the most prevalent field among all other fields. It belongs to the computer science and business world that interpret information and helps in decision making.

26.6.2 System optimization

'Linear programming' was the original term that was used for system optimization. System optimization was used for integration of the all elements of the system so that the system can work to its full extent.

26.6.3 Knowledge-based systems

Knowledge-based system can be defined as a computer program. This computer program solves the composite problems by using the knowledge base. These systems are built on AI and produce intelligent decisions. In knowledge-based systems, knowledge is represented using various rules, scripts, and frames.

26.6.4 Modeling

Modeling is a process of software engineering. Based on this process, data models are created. Different modeling tools are used including TheGMAX, CART, MARS, and TreeNet, etc. To create data models, different modeling methods and techniques are used. The model provides a facility to implement software in software engineering. Modeling helps us to understand various complex structures.

26.6.4.1 System architecture analysis

Architecture analysis involves the process of determining important system properties using the system's architectural model. It consists of four goals completeness, consistency, compatibility, and correctness. System architecture analysis uses the conceptual model to explain the structure, properties, and behavior of the system. A system's complete structure is represented using the architecture and it also concerns with the internal interface of the system's hardware components.

26.6.4.2 Algorithm architecture

An algorithm is a finite set of instructions that is used to do a specific task or to solve a problem. Algorithms are used for various purposes including calculations and data processing. While solving a real-time problem, algorithm is the main aspect that affects the time and other parameters such as cost, efficiency, and complexity. Various steps are elaborated in the growth of an algorithm including problem definition, modeling, specification, designing, checking, analyzing, implementation, testing, and documentation.

26.6.4.3 Intelligence agent systems (IAs)

IAs are software programs that use artificial intelligence to make decisions. They can learn and acquire knowledge from the surrounding environment. They can learn from this knowledge and attain their prescribed goals using this knowledge.

26.6.4.4 Dynamic prediction-based approach

The dynamic prediction-based method can be defined as a mathematical model. This model is used for the modeling process. Two major approaches of dynamic prediction are—first one is the joint modeling of longitudinal and survival data and the second approach is landmark analysis. This approach is used in medical and stock markets.

26.6.4.5 *Artificial neural networks*

An ANN is also called a "neural network" (NN). It is a nonlinear mathematical model which learns through training and is based on a biological neural system. ANN is an adaptive structure that changes its construction based on internal or external information. Artificial neural networks have a learning phase or training phase in which all information flows [15]. Artificial intelligence techniques are widely used in DM such as machine learning, automation and robotics, neural networks, natural language processing, machine vision, pattern recognition, knowledge acquisition, knowledge representation, etc.

26.7 Applications of data mining methods

The most common applications of data mining are given as follow:

26.7.1 Text mining and web mining

Web mining is the method of searching and extracting knowledge from web documents using confident keywords and key phrases. By searching various documents, relationships among them are found and organized information can be directly used by the users [1,24]. Text mining includes text classification, text clustering, text summarization, and text association rule analysis. Text mining is divided into two types of methods known as information retrieval method and database method [24]. For example, if a manufacturer wants to see client observations on different products, it can be easily obtained by text mining rather than using time-consuming review questions [2]. Text mining is a part of web mining. Web mining is a process to learn and excerpt useful knowledge from web documents through data mining technology. Web mining consist of different types of data such as unstructured, semistructured, dynamic, heterogeneous, and wide-distributed data. Web mining can be divided into three parts—web structure mining (page mining, URL mining), web content mining (web text mining, web multimedia mining), and web usage mining (general access mode analysis, analyze and customize web site) [24].

26.7.2 Medical/pharmacy

There has been extreme growth in biomedical research, which includes the development of new pharmaceuticals for various diseases such as cancer therapies and AIDS detection. This is the most critical application that manages the mining and understanding of organic grouping and structures. Research has been performed on DNA analysis which leads to finding causes of various diseases and develop various approaches to detect, treat, and prevent disease [1,2,10].

26.7.3 Insurance and health care

Medical terminologies are referred to as authenticated phrases by famous organizations that accurately describe the human body, conditions, and processes in a specific science-based manner. The approach used for mining the medical data is local mining, in which firstly, a tri-stage framework is established then embedded noun phrases medical records, identify the medical concepts, and then normalize the medical concepts [20].

26.7.4 Finance

Financial institutions and banks propose a wide variety of banking services which include checking of data, saving, individual client exchange, and business exchange. Data mining finance application helps in the prediction of the stock market, bankruptcies, changing currency, consideration and management of all types of financial risk, loan management, credit rating, and money laundering analyses. It also provides various credit services to its customers such as business, mortgage, automobile loans, and investment services such as mutual funds. It also provides insurance services to face any uncertain condition. Financial data that is composed in banks and monetary institutions are often whole, consistent, and of high value which helps in the methodical analysis of data [1,2].

26.7.5 Telecommunications

The telecom business or telecommunication industry produces a tremendous amount of data. It has rapidly grown from providing long and short remoteness connectivity services to provide numerous other far-reaching correspondence services such as voice, pager, fax, wireless, cellular phone, pictures, e-mails, and the Internet and many other ways of communications are underway. The telecommunication industry is spreading hugely and is highly competitive. It needs a great demand of DM to completely understand the telecommunication patterns, to clasp any fraud activity, understand the business properly. This helps in making the proper use of available properties [1,2,8].

26.7.6 Education

Education is a crucial application of DM which uses algorithms, tools, and methods of DM to explore educational data from teachers, pupils, and other organizational staff. In educational data mining, exclusive and more large scale data that come from educational backgrounds are explored [33]. Increasing demand of education domain and various applications of the knowledge graph, led to offer a system called KnowEdu. This system excerpts the concept of subjects or courses and then identifies the relationships between these concepts. Knowledge graphs are also called concept maps [27].

26.7.7 Retail industry

Data mining performs a lot in this industry because the trade industry gathers a large quantity of data from various sources such as sales of products, the history that provides the complete description like what the customers are buying like are they buying some combination of products, transferring goods from one location to another and consumption of goods, and services provided by manufacturers. The quantity of data collected increase rapidly due to e-commerce, business on the web, and the popularity of the products. Therefore, the data collected in huge amounts need data mining for decision-making [1]. There are many reasons behind the increase in the retail industry—rise of superstores, customer relationship management, supply chain management, and the rise of online retailing [40].

26.8 Conclusion

Data is crucial for every kind of application therefore and this data is in huge amount so it is mandatory to manage it very carefully. DM provides great help in maintaining the data. Data mining uses various techniques to extract, process, and utilize data in an efficient manner, which have been explained in the paper. Various application of data mining has been explained and many more applications of data mining are taking place that needs to be explored. The main need of stored data on various sources is quality, privacy, and security of data that need further consideration. In the future, DM will be needed everywhere where a large amount of data will be handled. DM should have the capability to work on multiple inputs and produce desired results.

References

[1] A.N. Paidi, Data mining: future trends and applications, Int. J. Mod. Eng. Res. 2 (6) (2012) 4657–4663.

[2] D.V. Kumar, K.T. Basha, Applications, current and future trends of data mining in diverse sectors, Int. J. Mod. Eng. Res.—IJMER 3 (4) (2018) 295–302.

[3] Y. Elovici, D. Braha, A decision-theoretic approach to data mining, IEEE Trans. Syst. Man, Cybern. Part A Systems Humans. 33 (1) (2003) 42–51.

[4] J. Mennis, D. Guo, Spatial data mining and geographic knowledge discovery—an introduction, Comput. Environ. Urban. Syst. 33 (6) (2009) 403–408.

[5] H. Mannila, Data mining: Machine learning, statistics, and databases, in: Proc. of the 8th Int. Conf. Sci. Stat. Data Base Manag. SSDBM 1996, 1996, pp. 2–8.

[6] B. Kitchenham, O. Pearl Brereton, D. Budgen, M. Turner, J. Bailey, S. Linkman, Systematic literature reviews in software engineering—a systematic literature review, Inf. Softw. Technol. 51 (1) (2009) 7–15.

[7] Z. Huang, M.K. Ng, A fuzzy k-modes algorithm for clustering categorical data, IEEE Trans. Fuzzy Syst. 7 (4) (1999) 446–452.

[8] R.L. King, A.M. Architecture, Special section communications, Image (Rochester, NY) 45 (4) (2007) 875–878.

[9] N.O. Alsrehin, A.F. Klaib, A. Magableh, Intelligent transportation and control systems using data mining and machine learning techniques: a comprehensive study, IEEE Access 7 (2019) 49830–49857.

[10] L. Pecchia, P. Melillo, M. Bracale, Remote health monitoring of heart, I.E.E.E. Trans. Biomed. Eng. 58 (3) (2011) 800–804.

[11] L. Cao, Social security and social welfare data mining: an overview, IEEE Trans. Syst. Man Cybern. Part C Appl. Rev. 42 (6) (2012) 837–853.

[12] S.H. Liao, P.H. Chu, P.Y. Hsiao, Data mining techniques and applications—a decade review from 2000 to 2011, Expert Syst. Appl. 39 (12) (2012) 11303–11311.

[13] S. Ghani, B.C. Kwon, S. Lee, J.S. Yi, N. Elmqvist, Visual analytics for multimodal social network analysis: a design study with social scientists, IEEE Trans. Vis. Comput. Graph. 19 (12) (2013) 2032–2041.

[14] G. Siemens, R.S.J.D. Baker, Learning analytics and educational data mining: towards communication and collaboration, in: ACM Int. Conf. Proceeding Ser, 2012, pp. 252–254.

[15] P. G. Zhang, "Data mining and knowledge discovery handbook," Data Min. Knowl. Discov. Handb., no. July 2010, 2010.

[16] D. Walte, H. Reddy, V. Ugale, A. Unwane, Overview of algorithm ED mining for higher education: an application, India 3 (2) (2014) 521–523.

[17] S. Patodi, M. Jain, Data mining: the massive data set, Int. J. Comput. Sci. Inform. Technol.—IJCSIT 6 (5) (2015) 4667–4671.

[18] S.S. Ghuman, A review of data mining techniques, Ind. Manag. Data Syst. 3 (4) (2014) 1401–1406.

[19] S.S. Nikam, A comparative study of classification techniques in data mining algorithms, Int. J. Mod. Trends Eng. Res. 4 (7) (2017) 58–63.

[20] L. Nie, Y.L. Zhao, M. Akbari, J. Shen, T.S. Chua, Bridging the vocabulary gap between health seekers and healthcare knowledge, IEEE Trans. Knowl. Data Eng. 27 (2) (2015) 396–409.

[21] S. Vijayarani, M.A. Sakila, Multimedia mining research—an overview, Int. J. Comput. Graph. Animat.—IJCGA, 5 (1) (2015) 69–77.

[22] B. Srinivasan, K. Pavya, A study on data mining prediction techniques in healthcare sector, Evaluation (2016) 552–556.

[23] R. Bansal, N. Gaur, S.N. Singh, Outlier detection: applications and techniques in data mining, in: 2016 6th International Conference—Cloud System and Big DataEngineering (Confluence), 2016, pp. 373–377.

[24] Y. Li, Research on technology, algorithm and application of web mining, in: Proc. of the 2017 IEEE Int. Conf. Comput. Sci. Eng. IEEE/IFIP Int. Conf. Embed. Ubiquitous Comput. CSE EUC 2017, vol. 1, 2017, pp. 772–775.

[25] S. Anand, P. Padmanabham, A. Govardhan, R.H. Kulkarni, An extensive review on data mining methods and clustering models for intelligent transportation system, J. Intell. Syst. 27 (2) (2018) 263–273.

[26] M. Prodel, V. Augusto, B. Jouaneton, L. Lamarsalle, X. Xie, Optimal process Mining for Large and Complex Event Logs, IEEE Trans. Autom. Sci. Eng. 15 (3) (2018) 1309–1325.
[27] P. Chen, Y. Lu, V.W. Zheng, X. Chen, B. Yang, KnowEdu: a system to construct knowledge graph for education, IEEE Access 6 (2018) 31553–31563.
[28] K. Liu, H. Kargupta, J. Ryan, Random projection-based multiplicative data perturbation for privacy preserving distributed data mining, IEEE Trans. Knowl. Data Eng. 18 (1) (2006) 92–106.
[29] H. Agahi, A. Mohammadpour, S.M. Vaezpour, Predictive tools in data mining and k-means clustering: universal inequalities, Results Math. 63 (3–4) (Jun. 2013) 779–803.
[30] Y. Shao, W. Hong, Z. Li, A new method to compute ratio of secure summations and its application in privacy preserving distributed data mining, IEEE Access 7 (2019) 20756–20766.
[31] V. Thanuja, B. Venkateswarlu, G.S.G.N. Anjaneyulu, Applications of data mining in customer relationship management, J. Comp. Math. Sci. 2 (3) (2011) 423–433.
[32] L. Zhang, et al., Assessment of career adaptability: combining text mining and item response theory method, IEEE Access 7 (2019) 125893–125908.
[33] O. Moscoso-Zea, J. Castro, J. Paredes-Gualtor, S. Lujan-Mora, A hybrid infrastructure of Enterprise Architecture and business intelligence analytics for knowledge management in education, IEEE Access 7 (2019) 38778–38788.
[34] C. Liu, H. Duan, Q. Zeng, M. Zhou, F. Lu, J. Cheng, Towards comprehensive support for privacy preservation cross-organization business process mining, IEEE Trans. Serv. Comput. 12 (4) (2016) 639–653.
[35] https://www.semanticscholar.org/paper/Distributed-Data-mining-using-Multi-Agent-data-Nure/39ecaef6ae79837e32ebe1b672e455651c19a484/figure/1.
[36] https://docs.oracle.com/database/121/SPATL/spatial-information-and-data-mining-ations.htm#SPATL791.
[37] https://cmapspublic3.ihmc.us/rid=1040063256546_510755563_5629/07.%20The%20Logic%20of%20Phenomenal%20Data%20Mining.cma.
[38] https://cmapspublic3.ihmc.us/rid=1040063256546_510755563_5629/07.%20The%20Logic%20of%20Phenomenal%20Data%20Mining.cmap.
[39] https://www.semanticscholar.org/paper/Trend-Analysis-of-Time-Series-Data-Using-Data-Baheti-Toshniwal/d3701cfe387e1ad67453aeae28175110582e9103.
[40] H. Li, Applications of data warehousing and data mining in the retail industry, in: 2005 Int. Conf. Serv. Syst. Serv. Manag. Proc. ICSSSM'05, vol. 2, 2005, pp. 1047–1050.

Chapter 27

D-PPSOK clustering algorithm with data sampling for clustering big data analysis

C. Suresh Gnana Dhas[a], N. Yuvaraj[b], N.V. Kousik[c], and Tadele Degefa Geleto[a]

[a]*Department of Computer Science, Ambo University, Ambo, Ethiopia,* [b]*Research and Development, ICT Academy, Chennai, India,* [c]*School of Computing Science and Engineering, Galgotias University, Greater Noida, Uttar Pradesh, India*

27.1 Introduction

Big data is a term containing various types of confusing and comprehensive data sets that are hard to prepare with normal framework data management. Big data such as storage, transportation, viewing, searching, analysis, and protection breaches and sharing are created with a range of problems. The exponential data development in all domains calls for step-by-step procedures to be followed and accessed. In Ref. [1], the creators stressed the demand for analysis in the context of Big Data, taking into account the final objective of addressing the online biocompatible data path. The importance of big data in organic and biological research has been predicted. It has thus exposed that an administrative plan for true identifiable data has been minimized. This is possible by deconstructing the metadata and consolidating past discrete data sets by using foresight insights [2–4]. When efforts are made to select a specific stack, which controls the data to be stored or prepared, opponents of large data investigations arrive. Social database management systems do well with organized data and continue to decide on certain requirements. Whether it is from interpersonal organizations, sensor systems, and other combined data with replication for the exponential development of unstructured information in terabytes or even petabytes, big data is the answer to handling this data.

The enormous expansion of data nowadays results in a stunning increase in the number of gadgets located on the periphery, including implanted sensors, smart mobile devices, and tablet PCs. The organization's workforce, long-range interpersonal communication destinations, and various machines create these

System Assurances. https://doi.org/10.1016/B978-0-323-90240-3.00027-8

large amounts of data sets (distributed storage, meters, CC TV cameras, and so on). The data centers provide these comprehensive sets of data. The problems related to these large or large data collections are their storage, management, recovery, and analysis. This work is largely related to the problem of breaking down such huge data on average. The large data in a conservative configuration, which will be an instructional form for the complete data, is one technique that can be used to overcome this issue.

Clustering is an approach used for the analysis of exploratory data. Expansive data volumes are examined externally in this way and help to decide quickly. Clustering is an unmonitored method for the classification of huge data collections into corresponding collections. For data points or incidents, there is no established class name [5]. Clusters aggregate data events into subsets so as to assemble comparable cases together, while various cases are called clusters and different groups. For this reason, many academics are transforming big data sets into parallel processing.

The k-means algorithm is a well-known technique for clustering and has been linked with a variety of practical difficulties in clustering. However, the k means is not curved and may contain many neighborhood minima, as its selection of inductions gives it a few disadvantages. Evolutionary computing, for instance, genetic algorithms, and particle swarm optimization presents late headways in cluster algorithms [6,7]. Genetic algorithms usually begin with certain applicants' optimization responses and they are developed through determination, crossover, and mutation to achieve a superior result. The PSO notion was designed to replicate social behavior, in which real estate is a trade-in data and practical uses. Many research studies used PSO to group data in multidimensional areas and obtained outstanding results. However, when it comes to global optimum, the rate of mergers is insufficient. Kao et al. study the relevance of the k-means PSO (K-PSO) to data vector clustering.

The PSO encompasses an irregular start phase random sequence as a parameter and is dependent on this parameter to update particle locations and velocities. In any event, in particular with complex multitop search challenges, PSO commonly leads to premature convergence, like high-dimensional clustering. The present research proposes the distributed-parallel particle swarm optimization with k-means (D-PPSOK) algorithm with samples of huge data sets for fewer computing clusters. The proposed D-PPSOK with large-scale data sampling provides superior performance compared to the existing data sample data set based on experimental results.

27.2 Related work

Jain and Verma [8] provided a description of an approximate k-means method. It is a new, rapid, scalable, and highly precise big data analytics strategy. It eliminates the inconvenience of k-means for an indefinite number of iteration sequences without losing precision. In this study, we present a new approach to optimize hybrid particles that combines the best features of PSO and fuzzy C-means for efficient clustering of Web documents. This half-and-hybrid PSO method has been tried in several text report collections. The range of the documents varies between 512 and 1639 in the data set and extends between 12,367 and 19,851. Based on the exploratory findings, our approach to particle swarm optimization for fuzzy c-means clustering (PSOFCM) is better clustered than previous techniques.

Sanse and Sharma [5] provided an overview of the different categorization methods, and an investigation (from a hypothesis) is also provided for general work. The study shows the characteristics, points of interest, and drawbacks of the various approaches.

Bisht and Singh [9] presented a hybrid genetic algorithm with k-means for the grouping of large data. For a big data partition, the k-means approach is more frequent and straightforward. However, the handling of large amounts of information and the preparation of less memory space are negatives. A genetic algorithm for the processing of large data is a search and optimization method. The new algorithms use genetic and k-means algorithms known as GKA or Genetic k-means algorithms. However, the partitioning of data is not allowed.

Bisht and Singh [9] introduced a strategy for PSO and k-means (EPSO), where PSO will be updated with search systems in the neighborhood. By means enhanced hybrid and half PSO, it helps to not only escape the algorithm from optima's neighborhood, but also the inadequacy of the modest fusion rate of the PSO. Exploratory results on eight benchmarks revealed that several alternative data clustering algorithms are defeated by the proposed approach and that it is efficient and robust.

Khan et al. [10] provided important clustering approaches that need severe computational requirements. A series of bioinspired algorithms understood as swarm intelligence (SI) has emerged late to satisfy these requirements and has been effectively linked to several actual clustering issues. This work examines the use of cluster analysis for particle swarm optimization. The possibility of fuzzy C-clustering provides better performance and also allows the particles to be robust to follow the evolving environment.

27.3 Proposed work

The PSO technique is used to create K-introductive centroids. In document clustering, this approach is proposed. An optimization process in the context of the population is the particle swarm optimizing (PSO) algorithm. The optimum solution can be found. Two modules are included in the hybrid DPPSOK algorithm: the DPPSO module and k-means. The first is to decide the center of the clusters by using the DPPSO module.

The PSO algorithm is used to determine the nearby objectives of the optimal solution at the early stage. The following centroids are then used by the K-means module to perfect and provide the ideal solution with a certain end

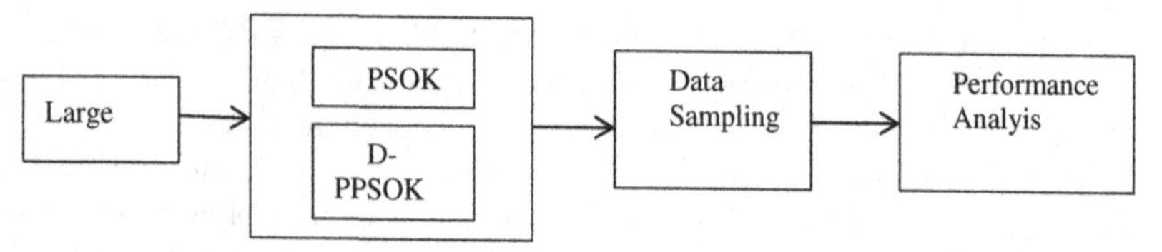

FIG. 27.1 An overview of proposed model.

objective. Finally, the conveyer and the parallel environment perform these two phases. The use of a distributed array by PPSOK reduces the computational time needed to cluster the larger files (Fig. 27.1).

27.3.1 Implementation of DPPSOK-means algorithm

PSO-based k-means clustering algorithm [11], perceived as PSOK and k-means clustering algorithm is consolidated with PPSO. In PSO, a resident of reasonable "particles" is instated among random positions X_i and velocities V_i, and work, f, is established, utilizing the particle's positional arranges as information qualities. In a n-dimensional analysis space, $X_i = (x_{i1}, x_{i2}, x_{i3}, \ldots, x_{in})$ and $V_i = (v_{i1}, v_{i2}, v_{i3}, \ldots v_{in})$: positions and velocities are balanced, and the function is assessed with the new coordinates at each timestep. The basic upgrade conditions for the dth measurement of the ith particle in PSO might be indicated as

$$V_{id}^{new} = V_{id}^{old} = +c_1, r_1, p_{id} x_{id} + c_2, r_2, p_{id}, x_{id} \tag{27.1}$$

$$X_{id}^{new} = X_{id}^{old} = V_{id}^{new} \tag{27.2}$$

Features r_1 and r_2 are random positive numbers, drawn from a uniform dispersion, and are classified by an extreme point of confinement Rmax, which is a parameter of the framework. In Eq. (27.2), c_1 and c_2 are called increasing speed constants though w is called idleness weight. Pb is the neighborhood's finest solution discovered so far by the ith particle, while P_g relates to the positional directions of the fittest particle discovered so far in the whole group. The PSOK algorithm in the PSOK-means algorithm, the global search of PSO and the speed of consultation of k-means are consolidated. The algorithm PSO is utilized as the initial step of finding the neighboring of the optimal solution. The got result is then utilized for the k-means algorithm, by utilizing a nearby search to produce the last result.

27.3.2 DPPSOK algorithm

A parfor (parallel for) loop is useful in the express that needs many loop cycles of a straightforward estimation. Specific techniques, for example, pcent ( ) and cent ( ) is overloaded and the return comes about that are themselves appropriated exhibits. It is informative to note this is like parallelism in the light of strings. The parallelism is covered up inside computational strategies and functions that are called from generally successive code.

Algorithm k-means (D, K)

Input: D data set, K-number of clusters,

Output: K overlapping clusters of data set

Step 1. At the first stage, each particle randomly chooses K different d vectors from the data set as the initial cluster centroids vectors.

Step 2. For each particle:

(a) Assign each vector in the data set to the closest centroid vector.

(b) Calculate the fitness value

$$f=\sum_{i=1}^{N_c}\{\sum_{j=1}^{P^i} d\left(\frac{O_iM_{ij}\}}{N_c}\right.$$

(c) Using the velocity and particle position update Eqs. (27.1) and (27.2) and generate the next solutions.

Step 3. Repeat step (2) until one of the following termination conditions is satisfied.

(a) The maximum number of iterations is exceeded.

This algorithm dispenses each position to the cluster whose center is adjacent. The center composes are the arithmetic mean calculated for each component one by one over every one of the points in the cluster. Assume it is as far as possible, the k-means algorithm.

.

27.3.3 Evaluation of the solutions

To evaluate the solutions utilize the recipe characterized by the condition (27.4). This function is utilized as a fitness function for the PSO algorithm and to evaluate the diverse results. It quantifies the normal separation between the archives of a cluster and its centroid. The smaller this value is, the more the cluster will be minimal. Thus, it can be utilized to assess the nature of the clusters. To discover the likeness between two reports, m_p and m_j regularly utilize a distance measure. The most utilized measures depend on the Minkowski equation given by

$$D_n(m_p, m_j) = \left(\sum_{i=1}^{d_m} |m_{i,p} - m_{i,j}|^n\right)^{1/n} \tag{27.3}$$

For $n = 2$, the formula describes the Euclidian distance. The standardized Euclidian distance between the archives m_p and m_j is given as takes after:

$$D_n(m_p, m_j) = \sqrt{\sum_{k=1}^{d_m} m_{pk} - m_{ik})^{2/} d_m} \qquad (27.4)$$

where m_p and m_j are two report vectors; d_m is the size of the vectors; and m_{pk} and m_{jk} speak to separate the weight estimations of the kth term in the gathering for the records m_p and m_j.

27.4 Experimental result and discussion

The goal of the experimental analysis is to calculate the accuracy and time taken by the proposed system. A parameter that can be used for evaluating D-PPSOK in this proposed model is the accuracy of the data sample produced by it. A detailed description of data set used for experimentation and analysis of an experiment, the results, and their discussions are mentioned below.

27.4.1 Test data

To find the efficiency of the proposed system, it should be tested by giving the input as data sets. The data sets have been taken from (http://archive.ics.uci.edu/ml/) and given input of data set I (Quarterly Income Tax-I)—which consists of 50,000 and data set II (Quarterly Income Tax-II)—which consists of 1,12,400 instances.

27.4.2 Experimental result

They compare the performance of our D-PPSOK with a data sampling algorithm based on distributed array technique against the traditional PSOK. The experimental results are conducted for various values of k taken for PSOK and D-PPSOK algorithms and the execution time taken by these approaches are shown in Table 27.1, Figs. 27.2 and 27.3.

TABLE 27.1 Parallel proposing elapsed time for various values.

Data set	No. of instance	K = 3		K = 5		K = 10	
		PSOK	*D-PPSOK*	*PSOK*	*D-PPSOK*	*PSOK*	*D-PPSOK*
Data set I	50,000	10.2	8.0	20.0	18.3	49.5	48.2
Data set II	1,12,400	26.4	16.9	54.2	42.6	135.5	112.4

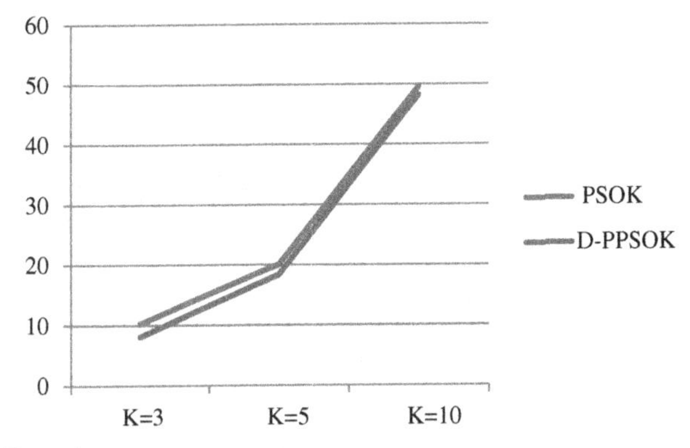

FIG. 27.2 Data set.

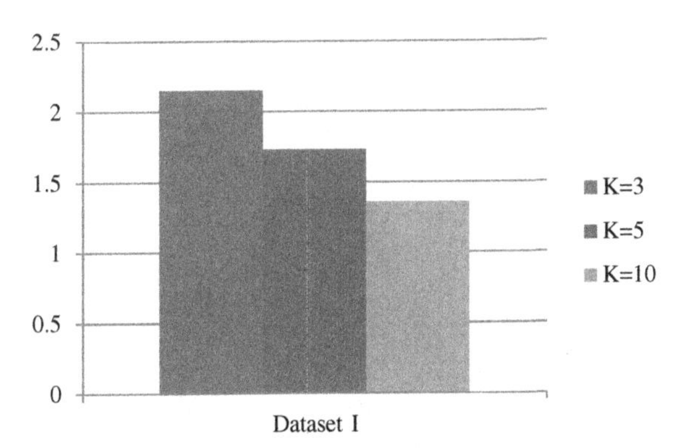

FIG. 27.3 Elapsed time variance of PSOK and D-PPSOK for data set I.

TABLE 27.2 PPSOK, D-PPSOK elapsed time variation of data set I and data set II.

Data set	K = 3	K = 5	K = 10
Data set I	2.15	1.73	1.36
Data set II	9.53	11.63	23.07

According to the experimental results, the proposed D-PPSOK with data sampling using distributed array technique is better than the PSOK clustering algorithm. Table 27.2, Figs. 27.3 and 27.4 shows the variance of elapsed time taken for data set I and data set II.

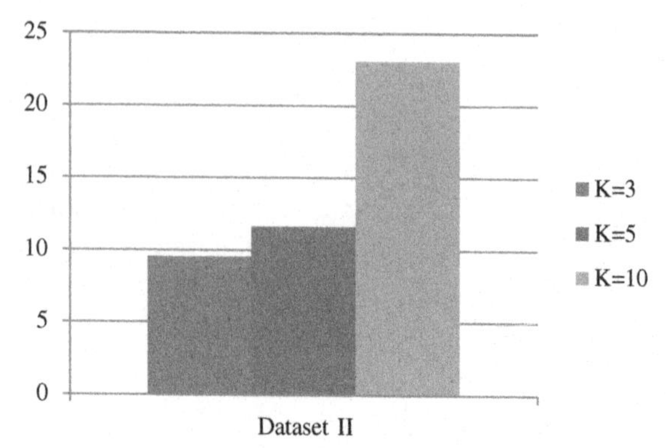

FIG. 27.4 Elapsed time variance of PSOK and D-PPSOK for data set II.

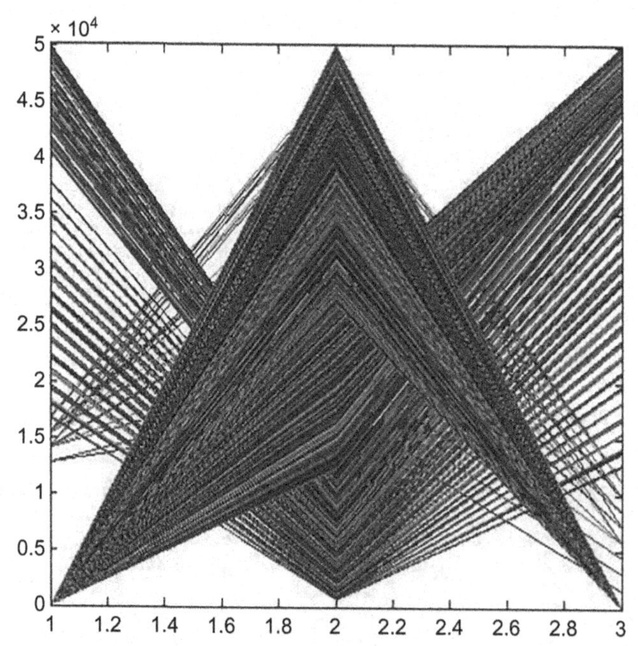

FIG. 27.5 A data sample of PSOK for data set II (K = 10).

Sampling is a valid and frequently used practice for statistical analysis so can apply data sampling algorithm to PSOK and proposed algorithm D-PPSOK. Fig. 27.5 and Table 27.3 show the experiment clearly that the proposed algorithm is better than the PSOK.

TABLE 27.3 PSOK algorithm best sample of data set II (K = 10).

0	0	0	0	0
0	0	0	0	0
43,235	0	46,914	27,663	0
0	0	0	0	0
0	0	0	0	0
80,663	100,564	83,394	105,666	100,859
0	0	0	84,883	0
102,706	0	106,920	102,932	0
0	0	0	37,239	0
0	0	0	110,037	0

27.5 Conclusion

Large data sets become difficult to cluster using traditional data clustering algorithms. Hence, efficient parallel clustering algorithms are required to resolve this difficulty. The proposed system D-PPSOK clustering algorithm with data sampling clusters the large data set efficiently and provides a convenient way of clustering with less computational steps.

References

[1] S. Justin Samuel, R.V.P. Koundinya, K. Sashidhar, C.R. Bharathi, A survey on big data and its research challenges, ARPN J. Eng. Appl. Sci. 10 (8) (2015).

[2] L.-Y. Chuang, Y.-D. Lin, C.-H. Yang, Member, IAENG, An improved particle swarm optimization for data clustering, in: International MultiConference of Engineers and Computer Scientists Vol I, IMECS, March 2012.

[3] P. Jaganathan, S. Jaiganesh, A particle swarm optimization based fuzzy c means approach for efficient web document clustering, Int. J. Eng. Technol. 5 (6) (2013–2014).

[4] D.C. Tran, Z. Wu, V.X. Nguyen, A new approach based on enhanced PSO with neighborhood search for data clustering, in: International Conference of Soft Computing and Pattern Recognition (SoCPaR), 2013.

[5] K. Sanse, M. Sharma, Clustering methods for big data analysis, Int. J. Adv. Res. Comput. Eng. Technol. 4 (3) (2015).

[6] S. Paterlini, T. Krink, Differential evolution and particle swarm optimisation in partitional clustering, Comput. Stat. Data Anal. 50 (2006) 1220–1247.

[7] S. Rana, S. Jasola, R. Kumar, A review on particle swarm optimization algorithms and their applications to data clustering, Artif. Intell. Rev. 35 (2011) 211–222.

[8] M. Jain, C. Verma, Adapting k-means for clustering in big data, Int. J. Comput. Appl. 101 (1) (2014). 0975–8887.

[9] P. Bisht, K. Singh, Big data mining: analysis of genetic k-means algorithm for big data clustering, Int. J. Adv. Res. Comput. Sci. Softw. Eng. 6 (7) (2016).
[10] A. Khan, N.G. Bawane, S. Bodkhe, An analysis of particle swarm optimization with data clustering-technique for optimization in data mining, Int. J. Comput. Sci. Eng. 02 (07) (2010).
[11] M. Thangarasu, H.H. Inbarani, DPPSOK algorithm for document clustering, Int. J. Control. Theory Appl. 9 (26) (2016).

Chapter 28

A review on optimal placement of phasor measurement unit (PMU)

Ashutosh Dixit[a], Arindam Chowdhury[b], and Parvesh Saini[a]
[a]*Department of Electrical Engineering, Graphic Era Deemed to be University, Dehradun, Uttarakhand, India,* [b]*Department of Electrical Engineering, Maryland Institute of Technology and Management, Jamshedpur, India*

28.1 Introduction

The burgeoning population around the globe has significantly increased electricity consumption. Consequently, there is also an increased demand for reliable electricity. The power from generating stations is transmitted and distributed to the consumers' end via a proper network. Recognition of the quality and stability of the electrical power transmission network monitoring and measuring equipment helps to avoid loss of electricity. In view of the GPS coordinated clock, the phasor estimation unit can gauge a tremendous measure of basic force network data, which incorporates voltage and current of the necessary bus, generator speed, power angle, and rate at which the recurrence changes. By receiving the real-time PMU measurement data from various areas, distribution's network quality under dynamic operating and static conditions can be recorded and analyzed by the operators in the central room. Setting up of PMUs in all substations can suggestively improve the electrical power network reliability but the installment of PMU devices at all locations is unaffordable as the price of a single PMU unit is quite high. To cut the maintenance fee and unit cost, optimal PMU placement (OPP) has been implemented for reducing the quantity of PMUs installed as well as to achieve the entire degree of observability. Thus, the trouble of the ideal PMU situation had arisen as another point of examination interest in the space of electrical power network, of late. Approaches for the solution of the problem of optimal PMU placement has been focused on as a research curiosity. The methods can be figured into three groups: heuristics methods, meta-heuristic optimization methods, and mathematical programming methods. This chapter presents an analysis on

System Assurances. https://doi.org/10.1016/B978-0-323-90240-3.00028-X

the optimal placement problem and the solution methodologies based on heuristic method, meta-heuristic method, and mathematical programing method, which will be very useful and serve as a guide for assisting researchers who intend to work in this field of research. The orientation of this analysis can be divided into various sections. In Section 28.2, optimal PMU placement (OPP) problem formulations are presented. Arrangement of OPP items relying on numerical programing techniques, heuristic strategies, and meta-heuristic techniques are portrayed in Sections 28.3–28.5 separately. The correlation of these calculations is introduced in Section 28.6. Section 28.7 recommends probable upcoming effort on PMU placement and lastly, Section 28.8 concludes this chapter.

28.2 Optimal PMU placement (OPP) problem formulation

PMU is a dynamic device that measures the phase value of the voltage of the bus where the installation has been done and similarly measures the phase value of current associated with that branch. Fig. 28.1 shows a layout of the wide area monitoring system. The GPS locks the measurement signals into time synchronism which are put into a phasor data concentrator with the help of PMU. The phasor data concentrator amasses and isolates the phasor estimations and the processor interprets information of PMUs into reasonable information, which is perceptible on Human-Machine Interface. The access of the dangerous data of the network condition is effortlessly done by a machinist.

The following assumptions are formulated for the placement of PMUs:

Condition 1: For PMU connected buses, the current and voltage phasor of all its incident branches are known. These are known as "direct estimations." Concerning the reason for the phasor estimation unit, a PMU arranged in the Bus D,

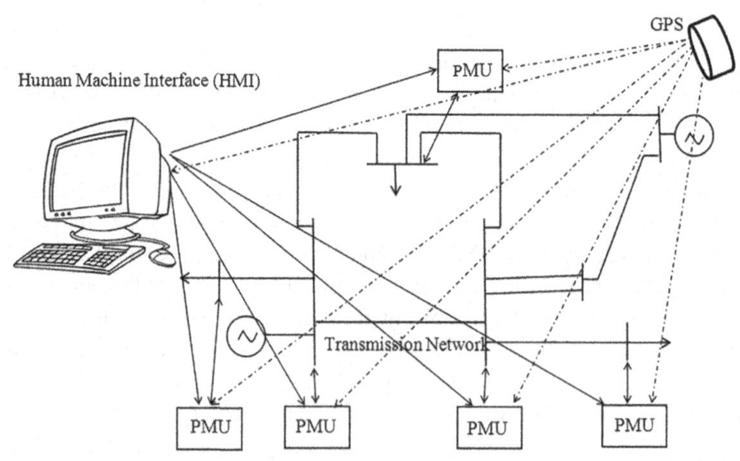

FIG. 28.1 Layout of wide area monitoring system (WAMS).

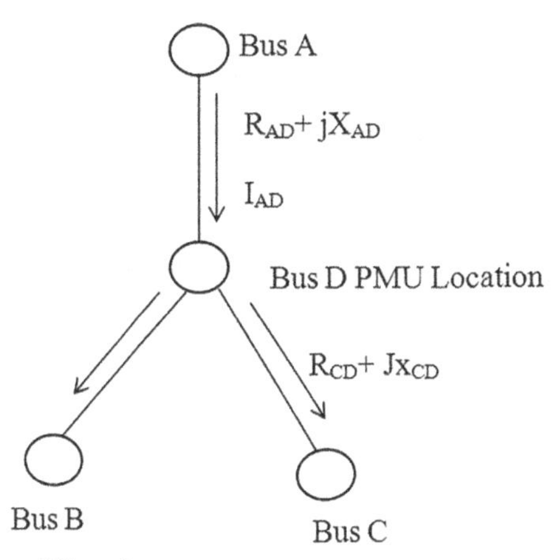

FIG. 28.2 First observability rule.

which is shown in Fig. 28.2, illustrates direct estimation of bus voltage is finished. Meanwhile, the branch flows associated with the hub are likewise estimated by the PMU. In Condition 1, the known parameters estimated by PMU are V_D, I_{AD}, I_{BD}, and I_{CD} and the characteristics of the transmission line are $R_{AD}+jX_{AD}$, $R_{BD}+jX_{BD}$, $R_{CD}+jX_{CD}$.

Condition 2: If phasors of voltage and current are identified at one end of a branch, then with the help of Eqs. (28.1–28.3) the phasor concerned with voltage at another termination of the branch is achieved. These are termed as "pseudo-estimations."

In view of the acknowledged values, for example, line impedance and branch flows, the voltage settles utilizing the accompanying condition (Fig. 28.3):

$$V_A = V_D + I_{AD}\left(R_{AD} + jX_{AD}\right) \tag{28.1}$$

$$V_B = V_D - I_{BD}\left(R_{BD} + jX_{BD}\right) \tag{28.2}$$

$$V_C = V_D - I_{CD}\left(R_{CD} + jX_{CD}\right) \tag{28.3}$$

Condition 3: On the off chance that voltage phasors of the two closures of a branch are recognized, the current phasor of this branch can be accomplished straightforwardly. These estimations are likewise named as "pseudo-estimations" (Fig. 28.4).

Beneath the present condition, expecting that the sizes of voltage in Buses A, B, and C are noticed, and then estimated, the line current in the part of BD, AD, and CD just as the voltage in Bus D can be determined. The necessary conditions are:

$$V_D = V_A - I_{AD}\left(R_{AD} + jX_{AD}\right) \tag{28.4}$$

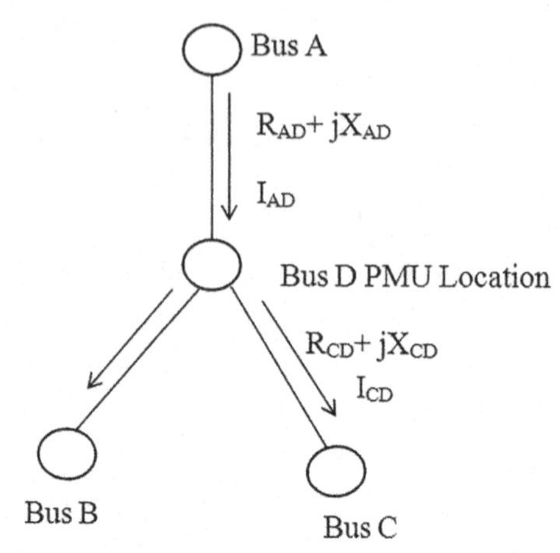

FIG. 28.3 Second observability rule.

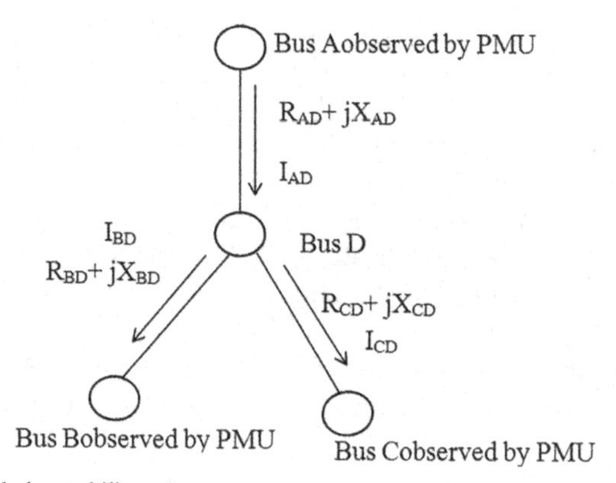

FIG. 28.4 Third observability rule.

$$V_D = V_B + I_{BD}\,(R_{BD} + jX_{BD}) \tag{28.5}$$

$$V_D = V_C + I_{CD}\,(R_{CD} + jX_{CD}) \tag{28.6}$$

$$I_{AD} = I_{BD} + I_{CD} \tag{28.7}$$

28.3 Mathematical programming method

28.3.1 Integer programming (IP)

Integer programing is one type of mathematical programming technique for resolving a problem on optimization having the entire design variable as an integer value. The constraints and objective function can be nonlinear, linear, or quadratic, resulting in integer nonlinear programing (INLP), integer linear programming (ILP), and integer quadratic programming (IQP) algorithms, respectively. A generalized optimal placement algorithm of PMUs is defined in Ref. [1] by using ILP. Cases that are considered include and exclude injection measurement and conventional power flow. The measurement placement problems, under those circumstances, are framed as an ILP which save computational time greatly. A simplified version of Ref. [1] is proposed in Ref. [2]. In Ref. [3] a mathematical execution of IP for power system analysis and price of PMU installation with mixed measurement sets, which involved conventional power flow and injection measurement are used. Approximations are used when dealing with zero-injection buses. A hybrid two-phase PMU placement technique is suggested in Ref. [4]. The first phase ensures topological observability by utilizing an ILP base approach. On the other hand, the second stage finds out the mathematical observability by considering whether the Jacobian matrix relating to voltage and current phasor measurement to bus voltage states in the linear model that is of the full rank. A new technique of PMUs placement which observes voltage and current phasor along with power system network branches has been discussed in Ref. [5]. Earlier studies on optimal placement of PMUs have anticipated that PMUs could be sited at a branch and would deliver branch voltage and current phasors along all branches directed to the bus. A multiobjective IQP that lessens the amount of PMUs is necessary for full system observability for normal operating conditions along with contingency conditions and a maximize the measurement redundancy at all buses [6]. An improved integer linear programming for optimal placement of PMU which is focused on enhancement in binary connectivity matrix of the system, to bring on the effect of zero-injection buses is discussed in Ref. [7]. The consequences of P1mg damage or a branch outage was not considered but its advantage is computation efficiency. A basic ILP model for network observability was developed in Ref. [8]. Thereafter, network contingencies such as measurement losses and line outages were considered. Finally, the consequences of restricted communication on PMU placement were taken into account. In Ref. [9] a novel integrated model is presented to take into consideration the consequence of conventional measurement and zero-injection buses with integer linear programming for PMU placement. The given model considers boundaries that are produced by zero-injection buses and conventional measurements. The basic contingencies such as PMU outage and the single branch are also separately and simultaneously considered. In Ref. [10] an integer linear programing approach for phasing of PMU placement over a time period horizon is used. The PMU placement for every phase helps in maximizing observation. The ultimate elucidation was similar to the optimum elucidation which was obtained without phasing. Two indices, Observability Redundancy Index and Bus Observability Index System are used to rank multiple results for PMUs observing a given bus.

28.4 Meta-heuristic methods

28.4.1 Simulating annealing (SA)

A genetic probabilistic meta-heuristic technique for the global optimization problem by trying a random variation on the current solution is considered as simulating mudding (SA). A bad change is acknowledged as the renewed answer with a condition that falls as the computation proceeds. The leisurelier cooling timetable, or the rate of decline is the most probable algorithm to discover an optimal solution. The novel simulating annealing technique in Ref. [11] presents a depth of unobservability technique in which the optimal location of the PMU is dependent on the incomplete observation capacity of the power system. Initially, to ascertain the minimum amount of PMUs so that the power system becomes entirely observable, the phasor measurement unit is placed considering spanning trees of power system graph. Then the novel theory of "depth-of-unobservability" is recommended that displays its effects on the number of PMUs placements. This technique also calculates the appropriate solutions to the problem of phasor measurement unit placement with communication limitations. In Ref. [12] a stochastic simulated annealing (SSA) technique is presented to solve the optimal PMU placement problem to observe topological sanity. Any single measurement of bad data can be detected by a critical measurement free system. The enhanced proposal [13] on optimal PMU placement problem result, which uses an improved bisecting search method to find out the amount of PMUs and a SA technique searches for a location set that steps into an observable power system for a fixed quantity of PMUs. A sensitivity constrained technique, which places phasor measurement unit in power system with greater sensitivities and considers the optimum PMU placement and the dynamic data of network for whole observation, is suggested in Ref. [14].

28.4.2 Genetic algorithm (GA)

GA is an optimization technique that duplicates the procedures of natural evolution. The working of GA is dependent on the population of individuals termed as "chromosomes," which are probable clarifications to the specified problem and are added to breed new individuals.

In Ref. [15] the GA method unravels the optimal PMU placement problem by different PMU location conditions, such as the absence of genuine estimations and basic sets from the framework, and the most extreme number of estimations got in contrast with the underlying one is the difference in the organization chart into a tree, maximum accuracy of estimates and minimum price of PMU placement.

In Ref. [16] a combination of the optimization method of a graph with a theoretic procedure that approximates separate optimum solutions of objectives and a test nondominated sorting genetic algorithm (NSGA) which treasures the optimum trade-offs between competing objectives are proposed. By utilizing an immunity genetic algorithm (IGA) that uses certain qualities and data of the issue for limiting the degenerative marvels during progress and expanding its productivity, the OPP issue arrangement is gotten in Ref. [17].

In Ref. [18] immunity genetic algorithm was established and its efficiency was proved on the traveling salesman problem (TSP) as a standard. It was noted that an immunity genetic algorithm is possible as well as effective and is conductive to improve the degeneration phenomenon in the original genetic algorithm, as a consequence, significantly increasing the converging speed. The principle thought to the back of this method is utilizing the past attention to the issue in the hunt technique.

28.4.3 Tabu search (TS)

By tracing and managing the search, one can solve combinational optimization problems through this process which is a meta-heuristic local search procedure. In Ref. [19], the Tabu Search strategy and a quick recognizability examination procedure reliant on expanded episode lattice, which just controls whole numbers, will take care of the issue of the ideal situation and the issue of PMUs. The solution of the combinatorial optimal PMU placement problem through the TS process has few computations. Therefore, this technique gives a timeless and precise solution related to orthodox observability investigation. In Ref. [20] a TS technique on PMU placement to exploit topological observability is presented.

28.4.4 Differential evolution (DE)

A meta-heuristic technique is applied for monitoring the mutation operation by utilizing the arbitrarily sampled pair differences of the objective vector rather than using probability distribution functions. The procedure recommended in Ref. [21], in reconciliation of Pareto nonruled arranging and differential advancement (NSDE), acknowledges calculation. This process understands worldwide multitarget advancement easily and quickly and this procedure has less computational and more noteworthy strength. Therefore, this technique provides a global and exact solution with complete observability.

28.4.5 Particle swarm optimization (PSO)

It is a populace-based pursuit technique that enhances the given issue utilizing entities (called particles) which change their position with time in the hunt space.

With the course of time, each particle guideline is impacted by its nearby situation as per its own insight and the experience of adjoining particles. This is required to push the multitude toward the best arrangement.

A BPSO-based technique for optimum location of PMUs is proposed in Ref. [22]. The objective of this process is to enhance the measurement redundancy and reduction of the total number of PMUs. Validation of the proposed method is done on IEEE 14, 30, 57, and 118 bus systems. A similar BPSO technique is also proposed in Ref. [23].

A binary PSO recommended in Ref. [24] fulfills the imperatives of PMU misfortune or branch blackout impact and its yields are contrasted and distinctive calculation.

The proposed technique in Ref. [25] is the combination of BPSO and immune mechanism, thus it has a higher converging speed. The effectiveness of the proposed technique was verified in IEEE 14 bus system and in the New England 39 bus system.

An improved discrete binary particle swarm optimization technique is proposed in Ref. [26].

To improve the convergence speed of the optimization process graph, a theoretic procedure is used. In Ref. [27], the genetic algorithm is meritoriously united with the particle swarm optimization technique to attain the optimum resolution. The cross and aberrance processes in the genetic algorithm are utilized for declining the searching possibility of the PSO technique and enhance the eminence of the inaugural PMU location, which results in quickening the solving procedure.

28.4.6 Ant colony optimization (ACO)

A probabilistic technique for solving computational problems can be condensed to find good paths with the help of graphs. A comprehensive Ant Colony Optimization technique is suggested in Ref. [28]. The method of adaptively changing the pheromone trajectory persistence coefficient and stochastic disquieting is presented for the betterment of the method on the capability to seepage from stagnation behavior and convergence speed.

28.5 Heuristic methods

28.5.1 Depth first search (DeFS)

This algorithm is noniterative and utilizes depth creation. It points out each and every vertex and partitions the graph as a forest. PSAT toolbox is utilized in Ref. [29], which relates the DeFS technique with Annealing Method, Direct Spanning Tree Method, and Graph Theoretic Procedure approaches.

28.6 Algorithm comparison

The least amount of PMUs positioned in IEEE bus system is used as a benchmark for algorithm comparison. From the works, the least amount of PMUs to empower complete observability is concised in Table 28.1.

TABLE 28.1 Minimum number of PMUs for test cases.

Case	Ideal minimum number of PMUs
IEEE 14-bus	3
IEEE 30-bus	7
IEEE 39-bus	8
IEEE 57-bus	11
IEEE 118-bus	28

Table 28.2 tabulates the algorithms with a required number of PMUs and their test system. In particular, the digits in the table denote the required amount of PMUs desired to accomplish complete observability. The optimal PMU placement problem is concerned with the required optimum number of phasor measurement units and optimum position net guaranteeing the observability of the entire system. The optimal PMU placement problem does not have a distinctive result. It depends on the starting point, optimization pattern, and contingency conditions. Different formulations for PMU placement problems with supplementary constrain have been presented in works and can be seen in Table 28.3. In Table 28.4 different contingency conditions previously discussed in the literature are tabulated and the lowest amount of PMUs necessary to complete observability of IEEE 14 and 30 bus systems is also presented.

Table 28.5 summarizes the advantages and disadvantages of various methods for optimum placement of PMUs. The selection of optimization method is reliant on the magnitude of the problem as well as the kind of application. Furthermore, the condition of the electrical power system must be taken into contemplation. The factors such as algorithm adaptability and user-friendliness also play an imperative role in the choice of optimization technique. The PMU placement technique will be accomplished offline and the accomplishment period is not prime anxiety. A method that is supported requires an insignificant amount of phasor measurement units when ideally found, has ability to coordinate possibilities, excellently models zero-infusions, computationally less demanding, advantageous, and simple to grasp.

28.7 Future scope

The previous section provides the comparison among three basic optimization algorithms but to implement the vision of smart grid, critical and costly technology needs to be deployed effectively in the electrical power system network.

TABLE 28.2 Algorithm assessment.

Method	Reference number	Minimum number of PMUs				
		Test system				
		IEEE 14-bus	*IEEE 30-bus*	*IEEE 39-bus*	*IEEE 57-bus*	*IEEE 118-bus*
Integer programming	[3]	3	7	—	11	28
Generalized integer linear programing	[2]	3	7	—	11	—
Nondominated sorting algorithm	[16]	—	—	8	—	29
Immunity genetic algorithm	[17]	3	7	8	11	28
Search tree and SA	[11]	3	7	—	11	—
Integer quadratic programming	[6]	2	10	—	17	32
Modified integer linear programming	[7]	—	7	8	—	—
Simulating annealing	[12]	4	10	—	19	34
Tabu search	[19]	3	—	10	13	—
Differential evolution	[21]	3	7	—	17	—
Binary particle swarm optimization	[22]	4	10	—	17	32
Binary particle swarm optimization and immunity algorithm	[25]	3	—	8	—	—
Contingency constrained	[8]	3	7	8	11	28
Optimal multistage	[10]	3	—	—	14	29

So, researchers should broaden their scope of the investigation beyond the basic optimization technologies. In this section, a new technological roadmap to achieve the optimization problem is developed.

(i) *Hybrid optimization techniques*: Using basic optimization techniques, there are possibilities to generate various hybrid algorithms which may solve the OPP problem efficiently. A large number of publications have proven that hybrid algorithm has great success in considering the different constraints in the OPP problem.

TABLE 28.3 Different formulation of OPP problem.

Reference number	Effect of zero injection buses	Effect of conventional measurement	Effect of conventional measurement	Single or multiple PMU outage	Single line outage or single PMU outage	Transmission line and transformer outage
[1]	√	×	×	×	×	×
[3]	√	×	×	×	×	×
[5]	×	×	×	√	×	√
[6]	×	×	√	√	×	×
[8]	√	×	×	×	√	×
[9]	×	×	×	√	√	×
[10]	√	×	×	√	×	×
[12]	×	√	×	×	×	×
[22]	√	×	×	×	×	×

TABLE 28.4 Minimum number of PMUs in IEEE 14 and IEEE 30 bus systems for different contingency condition.

Reference number	Method	Contingency consideration	Result			
					Minimum number PMU	
			Condition	*Observability condition*	*IEEE 14 bus*	*IEEE 30 bus*
[1]	Generalized integer linear programming	Effect of zero injection buses	Without zero injection bus	Complete observability	4	10
				Depth of 1 unobservability	2	4
				Depth of 2 unobservability	2	3
			With zero injection bus	Complete observability	3	7
				Depth of 1 unobservability	2	4
				Depth of 2 unobservability	2	3
[6]	Integer quadratic programming	Single PMU outage or single transmission line outage	Normal operating condition	Complete observability	—	10
			Under contingency condition	Complete observability	—	21

	programming	line outage, loss of measurement, line outage or loss of measurement	condition	observability		
			Line outage	Complete observability	7	13
			Loss of measurement	Complete observability	7	15
			Line outage or loss of measurement	Complete observability	8	17
			No PMU at zero-injection	Complete observability	3	7
[18]	Integer linear programming model with zero injection bus and conventional measurement	Single brunch outage, single PMU outage, single PMU, or single branch outage	Single brunch outage	Complete observability	7	13
			Single PMU outage	Complete observability	7	14
			Single PMU outage or single branch outage	Complete observability	8	16
[11]	Simulated annealing and tree search		Normal operating condition	Complete observability	3	7
				Depth of 1 unobservability	2	4
				Depth of 2 unobservability	2	3
				Depth of 2 unobservability	1	2
[22]	Binary particle swarm optimization (BPSO)	Effect of zero injection buses	Without zero injection bus	Complete observability	4	10
			Using zero injection bus	Complete observability	3	7

TABLE 28.5 Advantages and disadvantages of methods.

Technique	Reference number	Convergence speed	Advantages	Disadvantages	Whole convergence
Genetic algorithm	[15]	Slow depend on the fitness function	Robust, adaptable	Affinity to converge toward local optima instead of global optima Large-scale mutation size	Excellent
IGA	[17,18]	Fast	Provide optimal solution	Tougher to comprehend Restricted by dimensions of problem	Excellent
ILP	[1,2]	Very fast	Mathematical structure of ILPs is easier to understand and shows adaptability in term of phasing and modeling contingency	Constrains are nonlinear	Excellent
PSO	[24]	Fast	Simple concept, easy to implement, very efficient global search algorithm	Do not guarantee success	Poor
BPSO	[22,23,25]	Slow	Optimal solution reached	Does not mathematically offer themselves to malleability in term of measurement contingency, exclusively for large networks	Excellent
Tree search topology	[11]	—	Implementation of phasor measurement unit is very easy	Commonly appropriate for network previously fortified with a little observable islands	Poor
Contingency	[8]	Fast	Effective for large-scale systems	Constrains are nonlinear	—

Multistaging	[10]	Fast	Effective phasing. Does not upsurge least no. of PMUs after phasing Modeling is easy	Optimal no. not achieved Nonlinear constraints	Poor
DeFS	[29]	Fast	Computationally faster	Optimization criterion is stiff and unitary	Poor
SA	[11]	Very slow	Iterative improvement algorithm, memoryless algorithm	Continually annealing with a 1/log k schedule is very sluggish, particularly if the cost function is costly to calculate	Excellent
ACO	[28]	Very slow	Traveling salesman problem (TSP) relatively efficient retains memory of entire colony, less affected by poor initial solution	Owing to sequence of random decision, theoretically analysis is difficult Coding is difficult	Good

(ii) *Multiobjective optimization technique*: Researchers should not limit their objective to abate the total number of PMUs. Multiple objectives including redundancies performance, installation cost, and cyberattack to the PMU line may be considered to solve the OPP problem.
(iii) *Constrain consideration*: Outcome of zero-injection buses, a consequence of conventional measurement, single or multiple PMU outage, single line outage, transformer and transmission line outage, failure of the communication line, bad data and effect of environment, are some constraints that can be taken into account.
(iv) *Graph-based algorithm*: Other constructions of the optimal PMU placement problem, centered on the vertex covering or dominating set problems in graphs, are going to be an attention-grabbing path for future research.
(v) *Complicated network*: Most of the suggested methodologies are established in the IEEE 14 bus system to IEEE 300 bus system. The design and implementation of a novel robust optimization algorithm pertinent to the complicated power system should be reconnoitered.

28.8 Conclusion

An electrical power system network is an intricate and dynamic arrangement that should be monitored continuously for uninterrupted power supply. PMU is the most precise and advance synchronized technology available which provides information about the voltage and current phasor, phase angle and frequency which is then synchronized with great exactitude to a common reference, i.e., the global positioning system (GPS). The price of a PMU is very high so it is not cost-effective to mount PMU on each bus, therefore, to reduce the installation cost, optimal PMU placement (OPP) is implemented. In this chapter, various methods that are used to solve the OPP problems are discussed. The solution methodologies are divided into three categories: mathematical programing method, heuristic method, and meta-heuristic method. The most frequently used optimization techniques for optimal PMU placement are GA, PSO, and ILP. The various IEEE bus systems are used as a benchmark system for most of the research work. The review on optimization techniques discussed in this chapter will be beneficial for the investigators in order to determine and implement novel hybrid approaches for unraveling optimal PMU placement problems.

References

[1] B. Gou, Generalized integer linear programming formulation for optimal PMU placement, IEEE Trans. Power Syst. 23 (3) (2008) 1099–1104.

[2] B. Gou, Optimal placement of PMUs by integer linear programming, IEEE Trans. Power Syst. 23 (3) (2008) 1525–1526.

[3] B. Xu, A. Abur, Observability analysis and measurement placement for system with PMUs, in: Power Systems Conference and Exposition, New York, NY, USA, Vol. 2, 2004, pp. 943–946.

[4] R. Sodhi, S.C. Srivastava, S.N. Singh, Optimal PMU placement method for complete topological and numerical observability of power system, Electr. Pow. Syst. Res. 80 (9) (2010) 1154–1159.

[5] R. Emami, A. Abur, Robust measurement design by placing synchronized phasor measurements on network branches, IEEE Trans. Power Syst. 25 (1) (2010) 38–43.

[6] S. Chakrabarti, E. Kyriakides, D.G. Eliades, Placement of synchronized measurements for power system observability, IEEE Trans. Power Deliv. 24 (1) (2009) 12–19.

[7] G. Khare, N. Sahu, R. Sunitha, Optimal PMU placement using matrix modification based integer linear programming, in: International Conference on Circuit, Power and Computing Technologies [ICCPCT], Nagercoil, March 2014, pp. 632–636.

[8] F. Aminifar, A. Khodaei, M. Fotuhi-Firuzabad, M. Shahidehpour, Contingency-constrained PMU placement in power networks, IEEE Trans. Power Syst. 25 (1) (2010) 516–523.

[9] K.G. Khajeh, E. Bashar, A.M. Rad, G.B. Gharehpetian, Integrated model considering effects of zero injection buses and conventional measurements on optimal PMU placement, IEEE Trans. Smart Grid PP (99) (2015) 1–8.

[10] D. Dua, S. Dambhare, R.K. Gajbhiye, S.A. Soman, Optimal multistage scheduling of PMU placement: an ILP approach, IEEE Trans. Power Deliv. 23 (4) (2008).

[11] R.F. Nuqui, A.G. Phadke, Phasor measurement unit placement techniques for complete and incomplete observability, IEEE Trans. Power Deliv. 20 (4) (2005) 2381–2388.

[12] T. Kerdchuen, W. Ongsakul, Optimal PMU placement by stochastic simulated annealing for power system state estimation, GMSARN Int. J. 2 (2) (2008) 61–66.

[13] T.L. Baldwin, L. Mili, M.B. Boisen Jr., R. Adapa, Power system observability with minimal phasor measurement placement, IEEE Trans. Power Syst. 8 (2) (1993) 707–715.

[14] H.-S. Zhao, Y. Li, Z.-Q. Mi, L. Yu, Sensitivity constrained PMU placement for complete observability of power systems, in: Asia and Pacific Transmission and Distribution Conference and Exhibition, Dalian, 2005, pp. 1–5.

[15] A.Z. Gamm, I.N. Kolosok, A.M. Glazunova, E.S. Korkina, PMU placement criteria for EPS state estimation, in: International Conference on Electric Utility Deregulation and Restructuring and Power Technologies, Nanjing, April 2008, pp. 645–649.

[16] B. Milosevic, M. Begovic, Nondominated sorting genetic algorithm for optimal phasor measurement placement, IEEE Trans. Power Syst. 18 (1) (2003) 69–75.

[17] F. Aminifar, C. Lucas, A. Khodaei, M. Fotuhi-Firuzabad, Optimal placement of phasor measurement units using immunity genetic algorithm, IEEE Trans. Power Deliv. 24 (3) (2009) 1014–1020.

[18] L. Jiao, Wang 1., A novel genetic algorithm based on immunity, IEEE Trans. Syst. Man Cybern. Syst. 30 (5) (2000) 552–561.

[19] J. Peng, Y. Sun, H.F. Wang, Optimal PMU placement for full network observability using Tabu search algorithm, Int. J. Electr. Power Energy Syst. 28 (4) (2006) 223–231.

[20] H. Mori, Y. Sone, Tabu search based meter placement for topological observability in power system state estimation, in: IEEE Transmission and Distribution Conference, New Orleans, LA, 1999, pp. 172–177.

[21] C. Peng, H. Sun, J. Guo, Multi-objective optimal PMU placement using a non-dominated sorting differential evolution algorithm, Int. J. Electr. Power Energy Syst. 32 (8) (2010) 886–892.

[22] A. Ahmadi, Y. Alinejad-Beromi, M. Moradi, Optimal PMU placement for power system observability using binary particle swarm optimization and considering measurement redundancy, Expert Syst. Appl. 38 (2011) 7263–7269.

[23] S. Chakrabarti, G.K. Venayagamoorthy, E. Kyriakides, PMU placement for power system observability using binary particle swarm optimization, in: Australasian Universities Power Engineering Conference, Sydney, December 2008, pp. 1–5.

[24] M. Hajian, A.M. Ranjbar, T. Amraee, B. Mozafari, Optimal placement Of PMUs to maintain network observability using a modified BPSO algorithm, Int. J. Electr. Power Energy Syst. 33 (1) (2011) 28–34.

[25] C. Peng, X. Xu, A hybrid algorithm based on BPSO and immune mechanism for PMU optimization placement, in: Proceedings of the 7th World Congress on Intelligent Control and Automation, Chongqing, China, June 2008, pp. 7036–7040.

[26] M. Hajian, A.M. Ranjbar, T. Amraee, A.R. Shirani, Optimal placement of phasor measurement units: particle swarm optimization approach, in: International Conference on Intelligent Systems Applications to Power Systems, 2007, pp. 1–6.

[27] Y. Gao, Z. Hu, X. He, D. Liu, Optimal placement of PMUs in power systems based on improved PSO algorithm, in: IEEE International Conference on Industrial Electronics and Applications, 2008, pp. 2464–2469.

[28] W. Bo, L. Discen, X. Li, An improved ant colony system in optimizing power system PMU placement problem, in: Asia Pacific Power and Energy Engineering Conference, Wuhan, March 2009, pp. 1–3.

[29] G. Venugopal, R. Veilumuthu, P.A. Theresa, Optimal PMU placement and observability of power system using PSAT, in: International Joint Journal Conference on Engineering and Technology, Chennai, June 2010, pp. 67–71.

Chapter 29

Effective motivational factors and comprehensive study of information security and policy challenges

M. Arvindhan
Galgotias University, Greater Noida, India

29.1 Introduction

The authentication strategy by disguising identifying details for individuals throughout their encounters with each other is important. This unknown communicator can be used for malicious intent. It is indeed a serious security issue because governments cannot detect and kill keyhole hackers until they have been hacked. A safe and private mechanism is required to eradicate an entity's privacy. The disadvantages of these ventures are primarily high computing overhead, communication privacy problems, inability to trace and revoke malicious communication actors, and insecurity of roadside units (RSUs) to direct attacks. Private clouds apply to off-site data centers. Although cloud storage has made possible computing as a service, there are multiple parties in agreement over the protection and privacy of data as cloud customers, cloud providers, and third-party vendors. Therefore, it is appropriate to perform data monitoring to ensure data protection and protection in the cloud. Proprietorship of cars has risen as they became more available as wages have increased as they become more popular. That will also exacerbate high levels of air pollution and road congestion.

Planet Earth is estimated to host more than one billion cars by 2030, and this desire needs to be met by the advancement of more energy and infrastructure. According to the World Health Organization (WHO), 1.25 million people die each year due to traffic crashes, and the death toll will increase to 1.8 million by 2030. The automated driving industry has grown because of the need to provide efficient and efficient means of transportation as well as to be advanced. However, improvement in the popularity of autonomous vehicles has been hampered

System Assurances. https://doi.org/10.1016/B978-0-323-90240-3.00029-1

by several high-profile accidents in the past. According to the IEEE 802.11p protocol, the vehicle can exchange location data such as position, speed, acceleration, and control with its neighbors. There is also an immediate need for fine-grained input from other technology such as precise location information, vehicle sensing information, and effective navigation systems. Carmakers have introduced some new products in their vehicles because of the increasing competition in the industry.

More companies today know the value of security policies. According to PricewaterhouseCooper [1,2], 98% of big companies and 60% of small firms have a reported information management strategy. Compliance is not a problem.

Many organizations are also compelled to comply with security measures. In the E&Y Global Information Security Report, 57% of companies found their staff as the most possible source of an attack. Anxiety. UK's study on cyber security showed that 70% of all organizations with poorly understood cyber security policies had personnel problems, and the figure was just 41% for those with well-understood security policies. It would have a positive effect on the protection of an organization. The human aspect is also the lowest in crime-fighting link in the information technology chain that triggers more security threats. Many organizations remain ignorant of the need to handle the problem of cyber protection. Chan and Mubarak conducted a survey which showed that 50% of the employees in their sample did not know what the nature of information security policies was [3–5].

29.2 Key information security policies-related challenges

Security involves far more than just computers and it is human nature. Researchers have classified noncompliance conduct into three categories: intentional, negligent, and ignorant. The primary cause for transgression is venality.

Malicious intent is to carry a company's information infrastructure to a point where harm is imminent, while negligent intent is to breach an organization's security policy but not to cause harm to the organization. End-users remain unaware of their duties and the corporate principles of information security [6].

29.2.1 Security policy management and updating

Most organizations do not routinely revamp their information management procedures. Many organizations have not yet revised their procedures to be more related to the continuously and quickly changing knowledge [6]. A survey conducted by Protiviti showed that only 24% of the survey respondents had placed into effect appropriate use policies for cloud providers. This shows that many companies have obsolete practices and programs. Organizations must ensure that their strategies are still successful and do not go outdated. These staffs are able to create their own security solutions based on their knowledge and knowledge of security policies. For example, staff should use passwords that are more difficult to memorize. An individual may lack principles, ethics or policies, and honesty. This may contribute to security threats in the organization. Shadow protection could create a false sense of security. The company is at risk due to its "shadow protection strategy," which is not typically completely perceived by workers. Employees does not understand this and they cannot understand the danger that the company is in due to this behavior [7].

29.2.2 Illustration of security

The set of subjects S consists of U, C1, and C2, which represent the consumer, virtual private cloud, and public cloud alike. At first, the framework includes objects D1 and D2 stored on the private cloud and the public cloud, respectively. Recognize that the dictionary's first mark for a noun is (noun, n, {n}). The label of D1 says that it is owned by Gang, (ii) sensitive, and therefore readable only by Gang, and (iii) trustworthy, and hence can be affected by Gang alone. The label of D2 notes that it is owned by the public cloud, nonconfidential, and thus readable by both the public and private clouds, and thus untrustworthy because it can be manipulated by the public cloud. The subjects and artifacts go into constructing the initial state for this method. During the first clause, U demands his new job. In the second state, there is a new object denoted by Req, which represents an object U that is owned and affected by U only and all the subjects can read it. The second transformation C1 is followed by D1, and C2 is followed by D2. Remember that in this stage, if C2 attempts to read D1, the RWFM monitor will refuse this request because C2 is not assigned to be a reader of D1—note that the mark of D1 is (C1, {C1}, {C1}), and C1 is the only reader of D1 (second component). Other ways of modification and transformations of a structure may be interpreted in a similar way [8].

29.2.3 Open issues

As an initial stage in the design of a business operation, protection and privacy criteria and approaches are not considered. However, data on particular clients were obtained without actually being asked to offer a defined service (Fig. 29.1). Many current services implement protection and privacy schemes that allow the use of more recognizable data than required, while without preventing its use of data that are not really beneficial for finishing the technology solutions; corporations also do not know how much confidential information they carry [9].

Companies are ignorant of the confidentiality and privacy rules to be enforced on stored and transmitted data and are also unaware of the permission that the user gave when the data was provided. Social network analysis (SNA) will enhance anonymity. For population classification, machine learning-based

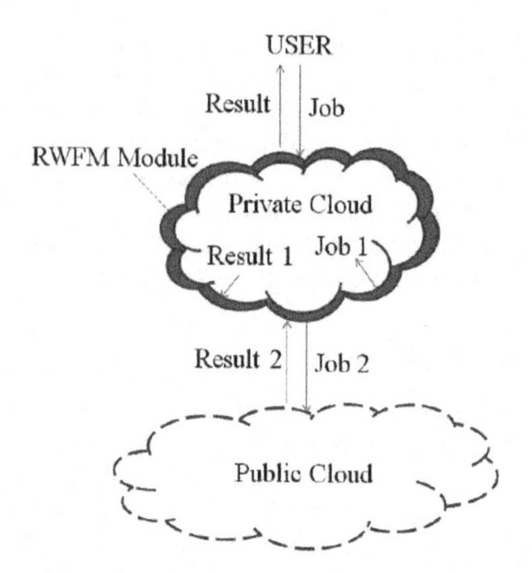

FIG. 29.1 Different level of job assignment on cloud.

community identification (ABCD algorithm) would be used to compare the performance of the different algorithms [10].

The ABCD algorithm approach outperforms other issues in terms of functionality and the number of groups in networks. Liu et al. [11] present their works on privacy-preserving matrix product-based role-based access restriction breach detection in an interoperable environment to fill the distance between two RBAC control domains. Since they have completed their efficiency study and experiment contrast, their construction is more effective and functional. This chapter explores a novel privacy-preserving mobile social network with the use of identity verification and private matching [12]. This scheme uses encryption and authentication for all correspondence, to prevent an attacker from accessing knowledge of a user's characteristics profile, and thus enhancing user privacy during a friendly matching phase. Healthy: This condition is for when an endpoint has been evaluated, analyzed, and found to follow a series of predefined laws (such as operating system (OS) patch level, antivirus software). Quarantined: This state is used to quarantine samples that do not adhere to security requirements (i.e., a risk to the network). The security analysis of our network is not something we are responsible for. This is a specialist function that is to be carried out by a specific device [13].

29.3 Organizational approaches to information security

In an SDN environment, the control plane specifies policies that are derived (i.e., "derived") and implemented in the application layer at the SDN switches and network devices. In this chapter, we explore the design of safe end-to-end communication paths in a distributed SDN environment. In wide networks of numerous autonomous systems crossing organizational borders, pathways are defined and chosen on the basis of the policy that is relevant to security. Being able to distribute defense capabilities intelligently as a service layer and the assignment of a complex security strategy is an important contribution of this chapter. We should have regulations that mandate how traffic would travel from one section of the grid to another. The unified control would be in control.

Controller, designed specifically for the purpose, with native apps for control of devices in a Software Specified Network: There are several third-party programs hosted in the controller which can be asked for based on various specifications. The manner in which the controller communicates with the programs identified as the north-bound interface (NBI) or the north-bound graphical user interface (GUI). The data plan consists of such networking systems and accompanying devices or hardware that link to each other [13].

Controller (which itself can be distributed in practice) with native applications for the management of devices in an ISDN network (Fig. 29.2). There can be several third-party applications that can be hosted (or access different services) in the controller. The interface between the controller and the applications is referred to as the NBI. The data plane consists of the networking devices and the interface between the controller and the networking devices is referred to as the South-Bound Interface.

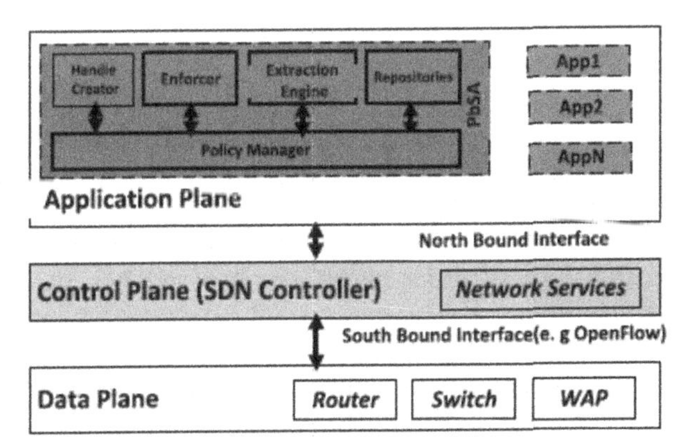

FIG. 29.2 Distributed security switch layer in SDN controller.

29.4 Network architecture and threat model

The network's architecture consists of several autonomous systems (AS) segregated by domain name servers (DNS), with each domain having its own routers, gateways, and switches (which are connected to users). In theory, there should be several SDN controllers, but for convenience, each domain would only have one. End hosts are connected to the public switched telecommunications network. In order to clearly explain each AS, we can presume that each AS has different entry and exit points. These are highly accessible hybrid forwarding devices. Assuming that OpenFlow switches are used, our implementation will also scale to accommodate legacy switches due to known paths of OpenFlow switches. The end hosts create traffic and forward the traffic to network devices. The security policy then controls the traffic. This is linked to the belief that threats and assaults are identical to device and operating system attacks. Popular methods of targeting websites include cross-site scripting, SQL injection, order injection, and buffer overflow. If an SDN program uses a web interface, the same application may be used to circumvent security on the controller to install a malicious program. If the attacker takes control of this machine, they will be able to carry out additional attacks including spying on the system traffic and removing or changing flow entries in the flow entry chart. Present protection methods are typically identical to those used in cyber security [14].

29.5 Policy-based SDN security architecture

Policies should be implemented and established in accordance with the applicable regulations of the respective countries. Also, the broadcast connection on SDN based mobile network area is also focused on OECD, which are the upholding:

(1) The acquisition of personal data should be restricted as far as possible and data owners should know what is being gathered as well as being able to give consent.

(2) The data must be collected for particular reasons as well as the consistency of the data must be ensured.

(3) Objective definition principle: At the time the data are obtained, the reasons for which they are obtained must be instantly specified and retained over time, particularly in the case of inconsistency with the purposes for which third parties may later request the data.

(4) Standards of data privacy: These should be ensured, and in particular physical security, access management, secrecy, and honesty should be protected.

(5) Transparency principle: Proper systems for identifying, in a short time, the existence, the scope, and the purposes of gathered personal information must be made available, to be always aware of changes, the procedures, and the policies implemented to personal/sensitive data.

(6) The need for appropriate governance mechanisms: These should be in operation to ensure that the values just mentioned are successfully translated into motion [15].

Cloud computing and cloud networking, which has been conceived as a means of exchanging services and data between devices on demand, will reap the benefits from the use of sticky policies for allowing users to explicitly monitor the access and the sharing rules surrounding their knowledge. Cloud processing and storage systems are third-party data centers situated in isolated locations that are not readily available. As a result, people will accept the treatment criteria and appropriate treatment protocols when using cloud services/applications or when supplying personal data. Cloud-based applications.

It covers the highest standard of protection and safety, and also detailed rules and processes to handle privacy. However, customers have to trust the service providers because, in exchange, the service providers can access the data that is in the cloud at any time, erase or change them (also accidentally), and exchange information with third parties, also without the users' permission. Since unauthorized access is unavoidable, users may encrypt the data before providing them to the server; however, this may be challenging and/or inadequate on its own. Adopted encryption and homomorphic encryption schemes are very weak, and, therefore, they can be easily broken by an attacker. [16].

Moreover, keys have to be kept on a separate storage block concerning the encrypted data, as well as, a system for the backup of the keys is needed, in case such a storage block is attacked. This is especially important because keys have to be stored away from the encrypted files, as well as a mechanism for backup of the keys is introduced. The sectors are required should they be targeted. Keys should be allowed to expire after a span of time and should be refreshed periodically at a periodic pace. These factors require more work, from a cloud management perspective. Last and not least, data obtained from a cloud may be often transmitted to users' mobile devices, which, in essence, needs to have encrypted messages, if the files are encrypted, in the way to attain them. This could not be advisable because mobile phones are vulnerable to hacking and open to everyone.

29.6 Trust

Trustworthiness is essential in cyber protection as it offers robust security measures against malicious hackers. In consideration of these aspects, we chose the suggested schemes of different writers as being especially apt in this particular issue. The developers argue that their confidence appraisal process proves to be the best at delivering what clients ask for. A more valuable trust in sensor clouds is established, with a hierarchical trust in mind, in which authorship certificates and chain of custody certificates are issued, with a digital signature and a digital signature on objects in mind (SCS). There are two primary elements of the

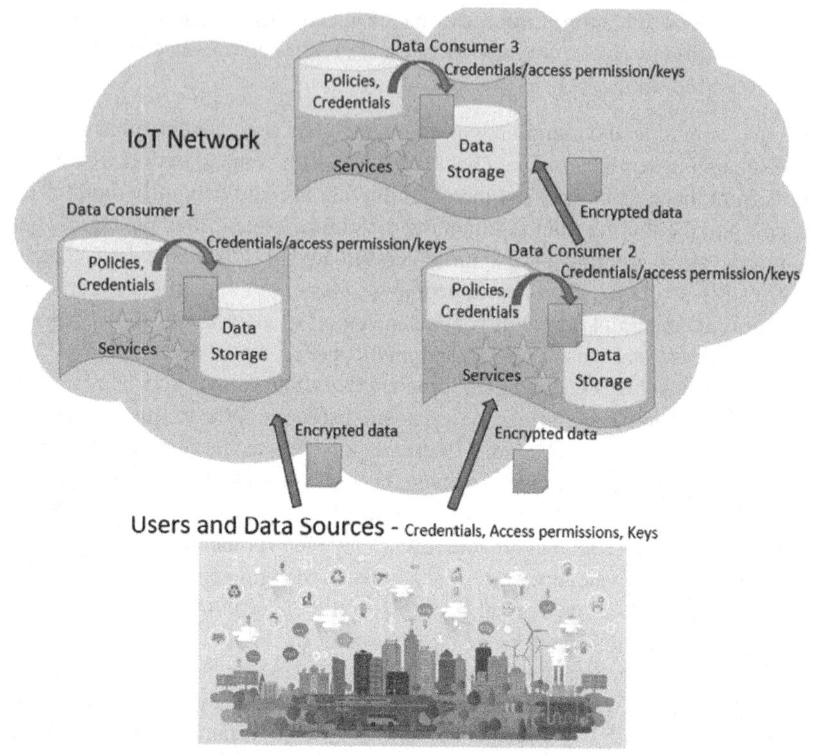

FIG. 29.3 Collection of data source for policy-based network.

hierarchical methodologies: confidence in the organization of the project, and trust in various tech vendors (SSPs). This is where the base level of reliability was built and where specific and refined in the testing layer, and only used on the finer levels of detail for more detailed and complicated data analysis. An interesting feature of this chapter, that is, explored is the use of WSNs to gather real-time information about service parameters and in real-time to track changes in other real-time, which serves as a differentiating factor in comparison to most organizations in that respect (Fig. 29.3). Data analytics provides a good understanding of service instances in the fog layer, which leads to the reliability of the entire network nodes. A data analytics-based approach monitors types of services connected to the edge node addresses, thereby keeping the network consistent [11,17].

29.7 Privacy

Data security is of the highest importance in a world that has made vast amounts of information freely available. There has been a large number of research efforts designed to protect the confidentiality and security of user data, particularly in the cloud. The current solution involves the transmission of location query data to a trustworthy location service provider (LSP) to achieve the location response. After using these findings, they are destroyed automatically. Such approaches can be costly as well as result in higher data loss to the consumer. This chapter uses a spatial K-anonym to provide its users with certain creative and practical approaches to online anonymity, as well as novel uses for online anonymity (CSKA). The plan they have put in place makes use of multiple caches to minimize the chances of intelligence leaks. From the results of a Markov model for motion, the privacy scores of the CSC scheme were predicted, and after that, the usability of the CSC was explored regarding the user's expectation of how mobile the user is and the cache rate of their cell. In modern vehicles, extensive information on destinations, previous locations, preferred routes, and private details are kept so users do not have to worry about missing them (e.g., such as call logs, contact lists, SMS messages, pictures, and videos), which prohibits the disclosure of several photos on social media at the same time so as well as posting more than one picture on social media platforms at a time (MIPP) [18,19]. Some companies such as Mayser and Camero address the challenge of this requirement by collecting usable photographs while also verifying the images from a variety of a single image source owner's privacy would not be disclosed to other image owners. Simulation research and real-world experiments show that MIPP can achieve retrieval precision and reliability simultaneously.

29.8 Privacy and security in cloud computing

Cloud computing allows multiple computational resources, software resources, and storage resources to be connected to form a public virtual source pool, where the users can procure services. With cloud computing's development in numerous fields, it has permeated into many different markets including scientific, manufacturing, education, usage, and entertainment (Fig. 29.4).

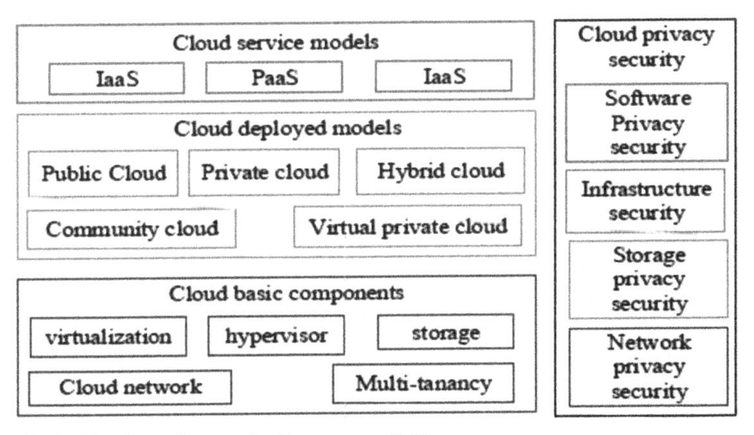

FIG. 29.4 Cloud security model with numerous fields.

The virtualization of cloud services offers a powerful infrastructure for end users. Cloud computing's features include reliability and performance, scalability, and accessibility. Cloud storage offers the benefits of economy, ease, off-the-shelf operation, versatility, and multitenancy [20,21].

29.9 Cloud computing framework

IaaS handles computing hardware (storage, virtual machine, computer, and network), as a service, and allows infrastructure optimization and provisioning, without needing additional resources or time. The field focuses on intrusion prevention, intrusion detection, data protection management, and other security-related fields.

PaaS is situated in the middleware of the business model and delivers services during the service creation lifecycle. PaaS faces numerous difficulties including partnerships with third parties, lifecycle management, and security architecture [22].

Software-as-a-Service is software that is deployed remotely, enabling third-party providers to deliver applications. A consumer can apply cloud services through the use of cloud technology.

Cloud storage is running and operates in the data center, which is related to a private cloud. It is easier for organizations to classify customers and suppliers in a public cloud given that the service is managed and run by the same entity.

A virtual private domain is a private cloud with lesser infrastructure, and a virtual private network (VPN) is a public network such as the Internet [23].

The defined cloud environment has a pooled resource pool assigned to the users. Virtualization is an essential component of an enterprise cloud deployment.

Virtualization enables networks to be built with several copies of applications, such as operating systems, server instances, networking configurations, and storage devices. An inter society will have different customers or users who could not look at or work on someone's records but could also obtain shared resources and make use of the resources provided to all of its members.

A policy outlines the organization's security strategy, defines and assigns roles, gives authority to security personnel, and determines how the organization can respond to an incident. Current control mechanisms have room for improvement but they can be changed by modifying policies based on Citrix Systems Inc.

29.9.1 Eligible

IT administrators have the discretion to allow users to carry and use mobile devices if security applications such as are installed antivirus.

29.9.2 Allowed systems

IT systems administrators can authorize only such devices on their networks. This would provide for flexibility when planning a security strategy that works on several platforms. The study recommended that network administrators discuss creating a list of appropriate devices and operating systems, and others that would not be allowed. For additional protection, you should only allow licensed mobile devices [24].

29.9.3 Service availability

You can have the services of BYO computers. Specific features and products can be offered at specific income levels, consumer styles, and platform types.

Cloud storage is a type of network-based storage that can be accessed via the Internet. Hypervisor: It allows a single hardware host to power several virtual machines. A hypervisor handles the different operating systems in the shared environment. A cloud network is a large cloud of interconnected data centers containing several servers. A cloud network necessities an Internet connection as well as (for greater security) a virtual private network. A large volume of data must be transferred into a cloud computing storage facility. The lack of power over resources has led consumers to seek greater privacy protection.

(1) A new law requires the use of lightweight and fine-grained data coding for cloud storage. In connection with cloud storage, multitrust domains are integrated with designated bodies, so there is no conventional need for data encryption and sharing (Fig. 29.5). Now, the design of a modern way is to encrypt many locations.

(2) Multiple data sourcing and cloud infrastructure will be explored in the chapter. When everything is seen to be an outsourcing project, there is a need to implement a mechanism to protect the data's secrecy [25].

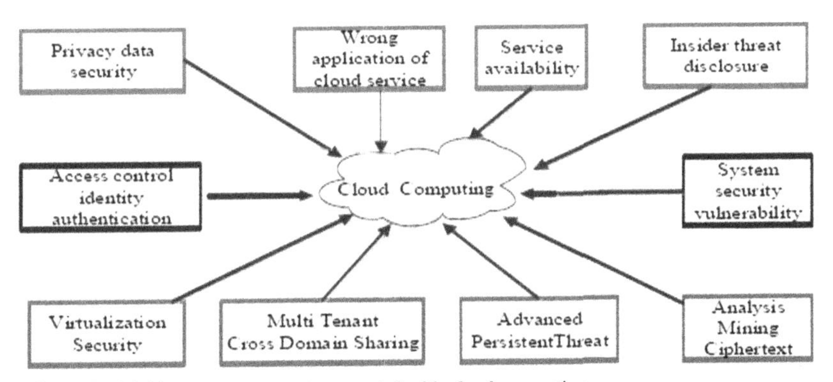

FIG. 29.5 Multiaccess component connected with cloud computing.

(3) Are computer security problems to do with Internet networks (on wide scales) with more complex network growing nowadays indulge in more complex problem. Due to the technical limitations of traditional telephone and Internet networking, as well as the physical restrictions of storage, neither privacy nor standard encryption techniques can be applied in the cloud computing setting [26,27].

29.9.4 Privacy security of cloud computing

(1) Privacy threats, such as privacy exposure, e.g., leaks and access protection, become more prevalent in a network-outsourcing strategy, particularly.
(2) For cloud computing, great effort is required in both managing who has access to the computing services and confirming who the person is who wants access to them.
(3) Virtualization has yielded better stability, but the conversion of virtual machines is a means of service isolation [28,29].
(4) The advanced persistent threat: An attack on the deep-seated appeal of hackers.
(5) Network manager insecurity: Due to different service providers, the security risk in the cloud can vary.
(6) Insider attacks on the agency systems have a high probability of damaging stability.
(7) Wrong utilization of cloud computing can generate issues for consumers, network operators, as well as unauthorized networks being affected [30,31].
(8) The attack on the cloud hosting unavailable systems leads to the distributed denial of service (DDoS) attacks being a critical security priority for cloud service providers.

29.10 ABE in cloud computing

ABE is a privacy security tool used in cloud computing that has the potential for international applications. Various privacy protective procedures have been suggested by academics to secure users' privacy sensitive data in a cloud storage environment [32,33].

ABE will have both a means of encryption and the ability to manage access to stored information. Attribute encryption offers four important benefits (Fig. 29.6):

(1) The owner of the data does not have to encode the data according to attributes, so it does not matter how many users there are:
(2) Only two users will decode (detect) ciphertext once they meet the required conditions.
(3) The secret of ABE is that a random number is used to deter collusion, which prevents the attack of users.
(4) ABE has quite good security controls.

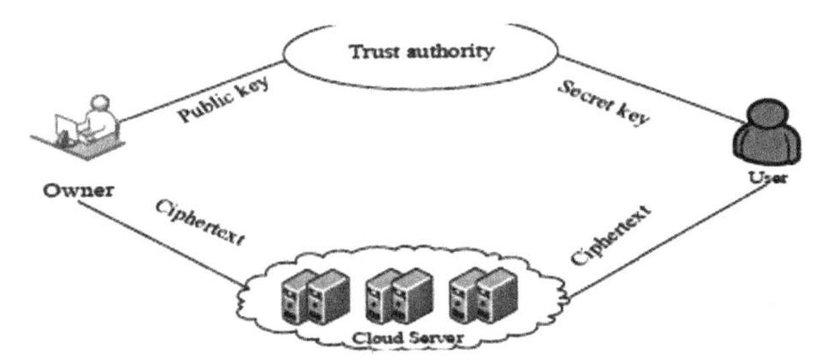

FIG. 29.6 Attribute-based encryption in cloud computing.

Due to ABE's ability to protect different users of the same attribute, the inability to revoke an attribute can influence certain users with the same attribute.

- The searchable encryption algorithm contains two algorithms: symmetric encryption and asymmetric encryption.
- The process of installing; preparation—the algorithm may generate a key and a public key depending on the provided security parameter and come up with additional secret keys such as the pseudorandom private key
- The algorithm extracts a password from the search terms provided in the written descriptions. The algorithm administrator is selected based on the situation and is owned either by the data owner or by the receiver.
- The data user can establish a keyword table of encrypted data and use a sortable function to generate it. In SAE, the private key should be used to encrypt files; in SSE, the symmetric key should be used to encrypt the key set [34,35].
- Query occurs automatically by the server. Search results and index data are used to determine the protocol preference. The program decides whether the input meets the criteria set by the user by deciding if the output matches the expected value. The search result is returned.

29.11 Conclusions

Defense, trust, and privacy are significant elements of cyber security because smart devices are growing in number, and the way we exchange information on the Internet is changing. There have been major advancements in artificial intelligence and machine learning, and these methods are used to provide better ways to protect against new cyber threats. Despite all this support from the academic community, hackers have still been able to do many wrong things with these tools, which opens up a new attack surface. Equipment such as routers, personal assistants, IoT, and other autonomous networks are the types of equipment that make up much of the Internet. Because they are the most prominent targets, it follows that they would be frequent targets. These three individuals' contributions, which are each among the many that have improved the research environment, show the trend and relevance of how writers improve it and help illustrate the potential importance of their work.

References

[1] Y. Qian, M. Chen, J. Chen, M. S. Hossain, and A. Alamri, 'Secure enforcement in cognitive internet of vehicles,'' IEEE Internet Things J., vol. 5, no. 2, pp. 12421250, Apr. 2018.

[2] K. Lin, J. Luo, L. Hu, M. S. Hossain, and A. Ghoneim, Localization based on social big data analysis in the vehicular networks, IEEE Trans. Ind. Informat., vol. 13, no. 4, pp. 19321940, Aug. 2017.

[3] Z. Zhou, D. Huang, and Z. Wang, ''Effcient privacy-preserving ciphertextpolicy attribute based-encryption and broadcast encryption,'' IEEE Trans. Comput., vol. 64, no. 1, pp. 126138, Jan. 2015.

[4] M. Nabeel, N. Shang, and E. Bertino, ''Privacy preserving policy-based content sharing in public clouds,'' IEEE Trans. Knowl. Data Eng., vol. 25, no. 11, pp. 260–2614, Nov. 2013.

[5] S. S. Karanki and M. S. Khan, ''SMMV: secure multimedia delivery in vehicles using roadside infrastructure,'' Veh. Commun., vol. 7, pp. 40–50, Jan. 2017.

[6] W. D. Hillis, 'Intelligence as an emergent behavior; or, the songs of Eden,'' Daedalus, vol. 5, pp. 17–189, Oct. 1988.

[7] T. Wolf, Analyzing and engineering self-organizing emergent applications, Ph.D. dissertation, Katholieke Univ. Leuven, Belgium, May 2007.

[8] S. Chavhan, P. Venkataram, Emergent intelligence: a novel computational intelligence technique to solve problems, in: Proc. 11th Int. Conf. Agents Artif. Intell, 2019, p. 93102.

[9] S. Chavhan and P. Venkataram, ''Emergent intelligence technique-based transport depot resource management in a metropolitan area,'' J. Vehicle Routing Algorithms, vol. 2, nos. 14, pp. 23–40, Dec. 2019.

[10] M.A. Rahman, M.S. Hossain, G. Loukas, E. Hassanain, S.S. Rahman, M.F. Alhamid, M. Guizani, Blockchain-based mobile edge computing framework for secure therapy applications, IEEE Access 6 (2018) 72469–72478.

[11] Y. Fan, S. Liu, G. Tan, F. Qiao, X. Lin, Fine-grained access control based on trusted execution environment, future Gener, Comput. Syst. 109 (2020) 551–561.

[12] D. Puthal, S. Nepal, R. Ranjan, and J. Chen, "Threats to networking cloud and edge datacenters in the internet of things," IEEE Cloud Computing, vol. 3, no. 3, pp. 64–71, 2016. [Online]. Available: doi:https://doi.org/10.1109/MCC.2016.63.

[13] T.D. Nguyen, C.E. Irvine, and J. Khosalim, "A multilevel secure mapreduce framework for cross-domain information sharing in the cloud," in Ground System Architectures Workshop, Mar 2013, pp. 18–21. [Online]. Available: http://gsaw.org/wpcontent/uploads/2013/06/2013s05nguyen.pdf.

[14] Y. Zhu, H. Hu, G.J. Ahn, Y. Han, S. Chen, Collaborative integrity verification in hybrid clouds, in: Collaborative Computing: Networking, Applications and Worksharing (CollaborateCom), 2011 7th International Conference on, Oct 2011, pp. 191–200.

[15] T. Wang, G. Zhang, M.Z.A. Bhuiyan, A. Liu, W. Jia, M. Xie, A novel trust mechanism based on fog computing in sensor-cloud system, Future Gener. Comput. Syst. 109 (2020) 573–582.

[16] S. Zhang, X. Li, Z. Tan, T. Peng, G. Wang, A caching and spatial k-anonymity driven privacy enhancement scheme in continuous location-based services, Future Gener. Comput. Syst. 94 (2019) 40–50.

[17] S.M. Elghamrawy, Security in cognitive radio network: defense against primary user emulation attacks using genetic artificial bee Colony (GABC) algorithm, Future Gener. Computing. Syst. 109 (2020) 479–487.

[18] H. Yin, Z. Qin, J. Zhang, L. Ou, F. Li, K. Li, Secure conjunctive multi-keyword ranked search over encrypted cloud data for multiple data owners, Future Gener. Comput. Syst. 100 (2019) 689–700.

[19] J. Shen, D. Liu, X. Sun, F. Wei, Y. Xiang, Efficient cloud-aided verifiable secret sharing scheme with batch verification for smart cities, Future Gener. Comput. Syst. 109 (2020) 450–456.
[20] M. Arvindhan, A. Anand, Scheming an proficient auto scaling technique for minimizing response time in load balancing on Amazon AWS Cloud, in: International Conference on Advances in Engineering Science Management & Technology (ICAESMT)-2019, Uttaranchal University, Dehradun, India, 2019. , March.
[21] D. Smyth, V. Cionca, S. McSweeney, and D. O'Shea, "Exploiting pitfalls in software-defined networking implementation," in Proc. Int. Conf. Cyber Secur. Protection Digit. Services, Jun. 2016, pp. 1–8.
[22] D. Estrin, G. Tsudik, Security issues in policy routing, in: Proc. IEEE Symp. Secur. Privacy, May 1989, pp. 183–193.
[23] K.K. Karmakar, V. Varadharajan, U. Tupakula, M. Hitchens, Policy based security architecture for software defined networks, in: Proc. 31st Annu. ACM Symp. Appl. Comput, 2016, pp. 658–663.
[24] P. Thai, J.C. de Oliveira, Decoupling policy from routing with software defined interdomain management: Interdomain routing for SDN-based networks, in: Proc. 22nd Int. Conf. Comput. Commun. Netw. (ICCCN), Jul. 2013, pp. 1–6.
[25] W. Lee, N. Kim, Security policy scheme for an efficient security architecture in software-defined networking, Information 8 (2) (2017) 65.
[26] R. Sahay, G. Blanc, Z. Zhang, K. Toumi, H. Debar, Adaptive policy-driven attack mitigation in SDN, in: Proc. 1st Int. Workshop Secur. Dependability Multi-Domain Infrastruct, 2017. p. 4.
[27] S. Hina, D.D. Dominic, Information security policies: Investigation of compliance in universities, in: 3rd Int. Conf. Comput. Inf. Sci. ICCOINS 2016 - Proc., 2016, pp. 564–569, https://doi.org/10.1109/ICCOINS.2016.7783277.
[28] H. Huang, N. Parolia, K.-T. Cheng, Willingness and ability to perform information security compliance behavior: psychological ownership and self-efficacy perspective, in: Pacific Asia Conf. Inf. Syst., 2016, https://doi.org/10.1186/1471-2334-12-S1-O4.
[29] G.C. Maphanga, O. Jokonya, The risk of users' negative Behaviours on information security compliance policy in organizations, Risk Gov Control Financ Mark Institutions 7 (2017) 30–40, https://doi.org/10.22495/rgc7i4art4.
[30] M.I. Merhi, J. Leighton, Top management can lower resistance toward information security compliance, Thirty Sixth Int. Conf. Inf. Syst. (2015) 1–11.
[31] M.M. O'Neill, S.R. Booth, J.T. Lamb, Using NVivo™ for literature reviews: the EightStep pedagogy (N7+1), Qual. Rep. 23 (13) (2018) 21–39. Available from: https://nsuworks.nova.edu/tqr/vol23/iss13/.
[32] N.S. Safa, R. Von Solms, An information security knowledge sharing model in organizations, Comput. Hum. Behav. 57 (2016) 442–451, https://doi.org/10.1016/j.chb.2015.12.037.
[33] S. Consolvo, M. Langheinrich, Identifying factors that influence employees' security behavior for enhancing ISP compliance, Priv. Secur. ACM SIGCAS Comput. Soc. 31 (2015) 8–23, https://doi.org/10.1145/503345.503347.
[34] J. Abed, G. Dhillon, S. Ozkan, Investigating continuous security compliance behavior: Insights from information systems continuance model, in: Twenty-second Am. Conf. Inf. Syst. San Diego, 2016, pp. 1–10.
[35] A. Alzahrani, C. Johnson, S. Altamimi S., Information Security Policy Compliance: Investigating the Role of Intrinsic Motivation Towards Policy Compliance in The Organisation, in: 2018 4th Int. Conf. Inf. Manag., IEEE, 2018, pp. 125–132, https://doi.org/10.1109/INFOMAN.2018.8392822.

Chapter 30

Integration of wireless communication technologies in internet of vehicles for handover decision and network selection

Shaik Mazhar Hussain[a], Kamaludin Mohamad Yusof[a], Afaq Ahmad[b], and Shaik Ashfaq Hussain[a]

[a]*Department of Communications Engineering and Advanced Telecommunications Technology, School of Electrical Engineering, Faculty of Engineering, Universiti Teknologi Malaysia (UTM), Johor Bahru, Malaysia,* [b]*Department of Electrical and Computer Engineering, Sultan Qaboos University, Muscat, Oman*

30.1 Introduction

The emergence of vehicular applications such as safety and nonsafety has brought immense challenges to the present vehicular communication and radio access technologies (RATs). No single technology can fulfill the requirements of vehicular applications [1]. For instance, 4G LTE offers a wide coverage range, high data rates, and high bandwidth but suffers from high transmission time intervals (HTTI). In contrast, dedicated short-range communication (DSRC) offers low latency but suffers from channel degradation and bandwidth limitations [2]. To maximize future generation wireless networks, 5G technology has been evolved to facilitate high data rates, ultrahigh bandwidth, and very low latencies. Table 30.1 shows a comparison of DSRC, 4G LTE, and 5G mm-wave wireless technologies in terms of latencies, data rate, frequency, and bandwidth.

Besides all these benefits, it suffers from short-range and line of sight propagation issues [2]. Traditional VANETs have been converted to the so-called Internet of vehicles (IoV) where the vehicles exchange information with other vehicles using different RATs. IoV permits communication of vehicles connecting with the internet. Vehicles are fast-moving on the road lane; this vehicle

System Assurances. https://doi.org/10.1016/B978-0-323-90240-3.00030-8

TABLE 30.1 Radio access technologies comparison.

Radio access technology (RAT)	Data rate	Latency	Bandwidth
Dedicated short-range communication (DSRC)	27 Mbps	150 ms	75 MHz
4G LTE (Long-term evolution)	DS—5 and 12 Mb/s US—2 and 5 Mb/s	1–3 s	Varying −1.4, 3, 5, 10, 15 and 20 MHz
5G mm-Wave	20 Gbps (Peak), 100 Mbps (Average)	<30 ms	30–300 GHz

communication is introduced to support the transmission of all types of data with each other during traveling. It also enables to transfer emergency data to all the other vehicles as a disseminator. As mentioned earlier, single technology such as DSRC, 4G LTE, and 5G might not suffice the requirements of vehicular technology because of their limitations. Hence, this raises a question—"How to Intelligently Cooperate among different radio access technologies (RATs) for appropriate network selection and optimal handover decisions in heterogeneous wireless networks". IoV integrates with 5G mm-Wave for reliable communication and also supports multimedia transmission. A vehicle is employed with multiple terminals such as DSRC, 5G (mm-Wave), and LTE for communication [1,3,4]. This combination has the common problem of vertical handover and network selection. To solve this, many research works have been designed to perform any one such as vertical handover and network selection. A set of metrics as channel constraints were taken into account to decide for handover [5,6]. As per the involvement of multiple radio access, it is essential to select a network since the signal strength and load of each network are not the same. The process of network selection for handover is also significant in this research [7] as an appropriate selection of network for handover could result in many concerns such as unnecessary handover, packet loss, handover failures, and degradation in the performance of quality of service (QoS). The process of handover requires especially for the vehicle to infrastructure communication since it demands long-distance traveling for connection. While the vehicles within the road lane can use DSRC for transmitting the data from one vehicle to another. For this, it requires selecting multiple relays in a path that is defined as routing to forward data [8,9]. Hereby, vehicle communication with DSRC, LTE, and mm-Wave involves the process of handover, network selection, and data forwarding. From this research field, the inefficient handover decision and network selection result in degraded performance of QoS. The network selection in handover was made by computing anyone of the following constraint such as channel metric and location. The use of anyone metric for handover decisions with a static threshold would not be able to hold connectivity for a long time after handover. Because of this, it leads to cause frequent handovers. Then the frequent handover of the vehicle will fail in data forwarding, i.e., routing. The strong connectivity of V2V takes place using a set of parameters such as channel metrics (signal to noise ratio (SNR), link quality), transmission metrics (delay, packet delivery ratio), vehicle metrics (speed, direction). The identified research gap in this field is the use of static threshold in a dynamic vehicular environment and the use of limited parameters for decision making. The following sections are organized as follows: existing works include the concepts, approach, methods, and issues. After this will be the proposed solution addressing the solutions to vertical handover, network selection, and routing issues. The later section will mention all the hardware and software requirements for the proposed work. In the next section, results are shown and compared with the existed method. In the end, our work is concluded followed by the references section.

30.2 Existing works

The number of handovers increases due to the absence of validating the vehicle regarding the need for handover. Hence, increases the number of unnecessary handovers. The multicriteria decision-making algorithm and machine learning algorithm were used for the handover decision, as it requires giving each decision for each vehicle since the metrics of each vehicle are not the same. For making the single decision of handover for a vehicle, the algorithm has to be executed on the RSU or the vehicle. In case, if several 20 vehicles need to perform handover, in this case, the decision is taken one after the other only. The previous works have discussed the use of multicriteria decision-making algorithms for data forwarding, i.e., routing that takes into account a limited number of metrics that fail to prefer an optimal path for transmission. If a path is selected from history metrics, then the current movement of the vehicle may be high. In this case, selecting such a vehicle in the path leads to breakage in the path and requires retransmission. In Ref. [2], the authors have used DSRC for V2V and mm-Wave for V2I communication using a reinforcement learning algorithm. To minimize the collision, this V2V uses a self-learning collision avoidance mechanism using Q-learning. From this reinforcement learning algorithm, the agent prefers the value of the contention window. For network selection, the TOPSIS algorithm considers several parameters into account. TOPSIS algorithm follows the distance principle that leads to abnormality in ranking results. On behalf of this problem, it performs very poor vertical handover decisions. As per the ranking result, the handover is performed by the vehicle. But the need for handover is not evaluated. In Ref. [10], a target discovery problem has been solved by using channel state information (CSI). The position of vehicles is determined using Kernel-based machine learning (ML) and handover decision using K-nearest neighbor (K – NN), which is also an ML. The handover decision also takes into account the history of handover data. For the identification of the position of a vehicle, the CSI is predicted between the vehicle and the unit. In this, the V2V data forwarding performs by estimating the probability of connectivity and interruption and SNR. K-NN algorithm for handover decision is not efficient as the accuracy depends on the quality of data. Also, it is slow processing when huge amounts of data are received and hence it takes time to make handover decisions. The data forwarding using these two metrics is not sufficient causing NLOS issues. In Ref. [1], a model is developed for vertical handover in a heterogeneous network. This paper proposes to use the k-partite graph to solve the criticality in selecting a network and then the Dijkstra algorithm is used for preferring the best path for transmission. This work aims to improve the performance of QoS. Here the handover decision is based on the cost function estimated from the graph. From these parameters, the handover decision is made and then the analytical hierarchy process (AHP) is used for estimating different criteria and select a path. Using AHP, weight is estimated and then the Dijkstra algorithm is applied to select a path, however, this results in frequent handovers. The maintenance of graphs in heterogeneous networks is complex since the devices are not the same at mobility, hence it requires larger resources and dynamic processing to manage the graph. The use of AHP is not efficient since it requires training of the data and then it can select the best path. But here as per the current situation of the vehicles, the path needs to be chosen, and also the movement of vehicles will not be the same in all the regions. Also, the addition of new criteria is difficult in this algorithm. In Ref. [2], the authors have proposed three different networks as WAVE, i.e., DSRC, WiFi, and LTE in which it employs the best interface election (BIS) algorithm. The BIS computes the following QoS parameters such as throughput, delay, and user preference. On estimating each parameter, it randomly selects a network concerning the QoS. The random selection of access networks with individual parameters leads to a poor selection of networks. Since the main constraints of QoS in this work is bandwidth or delay, i.e., it considers anyone from this, and hence the network selection is poor. Therefore, it minimizes QoS in the network. In Ref. [1], Mohamed Lahby et al. present the use of different terminals in vehicles and the terminals are LTE, mm-Wave, and DSRC (default in the vehicle for communication). The authors of this work give a basic idea about LTE and mm-Wave in-vehicle communication and also the illustration of the pros and cons of each network is highlighted. As a future direction, the selection of the best access for performing vehicle communication is studied in this work. In Ref. [3], the authors employ 5G, i.e., mm-Wave, and implement using network simulator and validate in both urban and rural environments. In this work, the vehicle network joins a software-defined network. It is evaluated only for video streaming data traffic. In Ref. [4], a network is deployed with an mm-Wave base station for communication purposes. However, this work uses static threshold and radio access differ for each network. In Ref. [5], a cluster-based handoff is proposed in which it introduces a dynamic edge-backup node

(DEBCK). The vehicles on the road lane are clustered and the backup node is the one that gets ready for handoff. The three main parameters that are taken into account for handoff are storage, communication, and energy. The failure of backup mobile edge-node or cluster head will lead to poor handoff or it may not perform a handoff. In Ref. [6], the authors have proposed a handoff protocol that takes into account LET for detecting the connectivity between vehicles. The partner selection protocols enable a selection of optimal partner nodes (PN). In Ref. [7], the authors have proposed a multitier heterogeneous network for network selection considering different parameters. In Ref. [8], the authors have proposed a cluster-based adept cooperative algorithm (CACA) focusing on the QoS metrics. As per this work, the process of clustering takes place and a cluster head is selected. This work uses the OLSR protocol with the Multi-Point Relay (MPR). The selection of MPR is not efficient, since the vehicles move at high speed. In Ref. [9], the authors have used the Analytic Hierarchy Process (AHP) to solve the line-of-sight (LOS) problem. However, it suffers from poor selection of relay and improper scoring value.

30.3 Research method

This research work focuses on integrating DSRC, 4G LTE, and 5G mm-Wave for heterogeneous IoV communication. This integration enables to provide reliable communication, especially for delay-sensitive applications. However, it subjects to the major problem of vertical handover and network selection. This research work aims to improve QoS for ongoing communication during data transmission and mitigating the issues related to handover and routing. In our proposed work, we have assumed that the vehicle is employed with three terminals DSRC, 4G LTE, and 5G. For vehicle-to-vehicle (V2V) communication, the DSRC is used and for Vehicle to infrastructure communication, LTE or mm-Wave access is used. Each network access differs in their coverage range and also they are the same in their characteristics with the vehicle. Fig. 30.1 shows the proposed architecture with the proposed algorithms.

30.3.1 Handover decision by dynamic Q-learning

In our proposed work, the threshold is set dynamically concerning the available network. Vehicle speed and signal strength are considered as input parameters and handover decision (Y/N) is the output. Both the input parameters are defined to be changing dynamic since the vehicle speed depends on the traffic density and the signal strength differs for each RAN that is in access. However, the vehicle speed will increase and decrease within a speed limit as per the vehicle's ability. Due to this, we set the threshold for signal strength in dynamic using Shannon entropy as

$$S(ss) = E[-\log(P(ss))] \tag{30.1}$$

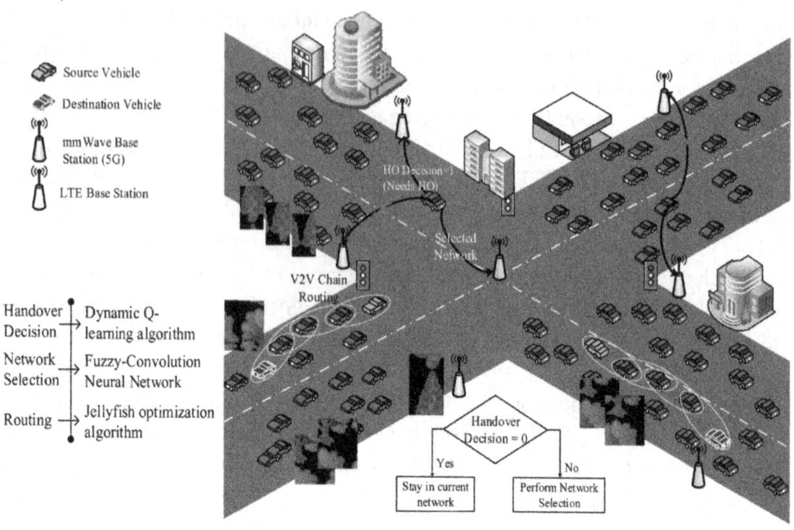

FIG. 30.1 Proposed system model.

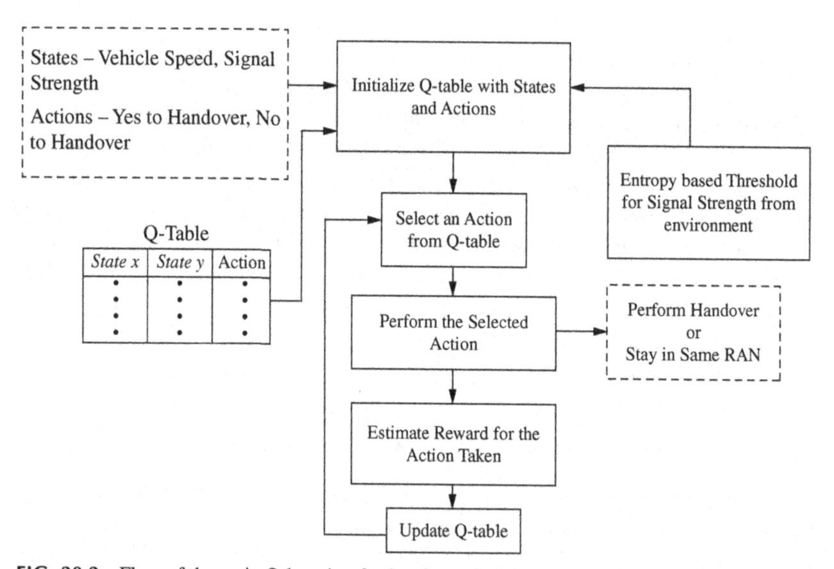

FIG. 30.2 Flow of dynamic Q-learning for handover decision.

$S(ss)$ is the Shannon entropy that composes of ss values for DSRC, mm-Wave, and LTE that ranges between −30 and − 70 dBm. In Eq. (30.1), $P(ss)$ denotes the probability of the signal strength, which is strong (Fig. 30.2).

The parameters signal strength and vehicle speed are the states that are taken into account to take an action for handover. Let $Q(A,B)$ represent state A and action B based on the Q-values. Each state A will have two parameters and this $Q(A,B)$ is determined and updated in the rule. The temporal difference update rule is as follows:

$$Q(A,B)+\alpha(R+\gamma Q(A',B')-Q(A,B))\rightarrow Q(A,B) \tag{30.2}$$

The term $Q(A',B')$ defines next state and action R is the reward given by the agent, γ is the discount factor, i.e., [0 – 1], α is the learning rate [0 – 1] that denotes the step length to estimate the (A,B). The action is taken using ϵ-greedy policy, the ϵ represents epsilon. The pseudocode for dynamic Q-learning is here to explain the processing of this algorithm to decide for handover (Fig. 30.3).

In ϵ-greedy policy, when the probability is $(1-\epsilon)$, then the action will be taken as per the value in the Q-table. If the handover request is agreed and the action is yes, then it will select a network.

30.3.2 Network selection using fuzzy CNN

Network selection is a process of selecting a network from the available radio access networks considering metrics such as signal strength, the distance between the base station and vehicle, vehicle density, data type, and line of sight. The definition for each metric is depicted below.

Pseudo Code 1: Dynamic Q-Learning

I – states (A), Q – *table*
O – Action (B)
1. start
2. V_a *(Request)* $\rightarrow$ *HandOver* //vehicle requests for handover
3. ini Q-table
4. ini Q (A,B)
5. for each $A \rightarrow ss,speed$ // vehicle 1 parameter
6. compute ss threshold from equ (1)
7. for (each step)
    apply ϵ – greedy policy
    obtain Q-value from Q-table
    perform action $A \rightarrow V_1$ // Action taken by vehicle 1
    compute R and next state A'
8. update Q-table using equ (2)
9. update A′ $\rightarrow A$
10. end

FIG. 30.3 Pseudocode of dynamic Q-learning.

Signal strength—signal strength defines the SNR which gives the number of signals. A channel is composed of noise as well as signal, the high the noise, the channel is unfit for transmission. The SNR (S_r) is determined from the signal power P_s and noise P_N, respectively. The formulation is

$$S_r = \frac{P_s}{P_N} \tag{30.3}$$

Distance between BS and vehicle—The distance between BS and a vehicle is estimated using Euclidean distance. This measure defines the stability of the link, as the distance increases the link will be unstable and when the distance decreases the link will be stronger. Euclidean distance is computed using the following equation:

$$D_{(L_{BS}, L_V)} = \sqrt{(x - x_1)^2 + (y - y_1)^2} \tag{30.4}$$

Vehicle density—The density of vehicle V_D denotes the number of vehicles that are connected with that particular BS.

$$V_D = \sum (N_{CL}, N_{NL}) \tag{30.5}$$

where N_{CL} and N_{NL} represents the number of connected links and the number of new links.

Data type—There are two types of data: safety (0) and nonsafety (1).

LoS—Line of sight defines the direct contact between the vehicle and BS without any obstacles that block the signals.

The above five metrics involves the development of fuzzy rules. The fuzzy logic deals with decision-making by the defined rules as high, medium, and low

$$(H, M, L) \rightarrow (mmWave, LTE, DSRC)$$

Fig. 30.4 shows the Fuzzy-CNN.

Fig. 30.5 explains the pseudocode for Fuzzy-CNN algorithm.

The use of CNN will give results for multiple vehicles at the same time by parallel processing. According to the selected network, the requested vehicle will be handed over from the present network to the target network.

30.3.3 Routing

V2V chain routing technique uses a chain method to form the chain between source vehicles to destination vehicles using jellyfish optimization algorithm and uses DSRC signals for communication. In this optimization algorithm, the objective function is defined using the consideration of three metrics channel metrics, vehicle metrics, and vehicle performance metrics.

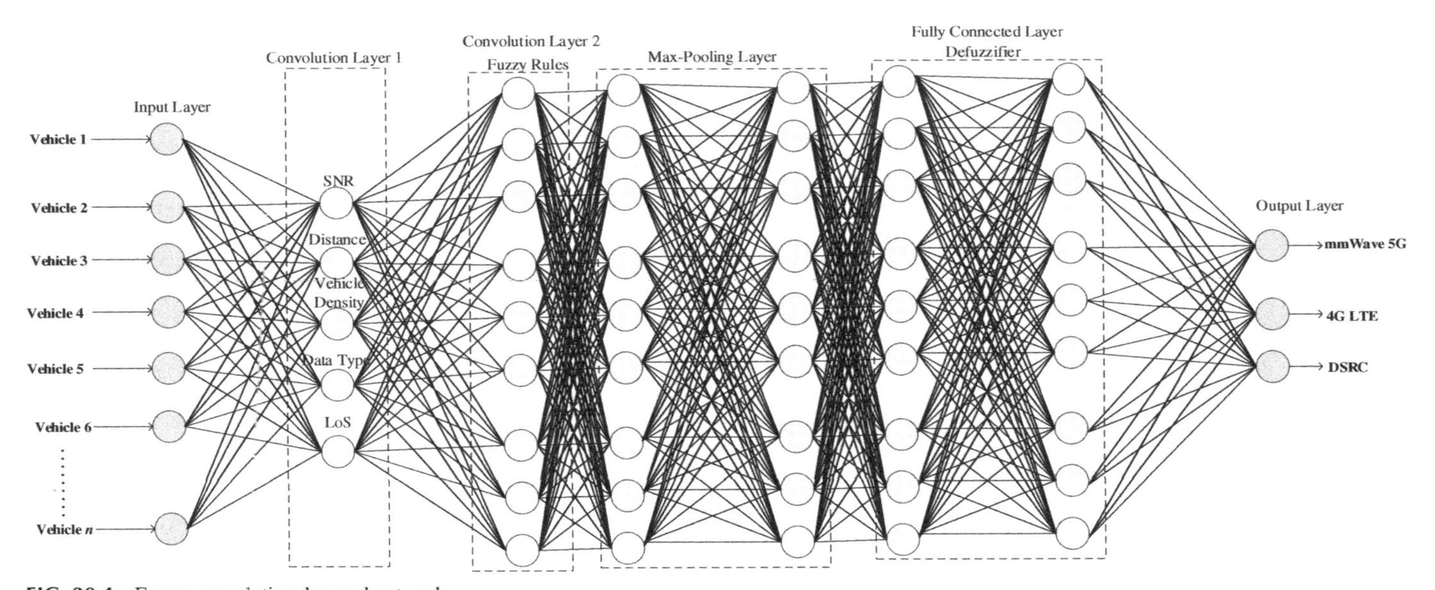

FIG. 30.4 Fuzzy-convolutional neural network.

```
Pseudo Code 1: Fuzzy-CNN
Input – Vehicle Req
Output – Network Selection (N_S)
1. begin
2. V(Req) → N_S                //vehicle requests for
selecting network
3. for each (V → SNR, D, Density, DT, LoS) //
convolution layer 1
4. compute SNR, D using equ (3) and equ (4)
5. determine the density, LoS with target
network
6. for (each V) do
    apply Fuzzy Rule           // convolution layer
2
7. if (V = R1)
    {
    select network (H) or (M) or (L)
    else
    go to next rule
        }
     end if
8. repeat step 7 until rule is satisfied
9. sum-up fuzzy values for each V   //max-
pooling layer
9. fuzzy set → crisp output     // fully connected
layer
10. return N_S              // output layer
11. end
```

FIG. 30.5 Pseudocode of Fuzzy-CNN.

The time control $c(t)$ is formulated from iteration and the random values as depicted in the following.

$$c(t) = \left| \left(1 - \frac{t}{\text{Max}_{ite}}\right) \times (2^* \text{rand}(0, 1) - 1) \right| \tag{30.6}$$

when $rand(0,1) > (1 - c(t))$, then passive motion

$$\text{rand}(0, 1) < (1 - c(t)), \text{then active motion} \tag{30.7}$$

The ocean current direction is represented as $\overrightarrow{OC}$ and it is mathematically given as

Let,

$$\overrightarrow{OC} = \frac{1}{V_p} = X^* - e_c\mu \tag{30.8}$$

Then,

$$\overrightarrow{OC} = X^* - d_{ff} \tag{30.9}$$

where V_p denotes the population of the vehicle as per this work. X^* indicates the best location, μ denotes mean location, and e_c is the attraction factor. Then the objective function is defined to select the best route. This function OF is formulated,

$$OF(Ms) = \sum (s_r, l_q)(s_p, R_d)(D_l, T_p) \tag{30.10}$$

Delay and speed should be minimum and other parameters can be maximum value to select the route.

$$l_q = \frac{1}{P_f \times P_r} \tag{30.11}$$

$$R_d = 2r \sin \sqrt{\sin^2\left(\frac{\Delta la}{2}\right) + \cos(la_v) * \cos\left(la_{np_i}\right) * \sin^2\left(\frac{\Delta ln}{2}\right)} \tag{30.12}$$

$$D_l = \frac{P_L}{b} \tag{30.13}$$

The criteria P_f, P_r denotes the number of the forwarded packet and the received packet in the same link between two vehicles, then (la, ln) represents the (latitude, longitude), so the vehicle location is (la_v, ln_v) and the next hop location is (la_{np}, ln_{np}) and r is the radius, i.e., coverage of the vehicle. The estimation of delay is computed from P_L, b that are packet length and bit rate.

30.4 Simulation setup

In our proposed work, we have created a vehicular network with 2 "LTE networks" (RSUs), 2 "5G Base Station" and 100 vehicles with the three terminal abilities of DSRC, mm-Wave, and LTE based on the sumo traffic model and configuration file. Vehicles use dynamic-Q learning algorithm for making handover decisions by taking speed and signal strength as inputs. The signal strength is set dynamically based on the Shannon entropy rule. Q-table is updated with states and actions and accordingly, actions are taken and the next state and actions are updated. It then selects a network from F-CNN. For network selection, the fuzzy rules are defined and used in CNN. Then routing takes place by using jellyfish algorithm that select V2V pairs between source to destination and so it is called V2V chain routing. We create a VANET network having 2-LTE networks (RSUs), 2 "5G Base Station," and 100 vehicles with the terminal ability of DSRC, mm-Wave, and LTE. Here, the vehicle is placed based on the sumo traffic model. Table 30.2 gives the simulation parameters that are considered in our work.

TABLE 30.2 Simulation specifications.

Parameter		Range/Value
Simulation area		2500 × 2500 m
Vehicle density		100
Number of 5G mm-Wave BSs		2
Number of 4G LTE BSs		2
Vehicle mobility type		Linear mobility
Vehicle speed		10–40 m/s
Transmission range	DSRC	300 m (Max)
	mm-Wave	~500 m
	LTE	100 km (Max)
Transmission rate		3–5 packets per second
Packet size		512 bytes
Simulation time		1000 s

30.5 Comparative analysis and results

A comparative analysis is shown for each case using the proposed approach and existing approach for vehicular movements of varying speeds between 20 and 100 km/h. Fig. 30.6 shows the number of unnecessary handovers performed, and Fig. 30.7 shows the comparative analysis of the handover failure rate between the proposed and existing method. It is clear from the figures that there is a good improvement seen with the proposed method with 25% reduction in unnecessary handover and 2% reduction in handover failure rate. Inappropriate selection of a network for handover activities could result in packet loss and call drops. This result shows that the decision by dynamic Q-learning and F-CNN have a better selection in handover. Figs. 30.8 and 30.9 show the comparison of throughput and delay of the proposed and existing approach. It is clear that the throughput is significantly better than the existing approach. The higher throughput is due to the fact of reduced handovers and less packet loss. From the proposed approach, there is a throughput improvement of 15%–20% and delay and packet losses are minimized. The figures below show the comparative analysis of our proposed approach with the existing approach. The performance metrics that are evaluated are a number of unnecessary handovers, HO success probability, throughput, packet loss, handover failure, and delay (Figs. 30.10 and 30.11).

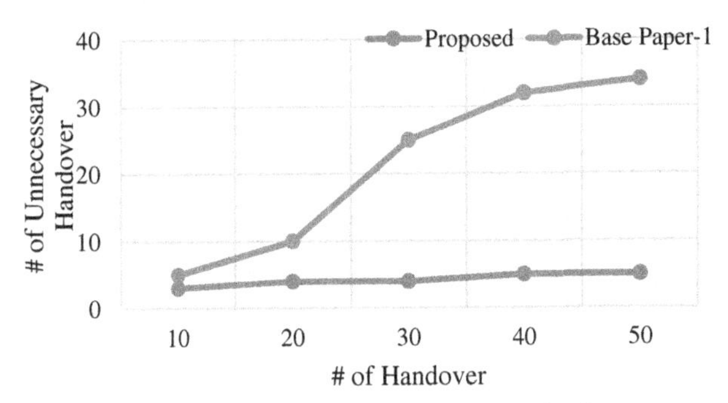

FIG. 30.6 Number of total handovers vs number of unnecessary handovers.

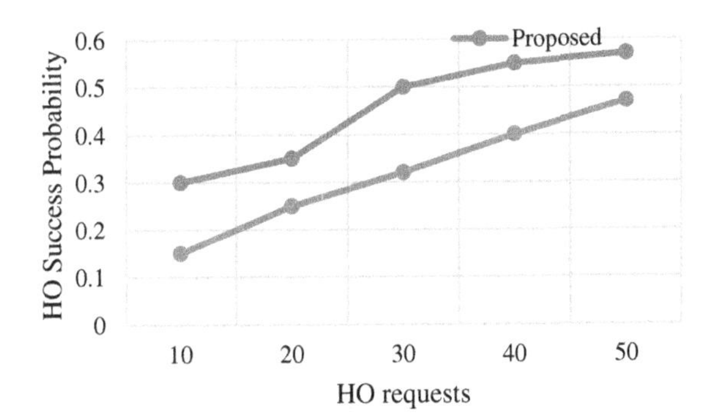

FIG. 30.7 HO request vs HO success probability.

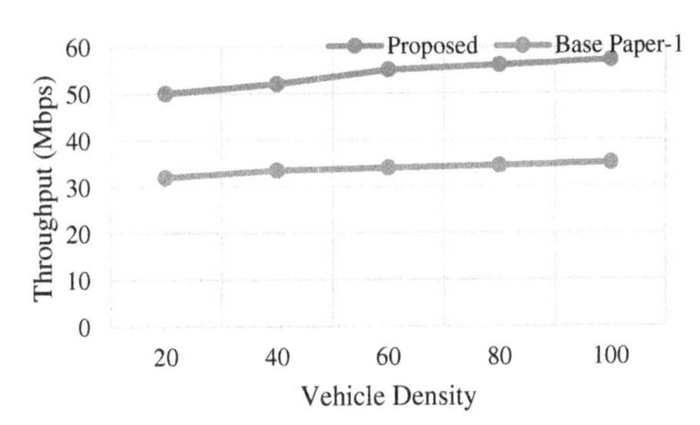

FIG. 30.8 Vehicle density vs throughput.

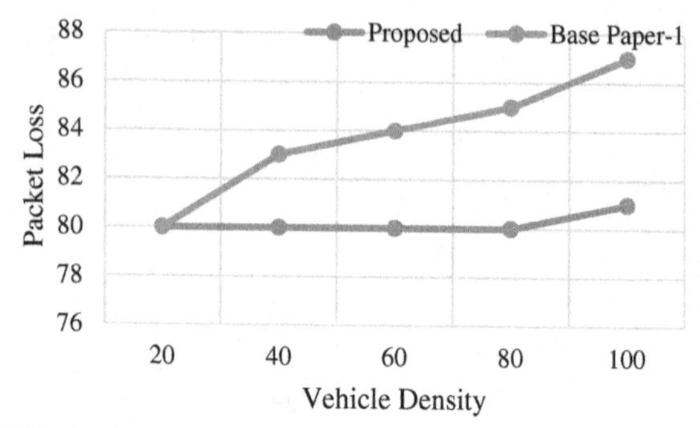

FIG. 30.9 Vehicle density vs packet loss.

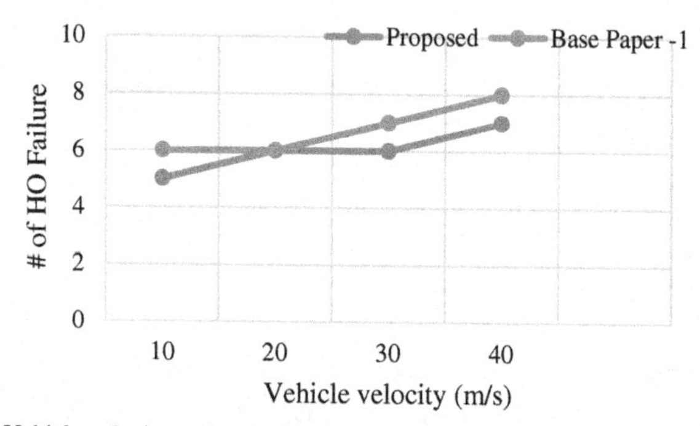

FIG. 30.10 Vehicle velocity vs handover failure rate.

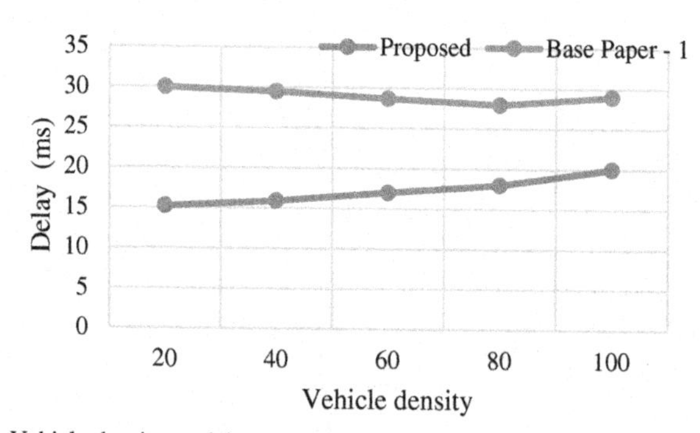

FIG. 30.11 Vehicle density vs delay.

30.6 Conclusion

In this chapter, we have highlighted the issues related to a single technology in terms of latency, bandwidth, and data rates, which are very critical in delay-sensitive applications. We have proposed three algorithms such as a solution to address handover decisions, network selection, and optimized routing paths.

We have constructed the IoV environment by assuming three RAT terminals equipped on vehicles. Dynamic Q-learning algorithm is used to take the decisions of handover with appropriate selection of networks using Fuzzy-convolutional neural networks. The jellyfish optimization algorithm is to select the shortest paths. Using a dynamic Q-learning algorithm, we have achieved a reduction in handover failure rates and improvements in handover success probability. Also, we have achieved higher throughputs and reduced delays using proposed algorithms. The implementation results of this work give better efficiency in terms of handover parameters as well as routing parameters. This article also claims that the single technology might be insufficient to meet the requirements of safety applications Hence, it is of paramount importance to integrate radio access technologies using appropriate algorithms to achieve better results when dealing with safety applications.

References

[1] M. Lahby, A. Essouiri, A. Sekkaki, A novel modeling approach for vertical handover based on dynamic k-partite graph in heterogeneous networks, Digit. Commun. Netw. 5 (4) (2019) 297–307.

[2] H.H.R. Sherazi, Z.A. Khan, R. Iqbal, S. Rizwan, M.A. Imran, K. Awan, A heterogeneous IoV architecture for data forwarding in vehicle to infrastructure communication, Mob. Inf. Syst. 2019 (2019).

[3] A. Tassi, M. Egan, R.J. Piechocki, A. Nix, Modeling and design of millimeter-wave networks for highway vehicular communication, IEEE Trans. Veh. Technol. 66 (12) (2017).

[4] K.M. Awan, M. Nadeem, A.S. Sadiq, A. Alghushami, I. Khan, K. Rabie, Smart handoff technique for internet of vehicles communication using dynamic edge-backup node, Electronics 9 (3) (2020).

[5] H. Ahmed, S. Pierre, A. Quintero, A cooperative road topology-based handoff management scheme, IEEE Trans. Veh. Technol. 68 (2018) 3154–3162.

[6] E. Ndashimye, N.I. Sarkar, S.K. Ray, A network selection method for handover in vehicle-to-infrastructure communications in multi-tier networks, Wirel. Netw 26 (2020) 387–401.

[7] N.M. Al-Kharasani, Z.A. Zukarnain, S.K. Subramaniam, M.H. Zurina, An adaptive relay selection scheme for enhancing network stability in VANETs, IEEE Access 8 (2020) 128757–128765.

[8] B. Fan, H. Tian, S. Zhu, Y. Chen, X. Zhu, Traffic-aware relay vehicle selection in millimeter-wave vehicle-to-vehicle communication, IEEE Wireless Commun. Lett. 8 (2) (2019).

[9] Z. Sheng, A. Pressas, V. Ocheri, F. Ali, R. Rudd, Intelligent 5G vehicular networks: an integration of DSRC and mmWave communications, in: 2018 International Conference on Information and Communication Technology Convergence (ICTC), Jeju, South Korea, 19 November 2018.

[10] L. Yan, H. Ding, L. Zhang, J. Liu, X. Fang, Y. Fang, M. Xiao, X. Huang, Machine learning-based handovers for sub-6 GHz and mmWave integrated vehicular networks, IEEE Trans. Wireless Commun. 18 (2019) 4873–4885.

Chapter 31

Modeling HIV-TB coinfection with illegal immigrants and its stability analysis

Rajinder Sharma
CR Department, University of Technology and Applied Sciences-Sohar, Sohar, Oman

31.1 Introduction

In the health sector, the problem of HIV-AIDS and tuberculosis (TB) pandemics is a challenge. It is commonly acknowledged that infectious diseases are one of the key root to human mortality, particularly in the developing countries. *Mycobacterium tuberculosis* is the causative agent of TB. Both human and animal populations are affected by the fatal disease. In agricultural dominant countries, it is probably transmitted from animals to humans. In addition to personal hygiene, environmental factors such as open water reservoirs, garbage dumps, and open drainage system in residential areas acts as catalyst to spread TB [1, 2]. It is present worldwide. Since the advent of AIDS pandemic, TB has been recognized as one of the most opportunistic infections in HIV-infected individuals. It is estimated that in non-HIV-infected individuals with latent *M. tuberculosis* has only 5%–10% of chance of developing active TB. The fatality rate is at least 50% if no treatment is available. The percentage of the development of active TB increases in the presence of factors that disturb immune function. An AIDS patient with latent TB has a 50% of life time chance of developing TB. Moghadas and Alexander [3] in their theoretical framework on TB pointed out that among the AIDS case reports so far, 60% of them had TB. The life span of the HIV-positive patients having TB decreases as it accelerates the progression of HIV. Several ecological investigations related to HIV/AIDS-TB coinfection have been conducted by using several mathematical models. For more details, one can refer to Refs. [4–17].

Illegal immigration is the relocation of individuals into a nation in transgression of the immigration laws of that dominion. As per UN DESA, 2020 available in the International Organization of Migrants (IOM) website (https://www.iom.int/), it is estimated that there are 280.6 million international immigrants

System Assurances. https://doi.org/10.1016/B978-0-323-90240-3.00031-X

worldwide, though it is difficult to make accurate estimates for illegal immigrants. Furthermore, most illegal immigrants arrive in a country legally, but overstay their authorized residence without the renewal of their residence visa. It is made of many reasons including family reunification, natural disaster, poverty, etc. The whole chapter is structured as follows. In Section 31.2, a brief mathematical description of the proposed model and the notations involved are given. The mathematical analysis comprises finding out of equilibrium points and its stability analysis are presented in Section 31.3. Finally, numerical analysis and discussion are provided in Section 31.4.

31.2 The mathematical model

To model the HIV-TB coinfection transmission dynamics with illegal migrants and foreigners without HIV-TB detection medical test, we partition the population of size $N(t)$ into four different categories. The susceptible $P(t)$, the TB infectives $T_1(t)$, the HIV infectives who are assumed to be infectious $T_2(t)$, AIDS patients $H(t)$, dot represents a differentiation with respect to time t and constructed the model as follows:

$$
\begin{aligned}
\dot{P} &= (1-b)Z_0 - \left(\frac{\gamma_1 P T_1 + \gamma_2 P T_2}{N}\right) - dP + \mu T_1; \\
\dot{T}_1 &= \frac{\gamma_1 P T_1}{N} - \frac{\gamma_3 T_1 T_2}{N} - (\mu + d) T_1; \\
\dot{T}_2 &= \left(\frac{\gamma_2 P T_2 + \gamma_3 T_1 T_2}{N}\right) - (\rho + d) T_2; \\
\dot{H} &= \rho T_2 - (\omega + d) H,
\end{aligned}
$$

where all the parameters taken into consideration are nonnegative;
Z_0 is the constant immigration rate of the susceptible population;
b is the illegal immigrants rate;
γ_1 is the rate at which the individuals from the susceptible class becomes infected with TB;
γ_2, γ_3 are the rates at which both susceptible and TB class individuals become infected with HIV;
ρ is the rate at which HIV-infected class individuals are getting full blown AIDS;
d is the natural death rate constant;
ω is the constant death rate related to disease; and
μ is the recovery rate.

Since $N(t) = P(t) + T_1(t) + T_2(t) + H(t)$. The modified system takes the following form:

$$\dot{N} = (1-b)Z_0 - dN - \omega H; \tag{31.1}$$

$$\dot{T}_1 = \frac{\gamma_1(N - T_1 - T_2 - H)T_1}{N} - \frac{\gamma_3 T_1 T_2}{N} - (\mu + d)T_1; \tag{31.2}$$

$$\dot{T}_2 = \frac{\gamma_2(N - T_1 - T_2 - H)T_2}{N} + \frac{\gamma_3 T_1 T_2}{N} - (\rho + d)T_2; \tag{31.3}$$

$$\dot{H} = \rho T_2 - (\omega + d)H. \tag{31.4}$$

In case of no infection, the steady-state value of population size approaches to $\frac{(1-b)Z_0}{d}$ which was otherwise $\frac{Z_0}{d}$ as shown in [18], that is, if there are no illegal migrants, $b = 0$, we will get the steady-state value of the population size taken under consideration in the mathematical model [18].

31.3 The mathematical analysis

In this segment, the mathematical investigation of the above-mentioned model has been done. The analysis mainly involves four nonnegative equilibria, stability analysis and can be easily followed by mathematicians and theoretical biologists:

(i) $P_0\left(\frac{(1-b)Z_0}{d}, 0, 0, 0\right)$;

(ii) $P_1\left(\frac{(1-b)Z_0}{d}, \frac{(1-b)Z_0(\gamma_1 - (\mu + d))}{d\gamma_1}, 0, 0\right)$;

(iii) $P_2(N, 0, T_2, H)$, the TB pathogen free, which exists if $\gamma_2 > (\rho + d)$ and $Z_0 > \frac{\omega d}{(1-b)(\omega + d)}T_2$, $b < 1$, where $N = \frac{1}{d}\left[Z_0(1-b) - \frac{\omega\rho}{(\omega + d)}T_2\right]$, $H = \frac{\rho}{\omega + d}T_2$,

$$T_2 = \frac{\dfrac{Z_0(1-b)}{d(\gamma_2 - (\rho + d))}}{\left[\dfrac{\gamma_2(\omega + d + \rho)}{\omega + d} + \dfrac{\omega\rho}{d(\omega + d)}(\gamma_2 - (\rho + d))\right]}.$$

(iv) $P_3{}^*(N^*, T_1{}^*, T_2{}^*, H^*)$, the coinfection equilibrium, where

$$N^* = \frac{1}{d}\left[Z_0(1-b) - \frac{\omega\rho}{(\omega + d)}T_2^*\right],\ H^* = \frac{\rho}{\omega + d}T_1^*,$$

$$T_1^* = \frac{\dfrac{Z_0(1-b)}{d(\gamma_1 - (\mu + d))} - \left[\dfrac{\gamma_1(\omega + d + \rho)}{\omega + d} + \gamma_3 + \dfrac{\omega\rho}{d(\omega + d)}(\gamma_1 - (\mu + d))\right]}{\gamma_1}T_2^*,$$

$$T_2^* = \frac{\dfrac{Z_0(1-b)}{d}\left[\gamma_2 - (\rho + d) + \left(\dfrac{\gamma_3 - \gamma_2}{\gamma_1}\right)(\gamma_1 - (\mu + d))\right]}{\left[\dfrac{\gamma_3(\omega + d + \rho)}{\omega + d} + \dfrac{\omega\rho}{d(\omega + d)}((\gamma_1 - (\mu + d)) + (\gamma_2 - (\rho + d))) + \gamma_3(\dfrac{\gamma_3 - \gamma_2}{\gamma_1})\right]}.$$

From above, it is clear that P_3* is positive only if, $Z_0 > \frac{\omega\rho}{(1-b)(\omega+d)}T_2^*$, $\gamma_1 > (\mu + d)$, $\gamma_2 > (\rho + d)$ and $\frac{Z_0(1-b)}{d(\gamma_1-(\mu+d))} > \left[\frac{\gamma_1(\omega+d+\rho)}{\omega+d} + \gamma_3 + \frac{\omega\rho}{d(\omega+d)}(\gamma_1 - (\mu+d))\right]$.
It is worth mentioning that in case of the disease to be endemic the disease-induced death rate diminished the population size at equilibrium from $\frac{(1-b)Z_0}{d}$ to $N^* = \frac{1}{d}\left[Z_0(1-b) - \frac{\omega\rho}{(\omega+d)}T_2^*\right]$, where

$$T_2^* = \frac{\frac{Z_0(1-b)}{d}\left[\gamma_2 - (\rho+d) + \left(\frac{\gamma_3-\gamma_2}{\gamma_1}\right)(\gamma_1 - (\mu+d))\right]}{\left[\frac{\gamma_3(\omega+d+\rho)}{\omega+d} + \frac{\alpha\rho}{d(\omega+d)}((\gamma_1 - (\mu+d)) + (\gamma_2 - (\rho+d))) + \gamma_3(\frac{\gamma_3-\gamma_2}{\gamma_1})\right]}.$$

Theorem 31.1

(I) System (31.1)–(31.4) with all positive parameters has a unique disease-free equilibrium P_0 and is locally asymptotically stable if $B_0 < 1$ (i.e., $\frac{\gamma_1}{(\mu+d)} < 1$), $B_1 < 1$ *(i.e.,* $\frac{\gamma_2}{(\rho+d)} < 1$) *and the second equilibrium point* P_1 *is unstable. Where* B_0 *and* B_1 *are the basic reproduction numbers.*

(II) Equilibrium point P_2 is locally asymptotically stable if $\gamma_1 < (\mu + d)$ and $\gamma_2 < (\rho + d)$ satisfies the Routh-Hurwitz criterion $k_1k_2 - k_3 > 0$, $k_1k_2k_3 > 0$, and unstable otherwise.

$$k_1 = \omega + 2d + \gamma_2\left(\frac{2T\hat{}_2}{N\hat{}} + \frac{H\hat{}}{N\hat{}}\right) - (\gamma_2 - (\rho+d));$$

$$k_2 = d(\omega+d) + \frac{\gamma_2\rho I\hat{}_2}{N\hat{}} + \gamma_2(\omega+d)\left(\frac{2T\hat{}_2}{N\hat{}} + \frac{H\hat{}}{N\hat{}}\right) - d(\omega+d)(\gamma_1 - (\rho+d));$$

$$k_3 = \left(\rho\left(\left(d + \omega\left(\frac{2T\hat{}_2}{N\hat{}} + \frac{H\hat{}}{N\hat{}}\right)\right)\right)\right)\frac{\gamma_2 T\hat{}_2}{N\hat{}} + \gamma_2(\omega+d)\left(\frac{2T\hat{}_2}{N\hat{}} + \frac{H\hat{}}{N\hat{}}\right)$$
$$-d(\omega+d)(\gamma_2 - (\rho+d)).$$

(III) The equilibrium point P_3 is locally asymptotically stable if it exists and satisfies the Routh-Hurwtiz criterion, that is, for all $c_i > 0 (i = 1, 2, 3, 4)$, $c_1c_2 - c_3 > 0$, and $c_1c_2c_3 - c_3^2 - c_1^2c_4 > 0$, *where* c_1, c_2, c_3, *and* c_4 *can be calculated easily as shown in [18].*

Similarly, global asymptotical stability of the equilibrium point P_3 can be easily verified on the lines of Lemma 1 and Theorem 2 of Naresh and Tripathi [18].

31.4 Numerical analysis and discussion

Numerical simulations for the system (31.1)–(31.4) are presented in this segment. We applied R-K method of fourth order with the following set of parameter values to integrate the formulated model. If there are no illegal immigrants, that is, $b = 0$, then we will get the simulation analysis as shown in [18]. Let us check the impact of illegal immigration on the proposed model with different Z_0 and initial values taken in [18]. We will set illegal immigration rate $b = 0$ in the beginning and then change/increase accordingly to see the impact.

$$\begin{aligned} Z_0 &= 3000, \ \ b=0, \ \ d=\frac{1}{50}, \ \ \gamma_1=0.925, \ \ \mu=0.3, \\ \gamma_2 &= 0.365, \ \ \gamma_3=1.15, \ \ \omega=1, \ \ \rho=0.2 \end{aligned}$$

and initial values

$$N(0)=30{,}000, \ \ T_1(0)=3000, \ \ T_2(0)=4000, \ \ H(0)=600.$$

The behavior of the model in the absence of illegal migration is same as that of shown in [18] (see Fig. 31.1). In subfigures (A–D), the susceptible population shows the increasing trend in the beginning and thereafter tends to its equilibrium position. It is also worth mentioning here that the growth in the susceptible population shows a decline as the illegal immigration rate increases before reaching the equilibrium position. In subfigure (E), the susceptible population falls sharply before reaching the equilibrium position. In subfigure (F), the susceptible population decreases continuously thereby giving a sharp edge to the infected population which is a clear indication of the eradication of the whole population with an increased time period. It has been observed that diseases such as AIDS and TB persist in the population due to the presence of illegal migrants as legal migrants can be easily approached, diagnosed, and given access to health-care facilities accordingly but different is the case with the illegal immigrants due to their nonresidential status. Thus, the illegal immigrants if remain unchecked put the susceptible population at risk to become infected with increased time period. Hence, if the illegal immigrants within the susceptible population can be checked/controlled through various means of law such as police crackdown, deportation, mercy schemes to exit the country without fine/penalty, awareness campaigns, etc., the chance of flow from susceptible to infected population can be controlled/minimized. Since, we focused on the impact of illegal immigrants on the susceptible population in this study, the behavior of the rest of the components is more or less same as in [18]. In our future study, we will explore the impact of illegal immigrants on rest of the components along with one of the control strategies as mentioned earlier in the discussion.

Remark 31.1 If we set $b = 0$, that is, there are no illegal immigrants in the population then our model reduces to [18].

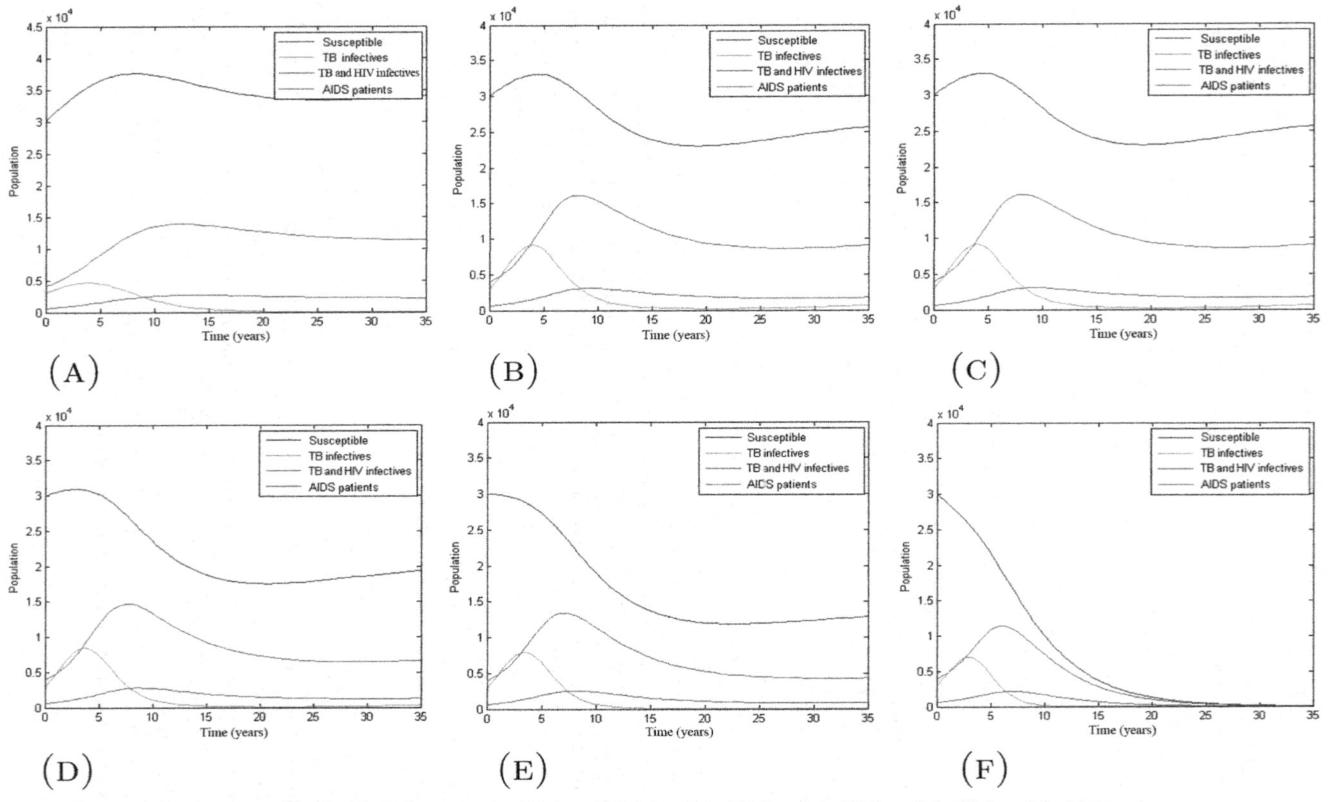

FIG. 31.1 Illegal immigration rate impact at different levels. (A) $b=0$, (B) $b=0.2$, (C) $b=0.4$, (D) $b=0.6$, (E) $b=0.8$, (F) $b=1$.

Acknowledgments

The author extends his sincere gratitude to the anonymous referees for their efficacious feedback which helps to a great extent in improving the quality of the chapter.

References

[1] M. Ghosh, J.B. Shukla, P. Chandra, P. Sinha, An epidemiological model for carrier dependent infectious disease with environmental effect, Int. J. Appl. Sci. Comp. 7 (3) (2000) 188–204.

[2] S. Singh, P. Chandra, J.B. Shukla, Modeling and analysis of the spread of carrier dependent infectious disease with environmental effects, J. Biol. Syst. 11 (3) (2003) 325–335.

[3] S.M. Moghadas, M.E. Alexender, Exogenous reinfection and resurgence of tuberculosis: a theoretical framework, J. Biol. Syst. 12 (2) (2004) 231–247.

[4] A.M. Tavares, I. Fronteira, I. Couto, D. Machado, M. Viveiros, A.B. Abecasis, S. Dias, HIV and tuberculosis co-infection among migrants in Europe: a systematic review on the prevalence, incidence and mortality, PLoS One 12 (9) (2017) e0185526, https://doi.org/10.1371/journal.pone.01-85526.

[5] M.J. Greenwood, W.R. Warriner, Immigrants and the spread of tuberculosis in the United States: a hidden cost of immigration, Popul. Res. Policy Rev. 30 (2011) 839–859.

[6] C. Castillo-Chavez, Z. Feng, W. Huang, On the computation of R_0 and its role on global stability, in: Mathematical Approaches for Emerging and Reemerging Infectious Diseases: An Introduction (Minneapolis, MN), IMA Vol. Math. Appl. 125, Springer, New York, 1999, pp. 229–250.

[7] C. Castillo-Chavez, B. Song, Dynamical models of tuberculosis and their applications, Math. Biosci. Eng. 1 (2) (2004) 361–404.

[8] L.I.W. Roeger, Z. Feng, C. Castillo-Chavez, Modeling TB and HIV co-infections, Math. Biosci. Eng. 6 (4) (2009) 815–837.

[9] S. Gakkhar, N. Chavda, A dynamical model for HIV-TB co-infection, Appl. Math. Comput. 218 (18) (2012) 9261–9270.

[10] TBFACTS.ORG, https://www.tbfacts.org/tb-hiv/.

[11] TBFACTS.ORG, https://www.tbfacts.org/tb-treatment/.

[12] TBFACTS.ORG, https://www.tbfacts.org/tb-statistics/.

[13] World Health Organisation, http://www.who.int/gho/mortality burden disease/life tables/situation trends/en/.

[14] World Health Organisation, http://www.who.int/mediacentre/factsheets/fs104/en/.

[15] World Health Organisation, http://www.who.int/mediacentre/factsheets/fs360/en/.

[16] F. Khamis, J.A. Noamani, H.A. Naamani, I.A. Zakwani, Epidemiological and clinical characteristics of HIV infected patients at a tertiary care hospital in Oman, Oman Med. J. 33 (4) (2018) 291–298.

[17] R. Singh, S. Ali, M. Jain, Rakhee, Epidemic model of HIV/AIDS transmission dynamics with different latent stages based on treatment, Am. J. Appl. Math. 4 (5) (2016) 222–234.

[18] R. Naresh, A. Tripathi, Modelling and analysis of HIV-TB co-infection in a variable size population, Math. Mod. Anal. 10 (3) (2005) 275–286.

Further reading

C. Castillo-Chavez, Z. Feng, To treat or not to treat: the case of tuberculosis, J. Math. Biol. 35 (6) (1997) 629–656.

Index

Note: Page numbers followed by *f* indicate figures, *t* indicate tables, and *b* indicate boxes.

A

Acceptance testing, 87
Active model inference, 68–69
AFSO (artificial fish swarm algorithm), 425–426
Agile procedures, 355–356
Anomaly detection
 angular subspace anomaly detection, 353
 challenges, 352–353
 data mining, 351
 dimensionality, 352–353
 outlier detection, 353
 statistical anomaly detection, 351–352
 techniques
  machine learning, 352
  statistical method, 352
 time series, 351
 volume and velocity, 352–353
ANT colony-based IoT systems, 420
Ant colony optimization (ACO), 420–421, 428, 520
Application layer business logic, 236
Artificial bee colony (ABC)-based IoT systems, 423–424
Artificial neural networks (ANN), 498
Aspect-oriented modeling (AOM)
 advice weaving, 450
  atomic join points weavers, 453–455, 454*f*
  EL-JP base model side weaver, 455*f*, 456
  LE-JP weaver base model side, 456–457, 456*f*
  nonatomic JP weaver, 455
  optimized EL-JP weaver, 457, 459*f*
  optimized LE-JP weaver, 457, 458*f*
  reduced nonatomic JP weavers, 456–457, 457*f*
 aspect weaving, 450
 atomic join points, 451–452, 453*f*
 base model, 450
 correctness
  aspect interference, 457
  correctness-by-construction modeling, 457
  at join points, 463–464
  noninterference of advices, 464–465
  verification and conditions, 457
  of weavers, 458–463, 460–462*f*
  weaving correctness verification effort, 472, 473*f*
 functionality, 450
 home rehabilitation system (*see* Home rehabilitation system (HRS))
 ideas, 448
 join points (JP), 450–452, 453*f*
 limitation, 474
 location and edge refinement, 451–452
 nonatomic join points, 452, 453*f*
 pointcut specifications, 450
 related work
  correctness criteria, 450
  UML-based approach, 448–449
  UPPAAL model, 449
  UTA models, 448–449
 UPPAAL timed automata, 450–451
 UTA modeling technique, 450
 verification, 471–472, 473*f*
 weaver, 451–453
Astrom-Hagglund technique (HA), 107
Automaton. *See also* UPPAAL automaton
 clustering, 76, 77*f*
 construction, 74–75, 74*b*

B

Bacterial foraging optimization (BFO)-based IoT systems, 424
BAT optimization (BO)-based IoT systems, 424–425
BCO (Bee colony optimization), 428
BFA (bacterial foraging algorithm), 425–426
Bias, 96
Big data, 503–504
Big data analytics (BDA)
 active data sources, 342
 big data security analytics model
  big data engines, 343
  data source, 344
  data warehousing, 344
  ETL (Extract, Transform, and Load) tool, 344
  monitoring services, 343
  security controls, 344

Big data analytics (BDA) *(Continued)*
big data security areas, 342–343, 343*f*
extract transform load, 342
passive data sources, 342
security analytics solution
business strategy, 346
data management infrastructure and platform, 346
network and suspicious layer, 347
security analytics programs, 346
streamlining, 346
threat detection
advanced persistent threat (APT), 344
credentials of users, 344–345
examples, 344–345
goal, 344
intellectual property, 345
mitigation, 345–346
personal information, 345
ransomware, 345
Biometric identification systems (BIS)
behavior qualities, 126
deep ANNs (artificial neural networks), 126
IoT
access and application, 128
analytics and data management, 128
attributes of environment, 126
connectivity, 127
devices and sensors, 126–127
structure, 126, 127*f*
Birth processes
branching processes, 356
definition, 357
failure detection
calibration method, 366–367
datasets, 365
linear contagion models, 365–366
MTBF calculation, 367–369, 368*f*
Musa-Okumoto and Goel-Okumoto models, 365–366
nonlinear least-squares fitting, 366–367
software reliability models, 365–366, 366*t*
probability mass function (PMF), 358–359
process classification, 357–358, 358*f*
Bitcoin
clients, 229–230, 244–245, 246*f*
components, 240–242
consensus
bitcoin network, 250–251, 251*f*
broadcasting transactions, 248
network discovery, 250–251, 251*f*
peer-to-peer (P2P) network, 250
propagating transactions, 248–250
vs. Ethereum, 229–230
miners, 251–252, 252*f*
nodes
decentralization, 247
full node, 247, 247*f*
network entity, 245–247
thin/light client, 248, 249*f*
types, 247
transaction
creation, 253
lifecycle, 252–253
structure, 253, 253–254*f*
wallets, 229–230
address formats, 243
advantage, 242
bitcoin address generation, 243, 243*f*
bitcoin address mapping, 242–243, 242*f*
desktop wallets, 245
deterministic wallets, 244
hardware wallets, 245
mobile wallets, 245
nondeterministic wallet, 244
types, 246*f*
working, 240, 241*f*
Bivariate software reliability growth model (SRGM)
bivariate assessment measures, 161
bivariate Weibull-type SRGMs
budget constraint, 167–168
detectable faults, 167
development scale, 169
estimated parameters, 169–170, 169*t*
goodness-of-fit comparison, 169–170, 170*t*
optimization problem, 167
parameter estimation, 165–167
sensitivity analysis, 170–173, 170–172*t*
testing-time functions, 164–165
Cobb-Douglas-type testing-time function, 155–156
constant elasticity of substitution (CES)-type production function, 156
data collection, 161–162
elasticity of substitution, 155–157, 156*f*, 162, 164*t*
extension method, 155
goodness of fit, 157, 162, 163*t*
mean squared errors (MSE), 161–162
nonhomogeneous Poisson process (NHPP) model
CD-DSS SRGM, 160
CD-EXP SRGM, 159

CD-ISS SRGM, 160
CES-DSS SRGM, 160
CES-EXP SRGM, 160
CES-ISS SRGM, 160
CES-type testing-time function, 159
Cobb-Douglas-type testing-time function, 159
detectable faults, 158
formulation, 158–159
probability mass function, 157–158
testing-time and testing-effort factors, 158–159
parameter estimation, 160–162, 162*t*
testing-time functions, 155, 157
Black-box testing, 67–68, 443
Blob antipattern, 281
Blockchain. *See also* Bitcoin; Ethereum
adequacy
architecting guidance, 240
checks, 236–238
data integrity, 238
data transparency, 238–239
drivers, 237*f*
reliability and availability, 239–240
scalability, 238
application, 231
decentralized network system, 231
definition, 230
emotional intelligence, 230
generalized architecture, 232, 233*f*
infrastructure layer
network, 232
nodes, 232
permissioned blockchain, 234
permissionless blockchain, 234
storage, 232
MIX Networks, 231
platform layer
application layer business logic, 236
consensus, 235
distributed computing layer, 234–235
JSON-RPC, 234
remote procedure calls (RPCs), 234
replication, 235
security, 235–236
transactions, 235
timestamping, 231–232
working, 231, 231*f*
Botnets, 335
Breaches, data, 143
Building automation system (BAS)
architecture, 210
availability model
joint maintenance strategy, 217–219, 218*f*
reliability and cybersecurity, 214–217, 215–216*f*
separate maintenance strategy, 219, 220*f*
contributions, 211
dependability, 210
evolution, 209–210
generalized model, 214, 214*f*
goal, 211
indicators, 210
input parameters, 219–220, 221–222*t*, 223*f*
multigoal maintenance
conditions, 211–212
ICS availability, 213, 213*f*
principles, 212
partial Amin index, 220–223
performance indicators, 220–222, 224*f*
results of model, 219–220, 223*f*, 225*f*
work-related analysis, 210–211

C

C-H-R set point regulation technique (CHR), 107
Classification technique, 493–494
Client-server architecture
advantages, 273
components, 272–273
illustration, 272, 272*f*
working, 272
Cloud computing
ABE, 542–543, 543*f*
allowed systems, 541
eligible, 540
IaaS, 540
multiaccess component, 541–542, 541*f*
PaaS, 540
privacy and security, 539–540, 539*f*
privacy security, 542
service availability, 541–542, 541*f*
software-as-a-service, 540
virtualization, 540
Cluster-based adept cooperative algorithm (CACA), 549–551
Clustering, 72, 73*b*, 76, 77*f*, 493, 504
Cobb-Douglas-type testing-time function, 155–156
Cohen-Coon technique (CC), 107
Colored petri nets (CPNs), 12
Common vulnerability scoring system (CVSS), 144–145

Complex fault modeling, 93–94
Condition-based maintenance (CBM), 391, 395–399, 398*t*
Condition monitoring, 287. *See also* Photovoltaic (PV) system; VRLA battery
Consensus mechanism, 235
Constant elasticity of substitution (CES)-type production function, 156
Constrained active machine learning (CAML), 69
Consumer internet of things (CIoT), 406–407
Contagion model, 355–356, 365–369
Continuous-time Markov chain (CTMC), 13
Core and structural analytics, 336
Correspondence analysis
  correlation coefficient, 2
  data visualization, 1–2
  estimated results
    assignee, 3, 4*f*
    component, 3, 4*f*
    edge-oriented OSS developers, 5–6
    factors, 3–5
    fault reporter, 4, 6*f*
    fault severity, 4, 6*f*
    fault status, 4, 7*f*
    hardware, 3, 5*f*
    OS, 3, 5*f*
    software version, 5, 7*f*
  estimation procedure, 3
  OSS fault large-scale data, 2
  variables, 2
Cyber-attacks, 439–440, 439*f*
Cyber physical systems (CPS), 412

D

Data mining (DM)
  algorithms, 478, 478*f*
  applications
    education, 499–500
    finance, 499
    insurance and health care, 499
    medical/pharmacy, 498–499
    retail industry, 500
    telecommunications, 499
    text mining and web mining, 498
  disciplines, 477
  information extraction, 477
  KDD process, 477–478
  knowledge discovery and, 477
  methodology
    data extraction strategies, 481, 485*t*
    description, 478, 479*f*
    digital resources, 480, 481*t*
    filtration process output, 481
    quality assessment rules, 480–481
    research questions identification, 478–480
    research strategy, 480
    study selection criteria, 480
    synthesis of extracted data, 481, 485–486*f*
  need of, 485–486
  pattern discovery, 478
  techniques
    algorithm architecture, 497
    artificial neural networks (ANN), 498
    categories, 496–498
    classification technique, 493–494
    clustering analysis, 493
    decision trees, 494–496, 494*f*
    dynamic prediction-based approach, 497–498
    information systems, 496
    intelligence agent systems (IAs), 497
    knowledge-based system, 497
    modeling, 497–498
    system architecture analysis, 497
    system optimization, 496
    types, 492, 493*f*
  types
    constraint-based data mining, 487–488
    distributed/collective data mining, 487
    educational data mining, 490, 491*f*
    hypertext data mining, 488
    multimedia data mining, 486–487, 487*f*
    phenomenal data mining, 489–490, 490*f*
    social security data mining, 490, 491*f*
    spatial and geographic data mining, 489, 489*f*
    time series/sequence data mining, 492, 492*f*
    ubiquitous data mining, 486
Dead time, 103–104
Decision trees
  association rules, 494
  data distillation for judgment, 496
  dynamic prediction-based approach, 495
  methods, 494
  outlier detection, 495
  prediction, 494–495
  process mining, 496
  relational mining, 495
  social network analysis (SNA), 496
  structure, 494, 494*f*
  text mining, 495
Denial-of-service (DoS) attack, 334
Dependability model

assessment, 210
concept, 210
factors, 210
indicators, 219–220, 223
information and control systems (ICS), 210
multigoal maintenance, 211–212
partial dependability index, 222–223
patching and proof testing, 214*f*
three-tier architecture, 215
Dependency inversion principle (DIP), 277
Depth first search (DeFS), 520
Design antipatterns
Blob antipattern, 281
functional decomposition antipattern, 282
patterns into antipatterns, 282–283
Poltergeist antipattern/Gypsy Wagons, 282
software quality, 281
Spaghetti Code antipattern, 281–282
Swiss Army Knife antipattern, 282
Design patterns
behavioral design patterns, 279
benefits, 279–281
classification, 278–279
creational design patterns, 278
essential elements, 277–278, 277*f*
gang of four (GoF) design patterns, 279, 280*t*
Mediator design pattern, 277–278
object-oriented software design, 277
structural design patterns, 278
Differential evolution (DE), 519
Digital systems modeling. *See also* Fault models
behavioral model, 308
categories, 308
fault avoidance, 317, 318*t*
fault coverage (FC) values, 317, 318*t*
fault detection efficiency (FDE), 317–318
fault diagnosis test procedures
data analysis techniques, 316–317
exhaustive testing pattern application time, 316, 316*t*
minimal test sequences, 316, 317*f*
test pattern-finding strategies, 316
typical test architecture, 315, 315*f*
fault tolerance, 318–319, 319*f*
functional model
binary decision diagrams, 309–310, 310*f*
primitive cube model, 309, 309*t*
register transfer levels (RTLs), 310
state table, 309–310
truth table, 309, 309*t*
Gajski-Kuhn Y-chart, 311, 311*f*
interconnectedness, 308, 320
level of modeling, 310
online test techniques, 319
structural models, 310
Dimensionality, 352–353
Distributed computing layer, 234–235
Distributed denial of services (DDoS) attack, 123, 131, 131*f*
Distributed-parallel particle swarm optimization with K-means (D-PPSOK) clustering algorithm
algorithm, 506–507
clustering analysis, 504
evaluation, 507–508
experimental analysis
elapsed time variation, 509, 509*t*, 510*f*
execution time, 508, 508*t*, 509*f*
goal, 508
parallel proposing elapsed time, 508, 508*t*
sampling, 510, 510*f*, 511*t*
test data, 508
implementation, 506
K-means algorithm, 504
modules, 505
overview, 505–506, 506*f*
particle swarm optimization (PSO), 504
related work, 504–505
Dynamic prediction-based approach, 497–498
Dynamic programming
characteristics, 202
classification, 203, 203*f*
computational procedure, 203
computation in forward and backward directions, 203–204
concepts, 201–202
disadvantage, 200
factor efficiency, 200
features, 202–203
history, 200–201
integrated efficiency model
load related reliability design, 206–207, 206*t*
number of factors and stage efficiency, 205–206, 205–206*t*
size related reliability design, 207, 207*t*
worth related reliability design, 206, 206*t*
multistage decision questions, 201
optimal decision policy, 204, 204*f*
phase, 201
principle of optimality, 201
representation, 201, 201*f*
state, 201
Dynamic Q-learning, 551–553, 552–553*f*

E

Economies of scale, 164
Elasticity of substitution, 155–157, 156*f*
Elliptical Curve Cryptography (ECC), 129
Emotional intelligence, 230
Ethereum
  vs. Bitcoin, 229–230, 254–255
  clients, 229–230
    Geth (Go-Ethereum), 258
    nodes, 261
    parity, 261
    software requirements, 261
    transactions, 258
  components, 257–258
  Ethereum accounts, 258
  Ethereum virtual machine (EVM)
    architecture, 256, 256*f*
    Ether and gas, 257, 257*t*
    smart contract, 257, 257*t*
    working, 256, 256*f*
  evolution, 255–256, 255*t*
  mining process, 258, 259–260*f*
  networks
    EthereumYellow Paper, 264–265
    hardware requirements for full node, 266
    implementation, 264–265, 264*f*
    local blockchain simulation, 265–266
    Mainnet, 264–265
    public Testnet, 265
    running an Ethereum client, 266
    solid state drive (SSD), 266
    Testnet, 264–265
    types, 264–265
  transactions
    binary message, 262
    externally owned account (EAO), 262
    gas, 263
    nonces, 262
    recipient, 263
    special transaction, 264
    structure, 262–264
    value and data, 263
  wallets, 229–230

F

Failure mode and effects analysis (FMEA), 51–52, 55, 65. *See also* Marine accidents
Fault detection efficiency (FDE), 317–318
Fault large-scale data. *See* Correspondence analysis
Fault models
  assumptions, 311
  bridging fault model, 313, 313*f*
  cell internal fault model, 314–315
  defect identification, 312
  multiple stuck-at faults model, 313
  open fault model, 313–314
  path delay fault model, 314, 314*f*
  physical faults, 311–312
  stuck-at fault model, 312, 312*f*
  transition fault model, 314
Fault tolerance, 318–319, 319*f*
Fog computing, 124–125, 130, 130*f*
Functional decomposition antipattern, 282
Functional model
  binary decision diagrams, 309–310, 310*f*
  primitive cube model, 309, 309*t*
  register transfer levels (RTLs), 310
  state table, 309–310
  truth table, 309, 309*t*
Fuzzy CNN, 553–554, 555–556*f*

G

Gajski-Kuhn Y-chart, 311, 311*f*
Gang of four (GoF) design patterns, 279, 280*t*
Genetic algorithm (GA), 518–519
Goodness of fit, 95–96, 99–100*f*
Grasshopper Optimization Algorithm, 323. *See also* Improved Grasshopper Optimization Algorithm (IGSD)
GWO (grey wolf optimization), 425–426

H

Handover decision
  dynamic Q-learning, 551–553, 552–553*f*
  multicriteria decision-making algorithms, 549–551
Hard fault modeling, 93
HIV-AIDS, 563. *See also* Illegal immigrants
Home rehabilitation system (HRS)
  aspect exercising quality models
    biometric characteristics, 467
    multiple exercise weaving, 467, 469*f*
    physical condition weaving, 467, 468*f*
    sampler weaving/exercising quality monitor, 467, 470*f*
  base model, 465–467, 466*f*
  weaving correctness, 467, 471

I

Illegal immigrants
  HIV-TB coinfection
    coinfection equilibrium, 565

human mortality, 563
illegal immigration rate, 567, 568*f*
mathematical analysis, 565–566
mathematical model, 564–565
numerical analysis, 566–567, 568*f*
Routh-Hurwitz criterion, 566
transmission dynamics, 564–565
transgression, 563–564
Image processing
Advanced Encryption Standard (AES) algorithm, 133–134
agricultural IoT, 134, 135*f*
cancer prediction system, 133–134
canny edge detection, 133
convolutional neural network (CNN), 132–133
encryption and decryption time, 123
F-RCNN model face mask detection, 132, 134*f*
infant cry system, 132, 133*f*
Multi-Context Fusion Network (MCFN), 134, 135*f*
YOLO object detection, 132–133
Improved Grasshopper Optimization Algorithm (IGSD)
crossover probability (CR), 325–326
gravity force, 324
hybridization, 323
limits, 327*t*, 329*f*
location of grasshopper, 324
loss assessment, 327–328*t*, 329*f*
mutation operation, 325
position of Grasshopper, 325
problem formulation, 323–324
real power loss, 329*t*
scaling factor (SF), 325–326
selection, 325
self-adaptive differential algorithm, 323
simulation study, 326
social communication, 324
wind advection, 324
Industrial IoT (IIoT), 407, 408*f*
Infectious diseases, 563
Information security governance (ISG)
ABCD algorithm, 534
authentication strategy, 531
cloud computing
ABE, 542–543, 543*f*
framework, 540–542, 541*f*
privacy and security, 539–540, 539*f*
network architecture and threat model, 536
objective definition principle, 536
open issues, 533–534, 534*f*
organizational approaches, 535, 535*f*
policy-based SDN security architecture, 536–537, 538*f*
privacy, 538–539
roadside units (RSUs), 531
SDN controller, 535, 535*f*
security illustration, 533
security policy management and updating, 532–533
social network analysis (SNA), 533–534
standards of data privacy, 536
transparency principle, 536
trust, 537–538
Inliers prone distributions
data analysis
critical values, 382, 382*t*
data description, 381–382
parameters estimates, 383, 383*t*
parametric functions, 383, 383*t*
density function, 374, 375*f*
early failure, 371–372
future research, 385–386
hypothesis, 379–381
inliers-prone models
early failure model-1, 373
early failure model-2, 373–374
instantaneous failure models, 372–373
model with inliers at zero and one, 374
instantaneous failures, 371–372
issues, 384–385
nonstandard mixture of distributions, 371–372
parameter estimation
maximum likelihood estimation, 375–376
unbiased estimation, 376–379
practical situations, 371–372
problems, 384
Integer programming (IP), 516–517
Integrated circuits (ICs), 307
Integrated efficiency model
cost restriction, 189
load related reliability design, 206–207, 206*t*
number of factors and stage efficiency, 205–206, 205–206*t*
size related reliability design, 207, 207*t*
unuseful systems, 189–190
worth related reliability design, 206, 206*t*
Integrated reliability model (IRM) system. *See* Parallel-series integrated reliability model (IRM) system
Integration testing, 87

Integration testing robotic systems, 67
Intelligence agent systems (IAs), 497
Interface segregation principle (ISP), 276
Internal model control (IMC), 107
Internet of everything (IoE), 412, 418–419
Internet of robotic things (IoRT), 413
Internet of services (IoS), 412
Internet of things (IoT). *See also* Image processing; Swarm intelligence (SI)
  advantages, 347
  AFSO (artificial fish swarm algorithm), 425–426
  ANT colony-based IoT systems, 420
  ANT colony optimization (ACO)-based IoT systems, 420–421
  applications, 403–404, 437
  artificial bee colony (ABC)-based IoT systems, 423–424
  assets, 404–405, 405*f*
  bacterial foraging optimization (BFO)-based IoT systems, 424
  BAT optimization (BO)-based IoT systems, 424–425
  BFA (bacterial foraging algorithm), 425–426
  biometric identification systems (BIS)
    access and application, 128
    analytics and data management, 128
    attributes of environment, 126
    behavior qualities, 126
    connectivity, 127
    deep ANNs (artificial neural networks), 126
    devices and sensors, 126–127
    structure, 126, 127*f*
  challenges, 443–444
  characteristics, 405–406, 405*f*
  connectivity, 437–438
  consumer internet of things (CIoT), 406–407
  crypto tools, 128–129
  cyber physical systems (CPS), 412
  data distributed storage process (D2SP), 129
  data integration, 349–350
  DDoS (distributed denial of services) attack, 123, 131, 131*f*
  definition
    Bosch, 407
    Cisco, 409
    Financial Times, 409
    Forbes Simply, 410
    Gartner, 407
    Google, 409
    Guardian, 409
    IBM, 407
    IERC, 409
    IoT-A with RFID technology, 411
    ITU IoT GSI, 410
    Margaret Rouse, 410
    McKinsey and Company, 411
    PC Magazine Encyclopedia, 410
    SAP, 410
    SAS, 410
    Tech World, 410
    Webopedia, 411
    Wikipedia, 411
  deterministic process (DP), 129
  devices, 437
  GWO (grey wolf optimization), 425–426
  heterogeneity, 404–405
  industrial internet, 411
  industrial IoT (IIoT), 407, 408*f*
  industry 4.0, 412
  interactive model
    application layer, 348
    components, 347–349
    network layer, 348
    perception layer, 347–348
    representation, 347–349, 348*f*
  internet of everything (IoE), 412, 418–419
  internet of healthcare things (IoHT), 413
  internet of medical things (IoMT), 413
  internet of robotic things (IoRT), 413
  internet of services (IoS), 412
  internet of things, services, and people (IoTSP), 413
  internet of things and services (IoTS), 413
  internet-related networks, 404
  IoTDS (Internet of things, data, and services), 413
  limitations, 404–405
  man in the middle (MitM) attack, 123, 131–132
  medical applications (healthcare), 129–130, 130*f*
  network and security, 128–129
  network connectivity, 406, 406*f*
  network privacy policies, 123
  ontology-based security modeling, 349–350
  particle swarm optimization (PSO)-based IoT systems, 421–423
  PIO (pigeon inspired optimization), 425–426
  security
    authentication, 443
    black-box testing, 443

commercialized secure hardware primitive designs, 439–440, 439–440*f*
cyber-attacks and mitigation techniques, 439–440, 439*f*
hardware attacks, 439
at hardware/silicon level, 439
integrated circuit (IC) design, 438
levels, 438, 438*f*
machine learning (ML) approaches, 444
malwares, 439–440
physical probing, 443
and privacy, 437
PUF, 443
random number generators, 443
reverse engineering, 443
secure hardware, 440, 440*f*
side-channel attack (SCA), 441–442, 442*f*
software attacks, 439
supply chain process, 438–439
trojans, hardware, 441, 441–442*f*
smart home
components, 350
home appliances control and management, 351
home conditions measurement, 350
services, 350–351
vehicular sensor networks
controller area networks (CAN), 124
data aggregation, 125
fog computing, 124–125
MSIP-IoT-A, 126
remote cloud center, 124–125
self-driving vehicle (SDV), 124
wired connection, 124
Internet of vehicles (IoV)
analytical hierarchy process (AHP), 549–551
cluster-based adept cooperative algorithm (CACA), 549–551
communication, 547–549
comparative analysis
delay, 558, 560*f*
handover failure, 558, 560*f*
HO success probability, 558, 559*f*
packet loss, 558, 560*f*
throughput, 558, 559*f*
unnecessary handovers, 558, 559*f*
Dijkstra algorithm, 549–551
handoff protocol, 549–551
handovers
dynamic Q-learning, 551–553, 552–553*f*
K-nearest neighbor (KNN), 549–551
multicriteria decision-making algorithms, 549–551
problem, 547–549
k-partite graph, 549–551
line-of-sight (LOS) problem, 547–551
network selection
fuzzy CNN, 553–554, 555–556*f*
problem, 547–549
TOPSIS algorithm, 549–551
quality of services (QoS), 547–551
radio access technologies (RAT), 547, 548*t*
simulation setup, 557, 558*t*
system model, 551, 552*f*
target discovery problem, 549–551
V2V chain routing technique, 554, 556–557
IP telephony. *See* Voice over Internet Protocol (VoIP)

J

JSON-RPC, 234

K

K-means algorithm, 504
KMeans clustering, 73
Knowledge-based system, 497

L

Latency, 34–35
Layered architecture
advantages, 274
components, 274
illustration, 273, 273*f*
Learning-based testing (LBT), 69
Lifecycle models of vulnerability, 337–338, 337*f*
Liskov substitution principle (LSP), 276

M

Maintenance strategies
advantages and disadvantages, 392*t*
application, 389–390
classification, 389–390
condition-based maintenance (CBM), 391
integrated approach, 391–392, 393*f*
reliability centered maintenance (RCM), 390–391
total productive maintenance (TPM), 390
Malware, 335, 439–440
Man in the middle (MitM) attacks, 123, 129, 131–132

Marine accidents
  failure probability, 52
  reason, 51–52, 52*t*
  risk assessment methodology
    aggregated experts' judgment, 59, 62*t*
    causes and effects of explosion, 59, 60–61*t*
    crisp values, 59, 63*t*
    data collection, 56
    detection, 58*t*, 59*f*
    failure modes identification, 56
    input parameter, 59, 61–62*t*
    membership functions, 56
    model flowchart, 59, 60*f*
    occurrence, 57*t*, 58*f*
    results, 62–63
    RPN of failure modes, 59, 63*t*
    sensitivity analysis, 63, 64*t*
    severity, 57*t*, 58*f*
    steps, 56–57, 59
  ship maintenance, 51–52
  starting air system (*see* Starting air system)
Marine propulsion system. *See also* Maintenance strategies
  condition-based maintenance (CBM), 395–399, 398*t*
  employees empowerment
    data collection, 400
    opinion of employees, 401
    overall equipment effectiveness, 400
    time loss reduction, 400–401
    training program, 400
  failure analysis worksheet, 395, 397*t*
  failure mode effect analysis, 395, 396*t*
  reliability centered maintenance (RCM) program, 395
  system selection and boundaries
    failure probabilities worksheet, 393–394, 394*t*
    schematic representation, 392–393, 393*f*
  task comparison, review and control
    cost-benefit analysis, 399, 399*t*
    overall equipment effectiveness, 399, 400*t*
    review, 400
  task selection, 395–399, 398*t*
Maximum likelihood estimation, 375–376
Mean-square fitting error MSE), 95–96
Mean time between failures (MTBF)
  failure detection
    calibration methods, 366–367
    intensity failure rate, 369
    interpolated MTBF curves, 367–369, 368*f*
    MTBF calculation, 367
    NHPP-based models, 367
    software reliability models, 365–366, 366*t*
  hardware reliability, 355–356
  and mean time to failure (MTTF)
    conditional MTBF, 365
    definition, 364
    theorem, 364–365
  reliability analysis, 355–356
  time domain analysis
    conditional/distribution function, 360–361
    definition, 360
    linear contagion process, 363–364
    nonhomogeneous Poisson process (NHPP), 364
    probability of *k*th failure, 361–363
Mean value function (MVF)
  definition, 359
  theorem, 359–360
Mediator design pattern, 277–278
Medical applications (healthcare), 129–130, 130*f*
MIMO PID controllers
  controller tuning
    Astrom-Hagglund technique (HA), 107
    C-H-R set point regulation technique (CHR), 107
    Cohen-Coon technique (CC), 107
    decoupling design, 106
    internal model control (IMC), 107
    MIMO control, 106
    RGA matrix, 105
    SISO system, 105
    transfer function, 105
    tuning techniques, 105, 106*t*
    Tyreus-Luyben technique (TL), 107
    Zeigler-Nichols technique (ZN), 107
  dead time, 103–104
  result analysis
    closed-loop step response, 108, 112–113*f*
    open-loop bode plots, 108, 111*f*
    open-loop step response, 108, 110*f*
    open-loop time and frequency response, 108, 109*t*
    performance indices, 108, 110, 115*t*, 116*f*
    robustness, 110
    stability margin, 110, 117–118*f*, 119*t*
    time response (step response) values, 108, 114*t*
    tuning technique comparison, 115–119, 119*t*, 119–120*f*
    variation of time response values, 108, 114–115*f*

Mobile robot testing
active model inference, 68–69
black-box testing, 68
learning-based testing (LBT), 69
model-based testing, 69–70, 70*f*
model inference, 68–69
passive model inference, 69
TestIt toolkit, 70, 71*f*
tool architecture
automaton construction, 74–75, 74*b*
clustering state space, 72, 73*b*
components, 70–71, 71*f*
KMeans clustering, 73
model checking, 75
model refinement, 75
postclustering analysis, 73
state-space exploration, 71–72, 72*b*
validation
automata clustering, 76, 77*f*
code coverage, 79–80, 80*f*, 81–82*t*
exploration, 76, 76–77*f*
intruder, 75
results, 79–80, 80*f*, 81–82*t*
UPPAAL automaton, 76, 78–79*f*, 79
Model inference, 68–69
Model-view-controller (MVC) architecture
advantages, 275–276
components, 275
illustration, 274–275, 275*f*
working, 274–275
Modularization techniques, 447. *See also* Aspect-oriented modeling (AOM)
Multigoal maintenance
conditions, 211–212
ICS availability, 213, 213*f*
principles, 212

N

Network and security, 128–129
Non-homogenous Poison process (NHPP), 87–90, 94–95, 157–158, 355–356

O

Ontological security, 349–350
Open-closed principle (OCP), 276
Open source software (OSS), 1. *See also* Correspondence analysis
ambiguity, 44
announcement planning, 43
entropy, 43–44
faults
announcement difficulty, 42
announcement planning, 43
arithmetical models, 43
defect identification and modification technique, 42
entropy measures, 43
fault estimation, 42
multiattribute efficiency investigation, 42
occurrence, 41–42
reserves distribution, 41
genetic and greedy algorithm, 43
Kapur entropy
envisaged bugs and predicted release time, 46, 47*t*
equation, 44
at time period, 46*t*
linear regression, 43–44, 46
multiple linear regression, 45–46
release planning (RP), 43
Shannon entropy
envisaged bugs and predicted release time, 46, 47*t*
equation, 44
at time period, 45*t*
simple linear regression (SLR) model, 45
software development, 41
OpenStack project, 3
Optimal PMU placement (OPP) problem
algorithm, 521, 523*t*
direct estimations, 514–515, 515*f*
methods, 513–514
pseudo-estimations, 515–516, 516*f*
wide area monitoring system (WAMS) layout, 514, 514*f*

P

Paper machine
PID controllers (*see* MIMO PID controllers)
schematics, 104, 104*f*
variables, 104
Parallel-series integrated reliability model (IRM) system. *See also* Dynamic programming
constant information, 198, 198*t*
efficiency design, 199–200, 199*t*
factor efficiency, 197
integrated efficiency model, 189–190
Lagrangean method
equality constraints, 190
exact and repetitive methods, 191
Lagrangean multiplier method, 191
model analysis, 192–197

Parallel-series integrated reliability model (IRM) system *(Continued)*
problem function, 192
unequal factors, 191
load constraint, 198, 198*t*
size constraint, 198–199, 198*t*
worth constraint, 198, 198*t*
Particle swarm optimization (PSO), 421–423, 504, 519–520
Passive model inference, 68–69
Patching
data breaches, 143
definition, 144
literature review, 146
mathematical model, 147–148
percentage of vulnerabilities dealt, 148–149, 150*t*
vulnerability correction process, 147–148
Permissionless blockchain, 234
Petri net (PN), 18
Phasor measurement unit (PMU)
algorithm comparison
advantages and disadvantages, 521, 526–527*t*
algorithm assessment, 521, 522*t*
contingency conditions, 521, 524–525*t*
minimum PMUs number for test, 520, 521*t*
OPP problem formulation, 521, 523*t*
complicated network, 528
constrain consideration, 528
depth first search (DeFS), 520
future research, 521–528
graph-based algorithm, 528
heuristic methods, 520
hybrid optimization techniques, 522
integer programming (IP), 516–517
mathematical programming method, 516–517
meta-heuristic methods
ant colony optimization (ACO), 520
differential evolution (DE), 519
genetic algorithm (GA), 518–519
particle swarm optimization (PSO), 519–520
simulating annealing (SA), 518
Tabu search (TS), 519
multiobjective optimization technique, 524
optimal PMU placement (OPP) problem
direct estimations, 514–515, 515*f*
methods, 513–514
pseudo-estimations, 515–516, 516*f*
wide area monitoring system (WAMS) layout, 514, 514*f*
trouble, 513–514
Phishing, 334
Photovoltaic (PV) system
aluminum-electrolytic capacitor and MOSFET
equivalent series resistance (ESR), 302
failure signature parametric values, 302
front panel VI, 303*f*
in-circuit condition monitoring, 302, 303*f*
in-circuit wear out condition, 301–302
on-state drain-source resistance, 302
thermal cycle, 302
infrared thermography (IRT), 287–288
need for, 287
VRLA battery (*see* VRLA battery)
Physical probing, 443
PID controllers. *See* MIMO PID controllers
PIO (pigeon inspired optimization), 425–426
Poltergeist antipattern/Gypsy Wagons, 282
Prediction error (PE), 96
Private permissioned blockchain, 234
Probability-based analytics, 336
Public permissioned blockchain, 234

R

Radio access technologies (RAT), 547, 548*t*
Random number generators, 443
Regression testing, 87
Reliability centered maintenance (RCM), 390–391, 395
Reliability engineering, 189
Remote procedure calls (RPCs), 234
Replication, 235
Reverse engineering, 443
Risk priority number (RPN), 55, 59, 63*t*
Robot operating system (ROS)-based robotic systems, 67–68. *See also* Mobile robot testing
Root mean square prediction error (RMSPE), 96
Routing technique, 554, 556–557

S

Satoshi client, 229–230
Security analytics
anomaly detection (*see* Anomaly detection)
architecture
attack graph model, 335–336, 339, 340*f*
common vulnerability scoring system (CVSS), 338–339
stochastic graph model, 339
big data (*see* Big data analytics (BDA))
classes

core and structural analytics, 336
lifecycle models of vulnerability, 337–338, 337*f*
probability-based analytics, 336
time-based analytics, 336–337
cyberattacks
botnets, 335
damage, 335
denial-of-service (DoS) attack, 334
malware attacks, 335
phishing, 334
spamming, 334
website threats, 335
Internet of things (IoT) (*see* Internet of things (IoT))
objectives, 335
representation of models
Markov model, 340, 341*f*
node rank analysis, 341
nonhomogeneous model, 341
situational awareness, 333, 334*f*
Self-driving vehicle (SDV), 124
Semi-Markov model
mean Sojourn times, 180
nonregenerative states, 177
notations, 176–177
performance measures
busy period analysis for server, 181
expected number of repairs to switch, 182
expected number of repairs to unit, 183
expected number of server visits, 183
expected number of treatments to server, 182
mean time-to-system failure, 180
profit, 184
reliability, 180
steady-state availability, 180–181
regenerative states, 177
server, 175
simulation study
availability behavior, 185, 185*f*
MTSF behavior, 184–185, 185*f*
profit analysis, 186, 186*f*
random variables, 184
standby system models, 176
states of system model, 177, 178*f*
switch, 175
transition probability, 177–179
Session Initiation Protocol (SIP)
colored petri nets (CPNs), 12
continuous-time Markov chain (CTMC), 13
contribution, 14
H.323 protocol, 11–12
INVITE and non-INVITE transaction, 11–12
layered structure, 15–16, 16*f*
model validation, 37–38, 38*f*
performance measures
badput, 33–34
goodput, 32–33
largeness problem, 31
latency, 34–35
SRN submodels, 31, 32*f*
state aggregation, 31
throughput, 32–34
responses, 14–15, 15*t*
results and discussion
goodput and badput *vs.* arrival rate, 35–36, 36*f*
goodput *vs.* arrival rate, 35, 36*f*
latency *vs.* arrival rate, 37, 37*f*
stochastic reward net (SRN) model
attributes, 18–20, 18*f*
components, 18*f*
continuous-time Markov chain (CTMC), 13
SIP INVITE transaction, 20–21, 21*f*, 22–25*t*, 25–31
timed colored petri nets (TCPNs), 12
transaction mechanism, 11–12, 16–18
working, 16–18, 17*f*
Ship maintenance, 51–52. *See also* Marine accidents
Side-channel attack (SCA), 441–442, 442*f*
Simple fault modeling, 92–93
Simulating annealing (SA), 518
Single responsibility principle (SRP), 276
Smart home
components, 350
home appliances control and management, 351
home conditions measurement, 350
services, 350–351
Software architecture and design
client-server architecture
advantages, 273
components, 272–273
illustration, 272, 272*f*
working, 272
design antipatterns
Blob antipattern, 281
functional decomposition antipattern, 282
patterns into antipatterns, 282–283
Poltergeist antipattern/Gypsy Wagons, 282
software quality, 281
Spaghetti Code antipattern, 281–282

Software architecture and design *(Continued)*
Swiss Army Knife antipattern, 282
design patterns
behavioral design patterns, 279
benefits, 279–281
classification, 278–279
creational design patterns, 278
essential elements, 277–278, 277*f*
gang of four (GoF) design patterns, 279, 280*t*
Mediator design pattern, 277–278
object-oriented software design, 277
structural design patterns, 278
design principles
dependency inversion principle (DIP), 277
interface segregation principle (ISP), 276
Liskov substitution principle (LSP), 276
object-oriented design, 276
open-closed principle (OCP), 276
single responsibility principle (SRP), 276
layered architecture
advantages, 274
components, 274
illustration, 273, 273*f*
model-view-controller (MVC) architecture
advantages, 275–276
components, 275
illustration, 274–275, 275*f*
working, 274–275
quality factors, 271
Software patching. *See* Patching
Software reliability engineering (SRE). *See also* Software reliability growth model (SRGM)
acceptance testing, 87
bugs, 85
definition, 85
error prevention, 86
fault/error detection, 86
operation, 86
software testing, 87
Software reliability growth model (SRGM), 355–356
bivariate (*see* Bivariate software reliability growth model (SRGM))
complex fault modeling, 93–94
concept, 87–88
criteria, 95–96
data description, 96, 97–98*t*, 99–100*f*
fault removal rate (FRR), 87–88
framework, 89–90
goodness of fit, 95–96, 99–100*f*
hard fault modeling, 93
modeling testing effort
exponential type curve, 91
logistic testing, 92
Rayleigh-type curve, 91
Weibull-type curve, 92
work, 91
model validation, 96, 97–98*t*, 99–100*f*
modules, 87–88
non-homogenous Poison process (NHPP), 87–89
notations for model, 90–91
parameter estimation, 95
proposed model assumptions, 90
simple fault modeling, 92–93
total fault removal phenomenon modeling, 94–95
Spaghetti Code antipattern, 281–282
Spamming, 334
Standby system model, 175–176. *See also* Semi-Markov model
StarlingX, 3
Starting air system
accessories, 52
block diagram, 53, 54*f*
bursting disc, 53
explosions, 54–55
failure mode and effects analysis (FMEA), 55
failure probability, 52
faults, 53
flame arrestor, 54
operation, 53
relief valve, 53
risk assessment
aggregated experts' judgment, 59, 62*t*
causes and effects of explosion, 59, 60–61*t*
crisp values, 59, 63*t*
detection chances of failure modes, 62*t*
occurrence probabilities of failure modes, 61*t*
results, 62–63
risk priority number (RPN) of failure modes, 59, 63*t*
sensitivity analysis, 63, 64*t*
severity of failure modes, 61*t*
safety devices and interlocks, 53–54
State aggregation technique, 31
State-space exploration, 71–72, 72*b*, 76, 76–77*f*
Statistical analysis. *See* Correspondence analysis
Stochastic modeling. *See also* Mean time between failures (MTBF)

birth processes
branching processes, 356
definition, 357
probability mass function (PMF), 358–359
process classification, 357–358, 358*f*
mean value function (MVF)
definition, 359
theorem, 359–360
Stochastic reward net (SRN) model
attributes, 18–20, 18*f*
components, 18*f*
continuous-time Markov chain (CTMC), 13
largeness problem, 31
petri net (PN), 18
SIP INVITE transaction
accepted state, 27, 29
badput, 33–34
calling state, 20–21, 25–26
completed state, 27–30
goodput, 32–33
guard functions, 25*t*
illustration, 20, 21*f*
immediate transition and their meaning, 25*t*
initiation, 20
places and their meaning, 24*t*
proceeding state, 26–29
terminated state, 26, 30–31
transitions and their meaning, 22–23*t*
state aggregation technique, 31
submodels, 31, 32*f*
validation, 37–38, 38*f*
Substitution Ceaser Cyphers (SCC), 129
Swarm intelligence (SI)
AFSO (artificial fish swarm algorithm), 425–426
algorithms, 415*f*
artificial intelligence (AI), 413, 416
augmented and quantified human, 418–419
BFA (bacterial foraging algorithm), 425–426
biotic development, 416
body-to-PC interface, 417–418
definition, 413–414
devices, augmentations and implants, 416–418
factors to be considered, 403
GWO (grey wolf optimization), 425–426
North Sense, 416–417
PIO (pigeon inspired optimization), 425–426
SI-based IoT systems
ACO (Ant colony optimization), 428
BCO (Bee colony optimization), 428
cloud computing, 430
connected cars, 429
data routing, 429–430
dynamic routing, 427
elasticity, 426
forecasting, 427
image processing, 427
interoperability, 426
radio frequency identification (RFID), 427
robustness, 426
scalability, 426
SI-based algorithms, 426–427
WSNs, 427
and systems intersections, 414
Swiss Army Knife antipattern, 282
System architecture analysis, 497
System complexity, 447
System optimization, 496

T

Tabu search (TS), 519
Testing process, 307. *See also* Digital systems modeling
TestIt toolkit, 70, 71*f*
Throughput
badput, 33–34
definition, 32
goodput, 32–33
Time-based analytics, 336–337
Timed colored petri nets (TCPNs), 12
TOPSIS algorithm, 549–551
Total fault removal phenomenon modeling, 94–95
Total productive maintenance (TPM), 390
Transaction-oriented protocol. *See* Session Initiation Protocol (SIP)
Transactions, 235
Transistor count trends, 308*f*
Trojans, hardware, 441, 441–442*f*
Tuberculosis (TB) pandemics, 563
Tyreus-Luyben technique (TL), 107

U

Unbiased parameter estimation, 376–379
UPPAAL automaton
manually created automaton, 79, 79*f*
refined model, 76, 78*f*
without labels, 76, 78*f*
UPPAAL timed automata, 449–451
UTA models, 448–450

V

Variation, 96
Vehicular communication, 547.
See also Internet of vehicles (IoV)
Vehicular sensor networks
applications, 124, 125f
controller area networks (CAN), 124
data aggregation, 125
fog computing, 124–125
MSIP-IoT-A, 126
remote cloud center, 124–125
self-driving vehicle (SDV), 124
wired connection, 124
Virtual machines (VMs), 123
Voice over Internet Protocol (VoIP), 11.
See also Session Initiation Protocol (SIP)
VRLA battery
aging mechanisms, 289–290
automatic fault diagnosis, 291, 292f
fault classification and analysis
de-fuzzification, 293, 296f
fault severity, 294
fuzzy inputs, 292–293, 293–294f
fuzzy output, 293, 294f
Fuzzy rule base, 293, 295t
fault detection, 291–292, 293f
fuzzy inference system (FIS), 289–293
infrared thermography (IRT), 289–292
LabVIEW 2015 software, 291
lead acid battery, 289–290
results
condition of battery, 298–300, 300f
fault classification, 296–298, 298f
fuzzy output values, 298–300, 300f
matching score at discharging, 294–296, 296–297f
system block diagram VI, 296–298, 299f
web-based condition monitoring, 300–301, 301f
Vulnerability
classification, 144
common vulnerability scoring system (CVSS), 144–145
definition, 143
disclosure, 144
discovered, 144
literature review, 145–146
mathematical model
vulnerability detection process, 146–147
vulnerability disclosure process, 147
vulnerability patching process, 147–148
model validation
comparison criteria, 150, 150t
parameter estimate values, 148, 149t
percentage of vulnerabilities dealt, 148–149, 150t
representation of estimated data, 148, 149t
need for, 151
notations, 146

W

Website threats, 335
White-box testing, 67–68

Z

Zeigler-Nichols technique (ZN), 107

Printed in the United States
by Baker & Taylor Publisher Services